THE GREEK ALPHABET

Greek name	Greek character		English equivalent	
alpha	Α	α	A	a
beta	Β	β	B	b
gamma	Γ	γ	G	g
delta	Δ	δ	D	d
epsilon	Ε	ϵ	Ĕ	ĕ
zeta	Ζ	ζ	Z	z
eta	Η	η	Ē	ē
theta	Θ	θ	Th	th
iota	Ι	ι	I	i
kappa	Κ	κ	K	k
lambda	Λ	λ	L	l
mu	Μ	μ	M	m
nu	Ν	ν	N	n
xi	Ξ	ξ	X	x
omicron	Ο	ο	Ŏ	ŏ
pi	Π	π	P	p
rho	Ρ	ρ	R	r
sigma	Σ	σ	S	s
tau	Τ	τ	T	t
upsilon	ϒ	υ	Y	y
phi	Φ	φ	Ph	ph
chi	Χ	χ	Ch	ch
psi	Ψ	ψ	Ps	ps
omega	Ω	ω	Ō	ō

THE CHEMIST'S COMPANION

A Handbook of Practical Data, Techniques, and References

ARNOLD J. GORDON
Pfizer, Inc.

RICHARD A. FORD
Montgomery College

A Wiley-Interscience Publication

JOHN WILEY & SONS, New York • Chichester • Brisbane • Toronto

The tables and data contained in this book are provided for informational and educational purposes only. While they are based on experience and information believed to be reliable, the authors and the publisher make no guarantee of their accuracy or completeness and assume no liability in connection therewith.

Library of Congress Cataloging in Publication Data

Gordon, Arnold J.
The chemist's companion: a handbook of practical data, techniques, and references.

Includes bibliographical references.
1. Chemistry—Handbooks, manuals, etc. I. Ford, Richard A., joint author. II. Title. [DNLM: 1. Chemical industry. 2. Chemistry. QD 31.2 G662c 1972]

QD65.G64 542 72-6660
ISBN 0-471-31590-7

Printed in the United States of America

20 19

For

Francesca

Shira and Ian

To Kathy

PREFACE

It is impossible for the chemist to remember or even to catalog handily more than a small portion of the information he regularly needs. Naturally, numerous textbooks, monographs, handbooks, data collections, and similar sources are available wherein specific data are sought. Nevertheless, our own experience has been that whether preparing a lecture, solving a problem, or working in the laboratory, there is a continuous and often immediate demand for certain information that seemingly should be, but usually is not, conveniently available. In addition to frequently called for physicochemical properties, there is an ever-present need for answers to such practical questions on how-to-do-it, what-is-it, what-do-I-use-and-when, and where-do-I-find-it-or-buy-it. An almost endless collection of workaday facts and figures are required, which cannot be found in any one place and often are not accessible at all.

The aim of this book is to provide, in an easily usable form, as much helpful information of this sort as possible. We have canvassed chemistry students and professionals for their pet needs, and have incorporated what we consider to be the most useful and often needed materials for a broad cross section of chemists. We cover almost all the properties mentioned in a 1965 survey of Americal Chemical Society members that asked, among other questions, "What physical, chemical, and mechanical properties of substances and systems do you seek most often in the literature?" [H. M. Wiseman, J. Chem. Doc., 7, 9 (1967)]. To our knowledge, the most reliable and recent data are used throughout. The art and methods of chemical synthesis are generally not discussed; this broad, separate area is, of course, the subject of many other books.

The choice and manner of presentation of some of the material included in this book reflect our interests and experience in organic chemistry. However, most of the information is applicable to nearly all chemical disciplines, and material of special interest to physical and inorganic chemists, as well as biochemists, is included. Topics discussed include properties of atoms and molecules, spectroscopy, photochemistry, chromatography, kinetics and thermodynamics, various experimental techniques, mathematical and numerical information, and a variety of hard-to-classify but frequently sought information. In addition to vital properties, we include important hints, definitions, and other material associated with the "art" as well as the state of the art of a particular subject. We have tried to incorporate as much useful information as possible — the sort that individuals keep tacked to the laboratory or office wall or in file-card collections. Many sections, in fact, are written as mini-reviews, containing the essence of a topic along with the most recent pertinent literature references. To some extent, finally, the book is meant to serve as a glossary of practical chemical lore.

In addition to a Subject Index, there is a Suppliers Index, which lists all suppliers and manufacturers mentioned in the book and cites the page on which an address appears. For pertinent topics, directions are given for obtaining special booklets or other material, many at no cost. A list of publishers' addresses can be found in the last chapter. We apologize in advance to any authors or suppliers we may have overlooked; we have tried to provide a representative collection, with coverage through July 1972.

Wherever possible, standard (IUPAC) chemical nomenclature is used; however, in the interests of clarity and custom, common names are used whenever deemed appropriate. The definitions and values for the newly adopted International System of Units are described (SI units, see p. 456), along with recommended rules of usage. We have knowingly violated two of those rules: those that prescribe the writing of a large number as (1 245.843 2) and the omission of a degree sign for degrees Kelvin. To our knowledge these are not yet in common usage, and we have retained the traditional practice (1,245.8432 and °K). We have also chosen to use mm Hg for pressure, although the equivalent unit torr is gaining wide acceptance. However, certain other recommendations have been adopted in this book. The micron ($1\mu = 10^{-6}$ m) is no longer considered to be a unit and is replaced by micrometer (μm), which is used, for example, for infrared wavelengths. Another change is the abolishment of the millimicron (mμ), formerly used for ultraviolet or visible spectra wavelengths; the official unit is now the nanometer (nm).

Although the book is purposely organized into separate, individualized sections, there is extensive cross referencing throughout. For example, in the section "Properties of Solvents and Common Liquids" each solvent is given a number; in many other sections of the book when a solvent is mentioned, this number is cited for easy reference to the more extensive data in the master table. The reader will notice in examining many of the tables that entries are missing; this is due to either the unavailability of the information or, in our judgment, its unreliability. It is hoped that readers will be able to fill in the missing data when they become available and that they will inform the authors of such additions.

We believe that this book can serve daily not only the professional chemist but also the student and the beginning practitioner. We welcome criticisms, corrections, and suggestions that will help to make future revisions of this book even more useful and accurate.

Arnold J. Gordon
Richard A. Ford

Riverside, Connecticut
Washington, D.C.
August 1972

ACKNOWLEDGMENTS

We owe our warm thanks for ideas and assistance in the preparation of this book to Dr. E. T. Becker (NIH) for his review of the sections on magnetic resonance; Daniel D. Traficante (MIT) for the use of unpublished material taken from the MIT Techniques Manual, MIT Chemistry Department; Drs. Sam Wilen (CUNY) and Ernest Eliel (University of North Carolina) for suggestions and data used in the section on optical resolution; Drs. Richard Timmons and Hao-Lin Chen (Catholic University of America) for their review of the photochemistry chapter; the numerous graduate students and colleagues everywhere who supplied ideas and information for various sections; the many companies and publishers who permitted liberal use of their printed material; the many typists who had to suffer through some very difficult material, especially to Mr. Joel Landry and Miss Tola Wilson for their patience and expertise in preparing the final, camera-ready copy of the entire manuscript.

Finally, we thank Francesca Gordon and Kathy Ford for their continued patience, encouragement, and assistance throughout all stages of the book's preparation; particular gratitude is owed to Francesca for help in the exhausting and seemingly endless task of proofreading nearly the entire final copy.

CONTENTS

The Chemist's Companion

A Handbook of Practical Data, Techniques, and References

PROPERTIES OF MOLECULAR SYSTEMS

I. PROPERTIES OF SOLVENTS AND COMMON LIQUIDS

Extensive data on organic solvents can be found in the definitive monograph, J. A. Riddick and W. B. Bunger, Organic Solvents, Volume II of Techniques of Chemistry (successor to the Technique of Organic Chemistry and the Technique of Inorganic Chemistry series), Wiley-Interscience, New York, 1971 (1041 pp; more than 5000 literature citations). Some of the discussion below was taken from this source.

Commercially available solvents come in a variety of purity grades, which are defined as follows:

ACS Reagent Grade. Meets standards established in Reagent Chemicals, ACS, Washington, D.C. 1960.

USP Grade and NF Grade. Meets standards established by the U.S. Pharmacopoeia (USP) or the National Formulary (NF) for exceptional purity of chemicals in drugs or in general pharmacological applications.

Spectrophotometric Grade. Spectrally "pure" in the uv, visible, near-ir, and mid-ir regions. For details, see page 167.

Gas Chromatographic Grade. Homogeneous by high-sensitivity gas liquid chromatography.

Electronic Grade. Extremely low amounts of metallic impurities and very low specific conductance; used in transistor industry and electronic circuitry.

Research Grade. High-Purity research and reference standards. (Bulletin 522, Chemical Department, Phillips Petroleum Co., Bartlesville, Okla.)
Pesticide Grade. Also known as "Nanograde" solvents. Used for pesticide residue analysis; meets ACS Reagent grade standards and has less than 10 ng/liter of chlorinated pesticide or other similar impurity.
Nonaqueous Titration Solvent Grade. Used for high-precision titration; very low in water and in acidic and basic substances. (Write for Titrants, Indicators, Solvents for Non-Aqueous Titrimetry, Eastman Organic Chemicals.)

The Burdick and Jackson Laboratories, Muskegon, Michigan 49442, sell some exceptionally high-purity solvents. Further discussion of purity and directions for custom purification of many common solvents appear on page 429.

This section contains tables of general constants and properties; alphabetical, mp, bp, dielectric constant, and azeotrope indexes to the entries; vapor pressure data for very common solvents (see also p. 30); a section on synonyms, structures, and properties of common name solvents and special liquids; and a section on empirical solvent parameters (Z-values, et al.). To determine if a given solvent forms a binary azeotrope, refer to the azeotrope index.

A. General Constants

Entries are arranged in the order employed by Chemical Abstracts: alphabetically for noncarbon-containing compounds, in the order of increasing number of carbons and hydrogens for carbon compounds, and then alphabetically for other elements. Compounds whose names are followed by a Z are those for which values of that empirical solvent parameter are given in the table on page 23.

The bp and mp are in °C at 760 mm Hg unless followed by another pressure in parentheses. Density (ρ) in g/cc is at 20°C, and dielectric constant (ε) is at 25°C for the pure liquid unless followed by another temperature in parentheses. Refractive index (n_D) is at 20°C unless followed by another temperature in parentheses. Dipole moments (μ) in debyes are for the gas phase unless followed by ℓ for neat liquid or by the solvent for solution measurement (B = benzene, D = 1,4-dioxane). Solubility in water at or near room temperature (S) is expressed as m = miscible, i = immiscible (insoluble) p = partly soluble (<10-15 g/100 g H_2O), v = very soluble (>25 g/100 g H_2O), and d = decomposes. Viscosity (η) is in millipoise (at 25° unless indicated otherwise).

Data for these tables were gathered principally from the following sources: L. Scheflan and M. B. Jacobs, The Handbook of Solvents, D. Van Nostrand, New York, 1953; J. J. Lagowski, Ed., The Chemistry of Non-Aqueous Solvents, Vols. 1, 2, and 3, Academic Press, New York, 1966, 1967 and 1970; T. C. Waddington, Non-Aqueous Solvent Systems, Academic Press, New York, 1965; I. Mellan, Handbook of Solvents, Vol. 1, Reinhold Publishing Corp., New York, 1957; Handbook of Chemistry and Physics, 50th ed., CRC, Cleveland, 1969; Organic Solvents, 2nd ed., Volume VII of Technique of Organic Chemistry, Interscience, New York, 1955; R. Nelson, Jr., D. Lide, Jr., and A. Maryott, Selected Values of Electric Dipole Moments for Molecules in the Gas Phase, NSRDS-NBS 10, U.S. Government Printing Office, Washington, D.C., 1967; A. L. McClellan, Tables of Experimental Dipole Moments, W. H. Freeman, San Francisco, 1963; A. Maryott and E. Smith, Table of Dielectric Constants of Pure Liquids, NBS Circular 514, U.S. Government Printing Office, Washington, D.C., 1951.

Compound	Mw	Bp	Mp
1. $AsBr_3$ (arsenic tribromide)	314.65	220	35
2. $AsCl_3$ (arsenic trichloride)	181.28	130	-13
3. AsF_3 (arsenic trifluoride)	131.92	63	-6
4. Br_2 (bromine)	159.81	59	-7
5. Br_3P (phosphorus tribromide)	270.70	173	-40
6. CCl_3F (trichlorofluoromethane)	70.01	24	-111
7. CCl_4 (carbon tetrachloride)	153.82	76.5	-23
8. CClFHBr (bromochlorofluoromethane)	147.38	36	-115
9. CCl_2HBr (bromodichloromethane)	163.83	90	-57
10. $CClHBr_2$ (dibromochloromethane)	208.29	119(748)	
11. $CHBr_3$ (bromoform)	252.75	150	8.3
12. CCl_3H (chloroform Z)	119.38	61.7	-63.5
13. CHN (hydrocyanic acid)	27.03	26	-14
14. $CClH_2Br$ (bromochloromethane)	129.39	68.1	-86
15. CH_2Br_2 (dibromomethane)	173.85	97	-52.6
16. CCl_2H_2 (dichloromethane Z)	84.93	40	-95.1
17. CH_2O (formaldehyde)	30.03	-21	-92
18. CH_2O_2 (formic acid)	46.03	101	8.4
19. CH_3Br (bromomethane)	94.94	3.6	-94
20. CH_3I (iodomethane)	141.94	42.4	-66.4
21. CH_3NO (formamide Z)	45.04	193	2.6
22. CH_3NO_2 (nitromethane)	61.04	101	-28.5
23. CH_4O (methanol Z)	32.04	64.5	-97.5
24. CH_5N (methylamine)	31.06	-6.3	-93.5
25. CS_2 (carbon disulfide)	76.14	46.2	-111.5
26. $C_2Cl_2O_2$ (oxalyl chloride)	126.93	63	-16
27. C_2Cl_4 (tetrachloroethylene)	165.83	121	-19
28. C_2Cl_3H (trichloroethylene)	131.39	87	-73
29. $C_2Cl_3HO_2$ (trichloroacetic acid)	163.39	198	58(α)
30. $C_2F_3HO_2$ (trifluoroacetic acid)	114.02	72.4	-15
31. $C_2Cl_4H_2$ (1,1,2,2-tetrachloroethane)	167.85	146	-36
32. $C_2H_2O_2$ (glyoxal)	58.04	50.4	15
33. C_2H_3Br (bromoethylene)	106.96	15.8	-140
34. C_2H_3OBr (acetyl bromide)	122.96	76	-96
35. C_2ClH_3 (chloroethylene)	62.50	-13.4	-154
36. C_2ClH_3O (acetyl chloride)	78.50	51	-112
37. $C_2Cl_3H_3$(1,1,1-trichloroethane)	133.41	74.1	-30
38. $C_2Cl_3H_3$ (1,1,2-trichloroethane)	133.41	113.8	-36
39. $C_2F_3H_3O$ (2,2,2-trifluoroethanol)	100.04	73.5	
40. C_2H_3N (acetonitrile Z)	41.05	81.6	-45.7
41. $C_2H_4Br_2$ (1,2-dibromoethane)	187.87	131	9.8
42. $C_2Cl_2H_4$ (1,1-dichloroethane)	98.96	57.3	-97
43. $C_2Cl_2H_4$ (1,2-dichloroethane)	98.96	83.5	-35
44. $C_2H_4I_2$ (1,2-diiodoethane)	281.86	200	83
45. C_2H_4O (acetaldehyde)	44.05	20.8	-121
46. C_2H_4O (ethylene oxide)	44.05	13.5(746)	-111
47. $C_2H_4O_2$ (acetic acid Z)	60.05	118	16.6
48. C_2H_5Br (bromoethane)	108.97	38	-119

S	ρ	n_D	ε	μ	$10^3\eta$	
d	3.33(50)		8.8(35)		54.1(35)	1.
d	2.16	1.621(14)	12.6(17)	1.59	12.25(20)	2.
d	2.67(0)		5.7(<-6)	2.59		3.
p	3.12	1.661	3.1(20)	0		4.
d	2.85(15)	1.697(27)	3.9(20)	1.7(D)		5.
i	1.46(30)		2.28(29)	0.45		6.
i	1.59	1.4601	2.23	0	9.69(20)	7.
i	1.98(0)	1.4144(25)				8.
i	1.98	1.4964				9.
i	2.45	1.5482				10.
p	2.89	1.5976	4.39(20)	0.99	21.5(15)	11.
i	1.48	1.4459	4.70	1.87	5.42	12.
m	0.7(22)	1.2675(10)	114	2.98	2.0(20)	13.
i	1.934	1.4838				14.
p	2.50	1.5420	6.7(40)	1.43		15.
i	1.33	1.4242	8.9	1.60	3.9(30)	16.
v	0.82(-20)			2.33		17.
m	1.220	1.3714	58(16)	1.41	19.66	18.
i	1.68	1.4218	9.82(0)	1.81	3.79	19.
i	2.28	1.5380	7	1.62	5.18(15)	20.
m	1.129(25)	1.44754	110	3.7	33.0	21.
p	1.137	1.3817	38.6	3.46	6.08	22.
m	0.791	1.3288	32.6	1.70	5.45	23.
v	0.66		9.4	1.31		24.
i	1.263	1.6319	2.64	0	3.76(20)	25.
d	1.478	1.4316	3.47	0.93(B)		26.
i	1.623	1.5053	2.5	0	9.32(15)	27.
i	1.464	1.4773	3.4(16)	0.77(ℓ)	5.32	28.
v	1.6	1.4603(61)	4.6(60)			29.
p	1.54(0)		39(20)	2.28	5.78	30.
i	1.595	1.4940	8.2(20)	1.32	18.4(15)	31.
m	1.14	1.3826				32.
i	1.49(ℓ)	1.4410		1.42		33.
d	1.66(16)	1.4538(16)	16.2(20)			34.
i	0.91	1.3700		1.45		35.
d	1.105	1.3898	15	2.72		36.
i	1.339	1.4379	7.5(20)	1.78		37.
i	1.44	1.4714		1.25(<70°)		38.
p	1.384(25)	<1.3	26.5			39.
m	0.777(25)	1.3441	36.2	3.92	3.45	40.
p	2.18	1.5387	4.78	1.01(35°)	18.8(15)	41.
i	1.176	1.4164	10(18)	2.06	5.05	42.
i	1.235	1.4448	10.4	1.44(35°)	8.0(20)	43.
i	3.325	1.871				44.
v	0.783	1.3316	21(20)	2.69		45.
v	0.88(10)	1.3597(7)	22(10)	1.89		46.
m	1.049	1.3716	6.19	1.74	11.6	47.
i	1.46	1.4239	9.39(20)	2.03		48.

Compound	Mw	Bp	Mp
49. C_2ClH_5 (chloroethane)	64.52	13.1	-139
50. C_2ClH_5O (2-chloroethanol)	80.52	128	-68
51. C_2FH_5O (2-fluoroethanol)	64.06	104	-26
52. C_2H_5NO (N-methylformamide)	59.07	180-5	
53. $C_2H_5NO_2$ (nitroethane)	75.07	115	-50
54. $C_2H_5NO_3$ (2-nitroethanol)	91.07	193	-80
55. C_2H_6O (ethanol Z)	46.07	78.3	-114.5
56. C_2H_6OS (dimethyl sulfoxide Z)	78.13	189(d)	18.4
57. $C_2H_6O_2$ (ethylene glycol Z)	62.07	198	-11.5
58. $C_2H_6O_4S$ (dimethyl sulfate)	126.13	189(d)	-32
59. C_2H_7N (dimethylamine)	45.09	7	-96
60. C_2H_7N (ethylamine)	45.09	16.6	-81
61. C_2H_7NO (2-aminoethanol)	61.09	170	10.3
62. $C_2H_8N_2$ (ethylenediamine)	60.11	116.5	8.5
63. C_3Cl_6O (hexachloroacetone)	264.75	203	-2
64. C_3F_6O (hexafluoroacetone)	166.02	-27.4	-125
65. $C_3F_6H_2O$ (hexafluoroacetone•H_2O)	184.04	55(80)	40
66. $C_3H_4O_3$ (ethylene carbonate)	88.06	248	39
67. C_3H_5N (propionitrile)	55.08	97.4	-93
68. C_3H_6O (acetone Z)	58.08	56.2	-95.4
69. C_3H_6O (oxetane)	58.08	46	
70. $C_3H_6O_2$ (ethyl formate)	74.08	54.5	-81
71. $C_3H_6O_2$ (methyl acetate)	74.08	57	-98
72. $C_3H_6O_3$ (dimethyl carbonate)	90.08	90	3
73. C_3ClH_7 (1-chloropropane)	78.54	46.6	-123
74. C_3ClH_7 (2-chloropropane)	78.54	35.7	-117
75. C_3H_7NO (N,N-dimethylformamide Z)	73.10	152	-61
76. C_3H_7NO (N-methylacetamide Z)	73.10	206	29.5
77. C_3H_8O (1-propanol Z)	60.11	97.4	-127
78. C_3H_8O (2-propanol Z)	60.11	82.4	-89.5
79. $C_3H_8O_2$ (2-methoxyethanol)	76.11	124	-85
80. $C_3H_8O_2$ (dimethoxymethane)	76.11	42	-105
81.[a] $C_3H_9O_3B$ (trimethyl borate)	103.91	68	-29
82. C_3H_9N (1-aminopropane)	59.11	47.8	-83
83. C_3H_9N (2-aminopropane)	59.11	32.4	-95
84. C_4H_4S (thiophene)	84.14	84.2	-38
85. C_4H_5N (pyrrole)	67.09	130	-15
86. $C_4H_6O_3$ (acetic anhydride)	102.09	139.6	-73
87. $C_4H_6O_3$ (propylene carbonate)	102.09	240	-70
88. C_4H_8 (cyclobutane)	56.12	12	-50
C_4H_4O (furan); see 225			
89. C_4H_8O (ethyl methyl ketone)	72.12	79.6	-86
90. C_4H_8O (tetrahydrofuran)	72.12	66	-65
91. $C_4H_8O_2$ (1,3-dioxane)	88.12	107	-42
92. $C_4H_8O_2$ (1,4-dioxane)	88.12	102	11.8
93. $C_4H_8O_2$ (ethyl acetate)	88.12	77.1	-83.6
94. $C_4H_8O_2S$ (sulfolane Z)	120.16	283	28.9
95. C_4H_9Br (1-bromobutane)	137.03	102	-112
96. C_4H_9Br (dl-2-bromobutane)	137.03	91.2	-112

[a] For $C_3H_8O_3$ (glycerol), see 226.

S	ρ	n_D	ε	μ	$10^3\eta$	
i	0.903(15)	1.3676	6.3(170)	2.05	2.79(10)	49.
m	1.200	1.4419	26	1.78	39.1(15)	50.
m	1.104	1.3639				51.
p	1.01	1.4319	182	3.83	16.5	52.
p	1.04(25)	1.3917	28.1(30)	3.6	6.61	53.
m	1.270	1.4434				54.
m	0.785(25)	1.3611	24.3	1.69	10.8	55.
m	1.101	1.4770	49	3.96	19.8	56.
m	1.109	1.4318	37.7	2.28	136(30)	57.
v	1.328	1.3874	42.6(20)			58.
v	0.68(0)		5.26	1.03		59.
m	0.683	1.3663	6.94(10)	1.22		60.
m	1.018	1.4541		2.6(D)		61.
m	0.9	1.4568	14.2(20)	1.99	15.4	62.
p	1.74(12)	1.5112	3.82(30)			63.
(m)	1.33(25ℓ)		1.96(-35)			64.
(m)		1.3179				65.
	1.32(39)	1.4158(50)		4.5		66.
	0.782	1.3655	27(20)	4.02	4.54(15)	67.
m	0.790	1.3588	20.7	2.88	3.16	68.
m		1.3961		1.93		69.
p	0.917	1.3598	7.2	1.93	3.58(30)	70.
v	0.933	1.3593	6.7	1.72	3.62	71.
i	1.069	1.3687		0.9		72.
i	0.891	1.3879	7.7	2.05	3.18(30)	73.
i	0.862	1.3777		2.17	2.86(30)	74.
m	0.945(25)	1.4303	36.7	3.86	7.96	75.
v	0.957(25)	1.4301	179(30)	3.73	38.9(30)	76.
m	0.804	1.3850	20.1	1.68	20.0	77.
m	0.786	1.3776	18.3	1.66	17.7(30)	78.
m	0.965	1.4024	16(30)	2.2(B)		79.
v	0.847(30)	1.3530	2.7(20)	0.74(35°)	32.5(30)	80.
d	0.915	1.3568	8.0(20)			81.
v	0.717	1.3870		1.17		82.
m	0.889	1.3742	5.5(20)			83.
i	1.065	1.5289	2.76(16)	0.55	6.21	84.
i	0.969	1.5085	7.48(18)	1.84		85.
v	1.082	1.3901	21(19)	2.8	78.3(30)	86.
	1.204	1.4189	65.1	4.98(B)	25.3	87.
i	0.72(5)	1.4260				88.
v	0.805	1.3788	18.5(20)	2.5(B)	36.5(30)	89.
m	0.889	1.4050	7.32	1.63		90.
m	1.034	1.4165		2.13(B)		91.
m	1.034	1.4224	2.21	0	10.87(30)	92.
p	0.900	1.3723	6.02	1.78	4.41	93.
v	1.262(30)		44(30)	4.7	98.7(30)	94.
i	1.276	1.4401	7.1(20)	2.08		95.
i	1.259	1.4366	8.64	2.23		96.

Compound	Mw	Bp	Mp
97. C_4H_9N (pyrrolidine)	71.12	89	
98. C_4H_9NO (N,N-dimethylacetamide Z)	87.12	165	-20
99. C_4H_9NO (morpholine)	87.12	128	-4.8
100. $C_4H_{10}O$ (1-butanol Z)	74.12	117	-90
101. $C_4H_{10}O$ (diethyl ether)	74.12	34.5	-116
102. $C_4H_{10}O$ (dl-2-butanol)	74.12	99.5	-115
103. $C_4H_{10}O$ (i-butyl alcohol)	74.12	108	-108
104. $C_4H_{10}O$ (t-butyl alcohol Z)	74.12	82	25.5
105. $C_4H_{10}O_2$ (1,1-dimethoxyethane)	90.12	64.5	-113
106. $C_4H_{10}O_2$ (2-ethoxyethanol)	90.12	135	
107. $C_4H_{10}O_2$ (ethylene glycol dimethyl ether Z)	90.12	83	-58
108. $C_4H_{10}O_3$ (diethylene glycol)	106.12	245	-10
109. $C_4H_{11}N$ (diethylamine)	73.14	56	-50
110. $C_4H_{12}Si$ (tetramethylsilane)	88.23	26.5	-102(α)
111. C_5H_5N (pyridine Z)	79.10	115.6	-41.8
112. C_5H_6O (γ-pyran)	81.09	80	
113. C_5H_8O (2,3-dihydro-γ-pyran)	84.13	86	
114. C_5H_{10} (cyclopentane)	70.14	49.3	-93.9
115. $C_5H_{10}O$ (tetrahydropyran)	86.14	88	
116. $C_5H_{10}O_3$ (diethyl carbonate)	118.13	126	-43
117. $C_5H_{11}N$ (piperidine)	85.15	106	-10.5
118. $C_5H_{11}NO$ (N,N-dimethylpropionamide)	101.15	175	-45
119. $C_5H_{12}N_2O$ (tetramethylurea)	116.16	167	-1
120. C_5H_{12} (neopentane)	72.15	9.5	-16.6
121. C_5H_{12} (pentane)	72.15	36.1	-130
122. $C_5H_{12}O$ (1-pentanol)	88.15	137	-79
123. $C_5H_{12}O$ (dl-2-pentanol)	88.15	119	
124. $C_5H_{12}O$ (3-pentanol)	88.15	116.1	<-75
125. $C_5H_{12}O$ (i-pentyl alcohol)	88.15	132	-117
126. $C_5H_{12}O$ (neo-pentyl alcohol)	88.15	113	52
127. $C_5H_{12}O$ (t-pentyl alcohol)	88.15	102	-8.4
128. $C_5H_{12}O_3$ (diethylene glycol monomethyl ether)	120.15	193	
129. C_6F_6 (hexafluorobenzene)	186.05	81(743)	$\sim$5
130. $C_6Cl_3H_3$ (1,2,4-trichlorobenzene)	181.45	213.5	17
131. $C_6Cl_2H_4$ (o-dichlorobenzene)	147.01	180.5	-17
132. C_6H_5Br (bromobenzene)	157.02	156	-30.8
133. C_6ClH_5 (chlorobenzene)	112.56	132	-45.6
134. C_6FH_5 (fluorobenzene)	96.11	85.1	-41
135. C_6H_5I (iodobenzene)	204.01	188	-31.3
136. $C_6H_5NO_2$ (nitrobenzene)	123.05	211	5.8
137. C_6H_6 (benzene Z)	78.12	80.1	5.5
138. C_6H_6O (phenol)	94.11	181.8	43
139. C_6H_7N (aniline)	93.13	184	-6.3
140. C_6H_7N (2-picoline)	93.13	129	-70
141. C_6H_7N (3-picoline)	93.13	144	-18
142. C_6H_7N (4-picoline)	93.13	145	3.6
143. C_6H_{12} (cyclohexane)	84.16	80.7	6.55

S	ρ	n_D	ε	μ	$10^3\eta$	
m	0.852	1.4431		1.58(B)		97.
m	0.937(25)	1.4384	37.8	3.81	9.2	98.
m	1.001	1.4548	7.33	1.5(B)		99.
p	0.810	1.3993	17.1	1.66	22.7(30)	100.
p	0.714	1.3526	4.34(20)	1.15	2.22	101.
p	0.806	1.3978	15.8	1.7(B)	42.1(20)	102.
p	0.794(30)	1.3955	17.7	1.64	39.1	103.
m	0.789	1.3878	10.9(30)	1.66(ℓ)	33.2(30)	104.
v	0.850	1.3668	3.49(20)			105.
m	0.930	1.4080		2.08		106.
m	0.863	1.3796			11(20)	107.
m	1.120	1.4472		2.7(D)	300	108.
v	0.71(18)	1.3873(18)	3.6(22)	0.92		109.
i	0.64	1.3587		0		110.
m	0.982	1.5095	12.3	2.19	9.45(20)	111.
		1.4559				112.
p	0.92	1.4399				113.
i	0.746	1.4065	1.97	0(ℓ)	4.16	114.
	0.881	1.4200				115.
i	0.975	1.3845	2.82(20)	1.10	7.48	116.
m	0.861	1.4530	5.8(22)	1.2(B)		117.
						118.
	0.969	1.4507	23.1	3.92(ℓ)		119.
i	0.614	1.3476(6)		0		120.
i	0.626	1.3575	1.84(20)	0	2.15	121.
i	0.814	1.4101	13.9	1.8(ℓ)	33.5	122.
v	0.810	1.4053				123.
p	0.821	1.4104	13.9			124.
p	0.809	1.4075	14.7	1.82(ℓ)	29.6(30)	125.
p	0.812					126.
p	0.806	1.4052	5.82	1.7(B)	28.1(30)	127.
m	1.027	1.4264				128.
i		1.3769				129.
i	1.454	1.5717		1.26(B)		130.
i	1.305	1.5515	9.93	2.50		131.
i	1.495	1.5597	5.40	1.70	9.85(30)	132.
i	1.106	1.5241	5.62	1.69	7.99(20)	133.
i	1.023	1.4684(40)	5.42	1.60		134.
i	1.831	1.6200	4.63(20)	1.70		135.
i	1.204	1.5562	35(30)	4.22	20.3(20)	136.
i	0.879	1.5011	2.28(20)	0	6.03	137.
v	1.072	1.5418(41)	9.78(60)	1.45	34.9(50)	138.
v	1.022	1.5863	6.89(20)	1.76(D)	37.7	139.
v	0.944	1.4957	9.8	1.9(B)		140.
m	0.967	1.5040		2.4(B)		141.
m	0.955	1.5037		2.6(B)		142.
i	0.778	1.4266	2.02(20)	0	8.98	143.

Compound	Mw	Bp	Mp
144. $C_6H_{12}O$ (cyclohexanol)	100.16	161	25.2
145. $C_6H_{12}O_3$ (paraldehyde)	132.16	128	12.6
146. C_6H_{14} (hexane)	86.18	69	-95
147. C_6H_{14} (2,2-dimethylbutane)	86.18	49.7	-99.9
148. C_6H_{14} (2,3-dimethylbutane)	86.18	58	-128
149. $C_6H_{14}O_2$ (ethylene glycol diethyl ether)	118.18	123.5	-74
150. $C_6H_{14}O_3$ (diethylene glycol monoethyl ether)	134.18	195	<-76
151. $C_6H_{14}O_3$ (diethylene glycol dimethyl ether)	134.18	161	
152. $C_6H_{15}O_3B$ (triethyl borate)	145.99	120	
153. $C_6H_{15}N$ (triethylamine)	101.19	89.4	-115
154. $C_6H_{15}O_3P$ (triethyl phosphite)	166.16	158	
155. $C_6H_{16}N_2$ (tetramethylethylenediamine)	116.21	121.5	
156. $C_6H_{18}N_3OP$ (hexamethylphosphoramide Z)	179.20	66(0.5)	
157. $C_7F_3H_5$ (benzotrifluoride)	146.03	102	-29
158. C_7H_5N (benzonitrile)	103.13	190.7	-13
159. C_7H_8 (toluene)	92.15	110.6	-95
160. C_7H_8O (anisole)	108.15	155	-37.5
161. C_7H_8O (benzyl alcohol)	108.15	205	-15.3
162. C_7H_9N (2,3-lutidine)	107.15	163	
163. C_7H_9N (2,4-lutidine)	107.15	159	
164. C_7H_9N (2,5-lutidine)	107.15	157	-16
165. C_7H_9N (2,6-lutidine	107.15	146	-6
166. C_7H_9N (3,4-lutidine)	107.15	164	
167. C_7H_9N (3,5-lutidine)	107.15	172	
168. C_7H_{14} (cycloheptane)	98.19	118.5	-12
169. C_7H_{14} (methylcyclohexane)	98.19	100.9	-126.6
170. C_7H_{16} (heptane)	100.21	98.4	-90.6
171. C_8H_{10} (ethylbenzene)	106.17	136.2	-95
172. C_8H_{10} (m-xylene)	106.17	139.1	-48
173. C_8H_{10} (o-xylene)	106.17	144.4	-25.2
174. C_8H_{10} (p-xylene)	106.17	138.4	13.3
175. $C_8H_{10}O$ (phenetole Z)	122.17	170	-29.5
176. $C_8H_{11}N$ (2,4,6-trimethylpyridine)	121.18	175	-44.5
177. $C_8H_{11}N$ (N,N-dimethylaniline)	121.18	194.2	2.45
178. $C_8H_{11}N$ (d-α-phenethylamine $[\alpha]_D^{25}$+40(MeOH))	121.18	187	
179. $C_8H_{11}N$ (dl-α-phenethylamine)	121.18	187	
180. $C_8H_{11}N$ (β-phenethylamine)	121.18	197	
181. C_8H_{18} (octane)	114.23	125.7	-56.8
182. C_8H_{18} (2,2,4-trimethylpentane Z)	114.23	99.2	-107.4
183. $C_8H_{18}O_3$ (diethylene glycol diethyl ether)	162.23	189	-44
184. $C_8H_{18}O_4$ (triethylene glycol dimethyl ether)	178.23	222	

S	ρ	n_D	ε	μ	$10^3\eta$	
p	0.962	1.4641	15.0	1.9	411(30)	144.
p	0.994	1.4049	13.9	1.43		145.
i	0.660	1.3751	1.89(20)	0.08	2.92	146.
i	0.648	1.3688				147.
i	0.662	1.3750				148.
	0.848	1.3860				149.
m	0.933	1.4300				150.
v		1.4073			20(20)	151.
d	0.855(28)	1.3749				152.
p	0.719(30)	1.4010	2.42	0.66	3.94(15)	153.
i	0.963	1.4127		1.82(D)		154.
		1.4177				155.
i	1.02	1.4579			35(60)	156.
d	1.188	1.4146	9.18(30)	2.86		157.
i	1.010	1.5289	25.2	4.18	14.47(15)	158.
i	0.867	1.4961	2.38	0.36	5.52	159.
i	0.996	1.5179	4.33	1.38	7.89(30)	160.
p	1.042	1.5396	13.1(20)	1.71	46.5(30)	161.
p	0.932	1.5057		2.20(B)		162.
v	0.931	1.5010		2.30(B)		163.
p	0.930	1.5006		2.15(B)		164.
m	0.923	1.4953		1.66(B)		165.
p	0.928	1.5096		1.87(ℓ)		166.
p	0.942	1.5061		2.58(B)		167.
i	0.810	1.4436				168.
i	0.769	1.4231	2.02(20)	0	6.85	169.
i	0.684	1.3878	1.92(20)	0.0	3.90	170.
i	0.867	1.4959	2.41(20)	0.59	6.37	171.
i	0.864	1.4972	2.37(20)	0.37(ℓ)	5.81	172.
i	0.880	1.5055	2.57(20)	0.62	7.56	173.
i	0.861	1.4958	2.27(20)	0	6.05	174.
i	0.967	1.5076	4.22(20)	1.45	11.58	175.
p	0.917(22)	1.495(25)		1.95(B)		176.
i	0.956	1.5582	4.9	1.68		177.
p	0.965(15)					178.
p	0.940	1.5238(25)				179.
p	0.958	1.5290				180.
i	0.702	1.3974	1.95(20)	0	5.14	181.
i	0.692	1.3915	1.94(20)	0(ℓ)	5.03(20)	182.
v	0.906	1.4115				183.
v		1.4224				184.

Compound	Mw	Bp	Mp
185. C_9H_7N (isoquinoline)	129.16	243(743)	26.5
186. C_9H_7N (quinoline)	129.16	238	-16
187. C_9H_{10} (indane)	118.18	178	-51
188. C_9H_{12} (mesitylene)	120.20	164.7	-44.7
189. C_9H_{20} (nonane)	128.26	150.8	-51
190. $C_{10}H_7Br$ (1-bromonaphthalene)	207.08	281	-6(α)
191. $C_{10}ClH_7$ (1-chloronaphthalene)	162.62	259(753)	-2.3
192. $C_{10}FH_7$ (1-fluoronaphthalene)	146.17	217	-9
193. $C_{10}H_7I$ (1-iodonaphthalene)	254.07	302	4.2
194. $C_{10}H_8$ (naphthalene)	128.19	218	80.6
195. $C_{10}H_{10}O_4$ (dimethyl phthalate)	194.19	284	0-2
196. $C_{10}H_{12}$ (tetralin)	132.21	208	-36
197. $C_{10}H_{18}$ (cis-decalin)	138.25	196	-43
198. $C_{10}H_{18}$ (trans-decalin)	138.25	187	-30
199. $C_{10}H_{22}$ (decane)	142.29	174	-30
200. $C_{12}H_{10}O$ (diphenyl ether)	170.21	258	26.8
201. $C_{14}H_{14}$ (m,m'-bitolyl)	182.27	280	9
202. $C_{14}H_{14}$ (o,o'-bitolyl)	182.27	256	20
203. $C_{16}H_{22}O_4$ (dibutyl phthalate)	278.35	340	-35
204. $C_{24}H_{38}O_4$ (dioctyl phthalate)	390.30	387	-50(gum)
205. Cl_2OS (thionyl chloride)	118.97	80	-105
206. Cl_3OP (phosphorus oxychloride)	153.33	108	1
207. Cl_3P (phosphorus trichloride)	137.33	74	-112
208. Cl_5Sb (antimony pentachloride)	299.02	∿140	5
209. D_2O (deuterium oxide)	20.031	101.42	3.82
210. D_2O_4S (sulfuric acid-d_2)	100.10		14.35
211. FH (hydrogen fluoride)	20.01	19.51	-89.4
212. FHO_3S (fluorosulfuric acid)	100.07	163	-89
213. F_5Sb (antimony pentafluoride)	216.74	150	7
214. F_6S (sulfur hexafluoride)	146.05	63.8	-50.5
215. HNO_3 (nitric acid)	63.01	82.6	-41.6
216. H_2O (water Z)	18.015	100.00	0
217. H_2O_2 (hydrogen peroxide)	34.01	150	-0.41
218. H_2O_4S (sulfuric acid)	98.08	305	10.371
219. $H_2O_7S_2$ (pyrosulfuric acid)	178.14	d	35
220. H_3N (ammonia)	17.03	-33.38	-77.7
221. H_3O_4P (ortho-phosphoric acid)	98.00	213	42.4
222. (H_2O) ("anomalous water")[a]		(>500d)	<-40
223. O_2S (sulfur dioxide)	64.06	-10.1	-75.5
224. O_3S (sulfur trioxide-α)	80.06	44.8	62.3
225. C_4H_4O (furan)	68.08	31.4	-85.7
226. $C_3H_8O_3$ (glycerol)	92.09	290d	20

[a]Not a true liquid, but probably a concentrated solution of salts. For recent discussions, see J. Chem. Educ., 48, 663, 667 (1971).

S	ρ	n_D	ε	μ	$10^3\eta$	
i	1.099	1.6148	10.7	2.73		185.
i	1.093	1.6228	9.00	2.29		186.
i	0.964	1.5378				187.
i	0.865	1.4994	2.28(20)	0(ℓ)		188.
i	0.718	1.4054	1.97(20)		6.67	189.
i	1.483	1.658	4.83			190.
i	1.19	1.6326	5.04	1.59(ℓ)	29.4	191.
i	1.13	1.5939		1.42(B)		192.
i	1.740	1.7026		1.44(B)		193.
i	1.03	1.5898(85)	2.54(85)	0(B)	7.80(99.8)	194.
i	1.191	1.5138	8.5	2.8(B)		195.
i	0.97	1.5414	2.76(20)	0.6(ℓ)	2.00(25)	196.
i	0.90	1.4810	2.20(20)	0(B)	33.8(20)	197.
i	0.870	1.4695	2.17(20)	0	21.3(20)	198.
i	0.730	1.4102	1.99(20)	0(ℓ)	8.54	199.
i	1.075	1.5787(25)	3.65(30)	1.3		200.
i	0.999	1.5946		0.5(CCl_4)		201.
i	0.991	1.5752		0.66(B)		202.
i	1.05	1.4911	6.44(30)		97.2(37.8)	203.
i	0.986	1.4853	5.1	3.1(CCl_4)		204.
d	1.655(10)		9.25(20)	1.45		205.
d	1.71(0)	1.460(25)	13.9(22)	2.4(B)	11.5	206.
d	1.56(21)	1.516(14)	3.43	0.78		207.
d	2.35	1.601(14)	3.2(21)			208.
m	1.105	1.33844	77.9	1.86		209.
m	1.857(25)					210.
m	1.123(-50)	1.1574(25)	84(0)	1.82	2.4(6)	211.
p	1.726(25)				15.6	212.
p	2.99(23)					213.
i	1.88(-50.5)					214.
m	1.504(25)	1.3970(24)		2.17	8.9(20)	215.
-	1.000(4)	1.33299	78.5	1.84	10.1(20)	216.
m	1.442(25)		84.2(0)	2.2		217.
m	1.827(25)		101		245	218.
d	1.9					219.
m	0.725(-70)	1.325	26.7(-60)	1.47	2.5(-33)	220.
v	1.83(18)				1780	221.
m	1.4	1.48				222.
v	1.46(-10)		15.4(0)	1.62	4.28(-10)	223.
d	1.97		3.11(18)	0		224.
i	0.951	1.4214	2.95	0.66		225.
m	1.261	1.4735(25)	42.5	2.56	9450	226.

B. Indexes to General Constants Table

1. Alphabetical Name Index

Name	No.
Acetaldehyde	45
Acetic acid	47
Acetic anhydride	86
Acetone	68
Acetonitrile	40
Acetyl bromide	34
Acetyl chloride	36
2-Aminoethanol	61
1-Aminopropane	82
2-Aminopropane	83
Ammonia	220
Aniline	139
Anisole	160
Anomalous water	222
Antimony pentachloride	208
Antimony pentafluoride	213
Arsenic tribromide	1
Arsenic trichloride	2
Arsenic trifluoride	3
Benzene	137
Benzonitrile	158
Benzotrifluoride	157
Benzyl alcohol	161
m,m'-Bitolyl	201
o,o'-Bitolyl	202
Bromine	4
1-Bromobutane	95
Bromobenzene	132
dl-2-Bromobutane	96
Bromochlorofluoromethane	8
Bromochloromethane	14
Bromodichloromethane	9
Bromoethane	48
Bromoethylene	33
Bromoform	11
Bromomethane	19
1-Bromonaphthalene	190
1-Butanol	100
dl-2-Butanol	102
i-Butyl alcohol	103
t-Butyl alcohol	104
Carbitol	150
Carbon disulfide	25
Carbon tetrachloride	7
Cellosolve	106
Chlorobenzene	133
Chloroethane	49
2-Chloroethanol	50
Chloroethylene	35
Chloroform	12
1-Chloronaphthalene	191
1-Chloropropane	73
2-Chloropropane	74
Collidine	176
Cyclobutane	88
Cycloheptane	168
Cyclohexane	143
Cyclohexanol	144
Cyclopentane	114
cis-Decalin	197
trans-Decalin	198
Decane	199
Deuterium oxide	209
Dibromochloromethane	10
1,2-Dibromoethane	41
Dibromomethane	15
Dibutyl phthalate	203
o-Dichlorobenzene	131
1,1-Dichloroethane	42
1,2-Dichloroethane	43
Dichloromethane	16
1,2-Diethoxyethane	149
Diethylamine	109
Diethyl carbitol	183
Diethyl carbonate	116
Diethylene glycol	108
Diethylene glycol ethers	
Diethyl	183
Dimethyl (Diglyme)	151
Monoethyl	150
Monomethyl	128
Diethyl ether	101

Name	No.
N-Methylformamide	52
Methylpyridines (picolines)	
2-	140
3-	141
4-	142
Morpholine	99
Naphthalene	194
Neopentane	120
Neopentyl alcohol	126
Nitric acid	215
Nitrobenzene	136
Nitroethane	53
2-Nitroethanol	54
Nitromethane	22
Nonane	189
Octane	181
Oxalyl chloride	26
Oxetane	69
Paraldehyde	145
Pentane	121
neo-Pentane	120
Pentyl alcohols	
n-	122
dl-2-	123
3-	124
i-	125
neo-	126
t-	127
d-α-Phenethylamine	178
dl-α-Phenethylamine	179
β-Phenethylamine	180
Phenetole	175
Phenol	138
ortho-Phosphoric acid	221
Phosphorus oxychloride	206
Phosphorus tribromide	5
Phosphorus trichloride	207
Picoline (see Methylpyridine)	
Piperidine	117
1-Propanol	77
2-Propanol	78
Propionitrile	67
Propylene carbonate	87
γ-Pyran	112
Pyridine	111
Pyrosulfuric acid	219
Pyrrole	85
Pyrrolidine	97
Quinoline	186
Sulfolane	94
Sulfur dioxide	223
Sulfur hexafluoride	214
Sulfuric acid	218
Sulfuric acid-d_2	210
Sulfur trioxide-α	224
1,1,2,2-Tetrachloroethane	31
Tetrachloroethylene	27
Tetrahydrofuran	90
Tetrahydropyran	115
Tetralin	196
Tetramethylethylenediamine (TMEDA)	155
Tetramethylene sulfone	94
Tetramethylsilane	110
Tetramethylurea	119
Thionyl chloride	205
Thiophene	84
Toluene	159
Trichloroacetic acid	29
1,2,4-Trichlorobenzene	130
1,1,1-Trichloroethane	37
1,1,2-Trichloroethane	38
Trichloroethylene	28
Trichlorofluoromethane	6
Triethylamine	153
Triethyl borate	152
Triethylene glycol dimethyl ether (triglyme)	184
Triethyl phosphite	154
Trifluoroacetic acid	30
2,2,2-Trifluoroethanol	39
Trimethyl borate	81
1,3,5-Trimethylbenzene	188
2,2,4-Trimethylpentane	182
2,4,6-Trimethylpyridine (collidine)	176
Water	216
m-Xylene	172
o-Xylene	173
p-Xylene	174

2. Melting Point Index in Increasing Order: Range (Compound Numbers)

<-130 (35, 33, 49, 121); -130 to -125 (148, 77, 169); -124.9 to -120 (73, 64, 45); -119.9 to -115 (48, 74, 125, 101, 8, 102, 153); -114.9 to -110 (55, 105, 36, 95, 96, 207, 25, 6, 46); -109.9 to -105 (103, 182, 80, 105); -104.9 to -100 (110); -99.9 to -95 (147, 71, 23, 42, 34, 59, 68, 16, 83, 146, 159, 171); -94.9 to -90 (19, 114, 24, 67, 17, 170, 100); -89.9 to -85 (78, 211, 212, 14, 89, 225, 79); -84.9 to -80 (93, 82, 60, 70, 54); -79.9 to -75 (122, 220, 223); -74.9 to -70 (149, 28, 86, 87, 140); -69.9 to -65 (50, 20, 90); -64.9 to -60 (12, 75); -59.9 to -55 (9, 107, 181); -54.9 to -50 (15, 187, 188, 214, 53, 88, 109, 204); -49.9 to -45 (172, 40, 133, 118); -44.9 to -40 (188, 176, 183, 111, 116, 197, 91, 215, 134, 5); -39.9 to -35 (84, 160, 31, 38, 196, 43, 203); -34.9 to -30 (58, 135, 132, 37, 198, 199); -29.9 to -25 (175, 81, 157, 22, 51, 173); -24.9 to -20 (7, 98); -19.9 to -15 (27, 141, 131, 120, 26, 164, 186, 161, 30, 85); -14.9 to -10 (13, 2, 158, 168, 57, 108); -9.9 to -5 (117, 192, 127, 4, 139, 3, 165, 190); -4.9 to 0 (99, 191, 63, 119, 217, 216); 0.1 to 5 (195, 206, 177, 21, 72, 142, 209, 193, 129, 208); 5.1 to 10 (137, 136, 143, 213, 11, 18, 62, 201, 41); 10.1 to 15 (61, 218, 92, 145, 174, 210, 32); 15.1 to 20 (47, 130, 86, 202, 226); 20.1 to 25 (--); 25.1 to 30 (144, 104, 185, 200, 94, 76); 30.1 to 40 (1, 219, 66, 65); >40 (221, 138, 126, 224, 194, 44)

3. Boiling Point Index in Increasing Order: Range at or Near 760 mm Hg (Compound Numbers)

-35 to 0 (220, 223, 64, 17, 35, 24); 0 to 9.9 (19, 59, 120); 10 to 14.9 (88, 49, 46); 15 to 19.9 (33, 60, 211); 20 to 24.9 (45, 6); 25 to 29.9 (13, 110); 30 to 34.9 (83, 225, 101); 35 to 39.9 (74, 8, 121, 48); 40 to 44.9 (16, 80, 20, 224); 45 to 49.9 (69, 25, 73, 82, 114, 147); 50 to 54.9 (32, 36, 70); 55 to 59.9 (109, 68, 71, 42, 148, 4); 60 to 64.9 (12, 3, 26, 214, 23, 105); 65 to 69.9 (90, 81, 14, 146); 70 to 74.9 (30, 39, 207, 37); 75 to 79.9 (34, 7, 93, 55, 89); 80 to 84.9 (112, 205, 137, 143, 121, 40, 104, 78, 215, 107, 43, 84); 85 to 89.9 (134, 113, 28, 115, 97, 153); 90 to 94.9 (9, 72, 96); 95 to 99.9 (15, 67, 77, 170, 182, 102); 100 to 104.9 (216, 169, 18, 22, 209, 92, 95, 127, 157, 51); 105 to 109.9 (117, 91, 103, 206); 110 to 114.9 (159, 126, 38); 115 to 119.9 (53, 111, 124, 62, 100, 47, 168, 10, 123); 120 to 124.9 (152, 27, 155, 149, 79); 125 to 129.9 (181, 116, 50, 99, 145, 140); 130 to 134.9 (2, 85, 41, 125, 133); 135 to 139.9 (171, 122, 174, 172, 86); 140 to 144.9 (208, 141, 173); 145 to 149.9 (142, 31, 165); 150 to 154.9 (11, 213, 217, 189, 75); 155 to 159.9 (160, 132, 164, 154, 163); 160 to 164.9 (144, 151, 162, 212, 166, 188); 165 to 169.9 (98, 119); 170 to 174.9 (61, 175, 167, 5, 199); 175 to 179.9 (118, 176, 187); 180 to 184.9 (52, 131, 138, 139); 185 to 189.9 (178, 179, 198, 135, 56, 58, 183); 190 to 194.9 (158, 21, 54, 128, 177); 195 to 199.9 (150, 197, 180, 29, 57); 200 to 204.9 (44, 63); 205 to 209.9 (161, 76, 196); 210 to 214.9 (136, 221, 130); 215 to 219.9 (192, 194); 220 to 224.9 (1, 184); 225 to 234.9 (--); 235 to 239.9 (186); 240 to 244.9 (87, 185); 245 to 249.9 (108, 66); 250 to 254.9 (--); 255 to 259.9 (202, 200, 191); 260 to 279.9 (--); 280 to 284.9 (201, 190, 94, 195); 290 (226); 302 (193); 305 (218); 340 (203); 387 (204).

4. Dielectric Constant Index in Increasing Order: ε Range (Compound Numbers)

1 to 1.99 (121, 146, 170, 182, 181, 114, 189, 199); 2 to 2.99 (143, 169, 198, 197, 225, 92, 7, 174, 6, 137, 188, 172, 159, 171, 153, 27, 194, 173, 25, 80, 84, 196, 116); 3 to 3.99 (4, 224, 208, 28, 207, 26, 105, 109, 200, 5); 4 to 4.99 (175, 160, 101, 11, 29, 135, 12, 41, 190, 177); 5 to 5.99 (191, 204, 59, 132, 134, 83, 133, 3, 117, 127); 6 to 6.99 (93, 47, 49, 203, 15, 71, 139, 60); 7 to 7.99 (20, 95, 70, 99, 85, 37, 73); 8 to 8.99 (81, 31, 195, 96, 1, 16); 9 to 9.99 (186, 157, 205, 48, 24, 138, 19, 140, 131); 10 to 10.99 (42, 43, 185, 104); 11 to 11.99 (103); 12 to 12.99 (111, 2); 13 to 13.99 (161, 122, 145, 206); 14 to 14.99 (62, 125); 15 to 15.99 (36, 144, 223, 102); 16 to 16.99 (79, 34); 17 to 17.99 (100); 18 to 18.99 (78, 89); 19 to 19.99 (--); 20 to 21.99 (77, 68, 45, 86); 22 to 23.99 (46, 119); 24 to 25.99 (55, 158); 26 to 27.99 (50, 39, 220, 67); 28 to 29.99 (53); 30 to 31.99 (--); 32 to 33.99 (23); 34 to 35.99 (136); 36 to 37.99 (40, 75, 57, 98); 38 to 39.99 (22, 30); 40 to 49.9 (58, 94, 56, 226); 50 to 59.9 (18); 60 to 69.9 (87); 70 to 79.9 (209, 216); 80 to 89.9 (211, 217); 90 to 99.9 (--); 100-115 (218, 21, 13); 179 (76); 182 (52).

5. Binary Azeotrope Index

The number underlined forms a binary azeotrope with each of the numbers in the parentheses. The azeotrope data themselves are located on page 24.

7 (18, 22, 23, 40, 43, 47, 55, 68, 77, 78, 89, 93, 103, 104, 216); 12 (23, 55, 68, 71, 89, 136, 216); 15 (47, 55); 16 (23, 25, 101, 216); 18 (7, 25, 159, 216); 19 (23); 20 (23, 55, 78); 22 (7, 25, 28, 55, 77, 78, 92, 100, 137, 146, 159, 216); 23 (7, 12, 16, 19, 20, 25, 40, 43, 68, 70, 71, 93, 121, 137, 143, 136, 159, 170, 181); 25 (16, 18, 22, 23, 28, 42, 55, 68, 70, 71, 77, 78, 89, 93, 101, 104, 121, 216); 27 (47, 77, 78, 100, 102, 103, 216); 28 (22, 23, 43, 47, 55, 77, 78, 100, 111, 216); 38 (216); 40 (7, 23, 43, 55, 93, 137, 216); 41 (47, 100); 42 (7, 25, 55, 68); 43 (23, 28, 40, 55, 78, 103, 137, 143); 45 (101); 47 (7, 15, 27, 28, 41, 92, 100, 111, 132, 137, 143, 159, 170); 53 (143); 55 (7, 12, 15, 20, 22, 25, 28, 40, 42, 43, 89, 93, 121, 137, 143, 146, 159, 170, 216); 57 (139); 68 (7, 12, 23, 25, 42, 71, 73, 114, 121, 143, 146, 170, 216); 70 (25, 23); 71 (12, 23, 25, 68, 121, 143, 146, 170, 216); 73 (68); 77 (7, 22, 25, 28, 27, 92, 137, 143, 146, 159, 216); 78 (7, 20, 22, 25, 27, 28, 43, 89, 93, 121, 137, 143, 146, 216); 89 (7, 12, 25, 55, 78, 137, 146, 216); 90 (146); 92 (22, 47, 77, 143, 216); 93 (7, 23, 25, 40, 55, 78, 143, 216); 100 (22, 27, 28, 41, 47, 111, 143, 146, 159, 170, 216); 101 (16, 25, 45, 121, 216); 102 (27, 137, 143, 159, 216); 103 (7, 27, 43, 137, 143, 159, 216); 104 (7, 25, 137, 143, 146, 216); 111 (28, 47, 100, 159, 170, 216); 114 (68); 121 (23, 25, 55, 68, 71, 78, 101, 216); 132 (47); 136 (12); 137 (22, 23, 40, 43, 45, 55, 77, 78, 89, 102, 103, 104, 143, 170, 216); 139 (57); 143 (23, 43, 47, 53, 55, 68, 71, 77, 78, 92, 93, 100, 102, 103, 104, 137, 216); 146 (22, 23, 55, 68, 71, 77, 78, 89, 90, 100, 104, 216); 159 (18, 22, 23, 47, 55, 77, 100, 102, 103, 111, 216); 161 (216); 170 (23, 47, 55, 68, 71, 100, 111, 137, 216); 181 (23, 216); 189 (216); 216 (7, 12, 16, 18, 22, 25, 27, 28, 38, 40, 55, 68, 71, 77, 78, 89, 92, 93, 100, 101, 102, 103, 104, 111, 121, 137, 143, 146, 159, 161, 170, 181, 189).

C. Selected Vapor Pressure Data for Some Liquids[a]

Compound	Mp	Bp	°C	mm Hg	°C	mm Hg	°C	mm Hg
Acetone	-95	56.5	-59	1	0	67	20	177
Acetonitrile	-41	81.8	-47	1	7.7	40	27	100
Benzene	5.5	80.1	-37	1	0	24.5	20	75
Carbon disulfide	-110.8	46.5	-74	1	-5	100	28	400
Carbon tetrachloride	-22.8	76.8	-50	1	0	32	20	87
Chloroform	-63.5	61.3	-58	1	10.4	100	43	400
Diethyl ether	-116.3	34.6	-78	0.7	0	183	20	436
p-Dioxane	10	101.1	-36	1	25	40	82	400
Ethanol	-112	78.4	-31	1	19	40	64	400
Hexane	-95.3	68.7	-54	1	-2.3	40	50	400
Methanol	-97.8	64.7	-44	1	0	30	20	94
Pyridine	-42	115.4	-19	1	38	40	96	400
Toluene	-95	110.6	-78	0.02	0	6.8	20	22
Water	0	100	-78	0.001	0	4.6	10	17.5

[a]Partly from Handbook of Chemistry and Physics, 50th ed., CRC Press, Cleveland, 1970, p. 148.

D. Commercial and Common Name Solvent Systems

The number in parentheses after the name refers to an entry in the table of solvent properties, page 4.

1. Carbitols: Diethylene Glycol Ethers -- $ROCH_2CH_2OCH_2CH_2OR'$

Name	Formula	R	R'	Bp
Methyl carbitol (128)	$C_5H_{12}O_3$	Me	H	193
Carbitol (150)	$C_6H_{14}O_3$	Et	H	195
Diethyl carbitol (183)	$C_8H_{18}O_3$	Et	Et	189

2. Cellosolves: Ethylene Glycol Ethers -- $ROCH_2CH_2OR'$

Name	Formula	R	R'	Bp
Cellosolve (106)	$C_4H_{10}O_2$	Et	H	135
Dimethyl cellosolve (107)	$C_4H_{10}O_2$	Me	Me	85
Diethyl cellosolve (149)	$C_6H_{14}O$	Et	Et	121
Methyl cellosolve (79)	$C_3H_8O_2$	Me	H	124
Cellosolve acetate	$C_6H_{12}O_3$	Et	$COCH_3$	156
Butyl cellosolve	$C_6H_{14}O_2$	n-Bu	H	171

3. Crownanes and Cryptates

Crownanes are cyclic polyethers containing from 9 to 60 atoms including 3 to 20 oxygen atoms per ring; many are solids at room temperature and are not necessarily used as solvents by themselves. Some form very stable complexes with a variety of metal and ammonium cations. See C. J. Pedersen, J. Amer. Chem. Soc., 89, 7017 (1967); 92, 386, 391 (1970); and H. K. Frensdorff, ibid., 93, 600 (1971). Somewhat related "solvents" are the bicyclic diamines (nitrogens at bridgeheads) containing oxymethylene bridges; such solvents are capable of forming encapsulated metal cations (cryptates). See J. Lehn, J. Sauvage, and B. Dietrich, ibid., 92, 2917 (1970). A recent review of ion binding properties of macrocyclic polyethers, polythioethers, polyamines, and related molecules has been published [J. J. Christensen, J. O. Hill, and R. M. Izatt, Science, 174, 459 (1971)]. A similar review on the polyethers is that by U. Takaki, T. E. Hogen Esch, and J. Smid, J. Amer. Chem. Soc., 93, 6760 (1971).*

4. Dowtherm A

A eutectic mixture of 26.5% biphenyl and 73.5% diphenyl ether; it is used as a high-temperature, inert solvent; mp 12°C, bp 258°C, stable to 385°C.

5. Freons: Fluorinated Hydrocarbons and Their Mixtures

Commercial Designation	Formula	Bp	Mp
11	CCl_3F (6)	24	-111
12	CCl_2F_2	-29.8	-158
13	$CClF_3$	-81.1	-181
21	$CHFCl_2$	9	-135
22	$CHClF_2$	-40.8	-146
112	CCl_2FCCl_2F	93	25
113	CCl_2FCClF_2	47.7	-36
114	$CClF_2CClF_2$	3.8	

6. Glymes: $CH_3O(CH_2CH_2O)_nCH_3$

Name	n	Formula	Bp
Glyme (107)	1	$C_4H_{10}O_2$	83
Diglyme (151)	2	$C_6H_{14}O_3$	161
Triglyme	3	$C_8H_{18}O_4$	222
Tetraglyme	4	$C_{10}H_{22}O_4$	119 (2 mm)

For information on glyme through heptaglyme, see K. Machida and T. Miyazawa, Spec. Chim. Acta, 20, 1865 (1964).

*Potassium permanganate is solubilized in benzene by complexing with "dicyclohexyl-18-crown-6" providing a convenient and efficient oxidant [D. J. Sam and H. E. Simmons, J. Amer. Chem. Soc., 94, 4024 (1972)].

7. Kerosene

A distillate mixture from crude petroleum, boiling range about 180-300°C. (Also called No. 1 Fuel Oil or Coal Oil.)

8. Naphthas

Synonym for "mineral spirits" or "petroleum spirits", naphthas are mixtures of hydrocarbons obtained as a distillate fraction from crude petroleum (e.g., that with bp range 152-204°C, sp. gr. 0.769 (20°C)). They are similar to ligroins and petroleum ethers.

9. Petroleum Ethers

Commercial mixtures of light paraffins. Those in the high boiling range (e.g., 65°-75°) are also called ligroins.

	Ordinary	ACS	ASTM I	ASTM II	ASTM III	
Bp Range	40-70	35-60	26-31	33-38	52-79	60-110
Sp.gr.(60°F)	0.635-0.660	0.64	0.63	0.63	0.67	0.69

10. Skellysolves: - saturated hydrocarbon mixtures

A: - mostly n-pentane, bp range 28-38°C.
B: - mostly n-hexane, bp range 60-71°C.
C: - mostly n-heptane, bp range 88-100°C.
D: - mixed heptanes, bp range 80-119°C.
F: - virtually same as ACS petroleum ether, bp range 35-60°C.

11. Superacids and Superbases

Fluorosulfuric acid (212) (FSO_3H) by itself and in mixtures with SO_3 and/or Lewis acids [e.g. SbF_5 (213)]. The combination FSO_3H-SbF_5 is also called "magic acid." [See R. J. Gillespie, Accts. Chem. Res., 1, 202 (1968) and G. A. Olah, Science, 168, 1298 (1970)]. Perchloric acid in chlorinated hydrocarbon solvents also has "superacid" qualities (e.g., K. Baum, ibid., 92, 2927 (1970)). Of similar interest is trifluoromethane sulfonic acid (CF_3SO_3H), said to be the strongest monobasic acid known; it is hydrolytically stable and nonoxidizing. Its properties are as follows: bp 162° (760mm); 54° (8 mm) [forms stable monohydrate, mp 34°, bp 96° (1 mm)]; ρ 1.696 g/cc at 25°; n_D^{25} 1.3250; dissociation constant (in HOAc), K = 1.26 X 10^{-5}.

For details, see Gramstad, Tidssker. Kjemi, Bergvesen Met., 19, 62 (1959) [C.A., 54, 12739 (1960)]; Tiera, U. S. 3,427,336; Hansen, J. Org. Chem., 30, 4322 (1965); and A. Streitweiser et al., J. Amer. Chem. Soc., 90, 1598 (1968). The compound is available from Pierce Chemical Co., Box 117, Rockford, Ill. 61105 (U.S.A.). It is manufactured under the name "Trimsylate" by the 3M Co.

Superacid reagents and related substances are commercially available from Cationics, Inc., 653 Alpha Drive, Cleveland, Ohio 44143.

Recently, potassium salts of primary and secondary amines have been shown to be extremely strong bases. These "superbases" are soluble in organic solvents [C. A. Brown, J. Amer. Chem. Soc., 95, 982 (1973)]. For other strong bases, see p. 67.

E. Empirical Solvent Parameters

1. Definitions

Empirical solvent parameters are used to correlate the effect of a change in solvent with a change in a reaction rate, equilibrium, or some other molecular property. Various numerical scales have been devised, using a standard reference reaction or property. For example, the most common parameter, known as the Z-value, is based on the influence a solvent has on the position of absorption of a particular transition for a pyridinium salt. If the parameter is a reliable one, the values should correlate well with similar or perhaps different measurements in some other system.

For more complete discussion of solvent parameters and medium effects, see E. J. Kosower, An Introduction to Physical Organic Chemistry, Part 2, Wiley, New York, 1968, and C. Reichardt, Angew. Chem., Intern. Ed. Engl., 4, 29 (1965). Most of the data given below are from these sources. For recent developments, including multiple correlations of solvent properties, see M. Dack, Chem. Britain, 5, 347 (1970).

Solvent Parameter	Definition
Z	The transition energy ($E = hc\nu = 2.859 \times 10^5/\lambda$, λ in Å) for the longest wavelength absorption band (a charge transfer band) for 1-ethyl-4-carbomethoxypyridinium iodide in the given solvent.
E_T	Same as Z, except for the use of a pyridinium N-phenolbetaine as the test substance.
Y	A quantitative measure of solvent ionizing power; Y-values are determined from the Winstein-Grunwald equation, $\log(k/k_0) = mY$, where k and k_0 are S_n1 rate constants at 25° in a given solvent and in 80% ethanol, respectively, and m = 1.000 for t-BuCl, the standard reactant (Y = 0 for 80% ethanol).
$\log k_{ion}$	Based on the k at 75° for the S_n1 reaction of p-methoxyneophyl tosylate [$CH_3OC_6H_4C(CH_3)_2CH_2OTs$]; parallels Y values.
Ω	The log of endo/exo ratio obtained in the Diels-Alder reaction of cyclopentadiene and methyl acrylate; of limited scope, especially in protic solvents.
δ	Called the solubility parameter (Hildebrand), it correlates solubility behavior of non-electrolytes and is a measure of the energy required to form a cavity to accommodate the transition state in a given solvent.
S	An empirical generalization of other solvent parameters, from $\log k_S / \log k_E = SR$, where k_S = a rate constant, equilibrium constant, or function of a spectral shift in a solvent, compared to that in absolute ethanol (k_E); R reflects the sensitivity of the system toward solvent change.

2. Z-Values (kcal/mole)

Name	Number[a]	Z
Benzene	137	54.0
Phenetole	175	58.9
2,2,4-Trimethylpentane	182	60.1
Glyme	107	62.1
Hexamethylphosphoramide	156	62.8
Chloroform	12	63.2
Pyridine	111	64.0
Dichloromethane	16	64.2
Acetone	68	65.7
N,N-Dimethylacetamide	98	66.9
N,N-Dimethylformamide	75	68.5
Dimethyl sulfoxide	56	71.1
Acetonitrile	40	71.3
t-Butyl alcohol	104	71.3
2-Propanol	78	76.3
Sulfolane	94	77.5
1-Butanol	100	77.7
N-Methylacetamide	76	77.9
1-Propanol	77	78.3
Acetic acid	47	79.2
Ethanol	55	79.6
95% Ethanol	-	81.2
Formamide	21	83.3
Methanol	23	83.6
Ethylene glycol	57	85.1
Water	216	94.6

[a]Refers to compound number in table of solvents, page 4.

3. Correlation of Various Empirical Solvent Parameters with Z-values at 25°C

The following equations and parameters are derived from (a) the E_T-value correlation of C. Reichardt, Angew. Chem., Intern. Ed. Engl., 4, 29 (1965), and (b) the δ-value correlation of H. Herbrandson and F. R. Newfeld, J. Org. Chem., 31, 1140 (1966).

(a) $y = aZ - b$			(b) $y = cZ - d$		
y	a	b	y	c	d
E_T	0.795	10.92	E_T	0.870	12.07
$\log k_{ion}$	0.142	14.31	δ	0.278	8.97
Ω	0.0122	0.207	Y	0.146	13.45
S	0.0107	0.851	S	0.0128	1.037

II. AZEOTROPIC DATA

The tables in this section give the properties of the various binary and ternary azeotropes at 760 torr (mm Hg) formed by the liquids included in the table of the properties of solvents and common liquids (p. 4). In the various sources for these data much discrepancy was found. The data on boiling points were much more reliable than those on percent composition but nevertheless contained enough discrepancies to justify the reporting of bp values only to the nearest degree. In some cases this discrepancy was resolved by finding agreement in two, three, or four of the sources, although usually it was impossible to determine if the data had come from the same original source. By far the most reliable figures appear in Azeotropic Data, Vols. I and II, Advances in Chemistry Series, Nos. 6 and 35, American Chemical Society, Washington, D.C., 1952 and 1962, from which most of the information appearing here was obtained. When a severe discrepancy could not be resolved, the azeotrope was not included in this table. A binary azeotrope index is given on page 18. The column entitled "No." in the tables below refers to entries in the main table of the properties of solvents and common liquids (p. 4). Composition of azeotropes is in weight percent.

A. Binary Azeotropes

No.	Components	Bp	%
45	Acetaldehyde	20	76
101	Ethyl ether		24
47	Acetic acid	80	2
137	Benzene		98
47	Acetic acid	118	95
132	Bromobenzene		5
47	Acetic acid	120	43
100	1-Butanol		57
47	Acetic acid	76	2
7	Carbon tetrachloride		98
47	Acetic acid	80	2
143	Cyclohexane		98
47	Acetic acid	114	55
41	1,2-Dibromoethane		45
47	Acetic acid	95	16
15	Dibromomethane		84
47	Acetic acid	119	80
92	1,4-Dioxane		20
47	Acetic acid	92	33
170	Heptane		67
47	Acetic acid	138	51
111	Pyridine		49
47	Acetic acid	107	39
27	Tetrachloroethylene		61
47	Acetic acid	101	28
159	Toluene		72
47	Acetic acid	86	4
28	Trichloroethylene		96
68	Acetone	39	33
25	Carbon disulfide		67
68	Acetone	56	88
7	Carbon tetrachloride		12
68	Acetone	65	21
12	Chloroform		79
68	Acetone	46	15
73	1-Chloropropane		85
68	Acetone	53	67
143	Cyclohexane		33
68	Acetone	41	36
114	Cyclopentane		64
68	Acetone	58	30
42	1,1-Dichloroethane		70
68	Acetone	50	59
146	Hexane		41
68	Acetone	56	90
170	Heptane		10
68	Acetone	56	88
23	Methanol		12

No.	Components	Bp	%
68	Acetone	56	48
71	Methyl acetate		52
68	Acetone	33	20
121	Pentane		80
68	Acetone	56	88
216	Water		12
40	Acetonitrile	73	34
137	Benzene		66
40	Acetonitrile	65	17
7	Carbon tetrachloride		83
40	Acetonitrile	79	49
43	1,2-Dichloroethane		51
40	Acetonitrile	73	44
55	Ethanol		56
40	Acetonitrile	75	23
93	Ethyl acetate		77
40	Acetonitrile	63	81
23	Methanol		19
40	Acetonitrile	77	84
216	Water		16
139	Aniline	181	76
57	Ethylene glycol		24
137	Benzene	79	85
102	2-Butanol		15
137	Benzene	74	63
104	t-Butyl alcohol		37
137	Benzene	79	92
103	i-Butyl alcohol		8
137	Benzene	78	55
143	Cyclohexane		45
137	Benzene	80	85
43	1,2-Dichloroethane		15
137	Benzene	68	68
55	Ethanol		32
137	Benzene	78	62
89	Ethyl methyl ketone		38
137	Benzene	80	99
170	Heptane		1
137	Benzene	58	61
23	Methanol		39
137	Benzene	79	87
22	Nitromethane		13
137	Benzene	77	83
77	1-Propanol		17
137	Benzene	72	66
78	2-Propanol		34
137	Benzene	69	91
216	Water		9
161	Benzyl alcohol	100	9
216	Water		91
100	1-Butanol	80	10
143	Cyclohexane		90
100	1-Butanol	115	44
41	1,2-Dibromoethane		56
100	1-Butanol	94	18
170	Heptane		82
100	1-Butanol	68	3
146	Hexane		97
100	1-Butanol	98	29
22	Nitromethane		71
100	1-Butanol	119	70
111	Pyridine		30
100	1-Butanol	110	32
27	Tetrachloroethylene		68
100	1-Butanol	106	28
159	Toluene		72
100	1-Butanol	87	3
28	Trichloroethylene		97
100	1-Butanol	93	57
216	Water		43
102	2-Butanol	76	18
143	Cyclohexane		82
102	2-Butanol	97	57
27	Tetrachloroethylene		43
102	2-Butanol	95	55
159	Toluene		45
102	2-Butanol	87	73
216	Water		27
103	i-Butyl alcohol	76	5
7	Carbon tetrachloride		95
103	i-Butyl alcohol	78	14
143	Cyclohexane		86
103	i-Butyl alcohol	83	6
43	1,2-Dichloroethane		94

No.	Components	Bp	%
103 159	i-Butyl alcohol Toluene	101	45 55
103 27	i-Butyl alcohol Trichloroethylene	85	9 91
103 216	i-Butyl alcohol Water	90	67 33
104 25	t-Butyl alcohol Carbon disulfide	45	7 93
104 7	t-Butyl alcohol Carbon tetrachloride	71	17 83
104 143	t-Butyl alcohol Cyclohexane	72	37 63
104 146	t-Butyl alcohol Hexane	64	22 78
104 216	t-Butyl alcohol Water	80	88 12
25 42	Carbon disulfide 1,1-Dichloroethane	46	94 6
25 16	Carbon disulfide Dichloromethane	36	35 65
25 55	Carbon disulfide Ethanol	42	91 9
25 93	Carbon disulfide Ethyl acetate	46	97 3
25 101	Carbon disulfide Ethyl ether	34	1 99
25 70	Carbon disulfide Ethyl formate	39	63 37
25 89	Carbon disulfide Ethyl methyl ketone	46	84 16
25 18	Carbon disulfide Formic acid	43	83 17
25 23	Carbon disulfide Methanol	38	86 14
25 71	Carbon disulfide Methyl acetate	40	70 30
25 22	Carbon disulfide Nitromethane	44	90 10
25 121	Carbon disulfide Pentane	36	11 89
25 77	Carbon disulfide 1-Propanol	46	95 5
25 78	Carbon disulfide 2-Propanol	44	92 8
25 216	Carbon disulfide Water	44	98 2
7 43	Carbon tetrachloride 1,2-Dichloroethane	75	83 17
7 55	Carbon tetrachloride Ethanol	65	84 16
7 93	Carbon tetrachloride Ethyl acetate	75	57 43
7 89	Carbon tetrachloride Ethyl methyl ketone	74	71 29
7 18	Carbon tetrachloride Formic acid	67	82 18
7 23	Carbon tetrachloride Methanol	56	79 21
7 22	Carbon tetrachloride Nitromethane	71	83 17
7 77	Carbon tetrachloride 1-Propanol	73	89 11
7 78	Carbon tetrachloride 2-Propanol	69	82 18
7 216	Carbon tetrachloride Water	66	96 4
12 55	Chloroform Ethanol	59	93 7
12 89	Chloroform Ethyl methyl ketone	80	17 83
12 136	Chloroform Hexane	60	72 28
12 23	Chloroform Methanol	53	87 13
12 71	Chloroform Methyl acetate	65	77 23
12 216	Chloroform Water	56	97 3
143 43	Cyclohexane 1,2-Dichloroethane	75	50 50
143 92	Cyclohexane 1,4-Dioxane	80	75 25
143 55	Cyclohexane Ethanol	65	69 31

No.	Components	Bp	%
143	Cyclohexane	72	44
93	Ethyl acetate		56
143	Cyclohexane	45	63
23	Methanol		37
143	Cyclohexane	55	17
71	Methyl acetate		83
143	Cyclohexane	70	73
53	Nitromethane		27
143	Cyclohexane	75	81
77	1-Propanol		19
143	Cyclohexane	69	68
78	2-Propanol		32
143	Cyclohexane	70	92
216	Water		8
15	Dibromomethane	76	62
55	Ethanol		38
42	1,1-Dichloroethane	55	86
55	Ethanol		14
43	1,2-Dichloroethane	71	63
55	Ethanol		27
43	1,2-Dichloroethane	60	65
23	Methanol		35
43	1,2-Dichloroethane	73	61
78	2-Propanol		39
43	1,2-Dichloroethane	82	61
28	Trichloroethylene		39
16	Dichloromethane	41	70
101	Ethyl ether		30
16	Dichloromethane	38	93
23	Methanol		7
16	Dichloromethane	39	98
216	Water		2
92	1,4-Dioxane	100	43
22	Nitromethane		57
92	1,4-Dioxane	95	45
77	1-Propanol		55
92	1,4-Dioxane	88	82
216	Water		18
55	Ethanol	72	26
93	Ethyl acetate		74
55	Ethanol	75	34
89	Ethyl methyl ketone		66
55	Ethanol	72	48
170	Heptane		52
55	Ethanol	58	21
146	Hexane		79
55	Ethanol	41	3
20	Iodomethane		97
55	Ethanol	76	71
22	Nitromethane		29
55	Ethanol	61	25
121	Pentane		75
55	Ethanol	77	68
159	Toluene		32
55	Ethanol	71	27
28	Trichloroethylene		73
55	Ethanol	78	96
216	Water		4
93	Ethyl acetate	62	51
23	Methanol		49
93	Ethyl acetate	76	75
78	2-Propanol		25
93	Ethyl acetate	71	92
216	Water		8
101	Ethyl ether	33	68
121	Pentane		32
101	Ethyl ether	34	99
216	Water		1
70	Ethyl formate	51	84
23	Methanol		16
89	Ethyl methyl ketone	64	30
146	Hexane		70
89	Ethyl methyl ketone	78	68
78	2-Propanol		32
89	Ethyl methyl ketone	73	89
216	Water		11
18	Formic acid	86	50
159	Toluene		50
18	Formic acid	100	74
216	Water		26
170	Heptane	59	46
23	Methanol		54
170	Heptane	57	4
71	Methyl acetate		96

No.	Components	Bp	%
170	Heptane	96	75
111	Pyridine		25
170	Heptane	79	87
216	Water		13
146	Hexane	50	74
23	Methanol		26
146	Hexane	52	39
71	Methyl acetate		61
146	Hexane	62	79
22	Nitromethane		21
146	Hexane	66	96
77	1-Propanol		4
146	Hexane	63	77
78	2-Propanol		23
146	Hexane	63	46
90	Tetrahydrofuran		54
146	Hexane	62	94
216	Water		6
20	Iodomethane	38	96
23	Methanol		4
20	Iodomethane	42	98
78	2-Propanol		2
23	Methanol	54	18
71	Methyl acetate		82
23	Methanol	63	68
181	Octane		32
23	Methanol	31	7
121	Pentane		93
23	Methanol	64	64
19	Tetrachloroethylene		36
23	Methanol	64	71
159	Toluene		29
23	Methanol	69	38
28	Trichloroethylene		62
71	Methyl acetate	34	22
121	Pentane		88
71	Methyl acetate	56	95
216	Water		5
22	Nitromethane	90	48
77	1-Propanol		52
22	Nitromethane	79	28
78	2-Propanol		72
22	Nitromethane	97	55
159	Toluene		45
22	Nitromethane	81	20
28	Trichloroethylene		80
22	Nitromethane	84	76
216	Water		24
189	Nonane	95	60
216	Water		40
181	Octane	90	74
216	Water		26
121	Pentane	36	6
78	2-Propanol		94
121	Pentane	35	99
216	Water		1
77	1-Propanol	94	48
27	Tetrachloroethylene		52
77	1-Propanol	93	49
159	Toluene		51
77	1-Propanol	82	17
28	Trichloroethylene		83
77	1-Propanol	88	72
216	Water		28
78	2-Propanol	82	81
27	Tetrachloroethylene		19
78	2-Propanol	76	30
28	Trichloroethylene		70
78	2-Propanol	80	88
216	Water		12
111	Pyridine	113	49
28	Tetrachloroethylene		51
111	Pyridine	110	22
159	Toluene		78
111	Pyridine	94	58
216	Water		42
27	Tetrachloroethylene	88	84
216	Water		16
159	Toluene	85	80
216	Water		20
38	1,1,2-Trichloroethane	86	84
216	Water		16
28	Trichloroethylene	73	95
216	Water		5

B. Ternary Azeotropes

No.	Components	Bp	%
68	Acetone		24
25	Carbon disulfide	38	75
216	Water		1
68	Acetone		24
12	Chloroform	63	65
55	Ethanol		11
68	Acetone		30
12	Chloroform	58	47
23	Methanol		23
68	Acetone		38
12	Chloroform	60	58
216	Water		4
68	Acetone		44
143	Cyclohexane	52	40
23	Methanol		16
68	Acetone		31
146	Hexane	47	60
23	Methanol		14
68	Acetone		6
23	Methanol	54	17
71	Methyl acetate		77
40	Acetonitrile		23
137	Benzene	66	69
216	Water		8
40	Acetonitrile		44
55	Ethanol	73	55
216	Water		1
40	Acetonitrile		21
28	Trichloroethylene	67	73
216	Water		6
137	Benzene		74
55	Ethanol	65	19
216	Water		7
137	Benzene		48
100	1-Butanol	77	4
143	Cyclohexane		48
137	Benzene		88
102	2-Butanol	38	5
216	Water		7
137	Benzene		42
103	i-Butyl alcohol	81	8
143	Cyclohexane		50
137	Benzene		71
104	t-Butyl alcohol	67	21
216	Water		8
137	Benzene		11
143	Cyclohexane	65	59
55	Ethanol		30
137	Benzene		65
89	Ethyl methyl ketone	68	26
216	Water		9
137	Benzene		82
77	1-Propanol	69	9
216	Water		9
137	Benzene		72
78	2-Propanol	66	20
216	Water		8
100	1-Butanol		8
170	Heptane	78	51
216	Water		41
100	1-Butanol		3
146	Hexane	62	78
216	Water		19
100	1-Butanol		15
181	Octane	86	25
216	Water		60
100	1-Butanol		12
111	Pyridine	109	21
159	Toluene		67
106	t-Butyl alcohol		12
7	Carbon tetrachloride	65	85
216	Water		3
25	Carbon disulfide		93
55	Ethanol	41	5
216	Water		2
7	Carbon tetrachloride		86
55	Ethanol	62	10
216	Water		4

No.	Components	Bp	%	No.	Components	Bp	%
7	Carbon tetrachloride		84	55	Ethanol		8
77	1-Propanol	65	11	93	Ethyl acetate	70	83
216	Water		5	216	Water		9
12	Chloroform		91	55	Ethanol		14
55	Ethanol	78	5	89	Ethyl methyl ketone	73	75
216	Water		4	216	Water		11
12	Chloroform		81	55	Ethanol		33
23	Methanol	53	15	170	Heptane	69	61
216	Water		4	216	Water		6
143	Cyclohexane		60	55	Ethanol		12
89	Ethyl methyl ketone	64	35	146	Hexane	56	85
216	Water		5	216	Water		3
143	Cyclohexane		33	55	Ethanol		37
23	Methanol	51	18	159	Toluene	74	51
71	Methyl acetate		49	216	Water		12
143	Cyclohexane		71	89	Ethyl methyl ketone		88
106	t-Butyl alcohol	65	21	78	2-Propanol	73	1
216	Water		8	216	Water		11
143	Cyclohexane		82	89	Ethyl methyl ketone		22
77	1-Propanol	67	10	146	Hexane	55	77
216	Water		8	216	Water		1
143	Cyclohexane		74	146	Hexane		59
78	2-Propanol	64	29	23	Methanol	45	14
216	Water		7	71	Methyl acetate		27
43	1,2-Dichloroethane		77	78	2-Propanol		38
55	Ethanol	68	16	159	Toluene	76	49
216	Water		7	216	Water		13
43	1,2-Dichloroethane		73				
78	2-Propanol	70	19				
216	Water		8				

III. EMPIRICAL BOILING POINT-PRESSURE RELATIONSHIPS

A. Approximate Changes in Boiling Point with Pressure for Unassociated Liquids

The table on pages 32 and 33 provides a rapid estimation of bp as a function of pressure. For example, a substance with bp 250° at 760 mm will distill at 93° under 2 mm.

B. Empirical Equation for Specific Substances

The following equation is useful for calculating vp at a given T, but its application is limited to those compounds for which a and b values are known.

$$\log_{10} p = -\frac{0.05223a}{T} + b$$

Calculate values of p (mm Hg) at T°K using values of a and b given, for example, in the CRC Handbook of Chemistry and Physics, 46th ed., p. D-124, or 50th ed., p. D-167.

In a recent innovation, a computer program (Fortran IV) can deduce the bp of organic compounds at reduced pressure. The program uses handbook data together with the Clausius-Clapeyron equation and prints out a table of bp as a function of vp. The program is available from A. G. Kreiger and C. K. Wallace, Jackson Community College, Jackson, Mich. 49201 [See J. Chem. Educ., 48, 457 (1971)].

C. Vapor Pressure-Temperature Nomograph

The figures on pages 34 and 35 are taken from Industrial and Engineering Chemistry, Vol. 38, 1946, p. 320. Copyright 1946 by the American Chemical Society; reproduced by permission of the copyright owner. The nomographs are used in conjunction with the following classification of structures by Groups.

Groups of Compounds Represented in Nomographs

Group 1

Anthracene
Anthraquinone
Butylethylene
Carbon disulfide
Phenanthrene
Sulfur monochloride
Trichloroethylene

Group 2

Benzaldehyde
Benzonitrile
Benzophenone
Camphor
Carbon suboxide
Carbon sulfoselenide
Chlorohydrocarbons
Dibenzyl ketone
Dimethylsilicane
Ethers
Halogenated hydrocarbons
Hydrocarbons
Hydrogen fluoride
Methyl ethyl ketone
Methyl salicylate
Nitrotoluenes
Nitrotoluidines
Phosgene
Phthalic anhydride
Quinoline
Sulfides

Group 3

Acetaldehyde
Acetone
Amines
Chloroanilines
Cyanogen chloride
Esters
Ethylene oxide
Formic acid
Hydrogen cyanide
Mercuric chloride
Methyl benzoate
Methyl ether
Methyl ethyl ether
Naphthols
Nitrobenzene
Nitromethane
Tetranitromethane

Group 4

Acetic acid
Acetophenone
Cresols
Cyanogen

(Continued)

Pressure (mm)	Temperature (°C)									
760	130	140	150	160	170	180	190	200	210	220
20	46	54	62	70	78	86	94	101	109	117
19	45	53	61	69	77	85	93	100	108	116
18	44	52	60	68	76	84	92	99	107	115
17	43	51	59	67	75	83	91	98	106	114
16	42	50	58	66	73	82	90	97	105	112
15	41	49	57	64	72	80	88	96	103	110
14	40	48	56	63	71	78	86	94	101	108
13	38	46	54	62	70	77	85	92	99	106
12	36	44	52	60	68	75	83	90	97	104
11	35	43	51	59	66	74	82	89	96	102
10	34	42	50	57	65	72	80	87	94	100
9	33	40	48	55	62	70	78	85	92	98
8	31	38	45	50	59	67	75	83	89	95
7	28	36	43	48	57	64	72	80	86	92
6	26	33	40	47	54	61	69	76	83	89
5	22	29	37	44	51	58	66	73	80	87
4	18	26	34	41	48	55	63	70	77	84
3	15	21	29	36	45	50	57	64	71	78
2	11	15	23	30	37	44	51	57	65	72
1	-1	6	13	20	27	33	40	47	54	60

Group 4

Dimethylamine
Dimethyl oxalate
Ethylamine
Glycol diacetate
Methyl formate
Nitrosyl chloride
Sulfur dioxide

Group 5

Ammonia
Benzyl alcohol
Methylamine
Phenol
Propionic acid

Group 6

Acetic anhydride
Isobutyric acid
Water

Group 7

Benzoic acid
Butyric acid
Ethylene glycol
Heptanoic acid
Isocaproic acid
Methyl alcohol
Valeric acid

Group 8

n-Amyl alcohol
Ethyl alcohol
Isoamyl alcohol
Isobutyl alcohol
Mercurous chloride
n-Propyl alcohol

With Pressure for Unassociated Liquids (see p. 30)

Temperature (°C)									Pressure (mm)
230	240	250	260	270	280	290	300	310	760
125	133	141	149	157	165	173	181	189	20
124	132	140	148	156	164	172	180	187	19
123	131	139	146	154	162	170	178	186	18
122	130	138	145	153	161	169	176	184	17
120	128	136	143	151	159	167	174	182	16
118	127	135	142	150	157	165	173	181	15
116	125	133	140	148	155	163	171	179	14
114	123	132	138	146	154	162	169	177	13
112	121	129	136	144	152	160	167	175	12
111	120	128	135	143	150	158	166	174	11
109	118	125	133	141	148	156	164	172	10
107	115	123	130	138	145	153	162	169	9
104	114	120	127	135	142	150	158	166	8
101	110	117	124	132	139	147	155	163	7
97	106	113	120	128	135	143	151	159	6
95	103	110	117	125	132	140	148	156	5
92	99	106	113	121	128	136	144	152	4
86	93	100	107	115	122	129	136	144	3
79	86	93	100	108	115	122	129	136	2
67	74	81	87	94	101	107	114	120	1

Vapor Pressure-Temperature Nomograph

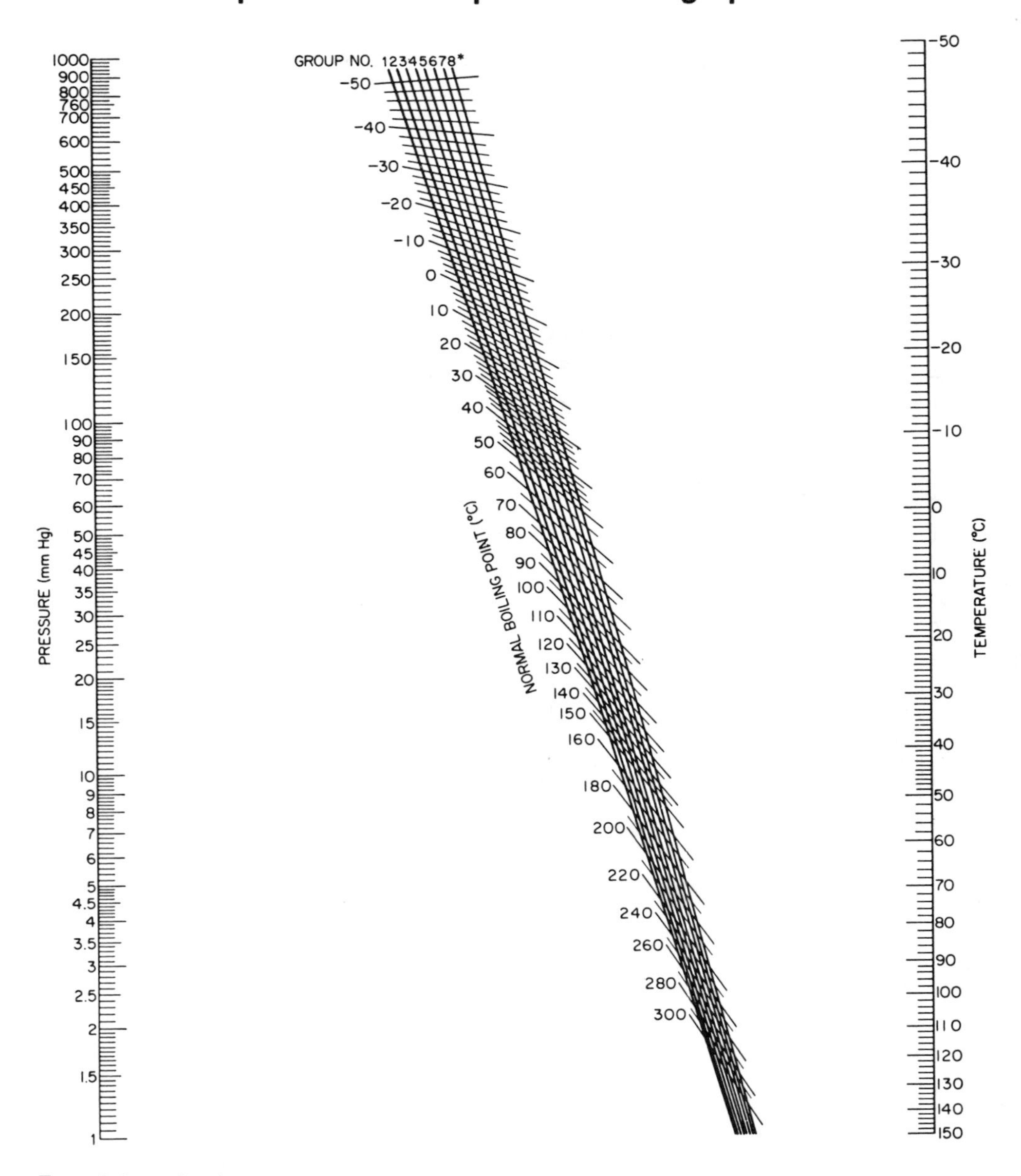

*For explanation of Group numbers, see page 31.

Vapor Pressure-Temperature Nomograph

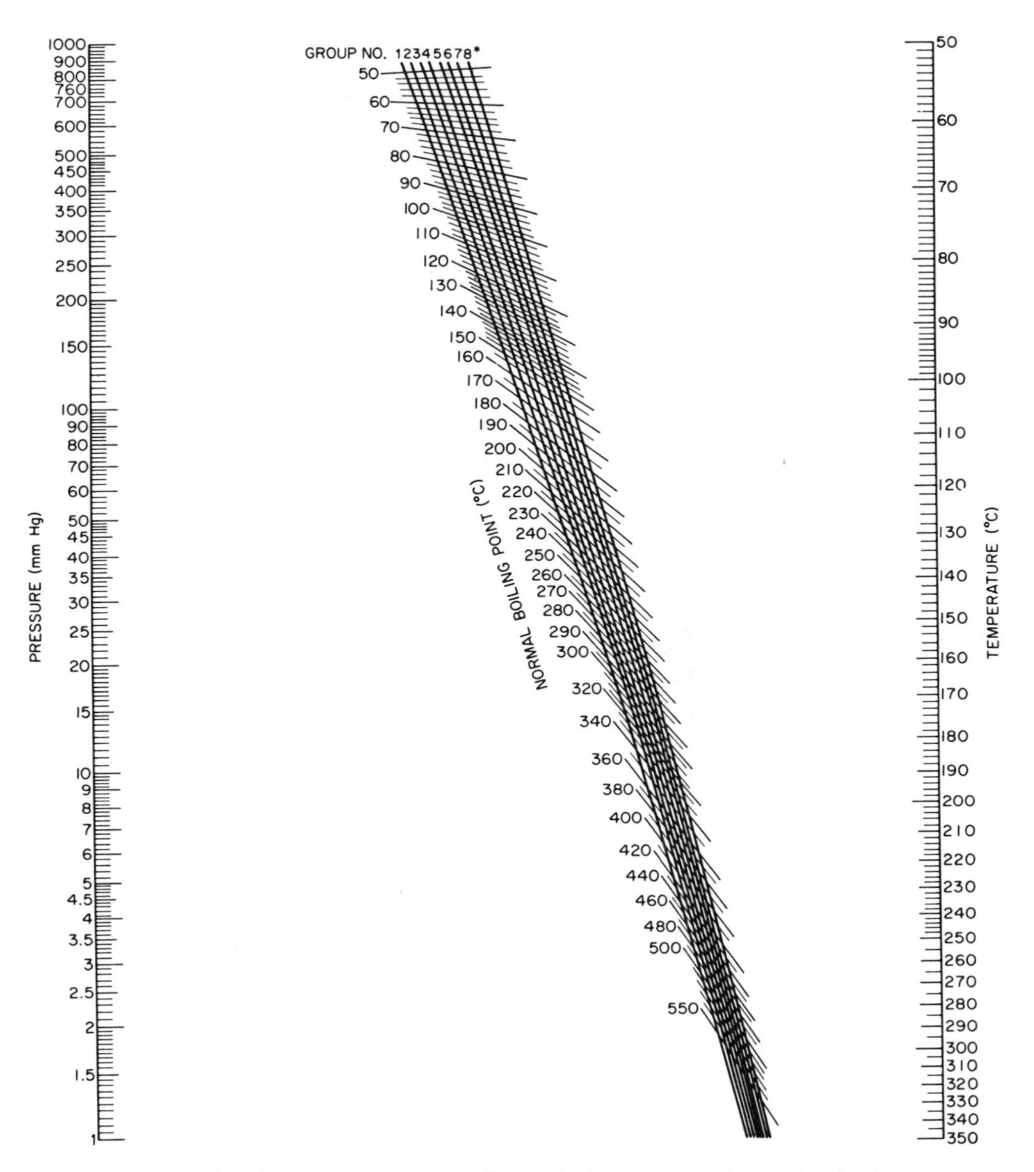

From *Industrial and Engineering Chemistry*, Vol. 38, p. 320, 1946. Copyright 1946 by the American Chemical Society and reproduced by permission of the copyright owner.

*For explanation of Group numbers, see page 31.

D. Pressure-Temperature Alignment Chart

The chart that follows is reproduced from one supplied by Maybridge Chemical Co., Ltd., Tintagel, Cornwall, England.

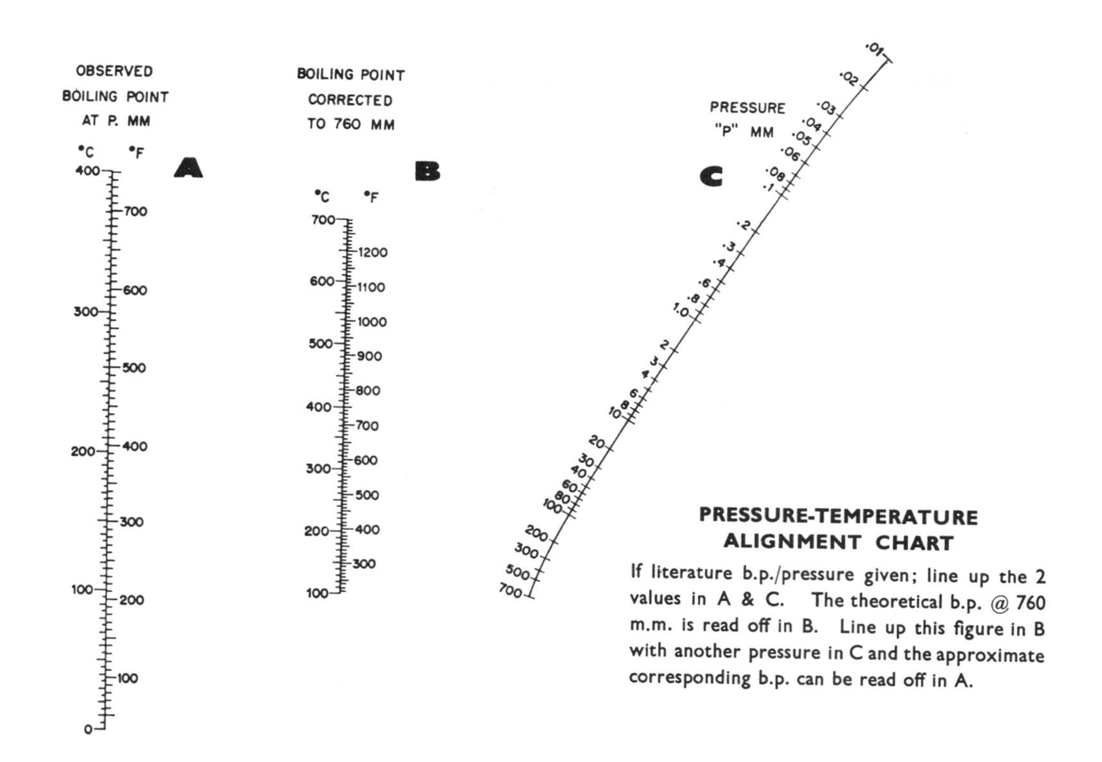

IV. PROPERTIES OF SELECTED GASES

The table entries were taken, and in many cases calculated, from the data in the following sources: Matheson Gas Data Book, 4th ed., The Matheson Company, Inc., New York, 1966; R. Nelson, D. Lide, Jr., and A. Maryott, Selected Values of Electric Dipole Moments for Molecules in the Gas Phase, NSRDS-NBS 10, U.S. Government Printing Office, Washington, D.C., 1967; L. Bergmann, Der Ultraschall, S. Hirzel Verlag, Stuttgart, 1954; Handbook of Chemistry and Physics, 50th ed., CRC, Cleveland, 1969; H. Stephen and T. Stephen, Solubilities of Inorganic and Organic Compounds, Macmillan, New York, 1963.

Melting and boiling points (°C) are at 1 atm (unless followed by mm Hg in parentheses; s = sublimes); density (ρ) in g/ℓ at 0°C and 1 atm unless indicated otherwise (g/cc followed by °C in parentheses); gas phase dipole moments (μ) are in debyes; thermal conductivity (TC) is in 10^{-6} cal/(sec-cm^2)/(°C/cm) at 100°F (37.8°C); sound velocity (V) is in m/sec at 0°C; solubility in water (S) is in g/100 cc H_2O at 1 atm at the temperature in parentheses (d = decomposition). For general information of gas solubilities in liquids see R. Battino and H. Clever, Chem. Rev., 66, 395 (1966). Viscosity and heat capacity data for some gases are found on page 394.

Gas	Formula Weight	Mp	Bp	ρ(0°)	μ	TC	V	S
Acetylene	26.04	-81.8	-83.6(s)	1.17	0	53.7		8.5(20)
Air	28.98	∿-217	-194	1.29	-	64.2	331	2.2(25)
Allene	40.07	-136	-34.5	0.76	0			
Ammonia	17.03	-77.7	-33.35	0.771	1.47	61.6	415	32(25)
Argon	39.95	-189.2	-185.9	1.784	0	44.2	319	0.006(25)
Boron trichloride	117.19	-107.3	12.5	1.349 g/cc(11)	0			d
Boron trifluoride	67.81	-127.1	-100.3	2.99	0			32.2(25d)
Bromomethane	94.95	-94	3.46	1.732 g/cc(0)	1.80			1.32(25)
Butane	58.12	-138.3	-0.5	0.601 g/cc(0)	0	40.9		1.96(20)
Carbon dioxide	44.01	-5(5 atm)	-78.5(s)	1.98	0	41.7	259	0.1(23)
Carbon monoxide	28.01	-207	-191	1.25	0.10	62.0	338	0.003(25)
Carbon tetrafluoride	88.01	-184	-128	3.034	0			0.001(25)
Carbonyl fluoride	66.01	-114	-83	1.14 g/cc(-114)	0.95			d
Carbonyl sulfide	60.07	-138	-50	1.24 g/cc(-87)	0.72			
Chlorine	70.91	-101	-34.6	3.214	0	21.9	206	0.63(25)
Chloroethane	64.52	-138.7	13.1	0.903 g/cc(15)	2.0	16		0.6(18)
Chloromethane	50.49	-97	-24.2	2.25	1.87	16		0.066(20)

(Continued)

Gas	Formula Weight	Mp	Bp	ρ(0°)	μ	TC	V	S
Cyanogen	52.04	-28	-21.17	0.95 g/cc(-21)	0			
Cyanogen chloride	61.48	-6.9	12.66	1.186	2.80			
Cyclobutane	56.10	-50	13	0.703 g/cc(0)				
Cyclopropane	42.08	-126.6	-33	1.88	0			
Deuterium	4.032	-254(121 mm)	-250	0.18	0	343	890	
Deuterium bromide	81.92			1.96 g/cc(0)	0.8			
Deuterium chloride	37.47	-115	-81.6		1.085	38(27)		
Deuterium fluoride	21.02		18.5					
Deuterium iodide	128.92				0.44			
Deuterium sulfide	36.10							
Diborane	27.67	-92.5	-165	0.47 g/cc(-120)	0			
Dimethylamine	45.08	-92	6.9	0.68 g/cc(0)	1.03			
Dimethyl ether	46.07	-141	-24.8	2.1	1.30			
Ethane	30.07	-183	-88.6	1.36	0	54.6	308(10)	0.001(25)
Ethylamine	45.09	-81	16.6	0.689 g/cc(15)	1.22	39.7		
Ethylene	28.05	-170	-104	1.26	0	52.1	317	0.02(21)
Ethylene oxide	44.05	-111	10.7	0.88 g/cc(10)	1.89			∞
Fluorine	38.00	-220	-188	1.7	0	59.2		
Fluoroform	70.02	-160	-84	1.52 g/cc(-100)	1.65			
Fluoromethane	34.03	-142	-78	0.58 g/cc(20)	1.85			
Helium	4.003	-272	-269	0.178	0	369	965	2 x 10^{-4}(20)
Hexafluoroacetone	166.02	-122	-27.4					
Hydrogen	2.016	-257	-253	0.090	0	446	1284	1.7 x 10^{-4}(25)
Hydrogen bromide	80.92	-86	-66	3.5	0.82	21.5	200	194(25)
Hydrogen chloride	36.47	-114	-85	1.00	1.081	35.1	296	71.9(20)
Hydrogen fluoride	20.01	-83	19.5	0.99 g/cc(19.5)	1.82			∞
Hydrogen iodide	127.92	-51	-36	5.66	0.44		157	25(-10)
Hydrogen selenide	80.98	-64	-42	2 g/cc(-42)				
Hydrogen sulfide	34.08	-83	-60	1.54	0.97	36.8	289	0.38(25)

Isobutene	56.11	-140	-6.9	2.58	0.5			
Krypton	83.80	-157	-153	3.65	0	23.6		0.014(25)
Methane	16.04	-183	-162	0.71	0	85.5	430	0.002(25)
Methylacetylene	40.07	-103	-23	1.82	0.781			
Methylamine	31.06	-92.5	-6.45	0.7 g/cc(-11)	1.31			40(60)
Methyl mercaptan	48.10	-121	5.96	0.87 g/cc(20)	1.52			2.4
Neon	20.18	-249	-246	0.90	0	118	430	0.001(20)
Nickel carbonyl	170.75	-25	43.2	1.32 g/cc(17)				
Nitric oxide (NO)	30.01	-164	-152	1.34	0.153	64.1	324(10)	0.006(25)
Nitrogen	28.02	-210	-196	1.251	0	64.1	334	0.002(25)
Nitrogen dioxide	46.01	-11.3	21.2	1.45 g/cc(20)	0.316	87(55)		
Nitrosyl chloride	65.47	-61.5	-5.8	2.99	1.9			
Nitrous oxide (N_2O)	44.02	-102	-90	1.98	0.167	36.1(0)	263	0.135(20)
Oxygen	32.00	-219	-183	1.43	0	65.9	316	0.004(25)
Ozone	47.998	-192	-112	2.14	0.53			0.001(27)
Phosgene	98.93	-128	7.6	1.38 g/cc(20)	1.17			
Phosphine	34.00	-133	-88	1.49	0.58			
Propane	44.10	-188	-42.1	0.59 g/cc(-45)	0.084	45.5		0.008(20)
Propylene	42.08	-185	-47.7	1.92	0.366			0.041(20)
Silane	32.12	-185	-112	1.4	0			
Silicon tetrafluoride	104.09	-95(s)	-65(1810)	4.69	0		167	
Sulfur dioxide	64.07	-75.5	-10.0	2.96	1.63	20.5(0)	213	8.3(25)
Sulfur hexafluoride	146.07	-50.8(32)	-63.8(s)	6.6	0			
Sulfur tetrafluoride	108.07	-121	-40	1.92 g/cc(-73)	0.632			
Sulfuryl fluoride	102.07	-137	-55	3.72	1.12			
Trimethylamine	59.11	-117	2.9	0.66 g/cc(0)	0.612			
Tritium	6.032				0			
Vinyl bromide	106.96	-140	15.8	1.49 g/cc(20)	1.42			
Vinyl chloride	62.50	-160	-13.9	0.92 g/cc(13)	1.45			
Vinyl fluoride	46.04	-161	-72	0.64 g/cc(19)	1.43			
Xenon	131.30	-112	-108	5.65	0			0.064(20)

V. PROPERTIES OF REPRESENTATIVE FUSED SALT SYSTEMS

Molten salts are good reaction media (or possibly constant temperature baths) for many materials, including inorganic and organic gases and solids, metals, and oxides. The variety available covers a temperature range greater than 1000°. More detailed discussions appear in the following references.

General

B. Sundheim, Ed., Fused Salts, McGraw-Hill, New York, 1964.
M. Blander, Ed., Molten Salt Chemistry, Interscience, New York, 1964.
W. Sundermeyer, Fused Salts and Their Use as Reaction Media, Angew. Chem., 77, 241 (1965), or Angew. Chem. Intern. Ed. Engl., 4, 222 (1965).
G. Charlot and B. Tremillon, Chemical Reactions in Solvents and Melts, Pergamon Press, New York, 1969 (includes 5000 references).
J. Corbett, in Survey of Progress in Chemistry, Vol. 2, Academic Press, New York, 1964, pp. 91 ff.
J. Braunstein, G. Mamantor, and G. P. Smith, Eds., Advances in Molten Salt Chemistry, Vol. 1, Plenum Publishing Corp., New York, 1971.

Organic Chemical Applications

Halogen exchange reactions of aryl halides and others: H. Fielding, et al., J. Chem. Soc. (C), 2142 (1966); D. Packham, Chem. Ind., 899 (1966); and J. Chem. Soc. (D), 207 (1965).
Fused organic salts: J. Gordon, et al., J. Amer. Chem. Soc., 87, 4347 (1965), and others in the series.
General Review: J. E. Gordon, Applications of Fused Salts, in Techniques and Methods of Organic and Organometallic Chemistry, D. Denney, Ed., Marcel Dekker, New York, 1969.

Fused Salt Spectrophotometry

D. Given, Quart. Rev., 19, 349 (1965).

A comprehensive compilation of fused salt data can be found in G. Janz, Molten Salts Handbook, Academic Press, New York, 1967. Most of the data given here were obtained from this very useful source.

A. Single Salts

1. Melting Points of Quaternary Ammonium Salts: $R_4N^+X^-$

X	R = n-Pr	n-Bu	n-Pentyl	iso-Pentyl	n-Hex
I	280d	147	135	147	105
Br	252	118	101		101
ClO_4	238	214	118	119	107
NO_2			97		
NO_3	260d	121	116	132	69
CNS		127	50	106	
Picrate	120		74	90	
Benzoate					25[a]

[a] C. Swain, et al., J. Amer. Chem. Soc., 89, 2648 (1967); this liquid salt has a solvent Y-value of -0.39 (Y = Winstein empirical solvent parameter).

2. Inorganic Salt Properties

The Mp and Bp are in °C at 1 atm; refractive index, n_D at t°C; C = cryoscopic constant (deg/mole-kg). Dielectric constants for some molten salts are known: AgCl (6.44), AgBr (5.91), alkali chlorides, bromides, iodides (∿3), $LiNO_3$ (2.5), $NaNO_3$ (1.9), KNO_3 (1.45), $AgNO_3$ (1.52). Dimethyl sulfone, $(CH_3)_2SO_2$ (mp 108°, bp 218°), is useful as a molten medium by itself and as a molten salt diluent [T. Griffiths, J. Chem. Soc. (D), 1222 (1967)].

Salt	Mp	Bp	$n_D(t)$	C
AgBr	430	1533		
AgCl	455	1557		
AgCN	350			
AgF	435	(1147)		
AgI	552	1504		
$AgNO_3$	210	d > 212	1.660(300)	27
$AlBr_3$	97.5	257		
AlI_3	191	386		
$AsBr_3$	31.2	221		
$BeCl_2$	405	547		
BI_3	43	210		
CaF_2	1418	(2509)		
$Ca(NO_3)_2$	561	d > 561		
$CsNH_2$	262			
$CuCl_2$	498	537d		
$GaCl_3$	77.5	200		
GaI_3	10	346		
KBr	734	1383	1.436(900)	39.3
KCl	772	1407	1.379(900)	25.4
$KClO_4$	610	400d		
KCN	635			
KCNS	360		1.537(300)	12.8
KF	856	1502	1.28(900)	21.9
KI	685	1324		52.4
KNH_2	330			
KNO_2	440	350d	1.356(600)	
KNO_3	337	d > 340	1.426(300)	31
KOAc	295			
KO_2CH	167.5			
KOH	360			22.5
K_2SO_4	1074		1.388(900)	68.8
LiBr	547	1310	1.60(570)	27.7
LiCl	610	1382	1.501(600)	13.8
$LiClO_4$	236	430d		
LiF	845	1681	1.32(950)	10.0
LiI	449	1171		41.8
$LiNO_3$	254	600d	1.467(300)	5.93
LiOAc	280			
LiOH	462	924d		5.27
Li_2SO_4	859		1.452(900)	142.0
Na_3AlF_6	1000		1.290(1000)	
NaBr	747	1392	1.486(900)	34.1
NaCl	808	1465	1.320(900)	20.0
$NaClO_4$	482			

(Continued)

Salt	Mp	Bp	$n_D(t)$	C
NaCN	562	1497		16.3
NaCNS	323			9.4
NaF	995	1704	1.25(1000)	16.7
NaI	662	1304		46.0
$NaNH_2$	208			
$NaNO_2$	271	d > 320	1.416(300)	
$NaNO_3$	310	380d	1.431(300)	16
NaO_2CH	300			
NaOH	318	1390		18.3
Na_2SO_4	884		1.395(900)	66.4
NH_4CNS	87.7			
NH_4NO_3	169			24.0
PI_3	61.5			
PtF_6	56.7			
RuO_4	25	(100d)		
$SbBr_3$	96.6	(288)		
$SbCl_3$	73.2	221		
SbF_3	290	376		
$SnBr_4$	30	205		
$SnCl_2$	247	623		
$TiBr_4$	38.5	230		
TiI_4	150	377		
$TlCl_3$	25	77d		
$ZnCl_2$	283	732	1.588(320)	
ZnF_2	872	1502		

3. Hydrate Melts

The data are from C. Angell, J. Electrochem. Soc., 112, 1224 (1965), and C. Moynihan, J. Chem. Educ., 44, 531 (1967).

Compound	Mp
$CaCl_2 \cdot 6H_2O$	30.2
$Ca(NO_3)_2 \cdot 4H_2O$	42.7
$LiNO_3 \cdot 3H_2O$	29.9
$Mg(NO_3)_2 \cdot 6H_2O$	89.9

B. Low-Melting Binary Eutectics

T (°C)	A-B	Mole %B	T (°C)	A-B	Mole %B
82.5	$AgNO_3$-$TlNO_3$	48	140	KNO_3-NaCNS	36.5
123.5	KCNS-NaCNS	70	145	KNO_3-$NaNO_2$	51.5
125	$LiNO_3$-KNO_3	44	150	KCl-CuCl	65
128	KCl-$AlCl_3$	67	164	$NaNO_3$-$TlNO_3$	76.5
131	KNO_3-$AgNO_3$	62	170	NaOH-KOH	50.6
133.5	$LiNO_3$-KNO_3	59	173	$LiNO_3$-$AgNO_3$	76

(Continued)

T (°C)	A-B	Mole %B	T (°C)	A-B	Mole %B
193	$LiNO_3-NaNO_3$	46	361	LiCl-KCl	41.5
217	$KNO_3-Pb(NO_3)_2$	23.4	388	$CdCl_2-KCl$	62
218	NaOH-LiOH	27	432	$ZnCl_2-KCl$	68.5
220	KNO_3-NaNO_3	50	450	LiCl-KCl	40
242	$CuCl-ZnCl_2$	88	453	LiBr-LiF	33
260	NaBr-NaOH	78	485	LiCl-LiF	29
285	KBr-AgBr	68	492	LiF-KF	48
287	$KNO_3-Ba(NO_3)_2$	12.4	500	$NaCl-CaCl_2$	52
295	$NaNO_3-NaCl$	6.4	535	$Li_2SO_4-K_2SO_4$	20
300	NaCNS-NaCl	6.0	710	$Na_2CO_3-K_2CO_3$	44
348	KBr-LiBr	60			

C. Low-Melting Ternary Eutectics

T (°C)	A-B-C	Mole %
86	$LiNO_3-NH_4NO_3-NH_4Cl$	25.8-66.7-7.5
120	$LiNO_3-NaNO_3-KNO_3$	30 - 17 - 53
142	$KNO_3-NaNO_3-NaNO_2$	44.2-6.9-48.9
188	KCl-TlCl-AgCl	7 - 37 - 56
203	$ZnCl_2-NaCl-KCl$	60 - 20 - 20
280	$PbCl_2-PbBr_2-AgBr$	66 - 7 - 27
318	RbCl-LiCl-NaCl	41.0-56.6-2.4
357	KCl-LiCl-NaCl	24 - 43 - 33
396	$MgCl_2-KCl-NaCl$	60 - 20 - 20
450	$CaCl_2-NaCl-BaCl_2$	47.0-38.5-14.5
454	NaF-KF-LiF	11.5-42.0-46.5
512	$Li_2SO_4-K_2SO_4-Na_2SO_4$	78 - 8.5 - 13.5

VI. STRUCTURE AND PROPERTIES OF NATURALLY OCCURRING α-AMINO ACIDS

Name (Abbreviation)[a]	Structure	Mw	Dec. Pt. (°C)	$[\alpha]_D^{25}(H_2O)$	pI[e]	pK[f]			
						A_1	A_2	B_1	B_2
Alanine (Ala)	$CH_3CH(NH_2)CO_2H$	89.1	297	1.8	6.11	2.35	-	4.13	-
Arginine (Arg)	$HN{=}C(NH_2)NH(CH_2)_3CH(NH_2)CO_2H$	174.2	238	12.5	10.76	2.01	-	1.52	4.96
Asparagine ($AspNH_2$)	$H_2NCOCH_2CH(NH_2)CO_2H$	132.1	236	33.2[b]	4.3				
Aspartic acid (Asp)	$HO_2CCH_2CH(NH_2)CO_2H$	133.1	270	5.0	2.98	2.10	3.86	4.18	-
Cysteine (Cys)	$HSCH_2CH(NH_2)CO_2H$	121.2	178	-16.5					
Cystine (Cys-Cys)	$[HO_2CCH(NH_2)CH_2S]_2$	240.3	261	-232[c]	5.02	1.04	2.05	3.75	6.00
Glutamic acid (Glu)	$HO_2CCH_2CH_2CH(NH_2)CO_2H$	147.1	249	12.0	3.08	2.10	4.07	4.53	-
Glycine (Gly)	$H_2NCH_2CO_2H$	75.1	290	-	6.20	2.35	-	4.22	-
Histidine (His)	(imidazol-4-yl)–$CH_2CH(NH_2)CO_2H$	155.2	277	-38.5	7.64	1.77	-	4.82	-
5-Hydroxylysine (Hylys)	$H_2NCH_2CH(OH)CH_2CH_2CH(NH_2)CO_2H$	162.2							
4-Hydroxyproline (Hypro)	(4-HO-pyrrolidine, N–H)–CO_2H	131.1	270	-76.0	5.82	1.92	-	4.27	-
Isoleucine (Ile)	$C_2H_5CH(CH_3)CH(NH_2)CO_2H$	131.2	284	12.4	6.04*	2.32	-	4.24	-
Leucine (Leu)	$(CH_3)_2CHCH_2CH(NH_2)CO_2H$	131.2		-11.0	6.04*	2.33	-	4.25	-
Lysine (Lys)	$H_2N(CH_2)_4CH(NH_2)CO_2H$	146.2	224	13.5	9.47	2.18	-	3.47	-
Methionine (Met)	$CH_3SCH_2CH_2CH(NH_2)CO_2H$	149.2	283	-10.0	5.74*	2.28	-	4.79	-
Phenylalanine (Phe)	$C_6H_5CH_2CH(NH_2)CO_2H$	165.2	284	-34.5	5.91*	2.58	-	4.76	-
Proline (Pro)	(pyrrolidine, N–H)–CO_2H	115.1	222	-86.2	6.3	2.00	-	3.40	-
Serine (Ser)	$HOCH_2CH(NH_2)CO_2H$	105.1	228	-7.5	5.68*	2.21	-	4.85	-

Threonine (Thr)	$CH_3CH(OH)CH(NH_2)CO_2H$	119.1	253	-28.5	5.59				
Tryptophan (Try)	(indolyl)-$CH_2CH(NH_2)CO_2H$	204.2	282	-33.7	5.88	2.38	-	4.61	-
Tyrosine (Tyr)	p-$HOC_6H_4CH_2CH(NH_2)CO_2H$	181.2	344	-10.0[c]	5.63	2.20	9.11	3.93	-
Valine (Val)	$(CH_3)_2CHCH(NH_2)CO_2H$	117.1	315	5.6	6.00*	2.29	-	4.28	-

All these amino acids are commonly found in naturally occurring proteins and are referred to as essential amino acids; other important natural amino acids include those given below.

α-Aminoadipic acid	$HO_2C(CH_2)_3CH(NH_2)CO_2H$	161.2		3.2	
α-Amino-n-butyric acid	$CH_3CH_2CH(NH_2)CO_2H$	103.1	292	20.6[c]	5.98*
3,4-Dihydroxyphenylalanine	$(HO)_2C_6H_3CH_2CH(NH_2)CO_2H$	197.2		-12.0[g]	
Homoserine	$HOCH_2CH_2CH(NH_2)CO_2H$	119.1		-8.8	
5-Hydroxylysine	$H_2NCH_2CH(OH)CH_2CH_2CH(NH_2)CO_2H$	162.2		9.2	
α-Methylserine	$HOCH_2C(CH_3)(NH_2)CO_2H$	119.1		4.5	
Ornithine	$H_2N(CH_3)_3CH(NH_2)CO_2H$	132.2	140	12.1	
Pipecolic acid	(piperidine-2-carboxylic acid: CO_2H, N-H)	129.2		-24.6(at 23°C)	
Sarcosine	$CH_3NHCH_2CO_2H$	89.1		-	6.00
Thyroxine	HO-(diiodophenyl)-O-(diiodophenyl)-$CH_2CH(NH_2)CO_2H$	776.9	235	-4.4[d]	

From Handbook of Chemistry and Physics, 50th ed., CRC, Cleveland, 1969, pp. C-741 to C-744, and Handbook of Biochemistry, CRC, Cleveland, 1968, p. B-20.

* D,L-Compound

[a] All have the L-configuration: (R, H_2N, CO_2H, H) Many exist also in the D-form, but not nearly as often. [b] In 3N HCl. [c] In 5N HCl. [d] In 0.13 N NaOH in 70% ETOH. [e] The pH at the isoelectric point. [f] Derived from the zwitterionic formulation of ionization at 25° in H_2O. [g] In 4% HCl.

VII. PROPERTIES AND APPLICATIONS OF LIQUID CRYSTALS

Certain liquids exhibit such unusual optical effects as double refraction. Although referred to as liquid "crystals," they are true liquids with flow properties and surface tension and do not have the usual crystalline properties. Such liquids are often called mesomorphic states. One interesting feature of such liquids is their ability as solvents to introduce orientation effects and selective ordering of solutes. Depending on chemical structure, liquid crystal compounds may exist in one or more phases: smectic, nematic, and cholesteric.

Smectic phases have a structure of parallel planes, moving in a gliding motion; they are unaffected by a magnetic field. Not many practical applications have yet been developed for this type. Nematic phases behave almost like true anisotropic liquids, yet have low viscosity and flow freely. They change polarization of light readily and are similar to birefringent crystals, with the additional feature that the direction of birefringence can be controlled by electric or magnetic fields. They also exhibit parallel needle-like structures in thick samples and are used as nmr solvents to obtain structural information of the solute. Cholesteric phases are formed by cholesteryl esters and some other optically active compounds. They have properties similar to those of nematic phases, but in addition show dramatic and extremely sensitive color effects under the influence of temperature and polarized light.

The following references are recommended for details on properties and applications of liquid crystals.

General

Structure and Physical Properties of Liquid Crystals, G. H. Brown, J. W. Doane, and V. D. Neff, CRC, Cleveland, 1970.

Transition Temperatures of Liquid Crystals, W. Kast, in Landolt-Börnstein, Vol. 6, Part 2a, pp. 266-335 (1960).

Ordered Fluids and Liquid Crystals, R. S. Porter and J. F. Johnson, Eds., Advances in Chemistry Series, No. 63, ACS, Washington, D.C., 1967.

Liquid Crystals and Ordered Fluids, J. F. Johnson and R. S. Porter, Eds., Plenum Press, New York, 1970.

Molecular Crystals and Liquid Crystals, a journal published by Gordon and Breach, New York.

Recent Results in the Field of Liquid Crystals, A. Saupe, Angew. Chem. Intern. Ed., Engl. 7, 97 (1968).

As Solvents

Liquid Crystals and Some of Their Applications in Chemistry, G. H. Brown, Anal. Chem., 41, No. 13, 26A (1969).

NMR Spectra in Liquid Crystals and Molecular Structure, S. Meiboom and L. C. Snyder, Accts. Chem. Res., 4, 81 (1971). (See also the article by Saupe above.)

NMR Studies of Molecular Oriented in The Nematic Phase of Liquid Crystals, P. Diehl and C. L. Khetrapal, in NMR, Basic Principles and Progress, Vol. 1, P. Diehl, E. Fluck, and R. Kosfeld, Eds., Springer-Verlag, New York, 1969, pp. 1-96. This chapter contains a list of 15 liquid crystals used as nmr solvents and a bibliography of compounds studied by this technique.

The Use of Liquid Crystals in Gas Chromatography, H. Kelker and E. Von Schivizhoffen, Adv. in Chromatog., 6, 247 (1968).

Mössbauer Investigation of the Smectic Liquid Crystalline State, D. Uhrich, J. Wilson, and W. Resch, Phys. Rev. Letters, 24, 355 (1970).

Polarized IR and Electronic Spectroscopy of Molecules Oriented in a Nematic Liquid Crystal, H. B. Gray et al., J. Amer. Chem. Soc., 91, 191 and 4017 (electronic spectra), 772 and 4017 (ir spectra) (1969).
Liquid Crystals Solvents as Reaction Media for the Claisen Rearrangement, W. Bacon and G. Brown, Mol. Cryst., Liquid Cryst., 6, 155 (1969), and Polymerization of a Nematic Liquid Crystal Monomer, C. M. Paleos and M. M. Labes, ibid., 11, 385 (1970).

Miscellaneous

Now that the Heat's off, Liquid Crystals Can Show Their Colors Everywhere, J. A. Castallano, Electronics, July 6, 1970 (a fine article on the use of liquid crystals for information storage, display systems, readout devices, etc.).
Aerospace Thermal Mapping Applications of Liquid Crystals, W. Woodmansee, Appl. Opt., 7, 1721 (1968).
Liquid Crystals and Their Roles in Inanimate and Animate Systems, G. H. Brown, Amer. Sci., 60, 64 (1972).

The major suppliers of liquid crystals and their synthesis intermediates are: Aldrich Chemical Company, 940 West St. Paul Avenue., Milwaukee, Wis. 53233 (will supply on request a compilation of articles on liquid crystals); Eastman Organic Chemicals, Rochester, N.Y. 14650 [also supplies solutions of cholesteric liquid crystal mixtures which change from red to blue over very narrow temperature ranges (1-5°)].; Princeton Organics, Route 206, Princeton, N. J. 08540 (high-purity liquid crystals).

The following table of properties is arranged according to the principal mesomorphic phase exhibited. Entries were chosen on the basis of availability and common usage as spectral solvents, glc phases, and so on. Generally, the most useful in these applications have been the nematic 4,4'-dialkoxy substituted azo- and azoxybenzenes [see, e.g., M. J. S. Dewar and J. P. Schroeder, J. Org. Chem., 30, 3485 (1965), and J. Amer. Chem. Soc., 86, 5235 (1964) on glc uses, and I. Morishima et al., ibid., 93, 1521 (1971) on nmr use]. The lowest temperature represents the point at which the solid crystal is transformed into a cholesteric (C), nematic (N), or smectic (S) phase; the highest temperature is that at which the substance becomes a truly isotropic liquid. For substances exhibiting more than one mesophase, the type (C, N, or S) is indicated in parentheses. The data were obtained from recent company catalogs and were checked with other sources, especially the compilation by Kast (see references above).

Name	Structure	Mesomorphic Range (°C)

Cholesteric Liquids

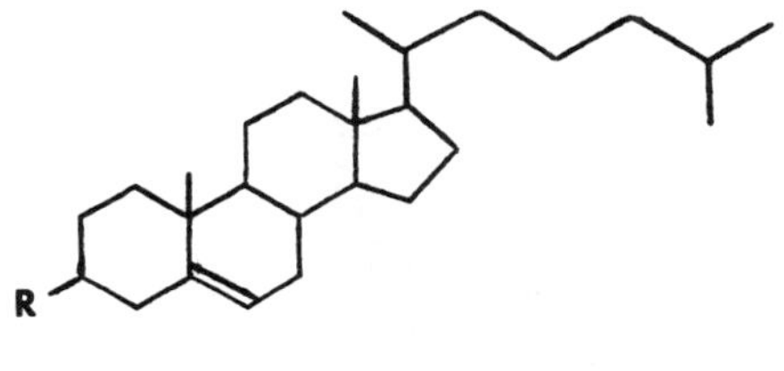

Name	Structure		Mesomorphic Range (°C)
Cholesteryl Derivative	R		
Acetate	CH_3CO_2		94.5-116.5
Benzoate	$C_6H_5CO_2$		147-178
Bromide	Br		67-94
n-Butyrate	n-$C_3H_7CO_2$		102-113
Carbonate	cholesteryl-CO_2		177-236
Chloride	Cl		62-97
Cinnamate	$C_6H_5CH{=}CHCO_2$ (trans)		161-210
2-Ethylhexyl carbonate	$CH_3CH(C_2H_5)(CH_2)_4OCO_2$		<r.t. (Eastman)
Myristate	$CH_3(CH_2)_{12}CO_2$		71(S)-81(C)-86.5
Oleate	$CH_3(CH_2)_7CH{=}CH(CH_2)_7CO_2$ (cis)		42-44.5
(-)-2-Methylbutyl-4-(4 cyanobenzylideneamino)cinnamate, (-)-4'-$NC{-}C_6H_4CH{=}NC_6H_4CH{=}CHCO_2CH_2CH(CH_3)CH_2CH_3$			91-109[a]
	Nematic Liquids*,[b]		
Butyl 4-(4'-ethoxyphenyl)carboxyphenyl carbonate	$C_2H_5OC_6H_4OCOC_6H_4OCO_2C_4H_9$		57.5-81
4-Methoxycinnamic acid	$CH_3OC_6H_4CH{=}CHCO_2H$		173.5-190
4-Octoxybenzoic acid	$C_8H_{17}OC_6H_4CO_2H$		101(S)-108(N)-147
LC-200	See footnote c		-10 to +59
R–⟨⟩–CH=N–⟨⟩–R'	R	R'	
4-(4'-Methoxybenzylideneamino) benzonitrile	CH_3O	CN	103-113.5
4-(4'-Ethoxybenzylideneamino) benzonitrile	C_2H_5O	CN	105-124.5
4-(4'-Cyanobenzylideneamino) methoxybenzene	NC-	OCH_3	115-125
4-(4'--Cyanobenzylideneamino) ethoxybenzene	NC-	OC_2H_5	115-132
4-(4'-Methoxybenzylideneamino) phenyl acetate	CH_3O	$OCOCH_3$	82-107.5
C_2H_5O–⟨⟩–N=N–⟨⟩–R		R	
4,4'-Azodiphenetole		OC_2H_5	155-162
4-(4'-Ethoxyphenylazo)phenyl pentanoate		$OCOC_4H_9$	79-125
4-(4'-Ethoxyphenylazo)phenyl hexanoate		$OCOC_5H_{11}$	70-126
4-(4'-Ethoxyphenylazo)phenyl heptanoate		$OCOC_6H_{13}$	68-117

(Continued)

*W. R. Young, A. Aviram, and R. J. Cox report that certain non-planar trans-stilbenes (mps 14-35°) constitute a new class of room temperature nematic liquids [J. Amer. Chem. Soc., 94, 3976 (1972)].

Name	Structure	Mesomorphic Range (°C)
	Nematic Liquids[b]	
	R[d]	
4,4'-Dimethoxyazoxybenzene	CH_3	118.2-135.3
4,4'-Diethoxyazoxybenzene	C_2H_5	136.6-167.5
4,4'-Dipropoxyazoxybenzene	1-C_3H_7	115.5-123.6
4,4'-Dibutoxyazoxybenzene	1-C_4H_9	102.0-136.7
4,4'-Dipentoxyazoxybenzene	1-C_5H_{11}	75.5-123.2
4,4'-Dihexoxyazoxybenzene	1-C_6H_{13}	81.3-129.1
4,4'-Diheptoxyazoxybenzene	1-C_7H_{15}	74.4(S)-95.4(C)-124.2
4,4'-Dioctoxyazoxybenzene	1-C_8H_{17}	79.5(S)-107.7(C)-126.1

[a] This substance is cholesteric according to G. H. Brown, Anal. Chem., 41, No. 13, 26A (1969) and Princeton Organics; older literature (Kast) lists the substance as exhibiting S and N Phases.

[b] Mixtures are often used to achieve lower temperature ranges for the liquid. See Brown, Anal. Chem., op. cit.

[c] A proprietary item from Princeton Organics; a r.t. liquid crystal with excellent light scattering capability.

[d] Data for all azoxy compounds from H. Arnold, Z. Physik. Chem., 226, 146 (1964).

VIII. PROTOTROPIC TAUTOMERISM

A. Introduction

Much literature, especially before 1950, is available on keto-enol equilibria, but discrepancies (up to several percent) occur in the concentration data reported by different workers. For a recent review on keto-enol equilibria, see S. Forsen and M. Nilsson, The Chemistry of the Carbonyl Group, Vol. 2, J. Zabicky, Ed., Interscience, New York, N.Y., 1970, p. 157. Wheland (1) and Ingold (2) give extensive discussions and data on the general subject of prototropic systems. The following papers are also noteworthy: phenol tautomerism, reference 3; β-diketones, reference 4; ketones and esters, reference 5; and heterocyclic systems, reference 6. Katritzky and Lagowski have reviewed critically the methods for studying prototropic tautomerism, especially in heterocyclic systems (6a); a sharply worded essay has also appeared, admonishing the chemist to pay careful heed to the assignment of structure in such molecules (7).

The data given here are the most consistent available and are taken or calculated mainly from the already existing compilations mentioned above, which should be consulted for the original references. Hammond (8) has also collected much of the older data on keto-enol systems, as well as on ring-chain tautomerism (2,8). It should be borne in mind that keto-enol compositions are solvent and temperature-dependent, and in some cases concentration-dependent (1,2,4). Enol acidity is discussed in references 8 and 9. For a review of ring-chain tautomerism, see reference 22.

Refs. p. 54.

B. Carbonyl Compounds

R	R'	X	Per Cent Enol In[a] Vapor	Pure Liquid	Water	Other	Ref.
			Simple Ketones: RCOR'				
CH_3	CH_3			0.00025		0.00015[b]	1,2
CH_3	C_2H_5			0.012			5
CH_3	1-C_3H_7			0.0086			5
CH_3	1-C_4H_9			0.11			5
C_2H_5	C_2H_5			0.067			5
C_2H_5	1-C_3H_7			0.047			5
2-C_3H_7	2-C_3H_7			0.0037			5
CH_3	C_6H_5			0.035			5
1-C_3H_7	C_6H_5			0.010			5
Camphor				0.14		100[i]	5
$-(CH_2)_4-$				0.0048		0.088[b]	1,2
$-(CH_2)_5-$				0.020		1.2[b]	1,2
			β-Diketones: RCOCHXCOR'				
CH_3	CH_3	H	66±5[q]	80[c]	15.5	92[d]	1,2
CH_3	CH_3	CH_3	44	31	2.8	59[d]	1,2
CH_3	CH_3	C_2H_5	35	27		26[d]	1,2
CH_3	CH_3	Br			8.1		1
CH_3	CH_3	$C_6H_5CH_2$	70	61		68[d]	1
CH_3	C_6H_5	H		99			9
CH_3	C_2H_5	H		72			2
C_2H_5	H	CH_3				100[i]	1
2-Formylcyclopentanone					41	98[e]	1,2
2-Acetylcyclopentanone					15		1,2
2-Formylcyclohexanone				100	48	98[f]	1,2
2-Acetylcyclohexanone					29	100[g]	1,2
Dimedon[h]					95		1
1,3-Indanedione					1.6		1
2-Methyl-1,3-Indanedione					1.0		1
			β-Ketoesters: $RCOCHXCO_2R'$				
CH_3	C_2H_5	H	54	7.5		49[d]	1,2
CH_3	1-C_3H_7	H	51				1
CH_3	C_2H_5	CH_3	14	4		12[d]	1,2
CH_3	C_2H_5	2-C_3H_7	6	5		6[d]	1,2
CH_3	C_2H_5	C_6H_5	80	31		67[d]	1,2
C_6H_5	C_2H_5	H		17.7		14.6[j]	5
OC_2H_5	C_2H_5	H		0.008			5
H	C_2H_5	$CO_2C_2H_5$		94			2
CH_3	C_2H_5	$CO_2C_2H_5$		69			2
C_2H_5	C_2H_5	$CO_2C_2H_5$		44			2

(Continued)

Refs. p. 45.

Compound	Per Cent Enol In[a]				Ref.
	Vapor	Pure Liquid	Water	Other	
$(CH_2)_n$ ring with C=O and $CO_2C_2H_5$					
n = 3	27	5		0[k]	1
n = 4	90	74		100[l]	1,2
n = 5	55	18			1
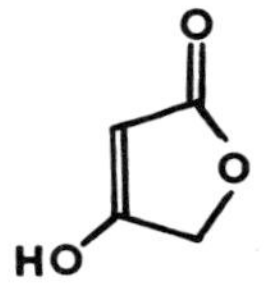 (tetronic acid)	100 (in solution and solid state)				13
α-Dicarbonyl Compounds					
CH_3CO-$COCH_3$		0.56			8
$C_6H_5CH_2CO$-$COCH_3$		60(100°C)			1
$C_6H_5CH_2CO$-COC_6H_5		30(100°C)			1
Cyclopentane-1,2-dione			100		8
Cyclohexane-1,2-dione			40		8
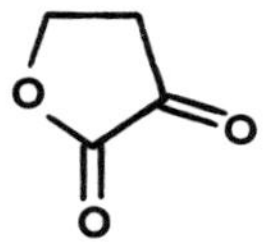 (dihydrofuran-2,3-dione)			100		8
Phenol Types					
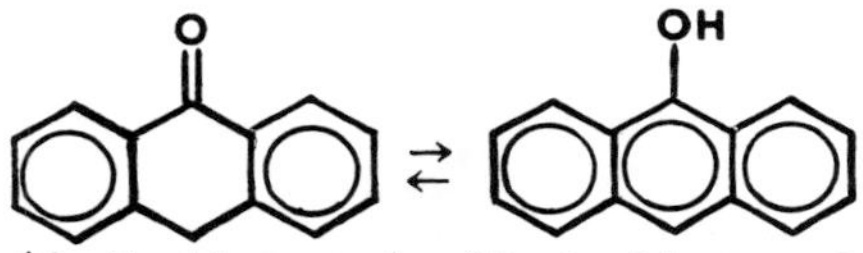 (9-Anthrone) ⇄ (9-Anthranol)	0.25[m]	50[n]			-
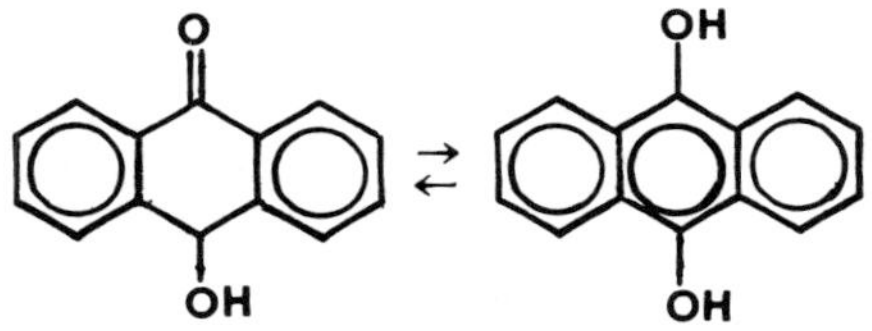 (Oxanthrone) ⇄ (Anthrahydroquinone)	5[o]	30[p]			12
1-, or 2-Naphthol	100 (in solution)				1
1-, or 2-Anthranol	100 (in solution)				1

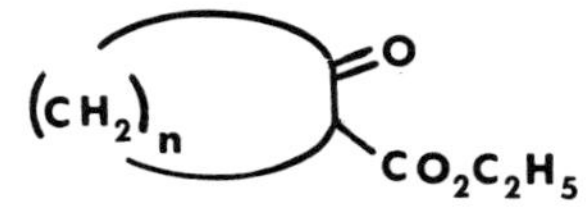

[a]In many cases, temperatures were not reported or controlled in the original studies and the data are assumed to refer to those at or near room temperature (≃25° ± 5°). [b]Pure liquid (5). [c]By nmr at 43°. W. B. Smith, J. Chem. Educ., 41, 97 (1964). [d]0.1M in hexane (1). [e]By nmr in DMSO or $CHBr_3$ at ≃40° (10). [f]In DMSO at 40°; 100% in $CHBr_3$ and $C_6H_5NO_2$ (by nmr) (10). [g]In DMSO, $C_6H_5NO_2$ or $CHBr_3$ at 40°, by nmr (10). [h]5,5-Dimethyl-1,3-cyclohexanedione. [i]Solid State (1). [j]Acetone solution. For pure liquid, value of 21 has been cited (2). [k]By nmr in $C_6H_5NO_2$, $CHBr_3$ and DMSO (10). [l]In $CHBr_3$; 80% in DMSO or $C_6H_5NO_2$ at 40° (10). [m]In isooctane or benzene at ≃30° (11). [n]In pyridine at 40° (12). Other values: 73, 20, 12, 8% in, respectively, DMSO, acetone, chloroform, DMF. [o]In DMSO at 40° (12). [p]In $CHBr_3$ at 40° (12). [q]At 105°. J. Karle, et al., J. Amer. Chem. Soc., 93, 6399 (1971).

Refs. p. 54.

C. Oxime ⇌ Nitroso Equilibria

Compound	Per Cent Oxime	Solvent or State	Ref.
RCH_2NO; R_2CHNO	100	Gas, liquid, solution	14, 15
$R_2C{=}CH\text{-}NO$	Probably 0	Liquid, solution	14
"p-Nitrosophenol"	83	Dioxane, ethanol, ether[a]	16
"p-Nitrosophenols"[b]			
2-Methyl	98	Dioxane	16
3-Methyl	97	Dioxane, ethanol	16
2,3-Dimethyl	99	Dioxane, ethanol	16
2,6-Dimethyl	≈100	Dioxane, ethanol	16
3,5-Dimethyl	93	Dioxane, ethanol	16
2-Chloro	97	Acetone, dioxane	17

[a]In acetone, 75% (17). [b]For data on naphthoquinone and anthraquinone monoximes, see A. Schors, A. Kraaijeveld, and E. Havinga, Rec. Trav. Chim., 74, 1243 (1955).

D. Nitro ⇌ Aci-nitro Equilibria: $R_2CH\text{-}NO_2 \rightleftharpoons R_2C{=}\overset{+}{N}(OH)O^-$

The following data (in H_2O at 25°) are found in (1), (15, p. 395), and (18). For discussion of nitronic acid esters ($R_2C{=}N(O)OR'$), see (19).

Compound	K = [Aci]/[Nitro]
CH_3NO_2	1.1×10^{-7}
$CH_3CH_2NO_2$	8.9×10^{-5}
$(CH_3)_2CHNO_2$	2.75×10^{-3}
$CH_3CH(NO_2)_2$	4×10^{-2}
$CH_3CH_2CH(NO_2)_2$	5×10^{-2}
$C_6H_5COCH_2NO_2$	10.3% aci-form in toluene; 2.7% in aq-CH_3OH (20)

E. Heterocyclic Systems

1. Pyridones ("Hydroxypyridines") and Analogs (6b,21)

Compound	log K[a]	Compound	log K[a]
2-Pyridone	2.96 (2.5)	4-Quinolone	4.2 (4.4)
Substituted 2-pyridones	m	9-Acridanone	7
2-Quinolone	3.88 (3.5)	3-Hydroxypyridine	-0.08[b]
1-Isoquinolone	4.85 (4.3)	3-Hydroxyquinoline	-1.08[b]
9-Phenanthridone	3.9	4-Hydroxyisoquinoline	0.46[b]

(Continued)

Refs. p. 54.

Compound	log K[a]	Compound	log K[a]
4-Pyridone	3.3	Pyridoxine[c]	0.90[b]
Substituted 4-Pyridones	m	Pyridoxal[d]	1.08[b]

[a]K = [NH-form]/[OH-form] in H_2O at 20°; m refers to oxo form as major (predominant) tautomer. The predominance of the oxo form is also found in other solvents (ethanol, benzene, etc.). [b]K = [Zwitterion form]/[OH-form]. [c]Vitamin B_6 (4,5-bis(hydroxymethyl)-3-hydroxy-2-methylpyridine). [d]4-formyl-3-hydroxy-5-hydroxymethyl-2-methylpyridine.

2. Pyrimidinones (6b)

The structure drawn represents the predominant form in solution and in the solid state.

4-Pyrimidinone: 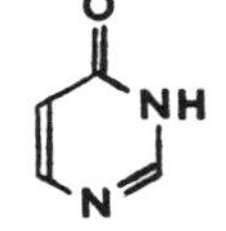

(same for 5,6-benzo compound, viz., 4-quinazolinone)

5-Hydroxypyrimidine: 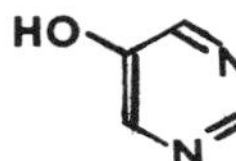

(2% [zwitterion: ⁻O, NH⁺, N] in solution)

Uracil: (same for 5,6-benzo compound, viz., 2,4-quinazolinedione)

Cytosine: 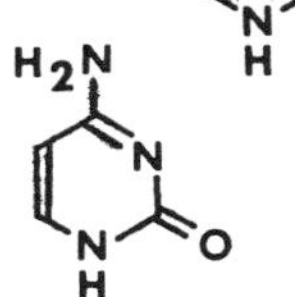

Barbituric acid: 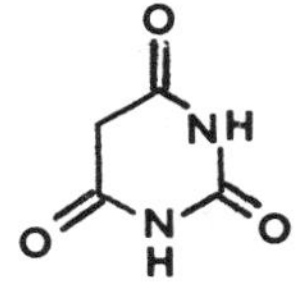

3. Amino Derivatives (6b)

The following exist predominantly in the amino form:

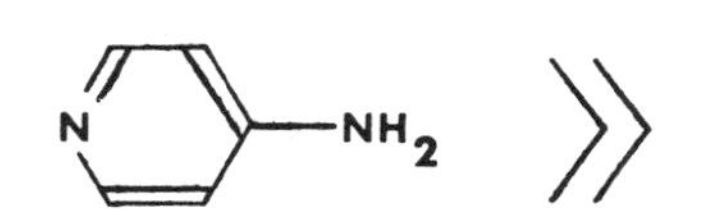

(Amino form) ≫ (Imino form)

Parent Compound	Amino Position
Pyridine	2-; 3-; 4-
Pyrimidine	2-; 4-; 5-
Acridine	1-; 2-; 3-; 4-; 5-
Phenanthridine	2-(9-Me); 6-; 9-
Melamine (2,4,6-amino-1,3,5-triazine: all-amino form)	

Refs. p. 54.

4. Other Heterocycles

Five-ring compounds with ≥1 heteroatom are covered in references 6c and 6d. General data and references on tautomerism of nucleic acid bases are found in reference 23.

F. References

1. G. W. Wheland, Advanced Organic Chemistry, 3rd ed., Wiley, New York, 1960, Ch. 14.
2. C. K. Ingold, Structure and Mechanism in Organic Chemistry, 2nd ed., Cornell University Press, New York, 1969, pp. 794-862.
3. R. H. Thomson, Quart. Rev., 10, 27 (1956). No composition data given.
4. G. Briegleb and W. Strohmeier, Angew. Chem., 64, 409 (1952). Detailed study of thermodynamic parameters for enolization (shown to be exothermic).
5. A. Gero, J. Org. Chem., 19, 1960 (1954).
6. (a) A. R. Katritzky and J. M. Lagowski, Advan. Heterocyclic Chem., 1, 311 (1963). (b) Ibid., 1, 341 (1963). (c) Ibid., 2, 1 (1963). (d) Ibid., 2, 28 (1963).
7. A. Katritzky, Chem. Ind., 331 (1965).
8. G. S. Hammond, in Steric Effects in Organic Chemistry, M. S. Newman, Ed., Wiley, New York, 1956, pp. 442-454; 460-470.
9. E. S. Gould, Mechanism and Structure in Organic Chemistry, Hold-Dryden, New York, 1959, pp. 376-384.
10. H. Sterk, Monatsh. Chem., 100, 1246 (1969).
11. H. Baba and T. Takemura, Tetrahedron, 24, 4779 (1968).
12. H. Sterk, Monatsh. Chem., 100, 916 (1969).
13. L. J. Haynes and J. R. Plimmer, Quart. Rev., 14, 292 (1960).
14. J. Boyer, in Chemistry of the Nitro and Nitroso Groups, Part I, H. Feuer, Ed., Wiley, New York, 1968.
15. P. A. S. Smith, Open Chain Nitrogen Compounds, Vol. 2, W. A. Benjamin, New York, 1966, Ch. 13.
16. R. K. Norris and S. Sternhell, Australian J. Chem., 19, 841 (1966).
17. H. Uffmann, Tetrahedron Letters, No. 38, 4631 (1966).
18. J. Belew and L. Hepler, J. Amer. Chem. Soc., 78, 4005 (1956).
19. N. Kornblum and R. A. Brown, ibid., 87, 1742 (1965).
20. Reference (1), p. 716.
21. For extensive data, see S. F. Mason, J. Chem. Soc., 674 (1958).
22. P. R. Jones, Chem. Rev., 63, 461 (1963).
23. G. Lee and S. Chan, J. Amer. Chem. Soc., 94, 3218 (1972); Y. P. Wong, ibid., 95, 3511 (1973).

IX. ACIDS AND BASES

A. Introduction

1. Bronsted Acid-Base Concept (1,2)

The Bronsted acid-base concept is defined in terms of proton transfer equilibria. Hence an acid in the Bronsted sense must possess at least one

$$\underset{\text{Acid}}{HA} + \underset{\text{Solvent (base)}}{S} \rightleftharpoons \underset{\text{Conjugate acid}}{HS^+} + \underset{\text{Conjugate base}}{A^-} \qquad [1]$$

acidic hydrogen atom; a base is simply a proton acceptor. The strength of an acid is measured in terms of the equilibrium constant (K) for the proton

Refs. p. 79.

transfer reaction; the stronger the acid, the greater the value of K (see below). In discussing base strength, it is conventional to refer to the strength of the conjugate acid, that is the protonated base:

$$\underset{\text{Conjugate acid}}{HB^+} + \underset{\substack{\text{Solvent}\\ \text{(Conjugate base)}}}{S} \rightleftarrows \underset{\text{Acid}}{HS^+} + \underset{\text{Base}}{B} \quad [2]$$

In each case, distinction must be made between the thermodynamic equilibrium acidity constant, K_a, and the concentration equilibrium acidity constant, K_c. The former is the true measure of equilibrium, while the latter is the value usually measured experimentally, especially in solutions that are not very dilute. For the example in equation [1], under the usual conditions of small concentrations of acid (thus concentration or activity of solvent is virtually constant):

$$K_a = K_a'(a_S) = \frac{(a_{A^-})(a_{HS^+})}{(a_{HA})} = \frac{(c_{A^-})\ (c_{HS^+})}{(c_{HA})} \cdot \frac{\gamma_{\pm}^2}{\gamma_{HA}}$$

$$K_c = K_c'\ (c_S) = \frac{(c_{A^-})\ (c_{HS^+})}{(c_{HA})}$$

where a = activity, c = molar concentration, and γ = activity coefficient. Activity, a, may be considered an "effective" concentration, one that correctly represents the actual behavior of real species under the usual nonideal conditions, as contrasted to the measured concentration, c. By definition, γ = unity in the standard state, which is a solution of unit concentration with the same properties as a solution of infinite dilution (Debye-Hückel conditions). Thus it is possible to determine the classical constant, K_c at a series of concentrations as a function of ionic strength, μ, and extrapolate to $\mu = 0$ to obtain K_a ($\mu = 1/2\ (\Sigma c_i Z_i^2)$, c_i = molar concentration of an ion of charge Z_i, summation for all ions present).

Since K_a values vary over many orders of magnitude, it is more convenient to use pK_a values:

$$pK_a = -\log_{10}\ K_a$$

Clearly, the stronger the acid, the smaller the value of pK_a; conversely, for bases, the smaller the pK_a of the conjugate acid (BH^+), the weaker is the base (B). One can also discuss base strength directly, as in the following equation, using water as solvent-acid:

$$B + H_2O \rightleftarrows HB^+ + HO^-$$

$$K_b = \frac{(a_{HB^+})\ (a_{HO^-})}{(a_B)}$$

Refs. p. 79.

K_b values are less common than the K_a values of the conjugate acid; to convert pK_a to pK_b, it is necessary to know the autoionization constant for the solvent, K_w in this case:

$$pK_a + pK_b = pK_w$$

Values of pK_w are 14.944 (0°), 14.167 (20°), 13.997 (25°), 13.680 (35°), and 13.017 (60°).

Another measure of the Bronsted "acidity" of a dilute aqueous solution is the activity of hydrogen ions, commonly denoted as the pH (= $-\log_{10} a_{H_3O^+}$). In practice, it is impossible to determine independently the activity of hydronium ion in a given solution, and accurate values must be measured relative to a standard. The United States pH scale is based on the pH values of a series of standard solutions, measured with a Pt,H_2(g)|buffer, Cl^-|AgCl,Ag cell (3a). The British scale is based on potassium hydrogen phthalate solutions as the primary standard (4).

In the special case of acid catalyzed ionization of alcohols to carbonium ions, $H_3O^+ + ROH \rightleftarrows R^+ + 2H_2O$, the strength of the alcohol as a base is denoted by pK_{R+}, defined as the pH of the solution when the alcohol is half-ionized: pK_{R+} = pH when $(c_{R+})/(c_{ROH}) = 1$. (5)

2. Lewis Acid-Base Concept (1,6,7)

A Lewis acid is any substance that will accept an electron pair from an electron pair donor (Lewis base); a Lewis acid must necessarily possess an unfilled orbital. The types of Lewis acids (bases) are classified as follows:

(a) Protons and metal cations. As the charge and electronegativity of the metal increases, the acid strength increases; if a metal has more than one oxidation state, the highest is the most active (Sb^{+5}, Sn^{+4}, Fe^{+3}, etc.).

(b) Molecules with ≥1 vacant orbital on a central atom, such as compounds of Group II and III elements (BX_3, AlX_3, BeX_2, ZnX_2, etc.) are the most common Lewis acids. It is noteworthy that vigorously anhydrous $AlCl_3$ and other Lewis acids are relatively inactive as catalysts unless a "co-catalyst," such as small amounts of HX or H_2O, is present.

(c) Molecules with unsaturation. Double and triple bonds react as Lewis bases toward electrophilic (Lewis acid) species, as in Friedel-Crafts reactions.

Unlike the pK and pH scales for Bronsted acids, no satisfactory general quantitative scale of Lewis acid strength has been developed, especially since the relative order of strengths is highly dependent on the reference base (6,7). SbF_5 is generally considered to be the strongest Lewis acid. Recently an "optical scale" for Lewis basicity has been developed for oxide systems (J. A. Duffy and M. D. Ingram, J. Amer. Chem. Soc., 93, 6448 (1971)).

B. Solvent Effects (8,9)

One of the most important factors determining acid (base) strength is the solvent used, especially when the solvent is a participant in the equilibrium as is generally the case. Changing solvents results in changes not only in acidity constants, but often also in the order of relative acidities of a series of compounds. A dramatic example is provided by the recent data obtained on gas phase acidities (10), which showed for aliphatic alcohols an acid strength order of primary<secondary<tertiary, the reverse of typical solution measurements. In addition, toluene was found to have an acid

Refs. p. 79.

strength between that of methanol and ethanol. Similar findings have been reported for gas phase amine acidity and basicity (10).

There are many reasons why solvents other than water are used in acid-base determinations: (1) solubility problems; (2) leveling effect (all acids measured appear to have the same strength; this effect is more serious for more basic (acidic) solvents); (3) acid or base strength of substance is beyond the range measurable in water (the strongest base capable of existence in significant concentration in water is hydroxide ion; the conjugate base of any solvent is the strongest base existing in that solvent). Ideally, solvents for acid-base determinations should (a) be both weakly acidic and weakly basic, (b) have a high dielectric constant to promote ionization, (c) be unreactive with the substances involved, including indicators, and (d) be easily obtainable in pure form. The following lists of recommended solvents are taken from (9).

Determination of Acids

Inert solvents (in order of decreasing dielectric constant): acetonitrile, acetone, ethyl methyl ketone, i-butyl methyl ketone, chlorobenzene, toluene, benzene.

Basic or amphoteric solvents (in order of increasing basicity): propylene glycol, methyl cellosolve, methanol, ethanol, 1-butanol, 1-propanol, 2-propanol, t-butyl alcohol, diethyl ether, p-dioxan, DMF, pyridine, 1-butylamine, ethylenediamine.

Determination of Bases

Inert solvents (in order of decreasing dielectric constant): acetonitrile, acetone, ethyl methyl ketone, i-butyl methyl ketone, chlorobenzene, chloroform, benzene, carbon tetrachloride, p-dioxan, cyclohexane, hexane.

Acid or amphoteric solvents (in order of increasing acidity): 2-propanol, cellosolve, propylene glycol, ethylene glycol, nitrobenzene, nitromethane, acetic anhydride, propionic acid, acetic acid, formic acid.

Some specific examples of the ranges available for measurable pK_a values are as follows (referred to water): ammonia (2 to 22), water (0 to 14), ethanol (-4 to 18), diethyl ether (-7 to 40), acetic acid (-8 to 11), formic acid (-9 to 7), sulfuric acid (-16 to -10), heptane (-20 to 40).

The meaning of pH changes considerably in nonaqueous media, and even in aqueous media at high concentrations of electrolytes. For solutions whose ionic strength is >0.1M, pH as a measure of acidity breaks down. Other scales have been developed, such as the Hammet acidity function; these are described in a separate section below. Recently, Bates has discussed approaches to measurement of pH for mixed aqueous solvents (3b).

C. Acidity Constants

1. General Guide To The Literature

It is impractical to include an exhaustive collection of acidity data in this book. In addition to the key references given within the tables below, the following bibliography to specific compound types is offered.

Refs. p. 79.

Compound Type or Study	Ref.
Prediction of acid and base strengths of organic compounds using Hammett σ,ρ relationships	11
Basicity of carbonyl compounds	12
Weak organic bases (alkenes; alkynes; aromatics; halogen compounds; compounds of N, O, S, P, and other elements)	13
Acidity of phenols	14
Basicity of aromatics	15
Heterocyclic compounds	16
Hydrocarbon acidity	17, 18
Acidity of benzenesulfonamides	19
Equilibrium acidities in dimethyl sulfoxide	20
General collections of large amounts of acid-base data	2, 21

2. Dissociation Constants of Common Acids in Water

These data were taken primarily from references 2, 22, and 23. The first ionization of sulfuric acid is too high to measure accurately, as are those for HCl, HI, $HClO_3$, and so on. For information on extremely strong ("super") acids, see page 21.

Acid	T(°C)	K_a	pK_a
Acetic (CH_3CO_2H)	25	1.754×10^{-5}	4.75
Arsenic (H_3AsO_4) (1)	18	5.62×10^{-3}	2.25
(2)	18	1.70×10^{-7}	6.77
(3)	18	3.95×10^{-12}	11.60
Arsenious ($HAsO_3$)	25	6×10^{-10}	9.23
Benzoic ($C_6H_5CO_2H$)	25	6.312×10^{-5}	4.20
o-Boric (H_3BO_3) (1)	20	7.3×10^{-10}	9.14
(2)	20	1.8×10^{-13}	12.74
(3)	20	1.6×10^{-14}	13.80
n-Butyric ($C_3H_7CO_2H$)	25	1.515×10^{-5}	4.82
Carbonic (H_2CO_3) (1)	25	4.30×10^{-7}	6.37
(2)	25	5.61×10^{-11}	10.25
Chloroacetic ($ClCH_2CO_2H$)	25	1.359×10^{-3}	2.87
Chromic (H_2CrO_4) (1)	25	1.8×10^{-1}	0.74
(2)	25	3.20×10^{-7}	6.49
p-Cresol ($CH_3C_6H_4OH$)	25	5.5×10^{-11}	10.26
Formic (HCO_2H)	20	1.77×10^{-4}	3.75
Hydriodic (HI)	25	$\sim 1.0 \times 10^{+11}$	~ -11
Hydrobromic (HBr)	25	$\sim 1.0 \times 10^{+9}$	~ -9
Hydrocyanic (HCN)	25	4.93×10^{-10}	9.31
Hydrofluoric (HF)	25	3.53×10^{-4}	3.45
Hydrogen selenide (H_2Se)	25	1.0×10^{-4}	4.0
Hydrogen telluride (H_2Te)	25	1.0×10^{-3}	3.0

Refs. p. 79.

(Continued)

Acid		T(°C)	K_a	pK_a
Hydrogen sulfide (H_2S)	(1)	18	9.1×10^{-8}	7.04
	(2)	18	1.1×10^{-12}	11.96
Hypobromous (HOBr)		25	2.06×10^{-9}	8.69
Hypochlorous (HOCl)		18	2.95×10^{-8}	7.53
Hypoiodous (HOI)		25	1.0×10^{-10}	10
Iodic (HIO_3)		25	1.69×10^{-1}	0.77
Nitric (HNO_3)		25	$4.36 \times 10^{+1}$	-1.64
Nitrous (HNO_2)		12.5	4.6×10^{-4}	3.37
Octanoic ($C_7H_{15}CO_2H$)		25	1.28×10^{-5}	4.89
Perchloric ($HClO_4$)		25	$\sim 1.0 \times 10^{+8}$	$\sim$-8
Periodic (HIO_4)		25	2.3×10^{-2}	1.64
Phenol (C_6H_5OH)		20	1.28×10^{-10}	9.99
Picric (2,4,6-$(NO_2)_3C_6H_2OH$)		25	4.2×10^{-1}	0.38
o-Phosphoric (H_3PO_4)	(1)	25	7.52×10^{-3}	2.12
	(2)	25	6.23×10^{-8}	7.21
	(3)	18	2.2×10^{-13}	12.67
o-Phosphorous (H_3PO_3)	(1)	25	1.58×10^{-2}	1.8
	(2)	25	6.3×10^{-7}	6.2
Pyrophosphoric ($H_4P_2O_7$)	(1)	25	1.0×10^{-1}	1.0
	(2)	25	1.0×10^{-2}	2.0
	(3)	25	1.0×10^{-7}	7.0
	(4)	25	1.0×10^{-9}	9.0
Sulfanilic ($H_2NC_6H_4SO_3H$)		25	5.9×10^{-4}	3.23
Sulfuric (H_2SO_4)	(1)	25	$\sim 1.0 \times 10^{+3}$	$\sim$-3
	(2)	25	1.20×10^{-2}	1.92
Sulfurous (H_2SO_3)	(1)	18	1.54×10^{-2}	1.82
	(2)	18	1.02×10^{-7}	6.91
Trifluoroacetic (CF_3CO_2H)		25	5.9×10^{-1}	0.23

3. Dissociation Constants of Common Nitrogen Bases in Water

These data were taken from references 16 and 24. K_a and pK_a refer to the conjugate acid; a large value of pK_a means a strong base.

Base	T(°C)	K_a	pK_a
Ammonia	25	5.66×10^{-10}	9.247
Aniline	25	2.34×10^{-5}	4.630
1,8-Bis(dimethylamino) naphthalene[a]	25	4.27×10^{-13}	12.37
Benzylamine	25	4.67×10^{-10}	9.33
n-Butylamine	20	1.67×10^{-11}	10.777
Cyclohexylamine	24	2.19×10^{-11}	10.66
Diethylamine	20	8.13×10^{-12}	11.090
Diisopropylamine	21	7.41×10^{-12}	11.13
Dimethylamine	25	1.85×10^{-11}	10.732
Diphenylamine	25	1.62×10^{-1}	0.79
Ethylamine	20	1.56×10^{-11}	10.807

(Continued)

Base	T(°C)	K_a	pK_a
Ethylenediamine (1)	20	8.41×10^{-11}	10.075
(2)	20	1.04×10^{-7}	6.985
Hexamethylenediamine (1)	20	7.85×10^{-12}	11.105
(2)	20	9.68×10^{-11}	10.014
Isoquinoline	20	4.17×10^{-6}	5.38
Methylamine	25	2.70×10^{-11}	10.657
N-methylaniline	25	1.41×10^{-5}	4.848
Morpholine	25	4.67×10^{-9}	8.33
Piperazine (1)	23.5	1.48×10^{-10}	9.83
(2)	23.5	2.76×10^{-6}	5.56
Piperidine	25	7.53×10^{-12}	11.123
Purine	20	4.07×10^{-3}	2.39
Pyrazine (1)	27±2	2.24×10^{-1}	0.65
Pyridine	25	6.17×10^{-6}	5.21
Pyrimidine	20	2.24×10^{-1}	0.65
Pyrrolidine	25	5.37×10^{-12}	11.27
Quinoline	25	1.35×10^{-5}	4.87
o-Toluidine	25	3.55×10^{-5}	4.45
m-Toluidine	25	1.86×10^{-5}	4.73
p-Toluidine	25	8.32×10^{-6}	5.08
Triethylamine	18	9.77×10^{-12}	11.01
Trimethylamine	25	1.55×10^{-10}	9.81
Urea	21	7.94×10^{-1}	0.10

[a]Sold under the trade name "Proton Sponge." It has unusual properties for use in organic chemical reactions (Aldrich Chemical Company).

4. Approximate pK_a Values for Common Reagents

The values in this table are meant only to be representative and to give some information on the effect of structure on acidity. The values above $pK_a = 15$ and below -2 are quite approximate, as indicated by the significant figures shown. These values of course are dependent on solvent, and in some cases the order of relative acidities may even change. These data were taken primarily from references 22 and 23.

Conjugate Acid	Conjugate Base	pK_a
$SbF_5 \cdot FSO_3H$	$SbF_5 \cdot FSO_3^-$	<-20
HPF_6	PF_6^-	-20
$R-N^+(OH)=O$	RNO_2	-12
$HClO_4$	ClO_4^-	-10
$R-C \equiv N^+-H$	RCN	-10
HI	I^-	-10
$R-C(=O^+-H)-H$	$R-C(=O)-H$	-8
$ArC(=O^+-H)-OR$	$ArC(=O)-OR$	-7.4
$ArC(=O^+-H)H$	$ArC(=O)H$	-7
$ArSO_3H$	$ArSO_3^-$	-7
HCl	Cl^-	-7
protonated cyclohexanone ($C_6H_{10}=O^+-H$)	cyclohexanone ($C_6H_{10}=O$)	-7

(Continued)

Refs. p. 79.

Conjugate Acid	Conjugate Base	pK_a
ArS^+H_2	$ArSH$	-7
ArO^+H_2	$ArOH$	-7
$CH_3C(=O^+H)-OH$	$CH_3C(=O)-OH$	-6.5
$(CH_3)_2SH^+$	$(CH_3)_2S$	-5.2
$(CH_3)_2OH^+$	$(CH_3)_2O$	-3.5
$CH_3OH_2^+$	CH_3OH	-2
$t\text{-}BuOH_2^+$	$t\text{-}BuOH$	-2
HNO_3	NO_3^-	-1.4
$R-C(=O)-NH_3^+$	$R-C(=O)-NH_2$	-1
CF_3CO_2H	$CF_3CO_2^-$	0
$2,4,6\text{-}(NO_2)_3C_6H_2-OH$	$2,4,6\text{-}(NO_2)_3C_6H_2-O^-$	0.4
CCl_3CO_2H	$CCl_3CO_2^-$	0.9
$CHCl_2CO_2H$	$CHCl_2CO_2^-$	1.3
$(C_6H_5SO_2)_2NH$	$(C_6H_5SO_2)_2N^-$	1.45
CH_2ClCO_2H	$CH_2ClCO_2^-$	2.8
HF	F^-	3.17
HNO_2	NO_2^-	3.29
$CH_2(NO_2)_2$	$^-CH(NO_2)_2$	3.6
HCO_2H	HCO_2^-	3.7
$2,4\text{-}(NO_2)_2C_6H_3-OH$	$2,4\text{-}(NO_2)_2C_6H_3-O^-$	4
$ArNH_3^+$	$ArNH_2$	4
$HC(=O)-CH_2-CH(=O)$	$HC(=O)-\overline{C}H-CH(=O)$	5
CH_3CO_2H	$CH_3CO_2^-$	5
$C_5H_5NH^+$ (pyridinium)	C_5H_5N (pyridine)	5.2
H_2CO_3	HCO_3^-	6.35
H_2S	HS^-	7.00
$p\text{-}O_2N-C_6H_4-OH$	$p\text{-}O_2N-C_6H_4-O^-$	7.2
$ArSH$	ArS^-	8
$p\text{-}(CH_3)_3N^+-C_6H_4-OH$	$p\text{-}(CH_3)_3N^+-C_6H_4-O^-$	8
$CH_3CH_2NO_2$	$CH_3\overline{C}HNO_2$	8.5
HCN	^-CN	9.1
NH_4^+	NH_3	9.24
$m\text{-}O_2N-C_6H_4-OH$	$m\text{-}O_2N-C_6H_4-O^-$	9.3
succinimide (N–H)	succinimide anion (N^-)	9.6
C_6H_5-OH	$C_6H_5-O^-$	9.98
RNH_3^+, $R_2NH_2^+$, R_3NH^+	RNH_2, R_2NH, R_3N	∿10
$C_6H_5SO_2NH_2$	$C_6H_5SO_2NH^-$	10.1
CH_3NO_2	$\overline{C}H_2NO_2$	10.2
HCO_3^-	CO_3^{--}	10.33
$o\text{-}HO-C_6H_4-CO_2^-$	$o\text{-}^-O-C_6H_4-CO_2^-$	13.4
$(RO_2C)_2CH_2$	$(RO_2C)_2\overline{C}H$	13
C_5H_6 (cyclopentadiene)	$C_5H_5^-$ (cyclopentadienide)	15
H_2O	^-OH	15.7
CH_3OH	CH_3O^-	16

(Continued)

Refs. p. 79.

Conjugate Acid	Conjugate Base	pK_a
CH_3CH_2OH	$CH_3CH_2O^-$	18
$PhC{\equiv}C{-}H$	$PhC{\equiv}C^-$	18.5
t-BuOH	$t{-}BuO^-$	19
$CH_3\overset{O}{\overset{\|}{C}}CH_3$	$CH_3\overset{O}{\overset{\|}{C}}CH_2^-$	20
$CH_3SO_2CH_3$	$CH_3SO_2CH_2^-$	23
$HC{\equiv}CH$	$HC{\equiv}C^-$	25 
$R\overset{O}{\overset{\|}{C}}NH_2$	$R\overset{O}{\overset{\|}{C}}NH^-$	25
$Ar\overset{O}{\overset{\|}{C}}NH_2$	$Ar\overset{O}{\overset{\|}{C}}NH^-$	25
Ph_3CH	Ph_3C^-	31.5
$PhCH_3$	$PhCH_2^-$	35
NH_3	NH_2^-	36
$CH_2{=}CH_2$	$CH_2{=}\overline{CH}$	36.5
	$C_6H_5^-$	37
CH_4	CH_3^-	40

5. Approximate Acidities of Carbon Acids

A carbon acid has been defined as "an organic substance that, when tested with a suitable base, donates a proton to that base by fission of a carbon-hydrogen bond"(18). Because of the very weak nature of these acids (pK > 15) direct measurement of equilibrium (thermodynamic) acidities is subject to large error. Several methods have been used to make such measurements; for specific details consult the references shown. The values reported here are meant only to be guides to the relative acidities of the different compounds and might change considerably with solvent. Reference 17 is a general review of the field.

Compound	pK_a	Ref.
$O_2NC\underline{H}_2NO_2$	3.6	25
$O_2NC\underline{H}_2CO_2CH_3$	5.8	25 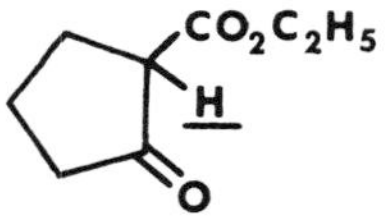
$COCH_3$, $\underline{H}$, O	8	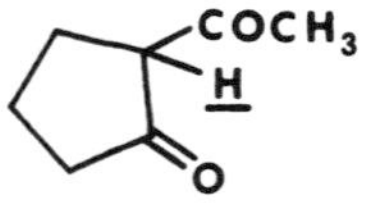 25
$CH_3C\underline{H}_2NO_2$	8.6	25 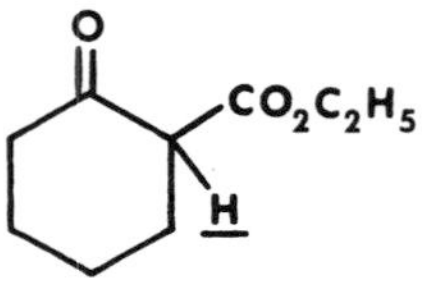
$NCC\underline{H}_2CO_2C_2H_5$	9	25
$CH_3\overset{O}{\overset{\|}{C}}C\underline{H}_2\overset{O}{\overset{\|}{C}}CH_3$	9	25
O, $COCH_3$, $\underline{H}$	10	25
CH_3NO_2	10.6	25
$CO_2C_2H_5$, $\underline{H}$, O	10.5	25
$CH_3\overset{O}{\overset{\|}{C}}C\underline{H}_2\overset{O}{\overset{\|}{C}}OR$	11	25
O, $CO_2C_2H_5$, $\underline{H}$	11.5	25
$NCC\underline{H}_2CN$	11-12	25
$RO_2CC\underline{H}_2CO_2R$		25
H, $\underline{H}$	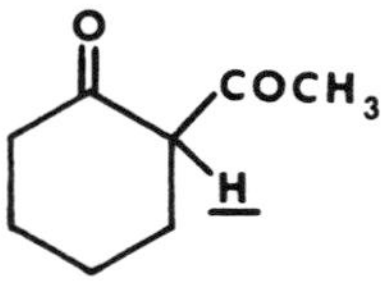  14-15	26

Refs. p. 79

(Continued)

Compound	pK_a	Ref.
$PhC{\equiv}C\text{-}\underline{H}$	18.5	27
$PhC(=O)C\underline{H}_3$	19	27
$CH_3C(=O)C\underline{H}_3$	20	25
	21	27
$C\underline{H}_3CO_2^-$	24	25
$RO_2CC\underline{H}_2R$	24.5	25
$C\underline{H}_3CN$	25	25
$CH_3C(=O)N\underline{H}_2$	25	25
$CH_3S(=O)C\underline{H}_3$	31.3	18
$Ph_3C\underline{H}$	31.5	28

Compound	pK_a	Ref.
$Ph_2C\underline{H}_2$	33	28
$PhC\underline{H}_3$	35	18
$CH_2{=}CH\text{-}C\underline{H}_3$	35.5	18
$H_2C{=}CH_2$	36.5	18
Benzene	37	18
Cyclopropane	39	18
CH_4	40	18
CH_3CH_3	42	18
Cyclobutane	43	18
$CH_3CH_2C\underline{H}_3$	44	18
Cyclopentane	44	18
Cyclohexane	45	18

6. Hammett and Other Acidity Functions (31,32,33)

a. Theory: The activity and concentration of hydrogen ion are routinely determined via pH measurements, but such measurements only have accurate meaning in very dilute solutions, in a single solvent, usually water. In concentrated solutions or in nonaqueous or mixed solvent systems some other parameter is necessary to measure the acidity of a solution. An acidity function is thus a measure of the proton donating ability of a medium, and is important in kinetic investigations of acid (or base) catalyzed reactions.

The best known and most widely used such function is the Hammett acidity function, H_o, which is the best function for comparing acidities of different media and is far superior to simple stoichiometric acid concentration:

$$H_o = pK_{HB^+} - \log\frac{c_{HB^+}}{c_B} = -\log\frac{a_{H^+}\gamma_B}{\gamma_{HB^+}} = -\log h_o$$

Ideally, H_o represents a unique value for a particular acid solution, say 85% H_2SO_4 (a comparable D_o is also possible for D_2SO_4/D_2O); further, an H_o scale should exist that would be a function of H_2SO_4 concentration. Note that $H_o \rightarrow pH$ for dilute, aqueous solutions. Such a scale was developed by using a series of so-called indicator bases, B, weak bases for which the extent of protonation (or deprotonation) in strong acids could be measured easily (e.g., spectrophotometrically). Then -- again ideally -- having measured H_o with a series of weak bases of different strengths, the extent of

Refs. p. 79.

ionization for any other weak base should correspond to the equation above; thus a plot of $\log(c_{HB^+}/c_B)$ against H_o should be linear with slope of -1.00 for all neutral bases (proton acceptors). If this were true, an H_o scale would depend only on the specific acid solution, and would be independent of the indicators (bases) used to measure it; for this to be true, the ratio (γ_B/γ_{HB^+}) and its dependence on acid concentration must be the same for all neutral bases in a given acid solution. In practice these assumptions are not met with all bases, and the "Hammett acidity function" applies only to any H_o scale derived originally from measurements of ionization of certain primary amine indicators (mainly substituted anilines). By definition, a Hammett base is one whose values of $\log c_{HB^+}/c_B$ plotted against H_o do give a straight line with a slope of unity; they need not be amines and can have any structure. Acidity functions defined by non-Hammett bases have also been derived and have other symbols, as shown by the following examples.

Function	System
$H_- = -\log \dfrac{(a_{H^+})(\gamma_{B^-})}{\gamma_{HB}}$	When base is negatively charged (not neutral), as for $BH \rightarrow B^- + H^+$, where BH is a neutral, acid indicator
$J_o = -\log \dfrac{(a_{H^+})(\gamma_{ROH})}{(a_{H_2O})(\gamma_{R^+})} = H_o + \log a_w$	This Gold and Hawes function (also referred to as C_o and H_R) applies to equilibria between alcohols and carbonium ions
$H_+ = -\log \dfrac{(a_{H^+})(\gamma_{HR^+})}{\gamma_{H_2R^{+2}}}$	A measure of ability of medium to transfer a proton to a cationic base, HR^+ to form the conjugate acid, H_2R^{+2}

There are many others (H_o''', H_A, H_I, etc.) introduced for specific reaction systems; for details and data consult references 31 to 33.

b. Some H_o Data for Aqueous Acids at 25°: The following tables were taken mainly from references 31 and 32, where considerably more data are available. Temperature dependence data of H_o for aq H_2SO_4 may be found in reference 34. The table headings have the following meaning: % = weight percent acid (g/100 g solution), M = acid molarity (moles/l), m = acid molality (moles/1000 g solvent), a_w = activity of water. Some recent determinations have been made of H_o values for "super-acid" systems (acidity greater than 100% H_2SO_4); see reference 37.

%	$-H_o$	M	m	a_w
		Hydrochloric Acid		
3	0.13	0.83	0.85	0.961
6	0.58	1.69	1.71	0.919
9	0.90	2.57	2.71	0.865
12	1.21	3.48	3.74	0.801
15	1.54	4.41	4.83	0.730
18	1.87	5.37	6.04	0.649
22	2.35	6.69	7.75	0.531
26	2.87	8.05	9.65	0.419
30	3.39	9.45	11.76	0.317
34	3.95	10.90	14.14	0.225
36		11.64	15.40	0.186
40		13.15	18.32	0.121
		Sulfuric Acid		
5	0.02	0.520	0.540	
10	0.31	1.085	1.13	0.956
15	0.66	1.685	1.79	0.923
20	1.01	2.32	2.55	0.879
25	1.37	3.00	3.39	0.824
30	1.72	3.73	4.37	0.752
35	2.06	4.49	5.48	0.666
40	2.41	5.31	6.80	0.564
45	2.85	6.18	8.34	0.459
50	3.38	7.11	10.19	0.352
55	3.91	8.095	12.45	0.251
60	4.46	9.16	15.30	0.160
65	5.10	10.28	19.12	0.091
70	5.80	11.48	23.5	0.050
75	6.56	12.75	30.6	0.018
80	7.34	14.07	40.8	0.005
85	8.14	15.40	57.7	0.002
90	8.92	16.65	91.6	0.0003
95	9.85	17.76	193.5	
98	10.27			
99	10.57			
99.5	10.83			
99.9	11.42			
100.0	11.94			
		Perchloric Acid		
5	-0.02	0.51	0.523	0.980
10	0.35	1.05	1.100	0.956
15	0.63	1.62	1.765	0.925
20	0.95	2.23	2.49	0.890
25	1.25	2.88	3.32	0.840
30	1.60	3.58	4.26	0.770
35	1.97	4.34	5.36	0.685

Refs. p. 79

(Continued)

%	$-H_o$	M	m	a_w
		Perchloric Acid		
40	2.40	5.15	6.63	0.580
45	2.92	6.02	8.15	0.460
50	3.52	6.98	9.95	0.330
55	4.22	8.01	12.10	0.200
60	5.25	9.13	14.92	0.110
65	6.43	10.33	18.45	
70	7.72	11.60	23.20	
75	9.15			
77	9.96			
78.6	10.31			
		Phosphoric Acid		
6	-0.90	0.63	0.64	0.998
10	-0.64	1.08	1.13	0.980
16	-0.33	1.78	1.94	0.963
20	-0.15	2.26	2.55	0.949
26	0.09	3.06	3.59	0.922
30	0.26	3.62	4.38	0.898
35	0.48	4.34	5.50	0.862
40	0.72	5.13	6.81	0.815
45	0.94	5.95	8.35	0.762
50	1.17	6.81	10.20	0.700
55	1.42	7.74	12.45	0.625
60	1.66	8.73	15.30	0.540
65	1.97	9.78	18.95	0.443
70	2.29	10.89	23.80	0.344
75	2.64	12.08	30.6	0.246
80	3.05	13.32	40.8	0.160
85	3.48	14.66	57.9	0.095

Refs. p. 79.

c. Some H_ Data at 25°

Aqueous KOH and NaOH (35)			Methanolic $NaOCH_3$ (38)		Ethanolic DMSO[a] (36)	
	H_					
Molarity	KOH	NaOH	Molarity	H_	DMSO (mole %)	H_
1	14.0	13.9	0.10	15.97	0	13.99
2	14.6	14.4	0.25	16.42	1	14.07
3	14.9	14.7	0.50	16.82	5	14.25
4	15.2	14.9	1.00	17.32	10	14.45
5	15.5	15.2	1.50	17.69	20	14.92
6	15.8	15.4	2.00	18.01	30	15.40
7	16.2	15.6	2.50	18.31	40	16.11
8	16.5	15.8			50	17.03
9	16.8	16.0			60	17.75
10	17.3	16.2			70	18.45
11	17.7	16.5			80	18.97
12	18.2	16.8			90	19.68
13		17.1			95	20.68
14		17.5				
15		17.8				
16		18.1				

[a] 10^{-2}M in sodium ethoxide.

D. Properties of Commonly Used Strong Bases

The bases in this section have pK_a values (of the conjugate acid) generally above 15 and thus are classified as strong bases. More detailed information on any of these bases can be obtained from Fieser and Fieser, Reagents for Organic Synthesis, Vols. 1 and 2 (29) ; see also H. O. House, Modern Synthetic Reactions, W. A. Benjamin, New York, 1965, page 185. The references given are to the Fieser volumes (F for volume 1 and F2 for volume 2), and the comments are mostly paraphrases of their comments. These comments are not meant to limit possibilities of usage, but only to be interesting observations. Where a single use is noted, often the reagent can be used for many other types of reactions.

Refs. p. 79.

Base	Formula	Mw	Solvents[a]	Comments	Ref.
Choline	$[(CH_3)_3N^+CH_2CH_2OH]OH^-$	121.18	H_2O, MeOH	Similar to Triton B	F-142
Claisen's alkali	CH_3OH-KOH	--	--	Extracts phenols insoluble in aqueous alkali	F-153
Lithium amide	$LiNH_2$	22.97	EtOH(s), ℓ-NH_3(s), $PhCH_3$(s)	Used for aldol condensations. Easier to handle than sodium amide	F-600
Lithium diethylamide	$LiN(C_2H_5)_2$	79.07	$HN(C_2H_5)_2$, DME, ether, THF	Used for opening epoxides	F-610 F2-247
Lithium ethoxide	$LiOC_2H_5$	52.00	EtOH	Generates carbenes from nitroso ureas; used in Wittig reactions	F-612
Lithium hydride	LiH	7.95	DME, ether(s)		F2-251
Lithium nitride	Li_3N	34.83	Diox(s), diglyme(s)	Both solvents strongly enhance reactivity; converts acid chlorides to triamides	F-618 F2-251
Naphthalene sodium	$[C_{10}H_8]^{\dot{-}}Na^+$	151.16	DME, ether, THF	Better than sodium in THF for metalation of terminal acetylenes and active methylene compounds	F-711 F2-289
Phenyl lithium	C_6H_5Li	84.04	Ether	Used in halogen lithium exchange reactions	F-845
Phenyl potassium	C_6H_5K	116.20	Ether	Metalates active methylenes	F-848

Refs. p. 79.

Phenyl sodium	C_6H_5Na	100.10	Ether	Used in benzyne production	F-848
Potassium amide	KNH_2	55.13	ℓ-NH_3, ether(s)	Flammable on contact with moisture; nitrates active methylene compounds in presence of alkyl nitrates; useful in dehydrohalogenations and Hoffmann degradations	F-907 F2-336
Potassium t-Butoxide	$(CH_3)_3COK$	112.21	t-BuOH, DME(s), DMF, DMSO, ether(s), PhH(s), Xy(s)	Good general base; probably most powerful alkoxide known	F-911 F2-336
Potassium ethoxide	KOC_2H_5	68.16	Ether(s), EtOH, $PhCH_3$(s)	Used as condensation catalyst	F-928
Potassium hydride	KH	40.11	Mineral oil suspension	More basic than sodium hydride	F-935 F2-346
Potassium hydroxide	KOH	56.10	EtOH, MeOH, H_2O	General base often used in ethanolic solution	F-935 F2-346
Potassium-2-methyl-2-butoxide	$CH_3CH_2C(CH_3)_2$-OK	126.24	t-AmylOH, ether, PhH, $PhCH_3$	Good for formation of oximes with hindered acetophenones	F-939
Sodium amide	$NaNH_2$	39.02	DME(s), ether(s), ℓ-NH_3(s), PhH(s), $PhCH_3$(s)	Good general base often used in dehydrohalogenations; once opened, it should all be used to prevent explosive decomposition products from forming	F-1034 F2-373
Sodium ethoxide	$NaOC_2H_5$	68.06	EtOH, ether(s)	General base often used in condensations and alkylations	F-1065

Refs. p. 79.

Refs. p. 79.

Base	Formula	Mw	Solvents[a]	Comments	Ref.
Sodium hydride	NaH	24.01	DME(s), DMF(s), ether(s), PhH(s) $PhCH_3$(s), Xy(s)	Good for condensations and alkylations; forms benzyl tosylates from benzyl alcohols and tosyl chloride	F-1075 F2-382
Sodium hydroxide	NaOH	40.00	H_2O	General base	F-1083
Sodium methoxide	$NaOCH_3$	54.03	DMF, DMSO, ether(s), MeOH, Pentane(s)	Yields dichlorocarbene from ethyl trichloro-acetate; good general base	F-1091 F2-385
Sodium-2-methyl-2-butoxide	$CH_3CH_2-C(CH_3)_2-ONa$	94.14	t-AmylOH, PhH, $PhCH_3$, Xy, ether	More soluble in organic solvents, especially aromatics, than t-butoxide	F-1096 F2-386
Sodium methylsulfinyl methide (dimsyl sodium)	CH_3SOCH_2Na	100.11	DMSO (prepare Na salt from NaH and excess DMSO)	Often reagent of choice for Wittig reaction; several general and specific uses	F-310
Tetraethylammonium hydroxide	$(C_2H_5)_4\overset{+}{N}OH^-$	147.26			F-1138
Triton B	$C_6H_5CH_2N(CH_3)_3OH$	167.26	EtOH, MeOH, Py	Good condensation catalyst	F-1252
Trityl lithium	$(C_6H_5)_3CLi$	250.25	DME, ether, THF	Used to study compositions of enolate anions	F-1256 F2-454

Trityl potassium	$(C_6H_5)_3CK$	282.43	DME, ether(s), ℓ-NH_3	Useful in acylations, alkylations, and carbonations	F-1258
Trityl sodium	$(C_6H_5)_3CNa$	266.32	DME, ether, PhH, $PhCH_3$, ℓ-NH_3	General condensation catalyst	F-1259

[a] The following shorthand for solvents is used: t-BuOH = t-butyl alcohol; Diox = dioxane; DME = 1,2-dimethoxyethane; DMF = dimethylformamide; DMSO = dimethylsulfoxide; ether = diethylether; EtOH = ethanol; Me_2CO = acetone; MeOH = methanol; ℓ-NH_3 = liquid ammonia; $PhCH_3$ = toluene; PhH = benzene; Py = pyridine; Xy = xylene. A lowercase s in parentheses following the solvent indicates only slight solubility or suspension.

E. Standard Buffer Solutions

The data below provide information for the preparation of most standard and some special buffer solutions of specified pH values. More data are available in references 21 and 30.

1. Standard Buffer pH values[a]

T (°C)	0.1M HCl	0.05M K-Tetraoxalate	Satd.(25°) KH-Tartrate	0.05M KH-Phthalate	Phosphate No. 1	Phosphate No. 2	0.01M Borax	Satd.(25°) $Ca(OH)_2$
0	1.10	1.67		4.00	6.98	7.53	9.46	13.42
5	1.10	1.67		4.00	6.95	7.50	9.40	13.21
10	1.10	1.67		4.00	6.92	7.47	9.33	13.00
15	1.10	1.67		4.00	6.90	7.45	9.28	12.81
20	1.10	1.68	3.56	4.00	6.88	7.43	9.23	12.63
25	1.10	1.68	3.56	4.01	6.87	7.41	9.18	12.45
30	1.10	1.68	3.55	4.02	6.85	7.40	9.14	12.29
35	1.10	1.69	3.55	4.02	6.84	7.39	9.08	12.13
40	1.10	1.69	3.55	4.04	6.84	7.38	9.07	12.04
45	1.10	1.70	3.55	4.05	6.83	7.38	9.04	11.98
50	1.11	1.71	3.55	4.06	6.83	7.37	9.01	11.84
55	1.11	1.72	3.55	4.08	6.83	7.37	8.99	11.71
60	1.11	1.72	3.56	4.09	6.84		8.96	11.57

(Continued)

Refs. p. 79.

[a]Solutions are prepared as follows:

<u>0.1M hydrochloric acid</u>. Dilute reagent grade HCl and titrate with standard base. Adjust to 0.1000 ± 0.005M.
<u>Potassium tetraoxalate</u>. Dissolve 1.271 g $KH_3(C_2O_4)_2 \cdot 2H_2O$ in water and dilute to 100 ml.
<u>Potassium hydrogen tartrate</u>. Allow salt to saturate in water at 25 ± 3°.
<u>Potassium hydrogen phthalate</u>. Dissolve 1.021 g salt in water and dilute to 100 ml.
<u>Phosphate No. 1</u>. Dissolve 3.40 g (0.025m) KH_2PO_4 and 3.55 g (0.025m) anhydrous Na_2HPO_4 in water and dilute to 1 liter.
<u>Phosphate No. 2</u>. Dissolve 1.183 g (0.008695m) KH_2PO_4 and 4.320 g (0.03043m) Na_2HPO_4 in water and dilute to 1 liter.
<u>Borax</u>. Dissolve 3.81 g of $Na_2B_4O_7 \cdot 10H_2O$ in water and dilute to 1 liter.
<u>Calcium hydroxide</u>. Allow reagent to saturate in water at 25°.

2. Nomograms for Acetate and Phosphate Buffers

The phosphate buffer nomogram below is used to determine the molar concentration of both KH_2PO_4 and K_2HPO_4 for the required solution. It is taken from W. C. Boyd, J. Biol. Chem., 240, 4097 (1965). For example, suppose a phosphate buffer of pH 7.4 and ionic strength 0.10 is desired. Position the straight edge between 7.4 on the pH scale, and 0.10 on the μ scale. Read 0.0096 on the KH_2PO_4 scale; then $[K_2HPO_4]$ is calculated using the equation shown at the top of the nomogram and the values read from the 1/3 KH_2PO_4 and μ/3 scales:

$$[K_2HPO_4] = 0.033 - 0.0033 = 0.0297M$$

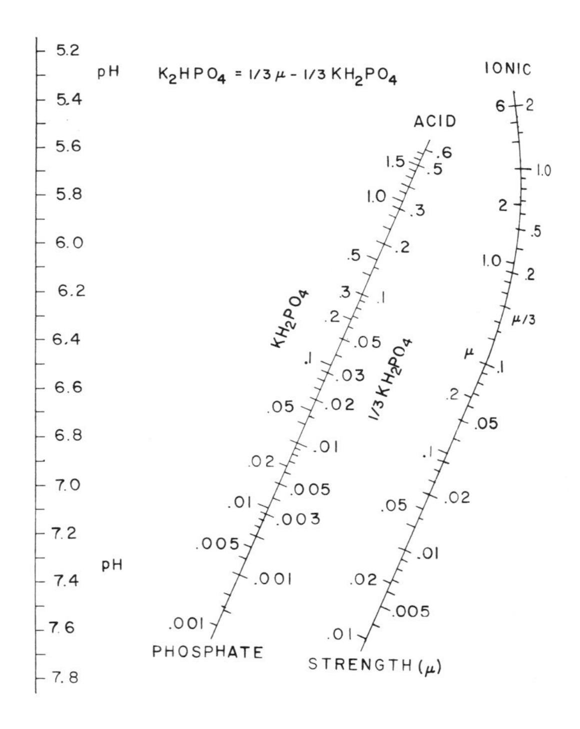

Refs. p. 79.

The acetate buffer nomogram below is taken from W. C. Boyd, J. Amer. Chem. Soc., 67, 1035 (1945). For example, an acetate buffer of pH 4.8 and ionic strength 0.01 can be obtained with a solution 0.01M in sodium acetate and 0.0087M in acetic acid.

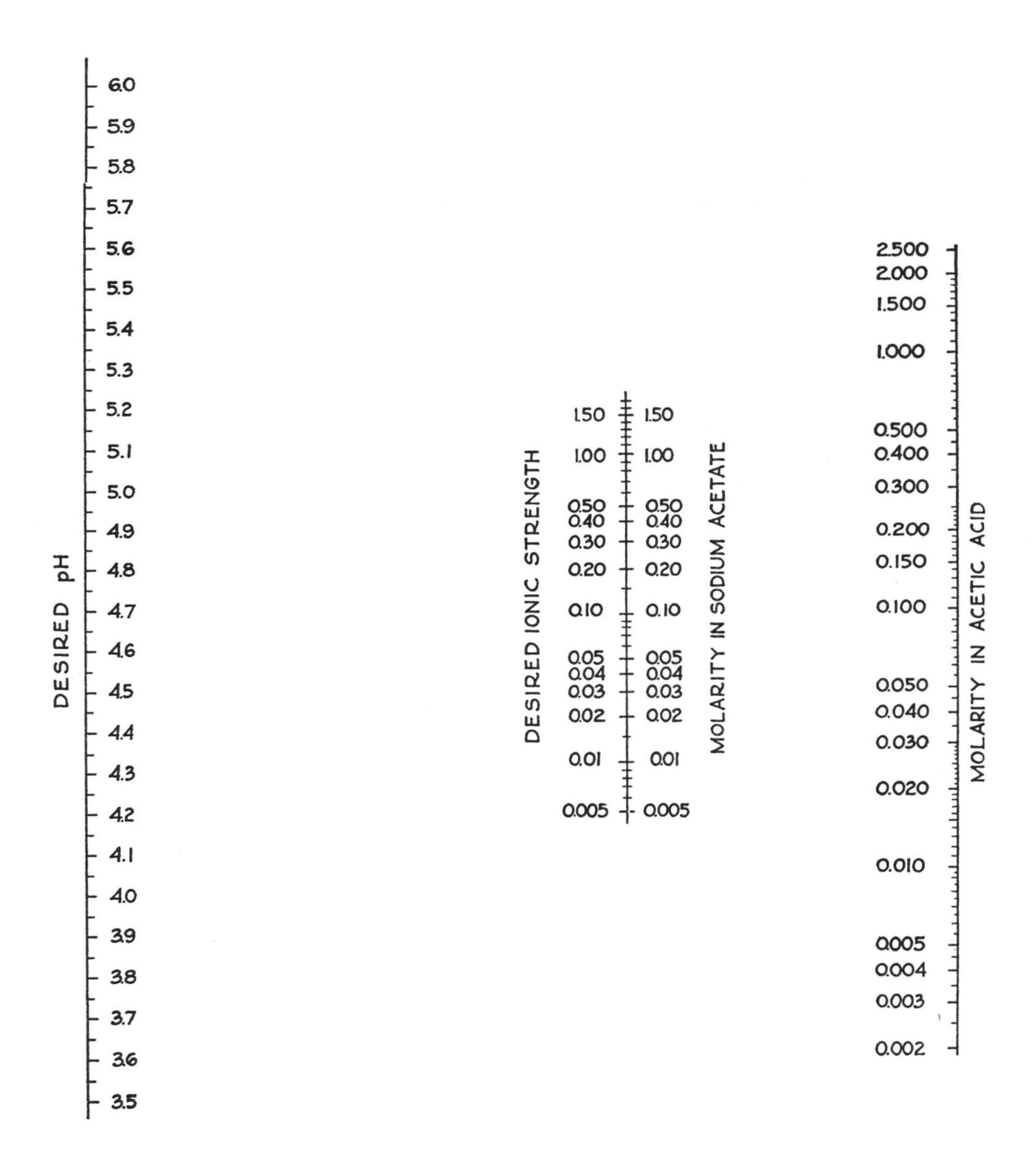

Refs. p. 79.

3. Miscellaneous Convenient Buffers

(a) Tris: made with HCl solutions of tris(hydroxymethyl)aminomethane (mw 121.14) and provide pH values from 7.55 to 9.67 (5°C), 7.00 to 9.10 (25°C), and 6.70 to 8.79 (37°C) depending on concentration. Sold as "Trizma" solutions or solid HCl salts by Sigma Chemical Co., 3500 DeKalb Street, St. Louis, Mo. 63118. For example, a 0.05M solution has pH values of 9.5, 8.9 and 8.6 at 5, 25, and 37°C, respectively.

(b) Succinate: a solution 0.05M in both NaHsuccinate and Na_2succinate has a pH of 5.34 (25°).

(c) Carbonate: a solution 0.05M in both $NaHCO_3$ and Na_2CO_3 has a pH of 9.93 (25°C).

(d) Na_3PO_4: a 0.05M Na_3PO_4 solution has a pH of 12.04 (25°C).

(e) Electrophoresis: see page 392 for electrophoresis buffers.

F. Primary Standards for Accurate Concentrations

A primary standard is a substance that can be used to make accurately known concentrations of active reagent. Often this solution is called the primary standard, while the substance used to make up the solution is called the primary standard substance. These substances must have a high degree of purity (>99.98%). The list below contains several substances that have been used as primary standards. The purity designation of these substances refers to the fact that Chem Sources, Directories Publishing Co., Flemington, N. J. lists them as being commercially available in very high purity (HP) or they are available from National Bureau of Standards as high purity reference substances (NBS) (see p. 496).

Constant-boiling hydrochloric acid has often been used as a primary standard; the concentration of HCl as a function of pressure is well known. A table showing this relationship appears below.

A secondary standard is one whose concentration (or purity) has been accurately determined by standardization with a primary standard. A sodium hydroxide solution for which the concentration had been accurately determined by titrating high purity benzoic acid is an example.

Name	Formula	Mw	Purity	Comment
Benzoic acid	$C_6H_5CO_2H$	122.123	NBS, HP	Dissolve about 0.5 g in about 50% ethanol and titrate using phenolphthalein as an indicator
Borax	$Na_2B_4O_7 \cdot 10H_2O$	381.360	NBS	Weak acid; dissolve in water and use methyl red as an indicator
Mercuric oxide	HgO	216.599	HP	Dissolve about 0.5 g with 15 g KBr in 25 ml water, excluding CO_2 in the solution process. Use bromthymol blue as an indicator

(Continued)

Refs. p. 79.

Name	Formula	Mw	Purity	Comment
Potassium bicarbonate	$KHCO_3$	100.116		Use bromcresol green as an indicator; titrate to first tint of green
Potassium biiodate	$KH(IO_3)_2$	389.912		Strong acid; low solubility in water; use bromthymol blue as an indicator
Potassium biphthalate	$KHC_8H_4O_4$	204.224	NBS	Weak acid; use phenolphthalein as an indicator
Potassium bitartrate	$KHC_4H_4O_6$	188.178	NBS	Solution subject to molding, thus useful for only a few days; use phenolphthalein as an indicator
Sodium carbonate	Na_2CO_3	105.988	HP	Slightly hygroscopic and should be dried; titrate to to first tint of green using bromcresol green

COMPOSITION OF CONSTANT-BOILING HCl(30)

Pressure (mm Hg)	Weight % HCl in Distillate	Weight of Distillate (g) Necessary to Make 1 Liter of 1 N HCl
770	20.197	180.407
760	20.221	180.193
750	20.245	179.979
740	20.269	179.766
730	20.293	179.555

Refs. p. 79.

Refs. p. 79.

G. Composition of Some Commercially Available Reagent-Grade Acids and Bases

These values, with the exception of the molecular weights, are approximate and are meant only to be representative of the reagents as commonly available in the United States. The more exact assay is usually printed on the bottle when purchased, but it should be kept in mind that concentrations change on exposure to air. This is particularly true for ammonium hydroxide. When making up a volumetric solution, therefore, it is better to use a little more than the calculated volume and then dilute if necessary. The millimeters of a concentrated reagent needed to make 1 liter of a solution of specified normality can be calculated with the formula:

$$V = \frac{100\ MN}{bpd}$$

where M is the molecular weight, N is the desired normality, b is the basicity (H_2SO_4 = 2, H_3PO_4 = 3, etc.), p is the percent acid (or base) in the concentrated reagent, and d is the specific gravity of that reagent. These data (p and d) are usually written on the reagent bottle.

Name	Formula	Mw	Normality of Conc. Reag.	% by Weight	Specific Gravity	Amount (mℓ) to Make 1 liter of 1N Solution
Acetic acid (glacial)	$HC_2H_3O_2$	60.052	17.4	99.8	1.05	57
Acetic acid (aqueous)			6.26	36	1.045	160
Ammonia	NH_3	17.031	14.8	29	0.90	65
Formic acid	HCO_2H	46.025	23.4	90	1.20	43
Hydriodic acid	HI	127.913	7.57	57	1.70	130
			5.51	47	1.50	180
Hydrobromic acid	HBr	80.912	8.90	48	1.50	120
Hydrochloric acid	HCl	36.461	12.0	37	1.19	83
Hydrofluoric acid	HF	20.006	28.9	49	1.16	35
Nitric acid	HNO_3	63.012	15.9	70	1.42	63
Perchloric acid	$HClO_4$	100.457	11.7	70	1.67	86
			9.5	60	1.54	110
Phosphoric acid	H_3PO_4	97.994	14.7	85	1.70	69
Potassium hydroxide	KOH	56.11	11.7	45	1.46	85
Sodium hydroxide	NaOH	40.00	19.1	50	1.53	52
Sulfuric acid	H_2SO_4	98.07	36	96	1.84	56
Sulfurous acid	H_2SO_3	82.09	0.7	6	1.02	-

H. Approximate pH Values For Various Concentrations of Selected Substances

The approximate pH values are given here for a number of common substances at various concentrations.

Compound	1N	0.1N	0.01N	0.001N
Acetic acid	2.4	2.9	3.4	3.9
Ammonia	11.8	11.3	10.8	10.3
Benzoic acid			3.1	
Citric acid		2.1	2.6	
Hydrogen chloride	0.10	1.07	2.02	3.01
Hydrogen cyanide		5.1		
Potassium hydroxide	14.0	13.0	12.0	11.0
Sodium bicarbonate		8.4		
Sodium carbonate		11.5	11.0	
Sodium hydroxide	14.05	13.07	12.12	11.13
Sulfuric acid	0.3	1.2	2.1	

Refs. p. 79.

Refs. p. 79

I. Acid-Base Indicators

The following common indicators cover the entire pH range (aqueous solutions). The data were taken from reference 30. For an extensive discussion of indicators for nonaqueous titrations, see reference 9.

Indicator	pH Range	Color		Preparation
		Acid	Base	
Methyl violet	0.1-1.5	Yellow	Blue	0.05% in water
Mentanil yellow	1.2-2.3	Red	Yellow	0.01% in water
Thymol blue (acid range)	1.2-2.8	Red	Yellow	0.1 g in 21.5 ml 0.01N NaOH + 229.5 ml water
Tropaeolin OO	1.4-2.6	Red	Yellow	1% in water
Methyl yellow	2.9-4.0	Red	Yellow	0.1% in 90% ethanol
Bromphenol blue	3.0-4.6	Yellow	Blue	0.1 g in 14.9 ml 0.01N NaOH + 235.1 ml water
Methyl orange	3.1-4.4	Red	Yellow	0.1% in water
Bromscresol green	3.8-5.4	Yellow	Blue	0.1 g in 14.3 ml 0.01N NaOH + 235.7 ml water
Methyl red	4.8-6.0	Red	Yellow	0.02 g in 60 ml ethanol + 40 ml water
Bromcresol purple	5.2-6.8	Yellow	Purple	0.02% in ethanol
Bromthymol blue	6.0-7.6	Yellow	Blue	0.1% in 50% ethanol
Phenol red	6.4-8.0	Yellow	Red	0.1% in ethanol
Neutral red	6.8-8.0	Red	Yellow	0.01 g in 50 ml ethanol + 50 ml water
Thymol blue (base range)	8.0-9.6	Yellow	Blue	0.1% in ethanol
Phenolphthalein	8.2-10.0	Colorless	Red	1% in ethanol
Thymolphthalein	9.4-10.6	Colorless	Blue	0.1% in ethanol
Alizarin yellow R	10.2-12.0	Yellow	Red	0.1% in water
Tropeolin O	11.0-13.0	Yellow	Orange	0.1% in water
Nitramine	10.8-13.0	Colorless	Brown	0.1% in 70% ethanol
1,3,5-Trinitrobenzene	11.5-14.0	Colorless	Orange	0.1% in ethanol

J. References

1. W. F. Luder and S. Zuffanti, The Electronic Theory of Acids and Bases, 2nd ed., Dover Publications, New York, 1961.

2. A. Albert and E. P. Serjeant, Ionization Constants of Acids and Bases, Methuen and Co., London, and Wiley, New York, 1962.

3. (a) R. G. Bates, Determination of pH, Wiley, New York, 1964; R. G. Bates, J. Res. Nat. Bur. Std., 66A, 179 (1962); B. R. Staples and R. G. Bates, Ibid., 73A, 37 (1969). (b) R. G. Bates, Pure Appl. Chem., 18, 419 (1969), and Solute-Solvent Interactions, J. F. Coetzee and C. D. Ritchie, Eds., Marcel Dekker, New York, 1969, p. 45.

4. "Specifications for pH Scale," British Standard 1647, British Standards Institution, London, 1961.

5. G. A. Olah and P. R. Schleyer, Eds., Carbonium Ions, Vol. 1, Wiley, New York, 1968.

6. G. A. Olah, Ed., Friedel-Crafts and Related Reactions, Vol. 1, Interscience, New York, 1963, pp. 169-201.

7. D. P. N. Satchell and R. S. Satchell, "Quantitative Aspects of Lewis Acidity of Covalent Metal Halides and Their Organo Derivatives," Chem. Rev., 69, 251 (1969).

8. M. M. Davis, "Acid-Base Behavior in Aprotic Organic Solvents," NBS Monograph 105, 1968.

9. I. Gyenes, Titration in Non-aqueous Media, D. Van Nostrand Co., New York, 1967.

10. J. I. Brauman et al., J. Amer. Chem. Soc., 93, 6360 (1971); 92, 5986 (1970); 93, 3911, 3914 (1971). See also D. Bohme, E. Lee - Ruff, and L. Young, Ibid., 93, 4608 (1971).

11. G. B. Barlin and D. D. Perrin, Quart. Rev., 20, 75 (1966) (acids), and J. Clark and D. D. Perrin, Ibid., 18, 295 (1964) (bases).

12. E. W. Arnett, R. Quirk, and J. Larsen, J. Amer. Chem. Soc., 92, 3977 (1970); V. A. Palm, V. Holdna, and A. Talvic, in The Chemistry of the Carbonyl Group, S. Patai, Ed., Wiley, New York, 1966, p. 421.

13. E. M. Arnett, "Quantitative Comparisons of Weak Organic Bases," Prog. Phys. Org. Chem., 1, 223 (1963).

14. C. H. Rochester and B. Rossall, Trans. Faraday Soc., 65, 1004 (1969).

15. H. H. Perkampus, "The Basicity of Unsaturated Compounds," Advan. Phys. Org. Chem., 4, 195 (1966).

16. A. Albert, in Physical Methods in Heterocyclic Chemistry, Vol. 1, A. R. Katritzky, Ed., Academic Press, 1963, p. 1.

17. A. Streitwieser and J. H. Hammons, Prog. Phys. Org. Chem., 3, 41 (1965).

18. D. J. Cram, Fundamentals of Carbanion Chemistry, Academic Press, New York, 1965.

19. G. Dauphin and A. Kergomard, Bull. Soc. Chim. Fr., 3, 486 (1961). See also F. A. Cotton and P. F. Stokely, J. Amer. Chem. Soc., 92, 294 (1970).

20. J. I. Brauman et al., J. Amer. Chem. Soc., 92, 6679 (1970).

21. Handbook of Biochemistry, CRC Press, Cleveland, 1968, pp. J-150 to J-189.

22. G. Kortum, W. Vogel, and K. Andrussow, Dissociation Constants of Organic Acids in Aqueous Solution, Plenum Press, New York, 1961.

23. H. A. Flaschka, A. J. Barnard, Jr., and P. E. Sturock, Quantitative Analytical Chemistry, Barnes and Noble, New York, 1969.

24. D. D. Perrin, Dissociation Constants of Organic Bases, Plenum Press, New York, 1965.

25. R. G. Pearson and R. L. Dillon, J. Amer. Chem. Soc., 75, 2439 (1953).

26. R. E. Dessy, Y. Okuzumi, and A. Chen, Ibid., 84, 2899 (1962).

27. W. K. McEwen, Ibid., 58, 1124 (1936).

28. A. Streitwieser, Jr., E. Cuiffarin, and J. H. Hammons, Ibid., 89, 63 (1967).

29. L. F. Fieser and M. Fieser, Reagents for Organic Synthesis, Vols. 1 and 2, Wiley, New York, 1967 and 1969.

30. Handbook of Chemistry and Physics, 47th ed., CRC Press, Cleveland, 1967; The Merck Index, 8th ed., Merck and Co., New York, 1968; Lange's Handbook of Chemistry, 10th ed., McGraw-Hill, New York, 1967.

31. For a brief introduction, see C. J. O'Connor, J. Chem. Educ., 46, 686 (1969).

32. C. H. Rochester, Acidity Functions, Academic Press, London and New York, 1970.

33. F. Coussement, Acidity Functions and Their Applications to Acid-Base Catalyses, Gordon and Breach, London, 1970.

34. P. Tickle, et al., J. Chem. Soc. (B), 65 (1970).

35. G. Schwarzenbach and R. Sulzberger, Helv. Chim. Acta, 27, 348 (1944).

36. K. Bowden and R. Stewart, Tetrahedron, 21, 261 (1965).

37. R. J. Gillespie, T. Peel, and E. Robinson, J. Amer. Chem. Soc., 93, 5083 (1971).

38. A. Streitwieser, Jr., C. J. Chang, and A. T. Young, ibid., 94, 4888 (1972).

PROPERTIES OF ATOMS AND BONDS

I. PROPERTIES OF THE ELEMENTS

The atomic weights are those adopted by the International Union of Pure and Applied Chemistry at its 1969 meeting and are based on a relative atomic mass of C^{12} = 12. For a full report, see Pure Appl. Chem., 21 (1), 91 (1970). These values apply to elements as they exist in materials of terrestrial origin and to certain isotopes; they are reliable to ±1 in the last digit or ±3 if that digit is in small type. A value in brackets denotes the mass of the isotope of longest known half-life with well known mass.

The other data are taken from several sources, including Handbook of Chemistry, 10th ed., N. A. Lange, Ed., McGraw-Hill, 1967; Handbook of Chemistry and Physics, 50th ed., Chemical Rubber Co., 1969-1970; L. P. Eblin, The Elements of Chemistry, 2nd ed., Harcourt, Brace and World, 1970. In cases where substantial disagreement or substantial error was reported, a "best" value was chosen and is reported in parentheses.

Element	Symbol	Atomic No.	Atomic Weight	Valence	Mp	Bp
Actinium	Ac	89	227.0278[a]		1050	3200
Aluminum	Al	13	26.9815	3	660.2	(2400)
Americium	Am	95	243.0614[a]	3,4,5,6	995	(2600)
Antimony	Sb	51	121.7_5	3,5	630.5	(1380)
Argon	Ar	18	39.94_8[b,c]	0	-189.2	-185.8
Arsenic	As	33	74.9216	3,5	817[e] (28 atm)	613(s)[e]
Astatine	At	85	210.9875[a]	1,3,5,7		
Barium	Ba	56	137.3_4	2	(725)	(1140)
Berkelium	Bk	97	247.0702[a]	3,4		
Beryllium	Be	4	9.01218	2	1278	(2970)
Bismuth	Bi	83	208.9806	3,5	271.3	1560
Boron	B	5	10.81[b,c]	3	(2200)	(3800)
Bromine	Br	35	79.904[b]	1,3,5,7	-7.2	58.78
Cadmium	Cd	48	112.40	2	320.9	764.9
Calcium	Ca	20	40.08	2	850	1490
Californium	Cf	98	249.0748[a]			
Carbon	C	6	12.011[c]	2,4	3500(s)	4830
Cerium	Ce	58	140.12	3,4	795	3468
Cesium	Cs	55	132.9055	1	28.7	690
Chlorine	Cl	17	35.453[b]	1,3,5,7	-100.98	-34.6
Chromium	Cr	24	51.996[b]	2,3,6	(1900)	(2500)
Cobalt	Co	27	58.9332	2,3	(1493)	(2900)
Copper	Cu	29	63.546[b,e]	1,2	1083	2595
Curium	Cm	96	245.0653[a]	3		
Dysprosium	Dy	66	162.5_0	3	1407	2600
Einsteinium	Es	99	254.0881[a]			
Erbium	Er	68	167.2_6	3	1497	2900
Europium	Eu	63	151.96	2,3	826	1439
Fermium	Fm	100	252.0827[a]			
Fluorine	F	9	18.9984	1	-219.6	-188.1
Francium	Fr	87	223.0198[a]	1		
Gadolinium	Gd	64	157.2_5	3	1312	3000
Gallium	Ga	31	69.72	2,3	29.78	(2400)

Melting and boiling points are reported in °C at 760 mm Hg, unless followed by another pressure in parentheses. An "s" after the value indicates sublimation. Densities are in g/cc (g/liter for gases) and are those of the element in its natural state at 20°C unless a temperature is noted in parentheses. Electronegativities are as reported by A. L. Allred and E. G. Rochow, J. Inorg. Nucl. Chem., 5, 264, 269 (1958); A. L. Allred, ibid, 17, 215 (1961). The atomic and covalent radii are in angstroms and electron affinities in electron volts. Ionization potentials are in volts and, for the most part, are from the United States Department of Commerce and National Bureau of Standards publication, Ionization Potentials, Appearance Potentials, and Heats of Formation of Gaseous Positive Ions, by J. L. Franklin, J. G. Dillard, H. M. Rosenstock, J. T. Herror, K. Droxl, and F. H. Field, Nat. Stand. Ref. Data Ser., Nat. Bur. Stand., 26 (1969). These values were determined spectroscopically.

For related data, see the table of isotopes below and a separate section on magnetic properties, p. 313.

Density	Electro-negativity	Radius		Electron Affinity	Ionization Potential	Symbol
		Atomic	Cov.			
10	1.00				6.9	Ac
2.702	1.47	1.43	1.18	0.52	5.986	Al
11.7					6	Am
6.684(25°)	1.82	(1.5)	1.38		8.641	Sb
1.7824[d]		1.91			15.759	Ar
5.72[e]	2.20	1.3	1.19	0.6	9.815	As
	1.96				9.5	At
3.51	0.97	2.2	1.98		5.211	Ba
						Bk
1.848	1.47	1.12	0.90	-0.19	9.322	Be
9.80	1.67	1.70	1.46		7.289	Bi
2.34	2.01	0.98	0.82	0.33	8.298	B
3.119[d]	2.74	1.13	1.14	3.36	11.85	Br
8.65	1.46	1.54	1.48		8.993	Cd
1.55	1.04	1.97	1.74		6.113	Ca
						Cf
3.51[e]	2.50	0.914	0.77	(1.2)	11.267	C
6.67[e]	1.06	1.81	1.65		6.54	Ce
1.873	0.86	2.67	2.25		3.894	Cs
3.214[d]	2.83		0.98	3.61	13.02	Cl
7.19	1.56	(1.27)			6.765	Cr
8.9	1.70	(1.3)			7.87±0.02	Co
8.96	1.75	1.28	1.38		7.726	Cu
7						Cm
8.536	1.10	1.77	1.59		5.80 ± 0.02	Dy
						Es
9.05	1.11	1.75	1.57		6.3 ± 0.1	Er
5.259	1.01	2.04	1.85		5.68 ± 0.03	Eu
						Fm
1.696(0°)[d]	4.10	0.68	0.72	3.45	17.426	F
	0.86					Fr
7.895[e]	1.11	1.79	1.61		6.16	Gd
5.907	1.82	(1.4)	1.26		5.997	Ga

(Continued)

Element	Symbol	Atomic No.	Atomic Weight	Valence	Mp	Bp
Germanium	Ge	32	72.5_9	4	937.4	(2830)
Gold	Au	79	196.9665	1,3	1063	2966
Hafnium	Hf	72	178.4_9	4	(2150)	5400
Helium	He	2	4.00260[b]	0	-269.7	-268.9
Holmium	Ho	67	164.9303	3	1461	2600
Hydrogen	H	1	1.0080[c]	1	-259.2	-252.7
Indium	In	49	114.82	3	156.17	2000
Iodine	I	53	126.9045	1,3,5,7	113.6	184.35
Iridium	Ir	77	192.22	3,4	(2410)	4500
Iron	Fe	26	55.847	2,3	1535	3000
Krypton	Kr	36	83.80	0	-157	-153
Lanthanum	La	57	138.9055	3	920	3470
Lawrencium	Lr	103	256[a]			
Lead	Pb	82	207.2[c]	2,4	327.4	(1744)
Lithium	Li	3	6.941[b,c]	1	180.5	(1326)
Lutetium	Lu	71	174.97	3	1652	3327
Magnesium	Mg	12	24.305[b]	2	650	1107
Manganese	Mn	25	54.9380	2,3,4,6,7	1245	(2097)
Mendelevium	Md	101	255.0906[a]			
Mercury	Hg	80	200.59	1,2	-38.87	356.9
Molybdenum	Mo	42	95.94	3,4,6	2610	5560
Neodymium	Nd	60	144.2_4	3	1024	3027
Neon	Ne	10	20.179[b]	0	-248.6	-246
Neptunium	Np	93	237.0482	4,5,6	640	
Nickel	Ni	28	58.71	2,3	1455	(2730)
Niobium	Nb	41	92.9064	3,5	2468	4930
Nitrogen	N	7	14.0067[b]	3,5	-209	-195.8
Nobelium	No	102	255[a]			
Osmium	Os	76	190.2	2,3,4,8	(3000)	(5000)
Oxygen	O	8	15.999_4[b,c]	2	-218.8	-182.97
Palladium	Pd	46	106.4	2,4,6	1550	(2927)
Phosphorous	P	15	30.9738	3,5	44.2[e]	280[e]
Platinum	Pt	78	195.09	2,4	1769	(3800)
Plutonium	Pu	94	242.0587[a]	3,4,5,6	640	3235
Polonium	Po	84	208.9825[a]	2,4,6	254	962
Potassium	K	19	39.102	1	63.7	774
Praseodymium	Pr	59	140.9077	3	935	3127
Promethium	Pm	61	145[a]	3	1035	2730
Protactinium	Pa	91	231.0359			
Radium	Ra	88	226.0254[a,c]	2	700	
Radon	Rn	86	222.0175[a]	0	-71	-61.8
Rhenium	Re	75	186.2		3180	(5625)
Rhodium	Rh	45	102.9055	3	1966	(3800)
Rubidium	Rb	37	85.4678[b]	1	38.9	688
Ruthenium	Ru	44	101.07	3,4,6,8	(2400)	(3900)
Samarium	Sm	62	150.4	2,3	1072	1900
Scandium	Sc	21	44.9559	3	1539	2727
Selenium	Se	34	78.96	2,4,6	217	685
Silicon	Si	14	28.086[c]	4	(1410)	2355
Silver	Ag	47	107.868[b]	1	960.8	2212
Sodium	Na	11	22.9898	1	97.8	890
Strontium	Sr	38	87.62[c]	2	770	1384

Density	Electro-negativity	Radius		Electron Affinity	Ionization Potential	Symbol
		Atomic	Cov.			
5.323(25)	2.02	1.37	1.22		7.89	Ge
19.32	1.42	1.44	1.50		9.223	Au
13.2	1.23		1.6		7.003	Hf
0.178[d]		0.9		0.19	24.586 ± 0.002	He
8.80	1.10	1.76	1.58		6.19 ± 0.02	Ho
0.0899[d]	2.20		0.37	0.747	13.598	H
7.31	1.49	1.66	1.44		5.786	In
4.93	2.21	1.3		3.06	10.457	I
22.42	1.55	1.36			9	Ir
7.86	1.64	1.26			7.90 ± 0.01	Fe
3.708		1.9			13.999	Kr
6.15[e]	1.08	1.86	1.69		5.615 ± 0.03	La
						Lr
11.34	1.55	1.75	1.47		7.417	Pb
0.534	0.97	1.5	1.34	0.82	5.392	Li
9.85	1.14	1.74	1.56		5.41 ± 0.02	Lu
1.738	1.23	1.60	1.30	-0.32	7.646	Mg
7.4[e]	1.60	1.26			7.434	Mn
						Md
13.546	1.44	1.57	1.49	1.54	10.44	Hg
10.22	1.30	1.37			7.099	Mo
7.00[e]	1.07	1.82	1.64		5.46	Nd
0.900(0°)[d]		1.60		-0.57	21.564	Ne
20.45[e]	1.22					Np
8.90	1.75	1.24			7.635	Ni
8.57	1.23	1.45			6.881	Nb
1.2506(0°)[d]	3.07	0.92	0.75		14.549	N
		1.45				No
22.5	1.52	1.35			8.7	Os
1.429(0°)[d]	3.50	0.6		1.47	13.618	O
12.02	1.39	1.37			8.336	Pd
1.82[e]	2.06	(1.2)		0.7	10.980	P
21.45	1.44	1.38			8.962	Pt
19	1.22				5.1 ± 0.5	Pu
9.3[e]	1.76	1.76			8.426	Po
0.862	0.91	(2.3)	1.96		4.341	K
6.77	1.07	1.83			5.76	Pr
						Pm
	1.14					Pa
(5)	0.97				5.277	Ra
9.73(0°)[d]					10.746	Rn
21	1.46	1.37			7.875	Re
12.4	1.45	1.34			7.463	Rh
1.53	0.89	2.45	2.11		4.176	Rb
12.2	1.42	1.33			7.366	Ru
7.54[e]	1.07	1.66			5.70 ± 0.02	Sm
3.0	1.20	(1.6)	1.44		6.561	Sc
4.79[e]	2.48	(1.3)		1.7	9.752	Se
2.33(25°)	1.74	1.32	1.11	(1.5)	8.151	Si
10.5	1.42	1.44	1.53		7.574	Ag
0.971	1.01	1.9	1.54	0.47	5.139	Na
2.6	0.99	2.14	1.92		5.694	Sr

(Continued)

Element	Symbol	Atomic No.	Atomic Weight	Valence	Mp	Bp
Sulfur	S	16	32.06[c]	2,4,6	119[e]	444.6
Tantalum	Ta	73	180.9479	5	2997	(5425)
Technetium	Tc	43	98.9062	6,7	(2200)	
Tellurium	Te	52	127.6_0	2,4,6	450	990
Terbium	Tb	65	158.9254	3	1356	2800
Thallium	Tl	81	204.37	1,3	304	1457
Thorium	Th	90	232.0381	4	(1700)	(4000)
Thulium	Tm	69	168.9342	3	1545	1727
Tin	Sn	50	118.69	2,4	231.9	2270
Titanium	Ti	22	47.9_0	3,4	(1675)	3260
Tungsten	W	74	183.85	6	(3400)	(5900)
Uranium	U	92	238.029[b,c]	4,6	1132	3818
Vanadium	V	23	50.9414[b]	3,5	(1890)	(3000)
Xenon	Xe	54	131.30	0	-111.9	-108
Ytterbium	Yb	70	173.04	2,3	824	1427
Yttrium	Y	39	88.9059	3	(1500)	2927
Zinc	Zn	30	65.37	2	907	419.5
Zirconium	Zr	40	91.22	4	1852	(4000)

[a] Longest-lived commonly available isotope.

[b] Atomic weight based on calibrated measurements.

[c] Variation of isotopic abundance in samples limits the precision; samples may have highly anomalous isotopic composition.

[d] Density for gases at 760 mm Hg and 20° in g/liter.

[e] Data for one allotropic form.

Density	Electro-negativity	Radius		Electron Affinity	Ionization Potential	Symbol
		Atomic	Cov.			
2.07[e]	2.44	1.27	1.02	2.07	10.360	S
16.6(0°)	1.33	1.45			7.883	Ta
	1.36	1.36			7.276	Tc
6.24	2.01	1.60	1.35	2.2	9.009	Te
8.27	1.10	1.77	1.59		6.74	Tb
11.85	1.44	1.71			6.108	Tl
(11.7)	1.11	1.80	1.65		6.95	Th
9.33	1.11	1.74	1.56		5.81 ± 0.02	Tm
7.31[e]	1.72	(1.5)	1.41		7.344	Sn
4.5	1.32	1.45	1.36		6.818	Ti
19.3	1.40	1.38			7.98	W
18.9	1.22	1.5			6.11 ± 0.05	U
6.1	1.45	1.3			6.740	V
5.88[d]		2.2			12.130	Xe
6.98[e]	1.06	1.93	1.70		6.22	Yb
4.47(0°)	1.11	1.8	1.62		6.528	Y
7.14	1.66	1.38	1.31		9.391	Zn
6.4	1.22	1.60	1.48		6.835	Zr

II. TABLE OF ISOTOPES

These data were taken from Handbook of Chemistry and Physics, 50th ed., CRC, Cleveland, 1969; C. M. Lederer, J. M. Hollander, and I. Perlman, Table of the Isotopes, 6th ed., Wiley, New York, 1968; and from the recent IUPAC tables [Pure Appl. Chem., 21 (1), 91 (1970)]. An attempt was made to include in this table all isotopes that occur naturally, are commercially available, or have an appreciable half-life. Abundance is in percent, mass is relative to C^{12} = 12.0000, and decay energy is in MeV. Symbols used are s = seconds, m = minutes, d = days, y = years, β^- = negative beta emission, β^+ = positron emission, α = alpha decay, EC = electron capture, IT = isomeric transition, and SF = spontaneous fission. The cross sections are from thermal neutron capture in barns (10^{-24} cm^2), microbarns (μb), millibarns (mb), or kilobarns (kb). If neutron capture can result in transformation to more than one isotope of an element, the cross section given is that for transformation to the isomer of longest half-life. For magnetic properties of the isotopes, see page 273.

Element	No.	Isotope	Abundance	Mass	Half-Life	Decay Mode	Decay Energy	Cross Section
Hydrogen(H)	1	1	99.984	1.007825	-	-	-	0.332 ± 2 mb
Deuterium(D)		2	0.0156	2.01400	-	-	-	0.51 ± 0.01 mb
Tritium(T)		3	-	3.01605	12.262y	β^-	0.01861	< 6.7 μb
Helium	2	3	0.00013	3.01603	-	-	-	< 0.1 mb
He		4	≃100	4.00260	-	-	-	≃ 0
Lithium	3	6	7.42	6.01512	-	-	-	45 ± 10 mb
Li		7	92.58	7.01600	-	-	-	37 ± 4 mb
Beryllium	4	7	-	7.0169	53.6d	EC	0.477	
Be		9	100	9.01218	-	-	-	9.2 ± 0.5 mb
		10	-	10.0135	2.5×10^6y	β^-	0.555	< 1 mb
Boron	5	10	19.7	10.0129	-	-	-	0.5 ± 0.2 b
B		11	80.3	11.00931	-	-	-	5 ± 3 mb
Carbon	6	12	98.892	12.00000	-	-	-	3.4 ± 0.3 mb
C		13	1.108	13.00335	-	-	-	0.9 ± 0.2 mb
		14	-	14.0032	5730y	β^-	0.156	< 10^{-6} b

Nitrogen	7	14	99.635	14.00307	-	-	-	1.81 b
N		15	0.365	15.00011	-	-	-	24 ± 8 μb
Oxygen	8	16	99.759	15.99491	-	-	-	0.178 ± 0.025 mb
O		17	0.037	16.9991	-	-	-	235 ± 10 mb
		18	0.204	17.9992	-	-	-	0.21 ± 0.04 mb
Fluorine	9	18	-		109.7m	β^+, EC	1.65	
F		19	100	18.99840	-	-	-	9.8 ± 0.7 mb
Neon	10	20	90.92	19.99244	-	-	-	38 ± 10 mb
Ne		21	0.257	20.99395	-	-	-	
		22	8.82	21.99138	-	-	-	36 ± 10 mb
Sodium	11	22	-	21.9944	2.62y	β^+		
Na		23	100	22.9898	-	-	-	0.4 ± 0.03 b
		24	-	23.99096	14.96h	β^-	5.51	
Magnesium	12	24	78.70	23.98504	-	-	-	52 ± 15 mb
Mg		25	10.13	24.98584	-	-	-	0.18 ± 0.05 b
		26	11.17	25.98259	-	-	-	0.03 ± 0.005 b
		28	-		21.2h	β^-	2.61	
Aluminum	13	26	-	25.98689	7.4×10^5y	β^+, EC	4.003	
Al		27	100	26.98153	-	-	-	232 ± 3 mb
Silicon	14	28	92.21	27.97693	-	-	-	160 ± 40 mb
Si		29	4.70	28.97649	-	-	-	280 ± 90 mb
		30	3.09	29.97376	-	-	-	100 ± 10 mb
		31	-	30.9753	2.62h	β^-	1.48	
		32	-	31.9740	≃650y	β^-	0.21	
Phosphorus	15	31	100	30.99376	-	-	-	190 ± 10 mb
P		32	-	31.9739	14.28d	β^-	1.710	
		33	-	32.9717	24.4d	β^-	0.248	
Sulfur	16	32	95.0	31.97207	-	-	-	
S		33	0.76	32.97146	-	-	-	15 ± 10 mb
		34	4.22	33.96786	-	-	-	200 ± 100 mb
		35	-	34.9690	87.9d	β^-	0.1674	
		36	0.014	35.96709	-	-	-	140 ± 40 mb

Table of Isotopes (Continued)

Element	No.	Isotope	Abundance	Mass	Half-Life	Decay Mode	Decay Energy	Cross Section
Chlorine Cl	17	35	75.53	34.96885	-	-		44 ± 2 b
		36	-	35.9797	3.08×10^{5}y	β^-,EC,β^+	0.712(β^-)	100 ± 30 b
		37	24.47	36.9658	-	-	-	430 ± 100 mb
Argon Ar	18	36	0.337	35.96755	-	-	-	6 ± 2 b
		37	-	36.9667	35.1d	EC	0.814	
		38	0.063	37.96272	-	-	-	0.8 ± 0.2 b
		39	-	38.964	269y	β^-	3.44	500 ± 200 b
		40	99.60	39.9624	-	-	-	650 ± 30 mb
Potassium K	19	39	93.22	38.96371	-	-	-	2.2 ± 0.2 b
		40	0.118	39.974	1.26×10^{9}y	β^-,EC,β^+		70 ± 20 b
		41	6.77	40.962	-	-	-	1.3 ± 0.2 b
		42	-	41.963	12.36h	β^-	0.60	
		43	-		22.4h	β^-	1.82	
Calcium Ca	20	40	96.97	39.96259	-	-	-	430 ± 40 mb
		42	0.64	41.95863	-	-	-	700 ± 160 mb
		43	0.145	42.95878	-	-	-	6.2 ± 1.1 b
		44	2.06	43.95549	-	-	-	1.1 ± 0.3 b
		45	-	44.956	163d	β^-	0.252	
		46	0.0033	45.9537	-	-	-	250 ± 100 mb
		47	-	46.954	4.535d	β^-	1.979	
		48	0.18	47.9524	-	-	-	1.1 ± 0.1 b
Scandium Sc	21	45	100	44.95592	-	-	-	(25 ± 2 b)
		46	-	45.955	83.9d	β^-	2.367	8.0 ± 1 b
Titanium Ti	22	44	-		48y	EC	0.16	
		46	7.93	45.95263	-	-	-	0.6 ± 0.2 b
		47	7.28	46.9518	-	-	-	1.7 ± 0.3 b
		48	73.94	47.948	-	-	-	8.3 ± 0.6 b
		49	5.51	48.94787	-	-	-	1.9 ± 0.5 b
		50	5.34	49.9448	-	-	-	140 ± 30 mb
Vanadium V	23	48	-	47.952	16.0d	β^+,EC	4.013	
		49	-	48.948	330d	EC	0.601	
		50	0.25	49.9472	6×10^{15}y	EC,β^-		100 ± 60b
		51	99.75	50.9440	-	-	-	4.8 ± 0.2 b

Chromium Cr	24	50	4.31	49.9461	-	-	-	16.0 ± 0.5 b
		51	-	50.945	27.8d	EC	0.752	
		52	83.76	51.9405	-	-	-	0.76 ± 0.06 b
		53	9.55	52.9407	-	-	-	18.2 ± 1.5 b
		54	2.38	53.9389	-	-	-	380 ± 40 mb
Manganese Mn	25	52	-					
		53	-	52.9413	1.9×10^6y	EC	0.598	≃170 b
		54	-	53.9402	303d	EC	1.379	<10 b
		55	100	54.9381	-	-	-	13.3 ± 0.1 b
		56	-		2.576h	β^-	2.7	
Iron Fe	26	54	5.82	53.9396	-	-	-	2.8 ± 0.4 b
		55	-	54.938	2.60y	EC	0.232	
		56	91.66	55.9349	-	-	-	2.5 ± 0.3 b
		57	2.19	56.9354	-	-	-	2.5 ± 0.2 b
		58	0.33	57.9333	-	-	-	1.23 ± 0.05 b
		59	-	58.935	45.6d	β^-		
		60	-		3×10^5y	β^-	0.14	
Cobalt Co	27	56	-	55.940	77.3d	EC,β^+	4.57	
		57	-	56.936	270d	EC	0.837	
		58	-	57.936	71.3d	EC,β^+	2.309	(1700 ± 200 b)
		59	100	58.9332	-	-	-	19.9 ± 0.91 b
		60	-	59.934	5.263y	β^-	2.819	(2.0 ± 0.2 b)
Nickel Ni	28	58	67.88	57.9353	-	-	-	4.4 ± 0.4 b
		59	-	58.934	8×10^4y	EC	1.072	
		60	26.23	59.9332	-	-	-	2.6 ± 0.2 b
		61	1.19	60.9310	-	-	-	2.5 ± 0.9 b
		62	3.66	61.9283	-	-	-	15 ± 2 b
		63	-	62.930	92y	β^-	0.067	
		64	1.08	63.9280	-	-	-	1.52 ± 0.24 b
Copper Cu	29	63	69.09	62.9298	-	-	-	4.5 ± 0.2 b
		64	-	63.930	12.80h	EC,β^-,β^+	0.573(β^-)	<6000 b
		65	30.91	64.9278	-	-	-	2.3 ± 0.3 b
		67	-					

Table of Isotopes (Continued)

Element	No.	Isotope	Abundance	Mass	Half-Life	Decay Mode	Decay Energy	Cross Section
Zinc Zn	30	64	48.89	63.9291	-	-	-	(0.82 ± 0.01 b)
		65	-	64.926	245d	EC, β^+	1.353	
		66	27.81	65.9260	-	-	-	
		67	4.11	66.9271	-	-	-	
		68	18.57	67.9249	-	-	-	0.09 ± 0.01 b
		70	0.62					
		72	-					
Gallium Ga	31	67	-					
		68	-	67.928	68.3m	β^+, EC	2.919	
		69	60.4	68.9257	-	-	-	1.8 ± 0.4 b
		71	39.6	70.9249	-	-	-	
		72	-					
Germanium Ge	32	68	-		275d	EC	≃0.7	
		69	-					
		70	20.51	69.9243	-	-	-	
		71	-	70.925	11.4d	EC	0.235	
		72	27.43	71.9217	-	-	-	0.98 ± 0.09 b
		73	7.76	72.9234	-	-	-	14 ± 1 b
		74	36.54	73.9219	-	-	-	0.14 ± 0.3 b
		76	7.76	75.9214	-	-	-	0.09 ± 0.02 b
		77	-					
Arsenic As	33	71	-					
		73	-					
		74	-	73.924	17.9d	β^+, EC, β^-	1.36	
		75	100	74.9216	-	-	-	4.3 ± 0.1 b
		76	-	75.922	26.4h	β^-	2.97	
		77	-	76.921	38.7h	β^-	0.686	
Selenium Se	34	74	0.87	73.9225	-	-	-	(55 ± 5 b)
		75	-	74.923	120.4d	β^-	0.865	
		76	9.02	75.9192	-	-	-	21 ± 2 b
		77	7.58	76.9199	-	-	-	42 ± 4 b
		78	23.52	77.9173	-	-	-	(0.2 ± 0.1 b)
		79	-	78.919	$\leq 6.5 \times 10^4$y	β^-	0.154	
		80	49.82	79.9165	-	-	-	
		82	9.19	81.9167	-	-	-	(6 ± 1 mb)

Bromine Br	35	76	-					
		77	-	76.921	57h	EC,β^+	1.365	
		79	50.54	78.9183	-	-	-	2.6 ± 0.2 b
		81	49.46	80.9183	-	-	-	
		82	-	81.917	35.34h	β^-	3.092	
		83	-					
Krypton Kr	36	78	0.35	77.9204	-	-	-	2.0 ± 0.5 b
		79	-					
		80	2.27	79.9164	-	-	-	14 ± 2 b
		81	-	80.917	2.1×10^5y	EC	0.29	
		82	11.56	81.9135	-	-	-	32 ± 15 b
		83	11.55	82.914	-	-	-	180 ± 40 b
		84	56.90	83.912	-	-	-	(0.1 ± 0.03 b)
		85	-	84.913	10.76y	β^-	0.67	
		86	17.37	85.911	-	-	-	0.06 ± 0.02 b
Rubidium Rb	37	83	-					
		84	-	83.914	33.0d	EC,β^+,β^-	2.68	
		85	72.15	84.9117	-	-	-	(0.45 ± 0.04 b)
		86	-	85.911	18.66d	β^-	1.78	
		87	27.85	86.909	4.8×10^{10}y	β^-	0.274	0.12 ± 0.03 b
Strontium Sr	38	83	-					
		84	0.56	83.9134	-	-	-	0.57 ± 0.05 b
		85	-	84.913	64.0d	EC	1.11	
		86	9.86	85.9094	-	-	-	
		87^m	-		2.83h	IT,EC		
		87	7.02	86.9089	-	-	-	
		88	82.56	87.9056	-	-	-	5 ± 1 mb
		89	-	88.907	52.7d	β^-	1.463	0.5 ± 0.1 b
		90	-	89.907	27.7y	β^-	0.546	0.9 ± 0.5 b
Yttrium Y	39	87	-	86.911	80h	EC,β^+	1.7	
		88	-	87.910	108.1d	EC,β^+	3.621	
		89	100	88.9054	-	-	-	1.3 ± 0.1 b
		90	-		64h	β^-	2.27	
		91	-	90.907	58.8d	β^-	1.545	1.4 ± 0.3 b

Table of Isotopes (Continued)

Element	No.	Isotope	Abundance	Mass	Half-Life	Decay Mode	Decay Energy	Cross Section
Zirconium Zr	40	88	-					
		89	-					
		90	51.46	89.9043	-	-	-	0.1 ± 0.07 b
		91	11.23	90.9053	-	-	-	(1.58 ± 0.12 b)
		92	17.11	91.9046	-	-	-	0.25 ± 0.12 b
		93	-	92.9057	1.5×10^{6}y	β^-	0.090	< 4 b
		94	17.40					
		95	-	94.9077	65.5d	β^-	1.121	
		96	2.80					
Niobium Nb	41	91	-					
		92	-					
		93^m	-		13.6y	IT		
		93	100	92.9060	-	-	-	
		94	-	93.9057	2.0×10^{4}y	β^-	2.06	(11 ± 2 b)
		95^m	-					
		95	-	94.9065	35.0d	β^-	0.925	< 7 b
Molybdenum Mo	42	92	15.84	91.9063	-	-	-	< 0.3 b
		93	-	92.9062	>100y	EC		
		94	9.04	93.9047	-	-	-	
		95	15.72	94.9046	-	-	-	14.4 ± 0.5 b
		96	16.53	95.9046	-	-	-	1.2 ± 0.6 b
		97	9.46	96.9058	-	-	-	2.2 ± 0.7 b
		98	23.78	97.9055	-	-	-	
		99	-	98.909	66.7h	β^-	1.37	
Technetium Tc	43	95^m	-					
		96	-					
		97^m	-					
		97	-		2.6×10^{6}y	EC	≃0.3	
		98	-	97.9072	1.5×10^{6}y	β^-	1.7	
		99^m	-		6.049h	IT	0.1427	
		99	-	98.908	2.12×10^{5}y	β^-	0.292	22 ± 3 b

Ruthenium Ru	44	96	5.51	95.9076	-	-	-	0.21 ± 0.02 b
		97	-		2.88d	EC	≃1.2	
		98	1.87	97.9055	-	-	-	< 8 b
		99	12.72	98.9061	-	-	-	10.6 ± 0.6 b
		100	12.62	99.9030	-	-	-	10.4 ± 0.7 b
		101	17.07		-	-	-	3.1 ± 0.9 b
		102	31.61	101.9037	-	-	-	1.4 b
		103	-		39.5d	β^-	0.74	
		104	58.58	103.9055	-	-	-	0.47 ± 0.2 b
		106	-		368d	β^-	0.0394	146 ± 45 mb
Rhodium Rh	45	101^m	-		4.4d	EC,IT	0.157(IT)	
		101	-		3.0y	EC	0.56	
		102^m	-		2.9y	EC	2.3	
		102	-		206d	EC,β^+,β^-	1.15(β^-)	
		103	100	102.9048	-	-	-	11 ± 1 b
		105			35.88h	β^-	0.565	
		106			30s	β^-	3.54	
Palladium Pd	46	102	0.96	101.9049	-	-	-	4.8 ± 1.5 b
		103	-		17.0d	EC	0.56	
		104	10.97	103.9036	-	-	-	
		105	22.23	104.9046	-	-	-	
		106	27.33	105.9032	-	-	-	279 ± 29 mb
		107	-		$\sim 7 \times 10^6$y	β^-	0.035	
		108	26.71	107.9030	-	-	-	12 ± 2 b
		109	-		13.47h	β^-	1.115	
		110	11.81		-	-	-	
Silver Ag	47	105	-		40d	EC		
		107	51.82	106.9041	-	-	-	35 ± 5 b
		108^m	-		>5y	EC,IT	0.110	
		109	48.18	108.9047	-	-	-	(4.2 ± 0.7 b)
		110^m	-		255d	β^-,IT		82 ± 11 b
		110	-		24.4s	β^-,EC	≃2.9	
		111	-		7.5d	β^-	1.05	

Table of Isotopes (Continued)

Element	No.	Isotope	Abundance	Mass	Half-Life	Decay Mode	Decay Energy	Cross Section
Cadmium Cd	48	106	1.22	105.907	-	-	-	1.0 ± 0.5 b
		108	0.88	107.9040	-	-	-	(1.5 ± 0.5 b)
		109	-		453d	EC	0.16	
		110	12.39	109.9030	-	-	-	
		111	12.75	110.9042	-	-	-	
		112	24.07	111.9028	-	-	-	
		113^m	-		13.6y	β^-,IT	0.58	
		113	12.26	112.9046	-	-	-	2 x 10^4 ± 300 b
		114	28.86	113.9036	-	-	-	(0.36 ± 0.007 b)
		115^m	-		43d	β^-,IT	1.62(β^-)	
		115	-		53.5h	β^-	1.45	
		116	7.58	115.9050	-	-	-	
Indium In	49	111	-		2.81d	EC		
		113^m	-		99.8m	IT	0.3916	
		113	4.28	112.9043	-	-	-	
		114^m	-		50.0d	IT,EC	0.191	
		115	95.72	114.9041	-	-	-	
Tin Sn	50	112	0.96	111.9040	-	-	-	0.8 ± 0.2
		113	-		115d	EC	1.02	
		114	0.66	113.9030	-	-	-	
		115	0.35	114.9035	-	-	-	
		116	14.30	115.9021	-	-	-	6 ± 2 mb
		117^m	-		14.0d	IT	0.317	
		117	7.61	116.9031	-	-	-	
		118	24.03	117.9018	-	-	-	
		119^m	-		≈250d	IT	0.089	
		119	8.58	118.9034	-	-	-	
		120	32.85		-	-	-	≈1 mb
		121^m	-		76y	β^-	0.45	
		121	-		27.5h	β^-	0.383	
		122	4.72	121.9034	-	-	-	1.0 ± 0.5 mb
		123^m	-		125d	β^-	1.42	
		124	5.94	123.9052	-	-	-	4 ± 2 mb
		125	-		9.4d	β^-	2.39	
		126	-		≈10^5y	β^-		

Antimony Sb	51	119	-		38.0h	EC		
		120^m	-		15.89m	β^+,EC	2.69	
		121	57.25	120.9038	-	-	-	6.2 ± 0.3 b
		122	-		2.80d	β^-,β^+,EC	1.972(β^-)	
		123	42.75	122.9041	-	-	-	3.4 ± 0.8 b
		124	-		60.4d	β^-	2.916	6.5 ± 1.5 b
		125	-		2.71y	β^-	0.764	
		126	-		12.5d	β^-	3.7	
		127	-		93h	β^-	1.60	
Tellurium Te	52	119^m	-		4.68d	EC	∿2.6	
		120	0.089	119.9045	-	-	-	0.34 ± 0.06 b
		121^m	-		154d	IT,EC,β^+	0.293(IT)	
		121	-		17d	EC	1.29	
		122	2.46	121.9030	-	-	-	1.7 ± 0.8 b
		123^m	-		117d	IT	0.247	
		123	0.87	122.9042	1.2×10^{13}y	EC	∿0.06	410 ± 30 b
		124	4.61	123.9028	-	-	-	6.8 ± 1.3 b
		125^m	-		58d	IT		
		125	6.99	124.9044	-	-	-	1.56 ± 0.16 b
		126	18.71	125.9032	-	-	-	125 ± 23 mb
		127^m	-		109d	IT,β^-	0.0887(IT)	
		127	-		9.4h	β^-	0.69	
		128	31.79	127.9047	-	-	-	14 ± 4 mb
		129^m	-		34.1d	IT,β^-	1.60(IT)	
		129	-		68.7m	β^-	1.48	
		130	34.48	129.9067				0.02 ± 0.01 b
		132	-		77.7h	β^-	0.50	
Iodine I	53	124	-		4.15d	EC,β^+	3.17	
		125	-		60.2d	EC	0.149	900 ± 90 mb
		126	-		12.8d	EC,β^-,β^+	2.150	
		127	100	126.9004	-	-	-	6.2 ± 0.2 b
		129	-		1.7×10^7y	β^-	0.189	9 ± 1 b
		130	-		12.3h	β^-	2.99	18 ± 3 b
		131	-		8.05d	β^-	0.970	∿ 0.7 b
		132	-		2.26h	β^-	3.56	

Table of Isotopes (Continued)

Element	No.	Isotope	Abundance	Mass	Half-Life	Decay Mode	Decay Energy	Cross Section
Xenon Xe	54	124	0.096	123.9061	-	-	-	100 ± 20 b
		126	0.090	125.9042	-	-	-	1.5 ± 1.0 b
		128	1.92	127.9035	-	-	-	
		129^m	-		8.0d	IT	0.236	
		129	26.44	128.9048	-	-	-	21 ± 7 b
		130	4.08	129.9035	-	-	-	
		131^m	-		11.8d	IT	0.163	
		131	21.18	130.9051	-	-	-	110 ± 20 b
		132	26.89	131.9042	-	-	-	50 ± 20 mb
		133	-		5.270d	β^-	0.427	
		134	10.44	133.9054	-	-	-	0.23 ± 0.02 b
		136	8.87	135.9072	-	-	-	280 ± 28 mb
Cesium Cs	55	131	-		9.70d	EC	0.35	
		132	-		6.59d	EC, β^+, β^-	2.09	
		133	100	133.9051	-	-	-	27.4 ± 1.5 b
		134	-		2.046y	β^-	2.062	134 ± 12 b
		135	-		3.0×10^6y	β^-	0.210	8.7 ± 0.5 b
		136	-		13.7d	β^-	2.54	
		137	-		30.0y	β^-	1.176	110 ± 33 mb
Barium Ba	56	130	0.101	129.9062	-	-	-	11 ± 3 b
		131	-		12.0d	EC	1.16	
		132	0.097	131.9057	-	-	-	8.5 ± 1.0 b
		133^m	-		38.9h	IT	0.288	
		133	-		7.2y	EC	0.488	
		134	2.42	133.9043	-	-	-	2 ± 2 b
		135	6.59	134.9056	-	-	-	5.8 ± 0.9 b
		136	7.81	135.9044	-	-	-	0.4 ± 0.4 b
		137	11.32		-	-	-	5.1 ± 0.4 b
		138	71.66	137.9050	-	-	-	0.35 ± 0.15 b
		140	-		12.80d	β^-	3.0	
Lanthanum La	57	137	-		6×10^4y	EC	≈0.5	
		138	0.089	137.9068	1.12×10^{11}y	EC, β^-		
		139	99.911	138.9061	-	-	-	8.8 ± 0.7 b
		140	-		40.22h	β^-	3.769	

Cerium Ce	58	134	-		72.0h	EC	≃0.16	
		136	0.193					
		138	0.250	137.9057	-	-	-	1.1 ± 0.3 b
		139	-		140d	EC	0.27	
		140	88.48	139.9053	-	-	-	0.58 ± 0.06 b
		141	-		32.5d	β^-	0.581	29 ± 3 b
		142	11.07	141.9090				0.95 ± 0.05 b
		144	-		284d	β-		1.0 ± 0.1 b
Praseodymium Pr	59	141	100	140.9074				3.9 ± 0.5 b
		142	-		19.2h	β^-	2.16	18 ± 3 b
		143	-		13.59d	β^-	0.933	
		144	-		17.27m	β^-	2.989	
Neodymium Nd	60	142	27.11	141.9075	-	-	-	18.8 ± 0.7 b
		143	12.17	142.9096	-	-	-	
		144	23.85	143.9099	2.4×10^{15}y	α	1.8	4.0 ± 0.5 b
		145	8.30	144.9122				50 ± 4 b
		146	17.22	145.9172	-	-	-	1.4 ± 0.2 b
		147	-		11.06d	β^-	0.91	
		148	5.73	147.9165	-	-	-	2.5 ± 0.2 b
		150	5.62	149.9207				1.3 ± 0.3 b
Promethium Pm	61	143		142.9110	0.73y	EC	1.1	
		144			0.96y	EC		
		145		144.9128	18.0y	EC	0.14	
		146			4.4y	EC,β^-	1.72(EC)	8400 ± 1680 b
		147		146.9152	2.62y	β^-	0.225	90 ± 30 b
		148^m			41.8d	β^-,IT	2.59	25 ± 4 kb
Samarium Sm	62	144	3.09	143.9117	-	-	-	≃0.7 b
		145	-		340d	EC	0.65	≃110 b
		146	-	145.9129	7×10^7y	α		
		147	14.97	146.9147	1.05×10^{11}y	α	2.314	75 ± 11 b
		148	11.24	147.9146	$>12 \times 10^{14}$y	α	2.001	2.4 ± 0.3 b
		149	13.83	148.9169	$>1 \times 10^{15}$y	α	1.90	41 ± 2 kb
		150	7.44	149.9170	-	-	-	102 ± 5 b
		151	-		93±8y	β^-	0.076	15 ± 1.8 kb
		152	76.72	151.9195	-	-	-	210 ± 10 b
		153	-		46.8h	β^-	0.801	
		154	22.71	153.9220	-	-	-	5.5 ± 1.1 b

Table of Isotopes (Continued)

Element	No.	Isotope	Abundance	Mass	Half-Life	Decay Mode	Decay Energy	Cross Section
Europium Eu	63	148	-		54d	EC,β^+	3.11	
		149	-		106d	EC	0.80	
		150	-		5y	EC	2.25	
		151	47.82	150.9196	-	-	-	5000 ± 300 b
		152	-		12.7y	EC,β^-,β^+	1.82(β^-)	
		153	52.18	152.9209	-	-	-	450 ± 100 b
		154	-		16y	β^-	1.97	1.5 ± 0.4 kb
		155	-		1.811y	β^-	0.248	14 ± 4 kb
Gadolinium Gd	64	148	-	147.9177	84y	α	3.27	
		150	-	149.9185	2.1×10^6y	α	2.80	
		152	0.200	151.9195	1.1×10^{14}y	α	2.24	<125 b
		153	-		242d	EC	0.243	
		154	2.15	153.9207	-	-	-	102 ± 7 b
		155	14.73	154.9226	-	-	-	61 ± 1 kb
		156	20.47	155.9221	-	-	-	
		157	15.68	156.9339	-	-	-	254 ± 2 kb
		158	24.87	157.9241	-	-	-	3.5 ± 1.0 b
		160	21.90	159.9071	-	-	-	0.77 ± 0.01 b
Terbium Tb	65	151	-		18h	EC,α		
		155	-		5.6d	EC	0.9	
		156	-		5.1d	EC,β^-		
		157	-		1.5×10^2y	EC	0.06	
		158	-		1.2×10^3y	EC,β^-		
		159	100	159.9250				30 ± 10 b
		160	-		72.1d	β^-	1.72	525 ± 100 b
		161	-		6.9d	β^-	0.58	
Dysprosium Dy	66	154	-	153.9248	$\approx 10^6$y	α	2.93	
		156	0.052	155.9238				
		158	0.090	157.9240	-	-	-	96 ± 20 b
		159	-		144d	EC	0.38	
		160	2.29	159.9248	-	-	-	55 ± 9 b
		161	18.88	160.9266	-	-	-	585 ± 50 b
		162	25.53	161.9265	-	-	-	200 ± 50 b
		163	24.97	162.9284	-	-	-	140 ± 30 b
		164	28.18	163.9288	-	-	-	
		165	-		139.2m	β^-	1.30	

Holmium Ho	67	163	-		$>10^3$y	EC	≃0.01	
		165	100	164.9303				
		166m	-		$1.2x10^3$y	β^-	1.847	
Erbium Er	68	162	0.136	161.9288	-	-	-	160 ± 30 b
		164	1.56	163.9293	-	-	-	13 ± 5 b
		166	33.41	165.9304	-	-	-	
		167	22.94	166.9320	-	-	-	700 ± 50 b
		168	27.07	167.9324	-	-	-	1.9 ± 0.2 b
		169	-		9.6d	β^-	0.34	
		170	14.88	169.9355	-	-	-	6 ± 1 b
		172	-		49.5h	β^-	0.91	
Thulium Tm	69	168	-		85d	EC	1.72	
		169	100	168.9344				115 ± 15 b
		170	-		134d	β^-	0.967	92 ± 4 b
		171	-		1.92y	β^-	0.098	4.5 ± 0.2 b
		172	-		63.6h	β^-	1.88	
Ytterbium Yb	70	168	0.135	167.9339	-	-	-	3.2 ± 0.4 kb
		169	-		31.8d	EC	≃1.2	
		170	3.03	169.9349	-	-	-	9.4 ± 0.9 b
		171	14.31	170.9365	-	-	-	50 ± 5 b
		172	21.82	171.9366	-	-	-	0.4 ± 0.1 b
		173	16.13	172.9383	-	-	-	19 ± 2 b
		174	31.84	173.9390	-	-	-	
		175	-		101h	β^-	0.467	
		176	12.73	175.9427	-	-	-	5.5 ± 1.0 b
Lutetium Lu	71	173	-		1.37y	EC	0.69	
		174^m	-		140d	IT,EC		
		175	97.41	174.9409				21 ± 3 b
		176	2.59		$2.2x10^{10}$y	β^-	1.02	7 ± 2 b
		177	-		6.74d	β^-	0.497	
Hafnium Hf	72	172	-		5y	EC	≃1	
		174	0.18	173.9403	$2.0x10^{15}$y	α	2.55	390 ± 55 b
		175	-		70d	EC	0.59	
		176	5.20	175.9435	-	-	-	15 ± 15 b
		177	18.50	176.9435	-	-	-	380 ± 30 b
		178	27.14	177.9439	-	-	-	
		179	13.75	178.9460	-	-	-	65 ± 15 b
		180	35.24	179.9468	-	-	-	12.6 ± 0.7 b
		181	-		42.5d	β^-	1.023	
		182	-		$9x10^6$y	β^-	~0.5	

Table of Isotopes (Continued)

Element	No.	Isotope	Abundance	Mass	Half-Life	Decay Mode	Decay Energy	Cross Section
Tantalum Ta	73	179	-		≃600d	EC	0.115	
		180	0.0123	179.9415	$>1x10^{12}y$			
		181	99.988	180.9480	-	-	-	22 ± 1 b
		182	-		115.1d	β^-	1.811	8.2 ± 0.6 kb
Tungsten W	74	180	0.14	179.9470				
		181	-		140d	EC	0.19	
		182	26.41	181.9483				20.7 ± 0.5 b
		183	14.40	182.9503				10.2 ± 0.3 b
		184	30.64	183.9510	-	-	-	1.8 ± 0.2 b
		185	-		75d	β^-	0.432	
		186	28.41	185.9543				37 ± 2 b
		187	-		23.9h	β^-	1.315	90 ± 40 b
		188	-		69.4d	β^-	0.349	
Rhenium Re	75	183	-		71d	EC	≃0.9	
		184	-		38d	EC	≃1.6	
		185	37.07	184.9530	-	-	-	110 ± 5 b
		186	-		88.9h	β^-,EC	1.071	
		187	62.93	186.9560	$4.3x10^{10}y$	β^-		
		188	-		16.7h	β^-	2.116	<2 b
Osmium Os	76	184	0.018	183.9526	-	-	-	<200 b
		185	-		96.6d	EC	0.982	
		186	1.59	185.9539	-	-	-	
		187	1.64	186.9560	-	-	-	
		188	13.3	187.9560	-	-	-	
		189	16.1	188.9586	-	-	-	
		190	26.4	189.9586	-	-	-	12 ± 6 b
		191	-		15.0d	β^-	0.310	
		192	41.0					
		193	-		31h	β^-	1.132	
		194	-		6.0y	β^-	0.097	

Iridium	77	188	-		41.5h	EC,β^+		
Ir		189	-		13.3d	EC	≃0.6	
		190	-		11d	EC	2.1	
		191	37.3	190.9609	-	-	-	300 ± 30 b
		192	-		74.2d	β^-,EC,β^+		
		193^m	-		11.9d	IT	0.08	
		193	62.7	192.9633	-	-	-	110 ± 15 b
Platinum	78	188	-		10.2d	EC	0.51	
Pt		190	0.0127	189.960	6.9×10^{11}y	α	3.18	
		191	-		3.0d	EC	≃2.0	
		192	0.78	191.9614	≃10^{15}y	α		
		193^m	-		4.3d	IT	0.148	
		193	-		<500y	EC	0.045	
		194	32.9	193.9628	-	-	-	1.2 ± 0.9 b
		195^m	-		4.1d	IT	0.259	
		195	33.8	194.9648	-	-	-	27 ± 2 b
		196	25.3	195.9650	-	-	-	0.9 ± 0.1 b
		197	-		18h	β^-	0.75	
		198	7.21	197.9675				4.0 ± 0.5 b
Gold	79	195	-		183d	EC	0.227	
Au		196	-		6.18d	EC,β^-	0.259(β^-)	
		197	100	196.9666	-	-	-	98.8 ± 0.3 b
		198	-		2.697d	β^-	1.374	25.8 ± 1.2 kb
		199	-		3.15d	β^-	0.46	30 ± 15 b
Mercury	80	194	-		1.9y	EC	≃0.4	
Hg		196	0.146	195.9658				
		197	-		65h	EC	0.42	
		198	10.02	197.9668	-	-	-	
		199	16.84	198.9683	-	-	-	2 ± 1 kb
		200	23.13	199.9683	-	-	-	<60 b
		201	13.22	200.9703	-	-	-	<60 b
		202	29.80	201.9706	-	-	-	4.9 ± 0.2 b
		203	-		46.9d	β^-	0.492	
		204	6.85	202.9735	-	-	-	0.43 ± 0.1 b
Thallium	81	201	-		74h	EC	≃0.41	
Tl		202	-		12.0d	EC	1.1	
		203	29.50	202.9723	-	-	-	10 ± 1 b
		204	-		3.81y	β^-,EC	0.763(β^-)	
		205	70.50	204.9745	-	-	-	0.5 ± 0.2 b

Table of Isotopes (Continued)

Element	No.	Isotope	Abundance	Mass	Half-Life	Decay Mode	Decay Energy	Cross Section
Lead	82	202	-		$\approx 3x10^5 y$	EC	0.05	
Pb		203	-		52.1h	EC	0.96	
		204	1.48	203.9731	-	-	-	0.66 ± 0.07
		205	-		$3.0x10^7 y$	EC		
		206	23.6	205.9745	-	-	-	30 ± 1 mb
		207	22.6	206.9759	-	-	-	0.71 ± 0.01 b
		208	52.3	207.9766	-	-	-	0.02 ± 0.01 b
		210	-		20.4y	β^-	0.061	
Bismuth	83	205	-		15.31d	EC, β^+	2.70	
Bi		207	-		30.2y	EC	2.40	
		208	-		$3.68x10^5 y$	EC	2.87	
		209	100	208.9804				19 ± 2 mb
		210^m	-		$\approx 2.6x10^6 y$	α, β^-		
		210	-		5.01d	β^-, α		54 ± 5 mb
Polonium	84	206		205.9805	8.8d	EC, α		
Po		208		207.9813	2.93y	α	5.21	
		209		208.9825	103y	α, EC	4.98(α)	
		210		209.9829	138.4d	α	5.408	
Astatine	85	211		210.9875	8.3h	α, EC		
At								
Radon	86	211		210.9906	15h	EC, α		
Rn		222		222.0175	3.82d	α	5.587	0.72 ± 0.07 b
Francium	87	212		211.996	19.3m	α, EC		
Fr		223		233.0198	22m	β^-, α		
Radium	88	223		223.0186	11.44d	α	5.977	
Ra		225			14.8d	β^-		
		226		226.0254	1600y	α		20 ± 3 b
		228			6.7y	β^-	0.055	
Actinium	89	225		225.0231	10.0d	α	5.931	
Ac		227		227.0278	21.8y	β^-, α		810 ± 20 b

Thorium	90	228		228.0287	1.910y	α	5.512	120 ± 15 b
Th		229		229.0316	7340y	α	5.167	
		230		230.0331	$8.0x10^4$y	α	4.767	33 ± 10 b
		232		232.0382	$1.41x10^{10}$y	α	4.08	7.4 ± <0.1 b
Protactinium	91	230			17.7d	EC,β^-,α		
Pa		231		231.0359	$3.26x10^4$y	α	5.148	200 ± 10 b
		233			27.0d	β^-	0.571	
Uranium	92	232	-	232.0372	72y	α,SF	5.414	75 ± 10 b
U		233	-	233.0396	$1.60x10^5$y	α	4.909	
		234	0.0057	234.0409	$2.47x10^5$y	α	4.856	
		235	0.72	235.0439	$7.0x10^8$y	α,SF	4.681	100.5 ± 1.4 b
		236	-	236.0457	$2.39x10^7$y	α,SF	4.573	6.0 ± 0.5 b
		238	99.27	238.0508	$4.5x10^9$y	α,SF	4.268	2.720 ± 0.025
Neptunium	93	235			410d	EC,α		
Np		236		236.0466	22h	EC,β^-		
		237		237.0480	$2.14x10^6$y	α	4.956	170 ± 5 b
Plutonium	94	236		236.0461	2.85y	α,SF	5.868	
Pu		237		237.0483	45.6d	EC,α,SF		
		238		238.0495	87y	α,SF	5.592	560 ± 25 b
		239		239.0522	24,300y	α,SF	5.243	265.7 ± 3.7 b
		240		240.0538	6600y	α,SF	5.255	290 ± 15 b
		241		241.0569	14.2y	β^-,α		360 ± 15 b
		242		242.0587	$3.86x10^5$y	α	4.98	20 ± 3 b
		244		244.0642	$8.2x10^7$y	α,SF	4.66	1.8 ± 0.3 b
Americium	95	241		241.0567	435y	α	5.640	70 ± 5 b
Am		242^m			152y	IT,α		2400 ± 500 b
		243		243.0614	$7.4x10^3$y	α	5.439	180 ± 20 b
Curium	96	242		242.0588	164d	α		20 ± 10 b
Cm		243		243.0614	32y	α	6.168	250 ± 50 b
		244		244.0629	18.1y	α	5.902	13 ± 5 b
		245		245.0653	$8.3x10^3$y	α	5.624	250 ± 50 b
		246		246.0672	$4.7x10^3$y	α,SF	5.476	8.4 ± 2.0 b
		247		247.0704	$1.6x10^7$y	α	∿5.3	
		248		248.0724	$3.5x10^5$y	α	5.161	6 ± 2 b
		250			$1.1x10^4$y	SF		

Table of Isotopes (Continued)

Element	No.	Isotope	Abundance	Mass	Half-Life	Decay Mode	Decay Energy	Cross Section
Berkelium Bk	97	247		247.0702	1.4×10^3y	α	5.86	
		249			310d	β^-,α,SF		1000 ± 500 b
Californium Cf	98	248		248.0724	360d	α,SF	6.37	
		249		249.0748	360y	α	6.295	300 ± 200 b
		250		250.0766	13y	α,SF	6.128	1500 ± 500 b
		251			900y	α	5.94	2.1 ± 1 kb
Einsteinium Es	99	252		252.0829	≃140d	α	6.75	
		253		253.0847	20d	α,SF	6.747	
		254		254.0881	270d	α	6.623	<40 b
Fermium Fm	100	252		252.0827	23h	α	7.17	
		253			3d	EC,α		
		257			80d	α	6.871	
Mendelevium Md	101	255		255.0906	0.6h	EC,α		
		258			54d	EC,α,SF		
Nobelium No	102	253			95s	α	8.2	
		255			185s	α,EC		
Lawrencium Lw	103	256			35s	α		

III. SELECTED BOND LENGTHS

These data were taken from Tables of Interatomic Distances and Configuration in Molecules and Ions. Supplement, Special Publication No. 18, The Chemical Society, London, 1965, pp. S3s-S23s, and Handbook of Chemistry and Physics, 50th ed., CRC Press, Cleveland, 1969, pp. F-154 to F-157. A collection of papers on structural effects is contained in "Epistologue on the Effect of Environment on Properties of Carbon Bonds", Tetrahedron, 17, 125-225 (1962). Each value in the table is an average based on determinations of many related molecules except for unique values, such as HCl; values in parentheses were derived from only one or two molecules. The symbol R is used mainly to illustrate the coordination (valence) of the atoms concerned, and may represent a variety of structural groups, organic and inorganic.

Bond Type	Structural System	Length (Å)
Bonds Not Involving Carbon		
H-H	H_2	0.74130±0.00006
H-F	HF	0.917
H-Cl	HCl	1.274
H-Br	HBr	1.408
H-I	HI	1.609
B-H	RBH_2	1.19±0.01
B-H	Bridged hydrides	1.32±0.02
B-F	R_2BF	1.29±0.01
B-Cl	R_2BCl	1.74±0.01
B-N	Borazoles	1.42±0.01
B-O	$(RO)_3B$	1.36±0.01
F-F	F_2	1.418±0.005
Cl-Cl	Cl_2	1.988
Br-Br	Br_2	2.284
I-I	I_2	2.666
N-H	NH_3	1.012
	RNH_2	1.01
	$RCONH_2$	0.99±0.02
N-N	R_2NNH_2	(1.451±0.005)
N=N	RN_2R	(1.25±0.02)
N≡N	N_2	1.0976±0.0002
N-O	$RO-NO_2$	1.36±0.02
N O	RNO_2	1.22±0.01
N=O	RON=O	(1.22±0.02)
O-H	H_2O	0.958
	ROH	(0.97±0.01)
O-O	O_2	1.20741±0.0002
	O_3	1.278±0.005
	H_2O_2	1.48±0.01
P-H	PH_3	1.437±0.004

Bond Type	Structural System	Length (Å)
Bonds Not Involving Carbon		
S-H	H_2S	1.335
	RSH	(1.329±0.005)
S-O	R_2SO_2	1.43±0.01
	$SOCl_2$	1.45±0.02
S-S	RSSR	2.05±0.01
Si-H	R_3SiH	1.476±0.005
Si-F	R_3SiF	1.561±0.005
Si-Cl	R_3SiCl	2.019±0.005
Si-Br	R_3SiBr	2.16±0.01
Si-I	R_3SiI	2.46±0.02
Si-O	R_3SiOR	1.633±0.005
Bonds Involving Carbon		
C-H	RCH_3	1.096±0.005
	R_2CH_2	1.073±0.005
	R_3CH	1.070±0.007
	Olefinic	1.083±0.005
	Allenic	1.07±0.01
	Aromatic	1.084±0.006
	Acetylenic	1.055±0.005
	▷X (X=O,NH,S)	1.081±0.007
	$CH_3C≡X$	1.115±0.004
C-B	R_3B	1.56±0.01
C-F	Paraffinic	1.379±0.005
	R_2CF_2, RCF_3, etc.	1.333±0.005
	Olefinic	1.333±0.005
	Aromatic	1.328±0.005
	Acetylenic	(1.27)
C-Cl	Paraffinic	1.767±0.005

Bond Type	Structural System	Length (Å)
Bonds Involving Carbon		
C-Cl	Olefinic	1.719±0.005
	Aromatic	1.70±0.01
	Acetylenic	1.635±0.005
C-Br	Paraffinic	1.938±0.005
	Olefinic	1.89±0.01
	Aromatic	1.85±0.01
	Acetylenic	1.79±0.01
C-I	C_2H_5I	2.207±0.005
	CH_2CHI	2.092±0.005
	Aromatic	2.05±0.01
	Acetylenic	(1.99±0.02)
C-C	Diamond	1.54452±0.00014
	Paraffinic	1.537±0.005
	C-C=C	1.510±0.005
	C-Aryl	1.505±0.005
	Aryl-aryl	1.52±0.01
	C=C-C=C	1.465±0.005
	C-C≡C	1.459±0.005
	C=C-C≡C	(1.45±0.02)
	C≡C-C≡C	1.371±0.005
	C-C=O	1.506±0.005
	Aryl-C=O	1.47±0.02
	O=C-C=O	1.49±0.01
	C=C-C=O	1.44±0.01
	C-C≡N	1.464±0.005
	CH_2CH-CN	1.426±0.005
C=C	Simple	1.335±0.005
	C=C-C=C	1.336±0.005
	C=C=C	1.310±0.005
	C=C-C=O	1.36±0.01
	CH_2=C=O	1.31±0.01
	CH_2=CHCN	1.339±0.005
	Graphite	1.4210±0.0001
	Aromatic	1.394±0.005
C≡C	Simple	1.202±0.005
	$(C≡C)_n$	1.206±0.005
C-N	RNH_2, R_2NH, etc.	1.472±0.005
	RNH_3^+, R_3NBX_3	1.479±0.005
	C-N=X (C-NO_2 *e.g.*)	1.47±0.01
Bonds Involving Carbon		
C-N	O=C-N	1.333±0.005
	C_6H_5-$NHCOCH_3$	1.43±0.01
C=N	Pyridine, etc.	1.339±0.005
	Pyrrole	1.383
C≡N	C_2H_5CN, etc.	1.157±0.005
C-O	Aliphatic alcohol, ether	1.426±0.005
	▷O	1.435±0.005
	RCO-OR' (acid, ester)	1.358±0.005
	Ar-OH	1.36±0.01
C=O	Aldehydes, ketones	1.215±0.005
	O=C-C=C	1.215±0.005
	O=C=X	1.160±0.001
	RCOX (X=halo)	1.171±0.004
	RNCO	1.17±0.01
	O=C-NR_2	1.235±0.005
	O=C(R)OR' (acid, ester)	1.233±0.005
	$-CO_2^-$	1.26±0.01
	p-Quinones	1.15±0.02
	Furan	1.371±0.016
C≡O	CO	1.128
	$Ni(CO)_4$, etc.	1.14±0.01
C-S	R_2S	1.817±0.005
	Perfluoro: $(CF_3)_2S$	1.83±0.01
C=S	Thiophene, etc.	1.718±0.005
	S=C=X	1.555±0.001
	$S=CR_2$	1.71±0.01
C-P	$(CH_3)_3P$	1.841±0.005
C-Si	Alkyl silanes	1.870±0.005
	Aryl silanes	1.843±0.005
	Alkyl halosilanes	1.854±0.005

IV. EFFECTIVE van der WAALS RADII

Experimentally, van der Waals radii are determined from measurements of interatomic distances in the crystalline state. They represent the distance of closest approach (equilibrium position) of an atom (or group) in one molecule to an atom of the same element in a neighboring molecule. Values are defined as one-half the equilibrium distance between the nuclei of such atoms (groups) in van der Waals contact. They are larger than the covalent radii, but for nonmetallic elements are approximately equal to ionic radii. More discussion can be found in L. Pauling, The Nature of the Chemical Bond, 3rd ed., Cornell University Press, Ithaca, New York, 1960, pp. 257-264. The following values (in Å) are usually referred to as effective van der Waals radii of the atoms and are those given by Pauling.

H	1.2				
N	1.5	O	1.40	F	1.35
P	1.9	S	1.85	Cl	1.80
As	2.0	Se	2.00	Br	1.95
Sb	2.2	Te	2.20	I	2.15

Radius of a methyl (CH_3) or methylene (CH_2) group = 2.0. Effective thickness of plane of aromatic hydrocarbon rings ("π-cloud thickness") = 3.4, or half-thickness (van der Waals "radius") = 1.7.

The same van der Waals radii are often used to estimate the nature of nonbonded interactions (attractive or repulsive) between atoms (or groups) within a single molecule. In this connection, for example, it is said that a covalently bound bromine atom has about the same "size" as a methyl group in terms of space-filling capacity.

Applications of the use of van der Waals radii in stereochemical analysis (nonbonded interactions, conformational energies) may be found in E. Eliel, N. L. Allinger, S. J. Angyal, and G. A. Morrison, Conformational Analysis, Interscience, New York, 1966, Ch. 7, and in F. H. Westheimer, Chapter 12, in M. S. Newman, Ed., Steric Effects in Organic Chemistry, Wiley, New York, 1956.

V. BOND ANGLES AND HYBRIDIZATION

The bond angles usually cited are those determined from electron diffraction and x-ray measurements; since these measurements give only the relative positions of nuclei, the angle is more precisely defined as an internuclear angle. Such a value may not be identical to the interorbital angle, defined as "the angle subtended by the bonding orbital axes at the nucleus" (K. Mislow, Introduction to Stereochemistry, Benjamin, New York, 1965). The distinction between internuclear and interorbital angles becomes important only for cyclic, particularly strained molecules. For example, the internuclear CCC angle in cyclopropane is 60° while the interorbital angle for the bent ("banana" or τ-) bonds is estimated as ≈105°.

Angles are a function of the hybridization of the bonds in question. For σ-bonds to carbon (composed of s and p character) the hybrid is defined as $(s + \lambda^2 p)$, where λ is the mixing coefficient and λ^2 is referred to as the hybridization index. For each bonding atomic orbital, i:

$$\text{Fraction s-character} = \frac{1}{1 + \lambda_i^2}$$

$$\text{Fraction p-character} = \frac{\lambda_i^2}{1 + \lambda_i^2}$$

The total fraction s-character contained in all the bonds to that carbon atom must be 1 (i may be up to 4):

$$\sum_i \frac{1}{1 + \lambda_i^2} = 1$$

For the case of i-equivalent orbitals, say in CH_4, $\lambda^2 = 3$, and each C orbital is sp^3 or 25% s-character. Similarly, the sum of all p-character contributions must equal the total number of p-orbitals mixed:

$$\sum_i \frac{\lambda_i^2}{1 + \lambda_i^2} = 1, 2, \text{ or } 3$$

Now, the interorbital angle, Θ_{ij}, between any two hybrid orbitals is related to λ by $1 + \lambda_i \lambda_j \cos \Theta_{ij} = 0$; thus, for i = j in the pure sp^3 case, $\cos \Theta = -1/3$, $\Theta = 109.5°$. In a "nonpure" sp^3 case, the measured CCC angle in propane is ≈112°, in which case $\lambda_{CC}^2 = 2.7$; the orbital is thus an $sp^{2.7}$ hybrid. Continuing this example, the HCH angle for the 2-carbon can be calculated as follows:

$$2[1/(1 + \lambda_{CH}^2)] + 2[1/(1 + 2.7)] = 1$$

$$\lambda_{CH}^2 = 3.35$$

(Thus the C-orbital in the C-H bond is an $sp^{3.35}$ hybrid with $1/(1 + 3.35) = 0.23$ parts s-character). Then, $\cos \Theta_{HCH} = -1/3.35$ or $\Theta_{HCH} = 107°$ (observed 106°).

An additional measure of carbon orbital hybridization in hydrocarbons is provided by the empirical relationship: $J = 500/(1 + \lambda_i^2)$, where J = nmr ^{13}C-H coupling constant in Hz and λ_i^2 is as above (for a discussion, see F. A. Bovey, NMR Spectroscopy, Academic Press, New York, 1969, p. 233-235). For ^{13}C data, see page 286.

A. Idealized Internuclear Angles for Pure Hybrids

Hybridization of Central Atom	Known Example	Symmetry Point Group[a]	Bond Angle (deg)
sp(linear)	HCN	$C_{\infty v}$	180
sp^2(trigonal)	BF_3	D_{3h}	120
sp^3(tetrahedral)	CH_4, BF_4^-	T_d	109.47
p^2(orthogonal)	-	$C_{2v}(XY_2)$	90
p^3(orthogonal)	-	$C_{3v}(XY_3)$	90
dp(linear)	I_3^-	$D_{\infty h}$	180
dsp^2(square planar)	$Ni(CN)_4^{-2}$	D_{4h}	90
d^2p^2(square planar)	ICl_4^-	D_{4h}	90
dsp^3(trigonal bipyramid)	PCl_5	D_{3h}	90 (apical-equatorial) 120 (equatorial-equatorial) 180 (apical-apical)
d^2sp^3(octahedral)	$Cr(CN)_6^{-3}$, SF_6	O_h	90 and 180

[a]For details, see page 471.

B. Selected Bond Angles

Because angles are dependent on hybridization, it is difficult to compile "typical" values. The data given here are representative. For specific molecules, consult "Tables of Interatomic Distances and Configuration in Molecules and Ions," Special Publication No. 11 (1958), and its "Supplement 1956-1959," Special Publication No. 18 (1965), The Chemical Society, London. The influence of steric and electronic effects on bond angles is discussed by Lide in "Epistologue on the Effect of Environment on Properties of Carbon Bonds," Tetrahedron, 17, 125 (1962). For discussions concerning theoretically calculated angles and how they relate to experimental values in organic systems, see Hendrickson, J. Amer. Chem. Soc., 89, 7036 (1967); Allinger et al., ibid., 90, 1199 (1968), and 93, 1637 (1971); and references therein.

The values given below are experimentally determined (hence internuclear angles) and were taken from the usual sources. The obvious bond angles are not listed for totally symmetrical cases (109.5° for CX_4; 120° for BX_3; 180° for RCN; etc.). Unless a specific molecule is given, the angles are approximate values or ranges for the compound type (R = alkyl, Ar = aryl; C-atoms are sp^3 unless indicated otherwise).

Structure	Angle	Degrees	Structure	Angle	Degrees
n-Alkanes	CCC	112.6±0.2	C-C=C	CCC	122-125
	HCH	104±2	C-C(sp^2)-R(H)	CCC,CCH	116
C-C-H	CCH	107-108	C=C-H	CCH	119
C-C-C(sp^2)	CCC	110-111			
C-C-C(sp)	CCC	110	$C=CH_2$	HCH	122

Structure	Angle	Degrees	Structure	Angle	Degrees
Cyclohexane	CCC	111.3	R_3N	CNC	109
			$ArNO_2$	ONO	124
H_2O	HOH	104.45	RCONHR	NCO	122-125
ROH	COH	108-109		CCN	115
R_2O	COC	110 ± 3		HNC(O)	121
	CCO	111			
	HCO	107	H_2S	HSH	92.3
Ar_2O,ArOR	COC	120-124	RSH	CSH	100
C-CR=O	CCO	120-122	Ar_2S	CSC	109
H-CR=O	HCO	119	SO_2	OSO	119.5
			R_2SO	CCS	110
RCl	CCCl	107		CSC	96
	HCCl	108		CSO	109
NH_3	HNH	107.3		HCS	109
RNH_2	HNH	106			
	CNH	112	PX_3(X=halo)	XPX	100-101
R_2NH	CNC	111	PR_3	RPR	100

VI. SELECTED BOND STRENGTHS

The standard bond dissociation energy is the enthalpy change for the process in which 1 mole of a specified bond is broken homolytically, with reactants and products in their standard states of a hypothetical gas at 1 atm and 25°C. These bond strengths are expressed in kcal/mol. Most of the values below are taken from S. W. Benson, J. Chem. Ed., 42, 502 (1965). The starred values are from the CRC Handbook of Chemistry and Physics, 50th ed., 1970. For a recent compilation of bond dissociation energies, see B. Darwent, Nat. Stand. Ref. Data Ser., Nat. Bur. Stand., No. 31, 1971. A fairly comprehensive and recent monograph, Chemical Bonds and Bond Energy, by R. T. Sanderson, is also useful for detailed discussion and data (Academic Press, New York, 1971); Sanderson's approach is "highly original," as discussed in a review of the book by R. Howald, J. Chem. Educ., 48, A562 (1971).

A. Single Bonds

1. Diatomic Molecules

Bond	Energy	Bond	Energy
H-H	104.2	F-Cl	61
D-D	106.0	F-Br	60
F-F	38	F-I	58
Cl-Cl	58	Cl-Br	52
Br-Br	46.0	Cl-I	50
I-I	36.1		

2. Polyatomic Molecules

Bond	Energy	Bond	Energy	Bond	Energy
$H-CH_3$	104	$Br-CCl_3$	54*	$H-OH$	119
$H-CH_2CH_3$	98	$F-COCH_3$	119	$H-O_2H$	90
$H-CHCH_2$	103	$Cl-COCH_3$	83.5	$H-SH$	90
$H-C_6H_5$	103	$I-COCH_3$	52.5	$H-OCH_3$	102
$H-CCH$	$\sim$125			$H-OC_6H_5$	85
$H-CH_2C_6H_5$	85	$Cl-Na$	98	$H-O_2CCH_3$	112
$H-CH_2CHCH_2$	85				
$H-CH_2OH$	93	CH_3-CH_3	88	$HO-CH_3$	91.5
$H-CF_3$	104	$CH_3-CH_2CH_3$	85	$HO-CH_2CH_3$	91.5
$H-CCl_3$	96	CH_3-CH_2OH	83	$HO-C_6H_5$	103
$H-COCH_3$	87.5	CH_3-CF_3	100*	$HO-COCH_3$	109
$H-CN$	130	CH_3-CHCH_2	92		
		$CH_3-C_6H_5$	93	CH_3O-CH_3	80
$F-CH_3$	108	CH_3-CCH	117	$CH_3O-CH_2CH_3$	80
$Cl-CH_3$	83.5	$CH_3-CH_2C_6H_5$	72	$CH_3O-CHCH_2$	87
$Br-CH_3$	70	$CH_3-CH_2CHCH_2$	72	$CH_3O-C_6H_5$	91
$I-CH_3$	56	$CH_3CH_2-CHCH_2$	89	$CH_3O-COCH_3$	97
$F-CH_2CH_3$	106	$CH_3CH_2-C_6H_5$	90		
$Cl-CH_2CH_3$	81.5	$CH_2CH-CHCH_2$	100	$HO-OH$	51*
$Br-CH_2CH_3$	69	$HCC-CCH$	150	$HO-Br$	57*
$I-CH_2CH_3$	53.5	$C_6H_5-C_6H_5$	100	CH_3O-OCH_3	36*
$Cl-CHCH_2$	84	$CH_2CH-C_6H_5$	99		
$F-C_6H_5$	116			H_2N-H	103
$Br-C_6H_5$	72	CH_3-COCH_3	82	H_2N-CH_3	79
$I-C_6H_5$	65*	$CH_3CH_2-COCH_3$	79	$H_2N-CH_2CH_3$	78
$F-CF_3$	129*	CH_3-CN	122	$H_2N-C_6H_5$	91
$Cl-CF_3$	85*	$CH_2CH-COCH_3$	89	$H_2N-COCH_3$	$\sim$96
$Br-CF_3$	70*	CH_2CH-CN	128		
$I-CF_3$	54*	$CH_3CO-COCH_3$	83	O_2N-NO_2	13.6
$F-CCl_3$	106*	$NC-CN$	144	$O_2N-COCH_3$	97
$Cl-CCl_3$	73*	CF_3-CF_3	97*		

B. Multiple Bonds

Bond	Energy	Bond	Energy
$O=O$	119	$CF_2=CF_2$	76.3*
$O=CO$	128*	$CH_2=NH$	$\sim$154
$O=CH_2$	175	$C\equiv O$	257
$O=NH$	115	$N\equiv N$	226
$HN=NH$	$\sim$109	$N\equiv CH$	224
$CH_2=CH_2$	163	$HC\equiv CH$	230

VII. FORCE CONSTANTS

The force needed to restore a bonding configuration (bond length or angle) to its equilibrium position is proportional to the displacement from that equilibrium for small displacements (Hooke's law). The proportionality constant is called the force constant, which has the dimensions of force per unit displacement (or energy per square unit displacement). The force constant values given here are, as usual, those obtained from vibrational spectra, and are not corrected for anharmonicity. For discussions of these points and much stretching force constant data, see T. L. Cottrell, The Strengths of Chemical Bonds, Butterworth, London, 1954, Ch. 11.

The data in the table below were taken from G. Herzberg, Molecular Spectra and Molecular Structure, Van Nostrand, Princeton, 1968, Ch. II; E. L. Eliel, N. L. Allinger, S. J. Angyal, and G. A. Morrison, Conformational Analysis, Interscience, New York, 1966, pp. 435-460; J. B. Hendrickson, J. Amer. Chem. Soc., 89, 7036 (1967); N. L. Allinger et al., ibid., 90, 5773 (1968), 91, 337 (1969), and 93, 1637 (1971). The force constant values cited are the most recent in use and are considered approximate (≃ ± 10%). The following conversion factors may be useful.

Bond length: 1 dyn/cm = 10^5 mdyn/Å = 1.438×10^{13} kcal/mole-cm^2 = 1.438×10^{-3} kcal/mole-Å^2

Bond angle: 1 mdyn-Å/rad^2 = 10^{-11} erg/rad^2 = 144 kcal/mole-rad^2 = 0.0438 kcal/mole-deg^2

(Unless indicated otherwise, carbon atoms are sp^3-hybridized.)

Bond	Stretching Force Constant mdyn/Å	Stretching Force Constant kcal/mole-Å^2	Bond	Stretching Force Constant mdyn/Å	Stretching Force Constant kcal/mole-Å^2
C-H			C-Cl	3.64	524
sp^3	4.8	692	C-Br	3.12	448
sp^2	5.3	762	C-I	2.65	381
Aromatic	5.0	720	C-O	5.0	720
Carbonyl(sp^2)	5.3	762	C=O(sp^2-C)	12.1	1740
sp	5.9	849	C=O(sp-C)	15.5	2230
			C-NH_2	4.9	705
C-C			C-NO_2	4.7	675
			C≡N(sp-C)	17.73	2550
sp^3-sp^3	4.5	648	C=N(sp^2-C)	10.5	1510
sp^2=sp^2	9.7	1380			
sp≡sp	15.6	2240	C-S(O)R	3.64	524
sp^3-sp^2	4.8	692			
sp^3-sp	4.5(5.2)	648(748)	-O-H	7.0	1006
sp^3-carbonyl (sp^2)	4.8	692	⋝N-H	6.35	914
			⋝S→O	5.0	720
Benzene[a]	5.2-5.6	748-806			
Cyclopentadienide ion[a]	5.39	775			
C-F	5.96	858			

[a]R. West, A. Sadô, and S. W. Tobey, J. Amer. Chem. Soc., 88, 2488(1966).

Bond Angle	Bending Force Constant mdyn-Å/ rad^2	kcal/ mole-deg^2
CCC		
sp^3-sp^3-sp^3	0.8	0.0350
sp^3-sp^3-sp^2	1.10	0.0482
sp^3-sp^2-sp^3	1.10	0.0482
sp^3-sp^3-sp	1.10	0.0482
sp^3-sp^2=sp^2	1.10	0.0482
CCH		
sp^3-sp^3-H	0.55	0.0241
sp^3-sp^2-H	0.66	0.0290
sp^2-sp^3-H	0.66	0.0290
sp^2=sp^2-H	0.66	0.0290
Ar-H	0.86	0.0376
HCH		
H-sp^3-H	0.32	0.0140
H-sp^2-H	0.55	0.0241

Bond Angle	Bending Force Constant mdyn-Å/ rad^2	kcal/ mole-deg^2
H-C-Cl	0.71	0.0312
C-C-Cl	0.82	0.0360
C-C=O	0.65	0.0285
H-C=O	0.50	0.0219
H-C-O	0.5	0.0219
C-O-H	0.76	0.0334
C-C-O	0.98	0.0430
C-C≡N	0.35	0.0154
C-C-S(O)R	1.10	0.0482
C-S(O)-C	0.9	0.0395
C-S(R)-O	0.9	0.0395
H-C-S(O)R	0.64	0.0281

VIII. TORSION AND INVERSION BARRIERS

Many species undergo unimolecular stereomutations that are not due to bond breaking. Such stereochemical changes are broadly classified as inversion (at an atom), psuedorotation (at an atom), and torsion (about a single or multiple bond). Each of the following sets of data is classified by the net spatial change. When experimental data are unknown, nonempirical SCF calculated barriers are given when available. It should be noted that at least in one case, namely tert-butylbenzylmethylamine, N-inversion and N-$C(CH_3)_3$ rotation appear to share a common transition state, giving rise to a single barrier of $\Delta G^{\dagger} = -6.2 \pm 0.2$ kcal/mole in CD_2CDCl at - 138° (see Bushweller, et al., J. Amer. Chem. Soc., 93, 544 (1971)). Recent advances in the general subject of barriers are discussed in ref 1d.

Barriers are given as the Eyring free energy of activation, $\Delta G^{\dagger}$, and/or as the potential energy barrier, V (same as Arrhenius E); when possible, enthalpy and entropy values are included. Note that footnotes for all the tables appear together on page 122.

A. Pyramidal Inversion (1a,1c,1d,1e)

A recent paper by Mislow, et al. (1e) reviews the state of theoretical calculations of barriers to pyramidal inversion and introduces a highly successful semiempirical method for first and second row elements based on the Pople CNDO/2 theory; the paper also summarizes most of the known experimental data.

Refs. p. 122.

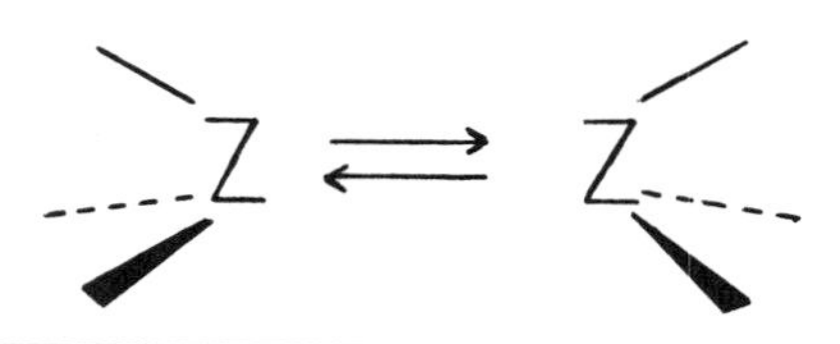

Species	$\Delta G^{\dagger}$ (°C)[a]	V[b]	Method[c]	Ref.
		At Carbon		
CH_3^-		3.8-5.9	T	2,41
▷⁻		20.8	T	3
▷•		Very low (-120°)	E	4a
$CF_3^{\bullet}$		27.4(21)	T	4b(41)
		At Nitrogen[d]		
NH_3		5.9(5.77)	M	5(41)
CH_3NH_2		4.8	V	6
$(CH_3)_2NH$		4.4	M	7
N-H		18.3(11.6)	T(M)	8(9)
O N—H		32.4(∿32)	T(N)	8(41)
N—H	17.0(52°)	11.0	N	10
N—X, X = CH_3		23-24;32	N	11
N—X, X = Cl	>23.5 (>180°)		N	12
N—X, X = CH_3	8.5 (-93°)		N	12
N—X, X = Cl	11.4 (-46°)		N	12
NF_3		56-59	V	13
$(CH_3)_2C(NF_2)CH(NF_2)CH_3$	⩾18(⩾147°)		N	14
H_2N-CHO		1.1	M	41

(Continued)

Refs. p. 122.

Species	$\Delta G^{\dagger}$ (°C) [a]	V^{b}	Method [c]	Ref.
At Oxygen				
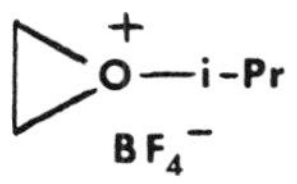 (cyclic oxonium, $\overset{+}{O}$—i-Pr, BF_4^-)	(T_c = -50°)	10 ± 2	N	15
At Silicon				
SiH_3^-		39.6	T	la
At Phosphorus				
PH_3		37.2	T	16,41
$R_1R_2PCH_3$ (R_1,R_2= alkyl, aryl)	30-36 (at 130°; <u>not</u> T_c)		R	17
For R_1=C_6H_5,R_2=n-Pr		30.7 (log A = 12.4)	R	18
$C_6H_5P(CH_3)PC_6H_5(CH_3)$		26±2 (log A = 14±1) [e]	N	19
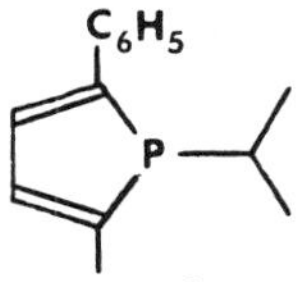 (C_6H_5, P, isopropyl ring compound)	$\Delta H^{\dagger}$ = 17.1 $\Delta S^{\dagger}$ = 3.1	$\Delta G^{\dagger}$ = 16 at 0.8° (T_c∿42°)	N	20
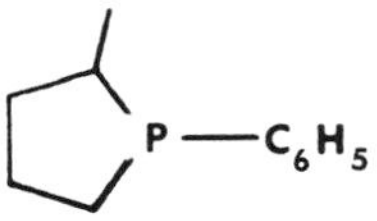 (P—C_6H_5 ring compound)	∿38 (∿170°)		N	20
At Sulfur				
H_3S^+		30.0	T	la
$CH_3(C_2H_5)\overset{+}{S}$(1-adamantyl)$ClO_4^-$		$\Delta H^{\dagger}$ = 26 $\Delta S^{\dagger}$ = 8	R	21
RS(O)(p-tolyl)				
R = CH_3		38.4 (log A = 11.7)	R	22
R = C_6H_5		37.2 (log A = 12.3)	R	22
R = 1-adamantyl		42.0 (log A = 14.3)	R	22

(Continued)

Refs. p. 122.

Species	$\Delta G^{\dagger}$ (°C)[a]	V[b]	Method[c]	Ref.
		At Arsenic		
$C_6H_5As(CH_3)AsC_6H_5(CH_3)$		27 ± 1 (log A = 14 ± 1)[e]	N	19

B. In-Plane Inversion[f]

Species	$\Delta G^{\dagger}$ (°C)[a]	V[b]	Method[c]	Ref.
		At Carbon		
$CH_2=CH^{\bullet}$		∿2(8)	E(T)	4a(41)
		At Nitrogen		
$CH_2=NH$		27.9(25-27)	T(N)	23(41)
HN=NH		50.1	T	23
HN=C=NH		8.4(6.7)	T(N)	23(41)
$R_2C=NC_6H_5$				
R = CH_3	20.3(126°)		N	24
R = Ac	19.5(105°)		N	24
R = p-$CH_3O\phi$	18.1(62.2°)		N	24
R = OCH_3	14.3(0°)		N	24
R = $N(CH_3)_2$	12.0(-36°)		N	24
R_2C =	22.2(140°)		N	24
$(CF_3)_2C=NCF(CF_3)_2$		13	N	25
		At Oxygen and Sulfur		
H_2O		>34	T	26
H_2S		92	T	27

Refs. p. 122.

C. Torsional Processes (28)[g]

1. Carbon Single Bonds in Simple Molecules (29)

Species	V[b]	Method[c]	Species	V[b]	Method[c]
CH_3CH_3	2.88(2.93)	TH(V)	CH_3NH_2	1.94	M
CH_3CF_3	3.48	TH	CH_3NHCH_3	3.28	V
CF_3CF_3	4.35	TH	$CH_3N(CH_3)_2$	4.41	V
CCl_3CCl_3′	10.8	ED	CH_3NO_2	0.006	M
CH_3CH_2F	3.30	M	$CH_3N{=}CH_2$	1.97	M
CH_3CH_2Cl	3.69	M	CH_3PH_2	1.96	M
CH_3CH_2Br	3.57	M	$CH_3As(CH_3)_2$	1.5-2.5	M
CH_3CH_2I	3.22	M	CH_3OH	1.07	M
CH_3BF_2	0.014	M	CH_3OCH_3	2.72	M
$CH_3CH_2CH_3$	3.3	TH	CH_3-OCHO	1.19	M
$CH_3CH(CH_3)_2$	3.87	TH	CH_3-ONO_2	2.32	M
$CH_3C(CH_3)_3$	4.80	TH	CH_3SH	1.26	M
$CH_3CH{=}CH_2$	1.98	M	CH_3SCH_3	2.13	M
$CH_3C(CH_3){=}CH_2$	2.21	M	CH_3SiH_3	1.70	M
$CH_3CH{=}CHCH_3$ (cis)	0.75	M	CH_3GeH_3	1.2	M
CH_3CHO	1.15	M	CH_3SnH_3	0.65	M
CH_3-CO_2CH_3	1.17	M			

2. Amides and Other Substances (30)

Process	$\Delta G^{\dagger}$ (°C)[a]	V[b]	log A

X=Y=CH_3, Z=H		22.0	13
X=Y=CH_3, Z=Cl		17-18	13-14
X=Y=CH_3, Z=t-Bu(31)	12.2 (40°)	11.5	12.3
X=Y=CH_3, Z=$OCH_2C_6H_5$	16 (-3°)		
X=Z=CH_3, Y=2,4,6-$(NO_2)_3C_6H_2$		21.0	14.3
Y=Z=CH_3, X=2,4,6-$(NO_2)_3C_6H_2$		19.2	13.5
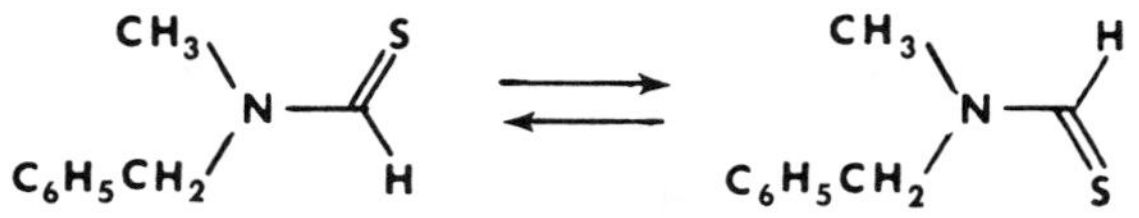	(Method R)	Forward: 25.2	14.2
		Reverse: 25.1	14.3

(Continued)

Refs. p. 122.

Process	$\Delta G^{\dagger}$ (°C)[a]	V^{b}	log A
R–C$_6$H$_4$–N=O ⇄ R–C$_6$H$_4$–N=O			
R = OC_2H_5 (32)	10.0 (-54°)		
R = $N(CH_3)_2$ (32)	12.4 (-3°)	[$\Delta H^{\dagger}$ = 14.9, $\Delta S^{\dagger}$ = 4.8 (33)]	
R = NH_2 (32)	12.4 (-4°)		
R = $O^{-}Na^{+}$ (32)	17.3 (+80°)		
R–O–N=O ⇄ R–O–N=O	(R = alkyl)	∿10	
Furfural (O-cis ⇄ O-trans)	Forward: $\Delta H^{\dagger}$ = 11.7, $\Delta S^{\dagger}$ = 3.8		
	Reverse: $\Delta H^{\dagger}$ = 10.7, $\Delta S^{\dagger}$ = 1.6		
Tetramethyl-bis(CH$_2$R)benzene, (R = t-butyl)		11.5	

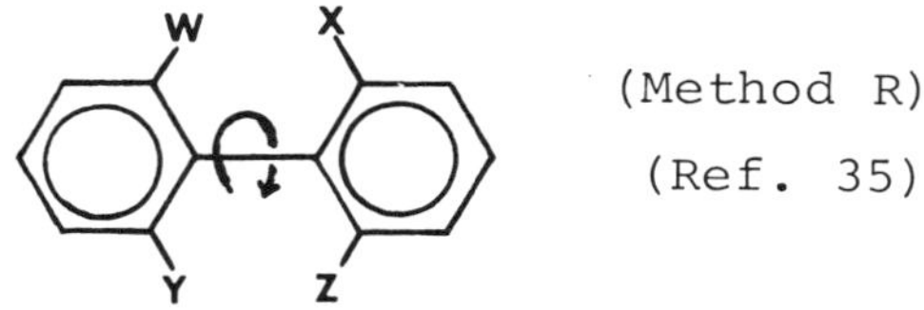

n = 4: $\Delta G^{\dagger}$ (155°) = $\Delta H^{\dagger}$ =35 (V=35.6) (34)
n = 5: $\Delta G^{\dagger}$ (-10°) = $\Delta H^{\dagger}$ =19 (V=20) (34)
n = 6: V = 10.7, log A = 11.7
(Method R for n = 4,5)

Biphenyl (W, X, Y, Z substituted) (Method R) (Ref. 35)

	V	log A	$\Delta H^{\dagger}$	$\Delta S^{\dagger}$
4,4'-$(CO_2H)_2$; W=X=H; Y=Z=Br	19.0	12.1	18.5	-4.9
Z=H; W=NO_2, X=OC_2H_5, Y=CO_2H	20.0	10.9	19.4	-10.5
W=X=OCH_3; Y=Z=$CONH_2$	27.8	10.4	27.1	-13.3
W=X=OCH_3; Y=Z=CO_2CH_3	25.0	10.1	24.3	-14.8
W=X=NO_2; Y,Z=$CH_2N(CH_3)CH_2$	30.0	11.5	29.2	-8.3
W=X=F; Y,Z=$CH_2\overset{+}{N}H(CH_3)CH_2$	27.8	12.7	27.1	-2.9
W=H; X=NO_2; Y,Z=$CH_2\overset{+}{N}(CH_3)_2CH_2$	25.1	15.0	24.5	+8.3

Refs. p. 122.

Process	$\Delta G^{\dagger}$ (°C)[a]	V[b]	log A
$(CH_3)_3C$–CH $(CH_2)_n$ (36)			
n = 4;5;6;7;8 : $\Delta G^{\dagger}$ (T_c) = ∿6(-155); ∿6.3(-150); 7.4(-126); 7.8(-118); 7.3(-130), resp.			
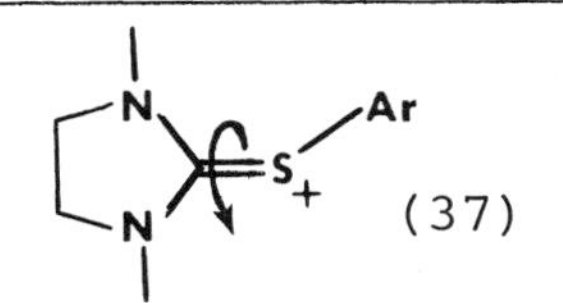 (37)	Ar = 2,4,6-R_3-C_6H_2; R = H; CH_3; i-Pr : $\Delta G^{\dagger}$ (T_c) = 7.6(-83); 8.0(-64); 9.3(-29)		

3. Ring Inversions (38) (Also Called Ring-Reversals)

	V	log A	$\Delta H^{\dagger}$	$\Delta S^{\dagger}$
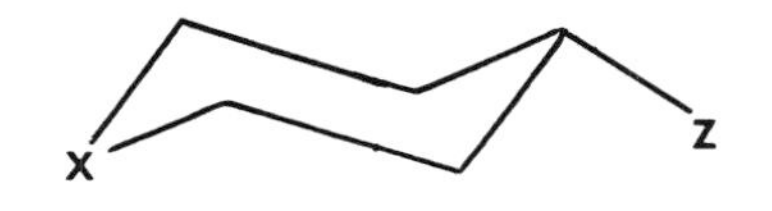				
X=CH_2; Z=H	10	12(uncertainty in values; see ref. 16)		
Perfluorocyclohexane	10.5	12.1		
X=CH_2; Z = F		Forward:	9.87	-1.46
		Reverse:	9.40	-2.51
X = C=CH_2; Z = H (39)			8.6	1.4
X = NH; Z = H	14.5	16.9		
X = NCH_3; Z = H	14.4	14.8		
X = S; Z = H (38)		$\Delta G^{\dagger}$ = 8.7 (T_c = -93°)		
X = O; Z = H (38)		$\Delta G^{\dagger}$ = 9.9 (T_c = -65°)		
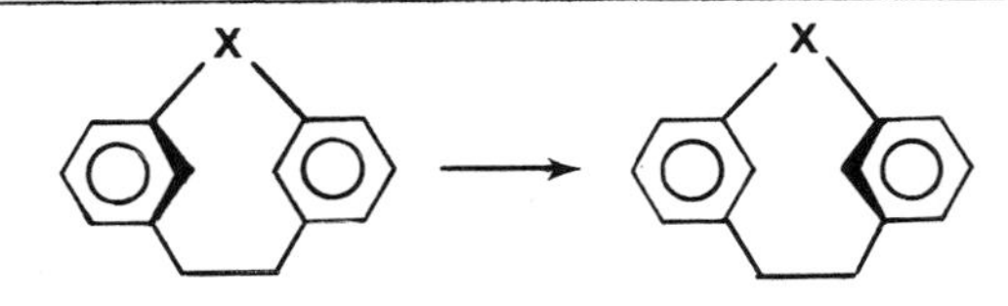				
X = CH_2CH_2		V>28		
X = CH_2SCH_2		$\Delta G^{\dagger}$ = 8.32 (T_c = 0.5°)		
X = $CH_2S(O_2)CH_2$		$\Delta G^{\dagger}$ = 13.7 (T_c = 57.5°)		
X = $CH_2CH(CO_2CH_3)CH_2$ (40)		$\Delta H^{\dagger}$ = 16.1, $\Delta S^{\dagger}$ = -2.4		

Refs. p. 122.

[a]For nmr determined barriers (in solution), $\Delta G^{\dagger}$ (kcal/mole) is at the temperature (°C) of signal coalescence of disastereotopic nuclei.

[b]For theoretically determined values, the potential energy barrier, V(kcal/mole), is comparable to the E_a value obtained experimentally.

[c]N = nmr (solution), M = microwave (vapor), R = thermal isomerism (racemization, e.g.), T = theoretical (nonempirical, SCF), E = esr, V = vibrational spectroscopy, TH = thermodynamic, ED = electron diffraction.

[d]Acyclic and cyclic amines with nonaxially symmetric heteroatomic substituents on N have been studied (e.g. RR'NOR"); however, the origin of the observed barriers (usually by nmr) is a subject of controversy and may be N-inversion, N-hetero bond rotation, or a combination. See references 1a and 1b.

[e]The observed barrier may involve restricted rotation as well as inversion for the meso-d,l equilibrium. See references 1 and 19.

[f]Also referred to as "two-dimensional" inversion (26) and the "lateral shift mechanism" (24). However, as with amines [d], the possible involvment of torsional processes has not as yet been ruled out in all cases examined (except for theoretical values). See reference 1.

[g]In this category are conformational changes involving restricted single bond rotations and ring inversions; for most cases, $\Delta S \sim 0$ (log A $\sim$ 13) (1b). For recent measurements of restricted rotation in radicals using esr, see Krusic and Kochi, J. Amer. Chem. Soc., 93, 846 (1971).

D. Pseudorotation*

This process has only recently been investigated and not much is known about the energetics. For recent studies, see D. Gorenstein and F. H. Westheimer, J. Amer. Chem. Soc., 92, 634 (1970), and D. Gorenstein, ibid., 92, 644 (1970).

E. References

1. For reviews, see (a) K. Mislow, L. Allen, and A. Rauk, Angew. Chem. Intern. Ed., Engl. 9, 400 (1970); (b) G. Binsch, Topics in Stereochemistry, Vol. 3, E. Eliel and N. Allinger, Eds., Interscience, New York, 1968, p. 146; (c) J. B. Lambert, Pyramidal Atomic Inversion, ibid., Vol. 6, 1971; (d) G. Chiurdoglu, Ed., Conformational Analysis, Scope and Present Limitations, Academic Press, New York, 1971; (e) A. Rauk, K. Mislow, et al., J. Amer. Chem. Soc., 93, 6507 (1971).
2. R. Kari and I. Csizmadia, J. Chem. Phys., 50, 1443 (1969). Calculations of potential surfaces have also been done for $HSCH_2^-$ (S. Wolfe et al., J. Chem. Soc.(D) 96 (1970); $H(O)SCH_2^-$ (A. Rauk et al., Can. J. Chem., 47, 113 (1969); and $H(O_2)SCH_2^-$ (A. Rauk et al., J. Amer. Chem. Soc., 91, 1567 (1969)).
3. D. Clark and D. Armstrong, J. Chem. Soc. (D), 850 (1969).
4. (a) R. Fessenden and R. H. Schuler, J. Chem. Phys., 39, 2147 (1963). Ordinary alkane radicals are planar according to H. Walborsky and C. Chen, J. Amer. Chem. Soc., 89, 5499 (1967). Cf. H. Walborsky et al., ibid., 90, 5222 (1968); M. J. S. Dewar and J. Harris, ibid., 91, 3652 (1969); R. W. Fessenden, J. Phys. Chem., 71, 74 (1967). (b) K. Morokuma, L. Pedersen, and M. Karplus, J. Chem. Phys., 48, 4801 (1968).

*Recent discussions on theoretical aspects of bonding and pseudorotation in PH_5 geometries are found in J. Amer. Chem. Soc., 94, 3035; 3040; 3047 (1972).

5. T. Snyder and C. Kenney, Microwave Spectroscopy of Gases, Van Nostrand, New York, 1965, pp. 162-190.
6. M. Tsuboi, A. Hirakawa, and T. Tamagake, J. Mol. Spectros., 22, 272 (1967).
7. J. Wollrab and V. Laurie, J. Chem. Phys., 48, 5058 (1968).
8. J. Lehn, B. Munsch, P. Millie, and A. Weillard, Theor. Chim. Acta, 13, 313 (1969).
9. M. Kemp and W. Flygare, J. Amer. Chem. Soc., 90, 6267 (1968).
10. T. Burdas, C. Szanaty, and C. Havada, ibid., 87, 5796 (1965).
11. M. Jautelat and J. D. Roberts, ibid., 91, 642 (1969); lower values of V (E_a) are for neat, and benzene and acetone solutions while the larger value is for $CDCl_3$ solution. For this molecule, $\Delta S^{\dagger}$ ($CDCl_3$) = 38 eu(log A $\sim$ 22) and T_c varies from 56 to 67°, depending on solvent.
12. J. Lehn and J. Wagner, J. Chem. Soc. (D), 148 (1968).
13. See footnote f of table 1 in M. Gordon and H. Fischer, J. Amer. Chem. Soc., 90, 2471 (1968).
14. S. Brauman and M. Hill, J. Chem. Soc. (B), 1091 (1969).
15. J. B. Lambert and D. H. Johnson, J. Amer. Chem. Soc., 90, 1350 (1968).
16. J. Lehn and B. Munsch, J. Chem. Soc. (D), 1327 (1969).
17. R. Baechler and K. Mislow, J. Amer. Chem. Soc., 92, 3090 (1970).
18. L. Horner and H. Winkler, Tetrahedron Letters, 461 (1964).
19. J. B. Lambert, G. F. Jackson, III, and D. C. Mueller, J. Amer. Chem. Soc., 90, 6401 (1968); 92, 3093 (1970).
20. W. Egan, R. Tang, G. Zon, and K. Mislow, ibid., 92, 1442 (1970).
21. R. Scartazzini and K. Mislow, Tetrahedron Letters, 2719 (1967).
22. D. Rayner, A. J. Gordon, and K. Mislow, J. Amer. Chem. Soc., 90, 4854 (1968). For a general review on sulfoxide stereomutation, see K. Mislow, Rec. Chem. Prog., 28, 217 (1967).
23. J. Lehn and B. Munsch, Theor. Chim. Acta, 12, 91 (1968).
24. H. Kessler and D. Leibfritz, Tetrahedron, 25, 5127 (1969); Tetrahedron Letters 427 (1969), and other papers in the series; N. Marullo and E. Wagener, ibid., 2555 (1969).
25. D. Y. Curtin, E. Grubbs, and C. McCarty, J. Amer. Chem. Soc., 88, 2775 (1966).
26. A. J. Gordon and J. P. Gallagher, Tetrahedron Letters, 2541 (1970).
27. A. Rauk and I. Csizmadia, Can. J. Chem., 46, 1205 (1968).
28. For a general treatment, see E. Eliel, N. Allinger, S. Angyal, and G. Morrison, Conformational Analysis, Interscience, New York, 1965, and J. Dale, Tetrahedron, 22, 3373 (1966). Also, J. P. Lowe in Progress in Physical Organic Chemistry, Vol. 6, A. Streitwieser, Jr., and R. W. Taft, Eds., Wiley-Interscience, New York, 1968, Ch. 1.
29. Taken from collections in Dale (28) and in Conformational Analysis, p. 140 (28) as well as in Lehn (41).
30. Unless indicated otherwise, data were determined by nmr and are those cited in reference 1b.
31. L. Graham and R. Diel, J. Phys. Chem., 73, 2696 (1969).
32. I. Calder and P. Garratt, Tetrahedron, 25, 4023 (1969).
33. P. Korver, P. van der Haak, and Th. de Boer, ibid., 22, 3157 (1966).
34. A. C. Cope, et al., J. Amer. Chem. Soc., 87, 3644, 3649 (1965).
35. For a review of biphenyl racemization studies, see D. M. Hall and M. M. Harris, J. Chem. Soc., 490 (1960); cf. D. M. Hall and T. Poole, ibid., (B), 1034 (1966). Similar data for 1,1'-binaphthyls may be found in A. Cooke and M. M. Harris, ibid., (C), 988 (1967).
36. F. Anet, M. St. Jacques, and G. Chmurny, J. Amer. Chem. Soc., 90, 5243 (1968).
37. H. Kessler and H.-O. Kalinowski, Angew. Chem., 82, 666 (1970); Angew. Chem. internat. edit. Engl., 9, 641 (1970).

38. For six-ring heterocycles, see R. Harris and R. Spragg, J. Chem. Soc. (B), 684 (1968); for dioxolanes (dioxanes) and dithiolanes, see H. Schmid, et al. Spectrochim. Acta, 22, 623 (1966), and E. Eliel, Accts. Chem. Res., 3, 1 (1970). For 5- and 6- ring heterocycles containing O and S, see C. Romers, et al., in Topics in Stereochemistry, Vol. 4, Wiley-Interscience, New York, 1969.
39. J. Gerig and R. Rimerman, J. Amer. Chem. Soc., 92, 1219 (1970).
40. R. Griffin, Jr., and R. Coburn, J. Amer. Chem. Soc., 89, 4638 (1967).
41. J. M. Lehn in reference 1d, pp. 129-155.

IX. BOND AND GROUP DIPOLE MOMENTS

A. Introduction

The dipole moment of a molecule can be considered to be the vector sum of individual bond and group moments, including contributions from any lone-pair electrons. If the geometry of a molecule is known or can be assumed, molecular dipole moments can be calculated as follows: establish a convenient set of reference axes, calculate the component (m) of each bond (group) moment along each axis, and obtain the resultant moment as the square root of the sum of the squares of the m_x m_y, m_z component sums along each axis (values always expressed in debyes (D), where 1 D = 10^{-18} esu-cm).

$$\mu = (m_x^2 + m_y^2 + m_z^2)^{1/2}$$

When only 2 bond (group) moments (m_1, m_2) are involved, the law of cosines is appropriate (Θ = angle between moments):

$$\mu = (m_1{}^2 + m_2{}^2 + 2m_1m_2\cos\,\Theta)^{1/2}$$

It has been pointed out that there are really three concepts of bond moments: intrinsic moments (theoretical), dynamic moments (from ir spectroscopic measurements), and static moments (empirically determined from measured dipole moments) [C. Cumper, Tetrahedron, 25, 3131 (1969)]. The data presented here are based on a static moment scale, the one of most general use. There are certain cautions to be heeded in the use of the data:

1. Although the magnitude of the C-H moment is usually taken as 0.4 D, its direction is as yet unknown; the choice of direction has a considerable influence on other bond and group moments calculated from dipole moments of molecules (see data below). Cumper has shown that the magnitude of the static C-H moment is virtually independent of the hybridization at C, and has assumed its value to be zero in establishing his moment data. However, the data herein are based on the more traditional practice (μ_{C-H}= 0.4 D).

2. Bond moments are not independent of the hybridization (sp^3, sp^2, etc.) of the atoms(s) involved or of the nature of the functional group; e.g., the C-Cl bond moment in R_2CCl_2 is not the same as in R_3CCl .

3. As with molecular dipole moments themselves, bond and group moments are not the same in the gas phase and in solution.

For details on measurement and interpretation of dipole moment data, consult the following references, from which most of the data below were taken:

L. E. Sutton, in Determination of Organic Structures by Physical Methods, E. A. Braude and F. C. Nachod, Eds., Vol. 1, Academic Press, New York, 1955, Ch. 9. (Some of the data therein have been superseded.)
C. P. Smyth, in Physical Methods of Organic Chemistry, 3rd ed., A. Weissberger, Ed., Vol. 1, Part 3, Interscience, New York, 1960, CH. XXXIX.
V. I. Minkin, O. A. Osipov, and Yu. A. Zhadanov, Dipole Moments in Organic Chemistry, Plenum Press, New York, 1970. A very readable and thorough treatment. (A translation of Dipol'nye Momenty, Khimiya Press, Leningrad, 1968.)

B. Moments of Carbon-Carbon Bonds

	Moment (D)	
Bond Type	Cumper[a]	Minkin (Compound Used)[b]
$sp^3 \rightarrow sp^2$	0.30	0.69, 0.67 (toluene, 0.37D; propylene, 0.35D)
$sp^3 \rightarrow sp$	0.70	1.48 (propyne, 0.75D)
$sp^2 \rightarrow sp$	0.40	1.15 (phenylacetylene, 0.73D)

[a]Average values, determined from dipole moments of many molecules, assuming μ(C-H) = 0 (Cumper, op. cit.). The values are increased by ≃ 0.10 D if a CH_3 is attached to the sp^3-C, and by ≈0.15 D for longer hydrocarbon chains.

[b]Based on C→H moments from ir measurements, as follows: sp^3-C, 0.31 D; sp^2 — C, 0.63 D; sp-C, 1.05 D. Minkin et al., op. cit., p. 89.

C. Moments of Various Bonds

Bond, as well as group moments are derived from model compounds, and the values are obviously strongly dependent on the value of the dipole moment used for the model. Examination of the books cited above (as well as others, including textbooks) shows some disagreement in bond moment values even for those derived from simple methyl or ethyl derivatives. The values given below were taken mainly from Minkin et al., op. cit., p. 88, and are derived from dipole moments of methyl compounds in benzene solution at 25°. Values in parentheses were calculated by the present author from the recommended gas phase dipole moment values of methyl compounds given by A. L. McClellan (Tables of Experimental Dipole Moments, Freeman, San Francisco, 1963). Values are listed for both assumed directions of the C-H bond moment (0.4 D). The atom to the left of the bond is the positive end of the moment, and, unless indicated otherwise, all multivalent atoms are sp^3.

Bond	Moment (D) C → H	Moment (D) C ← H	Bond	Moment (D) C → H	Moment (D) C ← H
C-N	1.26	0.45	C-S	1.6	0.9
C=N		1.4	C=S		2.0
C≡N	3.94(4.4)	3.1(3.6)	C-Se	1.5	0.7
C-O[a]	1.9	0.7	H-O	1.51	1.51
C-O[b]	1.5(1.2)	0.7(0.4)	H-N	1.31	1.31
C=O	3.2	2.4	H-S	0.7	0.7
C-F[c]	2.19(2.22)	1.39(1.42)	Si-H	1.0	1.0
C-Cl[c]	2.27(2.34)	1.47(1.54)	Si-C		1.2
C-Br[c]	2.22(2.19)	1.42(1.39)	Si-N	1.55	1.55
C-I[c]	2.05(2.04)	1.25(1.24)	N-:[d]	≃1.0	≃1.0

[a]Alcohols. [b]Ethers. [c]Values for vinyl C-halogen are about 0.7-0.8 D lower. [d]Lone pair on sp^3-nitrogen.

D. Moments of Coordination Bonds

The following, taken from Minkin et al., op. cit., p. 240, serve as a rough measure of polarity in "dative" bonds (as in amine oxides).

Bond	Moment (D)	Bond	Moment (D)
N→B	2.6	As→O	4.2
O→B	3.6	Se→O	3.1
S→B	3.8	Te→O	2.3
P→B	4.4	P→S	3.1
N→O	4.3	P→Se	3.2
P→O	2.9	Sb→S	4.5
S→O	3.0		

E. Moments of Various Groups

Values are given for substituents attached to alkyl and aryl groups. The data are mainly from Minkin et al., op. cit., p. 91, and Smyth, op. cit., p. 2602; solution values were obtained from measurements in benzene. Θ is the angle that the group moment makes with the bond from the functional group (X) to the carbon atom to which it is attached:

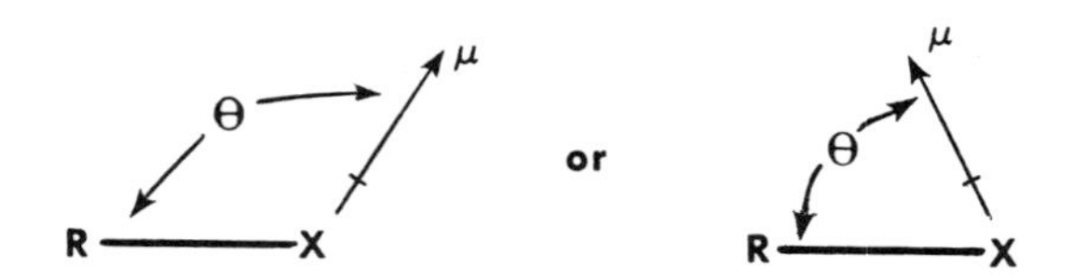

(Thus if X is the negative pole of the moment, Θ values are always >90°, and vice versa.)

	C_6H_5-X			CH_3-X			C_2H_5-X		
Group, X	Gas	Soln.	Θ	Gas	Soln.	Θ	Gas	Soln.	Θ
CH_3	0.37	0.37	0	0	0	-	0	0	-
CF_3	2.86	2.54	180	2.35	2.32	180			
CCl_3		2.04	180	1.77	1.57	180			
CN	4.39	4.05	180		3.4		4.00	3.57	180
CHO		2.96	146	2.72	2.49	125	2.73	2.5	
COOH		1.64		1.73	1.63	106	1.73	1.68	106
$COCH_3$	3.00	2.96	132	2.90	2.75	120	2.78		
CO_2CH_3		1.83	110	1.67	1.75	130	1.76	1.9	130
$CO_2C_2H_5$		1.9	118		1.8	89			
OH	1.40	1.6	90	1.70	1.7	118	1.69	1.7	118
OCH_3	1.35	1.28	72	1.30	1.28	124			
$OCOCH_3$		1.69	66						
OCF_3		2.36	160						
NH_2	1.48	1.53	48.5	1.28	1.46	91	1.2	1.38	100
$NHCH_3$		1.71	40						
$N(CH_3)_2$	1.61	1.58	30	0.61	0.86	109			
$NHCOCH_3$		3.69	100						
NO		3.09	149						
NO_2	4.19	4.01	180	3.50	3.10	180	3.68	3.3	180
N_3		1.44	140						
F	1.61	1.47	180	1.85	1.79	180	1.92		180
Cl	1.76	1.59	180	1.86	1.87	180	2.05	1.8	180
Br	1.64	1.57	180	1.82	1.82	180	2.01	1.9	180
I	1.71	1.40	180	1.70	1.65	180	1.8	1.8	180
SH		1.22	135	1.26	1.55				
SCH_3		1.34	77.5	1.50	1.40	57			
SCF_3		2.50	156						
SO_2CH_3		4.73	117						
SO_2CF_3		4.32	167						
$SOCF_3$		3.88	143						
SCN		3.59	127						
SeH		1.08	169						
$SeCH_3$		1.31	110		1.32				
$Si(CH_3)_3$		0.44	0		0	0			

X. AROMATICITY

A. Introduction

There are several recent monographs and important articles dealing with the concept of aromaticity and with the properties of aromatic molecules (1-9); reference 6b is the most recent, and presents thorough reviews of the current "state of the art." In the past, one of the criteria used to define a system as aromatic was its unexpected chemical reactivity -- for example, substitution rather than addition, and resistance to oxidation. However, it is

Refs. p. 133.

now generally accepted that chemical reactivity (which is based on transition state considerations) is a very poor guide. Rather, aromaticity is strictly a ground state phenomenon, manifested by unusual total molecular stability (lower ground state enthalpy) due to π-electron delocalization. Aromatic molecules can be carbocyclic or heterocyclic, including sydnones or meso-ionic compounds (5); they can be neutral or charged. Generally, aromatic molecules are classified as benzenoid (benzene-like structures: benzene, naphthalene, etc.) and nonbenzenoid (all others: azulene, cyclopentadiene anion, borazines, azepines, tropolones, etc.). Experimental criteria for aromaticity are described below.

B. Definitions

1. Aromaticity (1)

"An unsaturated cyclic or polycyclic molecule or ion (or part of a molecule or ion) may be classified as aromatic if all the annular atoms participate in a conjugated system such that, in the ground state, all the π-electrons (which are derived from atomic orbitals having axial orientation to the ring) are accommodated in bonding molecular orbitals in a closed (annular) shell." The original theoretical basis for aromaticity rests in Hückel's rule, namely, that to fill such a closed shell requires (4n + 2) π-electrons (n = 0, 1, ...); theoretically the rule applies to monocyclic systems, but it still works for polycyclic molecules. Recently (15) a "topological definition of aromaticity" has been introduced.

2. Antiaromaticity (8)

A closed ring having 4n conjugated π-electrons possesses antiaromatic character; that is, it is destabilized by resonance. In other words, an antiaromatic system is one in which electron delocalization raises the ground state energy considerably, compared to a noncyclic analog; for example, cyclopropenyl anion (as yet unprepared) and pentachlorocyclopentadienyl cation (a ground state triplet).

3. Homoaromaticity (6a)

Molecules designated as homoaromatic are charged species that have interruption of the σ-backbone joining π-electron centers, yet still exhibit enhanced delocalization energy:

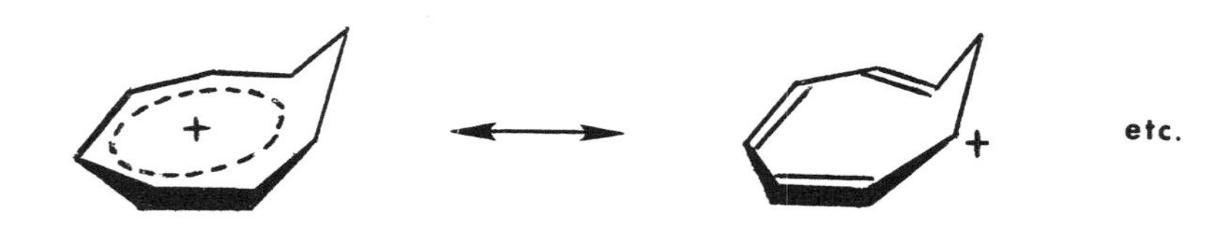

Homotropylium ion
(homoaromatic)

Refs. p. 133.

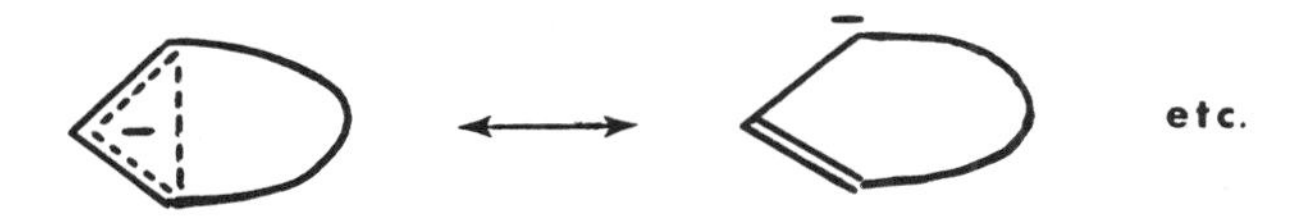

Homocyclopropenyl anion
(antihomoaromatic)

4. Alternant and Nonalternant Hydrocarbons

In alternant hydrocarbons, conjugated carbons can be divided into starred and unstarred sets, such that no two atoms of the same set are directly bonded to one another. For example, naphthalene (even alternant-10 atoms) and benzyl cation,anion, or radical (odd alternant-7 atoms). In alternant systems, for every bonding orbital of energy -E there is an antibonding orbital with energy +E (thus energy levels are symmetrically disposed about the zero level). Odd alternant systems (which must be charged or radicals) possess a nonbonding orbital ("zero" energy level); in addition, the cation, anion, and radical all have the same charge distribution (unpaired electron distribution) over the entire molecule. Nonalternant hydrocarbons do have atoms of the same set next to each other. The energies of bonding and antibonding orbitals in such molecules are not equal and opposite, and charge distributions are not the same in charged and radical species. Examples are azulene and fulvene:

There is no relationship between aromatic character and the alternant or nonalternant nature of a compound.

5. Craig's Rule

This rule is a scheme for predicting whether a cyclic, conjugated, nonbenzenoid hydrocarbon has aromatic character. It is based on a symmetry criterion, namely, that for the rule to be applied, the molecule must have an axis of symmetry passing through at least two π-electron centers; this symmetry axis, in effect, converts one Kekulé structure into another. The structure is labeled alternately (as far as possible) with spin symbols α and β. The sum is taken of the number (f) of symmetrically related pairs of π-centers off the axis, and of the number (g) of interchanges of α and

Refs. p. 133.

Pentalene

β by rotation about the axis. If the sum (f + g) is even, aromaticity is expected. For pentalene, f + g = 3 + 0 = 3 (nonaromatic); for fulvene, f + g = 0 + 2 = 2 (expected to have aromatic character). Molecules with (f + g) odd are often called pseudoaromatic.

6. Resonance Energy and Related Parameters (10,11)

When empirical data for a substance indicate that the molecule cannot be represented adequately by a single Lewis structure, it is said that the system exhibits resonance; the real structure is therefore some type of hybrid. Resonance is observed in aromatic as well as nonaromatic systems such as 1,3-butadiene and allyl cation, anion, or radical. In valence-bond theory, the vertical resonance energy is the difference between the calculated energy for the Lewis structure of lowest energy and the true energy of the molecule; obviously, since Lewis structures for molecules exhibiting resonance are not real, the so-called lowest energy value can only be estimated. The empirical resonance energy is determined by comparison of experimentally determined heats of combustion or hydrogenation, with values calculated from nonconjugated model compounds. The following energy cycle (10) illustrates the definitions for benzene (energies in kcal/mol):

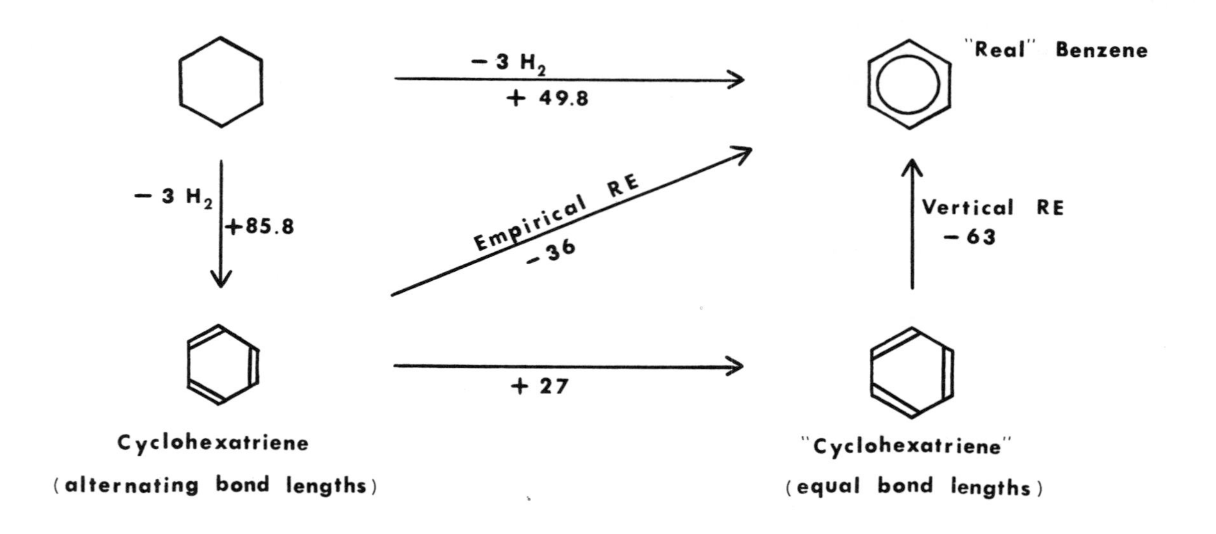

The compression energy (distortion energy) of +27 (a rough estimate) is that required to equalize the bond lengths of a hypothetical cyclohexatriene in order to create the Kekulé structure. The value 85.8 derives from three times the heat of hydrogenation of cyclohexene.

A more direct measure of π-electron conjugation is the delocalization energy, the calculated additional bonding energy that results from such delocalization (in contrast to isolated double bonds). Values are expressed in units of β, where β is the resonance (exchange) integral in the Hückel

Refs. p. 133.

molecular orbital (HMO) method for calculating π-energy levels. (Note. The "resonance integral" is a mathematical parameter used in the calculation for any molecule, and is not a measure of resonance in the sense discussed above.) The absolute value for β in kcal/mol is obtained by comparing empirical resonance energies with calculated delocalization energies. However, there is no one value of β since it changes with the nature of the system; for benzenoid aromatic molecules, a fairly accurate value of one β unit is -16 kcal/mole.

C. Values of Resonance and Delocalization Energies

The following table presents values of delocalization and empirical resonance energies for aromatic and some nonaromatic substances. Empirical resonance energies from heats of hydrogenation are considerably more accurate than those from heats of combustion. However, hydrogenation data are not always available and the values given are derived from combustion data; values in parentheses were obtained from heats of hydrogenation. The data are taken from references 1, 10, and 11. A review and collection of parameters used in HMO calculations is available (13). See reference 17 for recent data.

Substance	Empirical Resonance Energy (kcal/mole)	Delocalization Energy (β units)
	Benzenoid Aromatics	
Benzene	36.0 (36.0)	2.000
Naphthalene	61.0	3.683
Anthracene	83.5	5.314
Phenanthrene	91.3	5.448
Benz[c]phenanthrene	109.6	7.187
Pyrene	108.9	6.506
Naphthacene	110.0	6.932
Benz[a]anthracene	111.6	7.101
Chrysene	116.5	7.190
Triphenylene	117.7	7.275
Perylene	126.3	8.245
Indene	(39.4)	
Styrene	38.1 (36.9)	
trans-Stilbene	76.9 (79.0)	
Biphenyl	74	
o-Xylene	(35.1)	
	Heteroaromatics*	
Furan	15.8	
Pyrrole	21.2	
Thiophene	28.7	
Pyridine	23.0	
Indole	46.5	
Carbazole	73.8	
Quinoline	47.3	

(Continued)

*Resonance energies for heteroaromatics (and other systems) can be obtained from pK as well as thermochemical data (16).

Refs. p. 133.

Substance	Empirical Resonance Energy (kcal/mole)	Delocalization Energy (β units)
	Nonbenzenoid Hydrocarbons	
Azulene	33	3.364
Biphenylene	17.1	4.506
Cyclobutadiene	-	0
Cyclopentadienyl anion		2.48
Cyclooctatetraene	4.8	1.657*
Fulvalene		2.779
Fulvene	11	1.466
Heptafulvalene	29	4.004
Heptafulvene	14	1.994
Pentalene	-	2.46
Tropolone	21	
Tropylium		2.99
1,3-Butadiene	3.0 (3.5)	0.47
1,3-Cyclopentadiene	1.6 (2.9)	0.47
1,3-Pentadiene	(4.2)	0.47
1,3,5-Cycloheptatriene	(6.7)	0.99

*Calculated for planar cyclooctatetraene.

D. Criteria for Aromaticity

The traditional experimental criteria for aromaticity are anomalous chemical reactivity (a very poor criterion-see above) and the presence of a high degree of empirical resonance energy(also not the best criterion, since the resonance energy values obtained are not the consequence of π-delocalization alone, but include contributions from other factors involving σ-bonds). A recent, and more consistent, basis is the diamagnetic or paramagnetic anisotropy observed in the nmr (^{1}H and in special cases ^{13}C) chemical shifts of aromatic nuclei (2). This criterion is usually interpreted in terms of a "ring current" model for the delocalized π-system; however, although such anisotropy exists, its origin is still of some doubt, making a general correlation between chemical shift and aromaticity somewhat tentative (1).

More recently, Dauben and co-workers (12) have shown that diamagnetic susceptibility exaltation is a property unique to aromatic compounds. This property is defined as the difference between the measured molar susceptibility and the value estimated by neglecting the contribution of a ring current. A recent application of the magnetic susceptibility method has classified 2- and 4-pyrone as nonaromatic, at least by magnetic criteria (14).

Refs. p. 133.

E. References

1. G. M. Badger, Aromatic Character and Aromaticity, Cambridge University Press, London, 1969.
2. A. L. Jones, "Criteria for Aromatic Character", Rev. Pure Appl. Chem., 18, 253 (1968).
3. D. Ginsburg, Ed., Non-Benzenoid Aromatic Compounds, Interscience, New York, 1959.
4. D. M. G. Lloyd, Carbocyclic Non-Benzenoid Aromatic Compounds, Elsevier, Amsterdam, 1966.
5. J. P. Snyder, Ed., Nonbenzenoid Aromatics, Vol. 1, Academic Press, New York, 1969; Vol. 2, 1971.
6. (a) "Aromaticity. An International Symposium," Special Publication No. 21, The Chemical Society, London, 1967. (b) Aromaticity, Pseudo-aromaticity, Antiaromaticity (Vol. 3 of the Jerusalem Symposia on Quantum Chemistry and Biochemistry), E. D. Bergmann and B. Pullman, Eds., Academic Press, New York and London, 1971.
7. A. R. Katritzky and R. D. Topsom, "Distortions of the π-Electron System in Substituted Benzenes", Angew. Chem. intern. Ed., Engl., 9, 87 (1970).
8. R. Breslow, "Small Antiaromatic Rings," ibid., 7, 565 (1968).
9. C. A. Coulson and A. Streitwieser, Jr., Dictionary of π-Electron Calculations, W. H. Freeman and Company, San Francisco, 1965.
10. A. Streitwieser, Molecular Orbital Theory for Organic Chemists, Wiley, New York, 1967, Chapters 9 and 10.
11. A concise discussion and useful data are found in R. D. Gilliom, Introduction to Physical Organic Chemistry, Addison-Wesley, Reading, Mass., 1970, Ch. 5. See also N. C. Baird, J. Chem. Ed., 48, 509 (1971).
12. H. J. Dauben, Jr., J. D. Wilson, and J. L. Laity, J. Amer. Chem. Soc., 91, 1991 (1969); see 91, 4943 (1969) for a correction.
13. W. P. Purcell and J. A. Singer, "A Brief Review and Table of Semiempirical Parameters Used in the HMO Method", J. Chem. Eng. Data, 12, 234 (1967).
14. R. C. Benson, C. L. Norris, W. H. Flygare, and P. Beak, J. Amer. Chem. Soc., 93, 5591 (1971).
15. M. J. Goldstein and R. Hoffmann, J. Amer. Chem. Soc., 93, 6193 (1971).
16. For a recent discussion, see D. Lloyd and D. R. Marshall, Chem. Ind., 335 (1972).
17. W. C. Herndon, J. Amer. Chem. Soc., 95, 2404 (1973). This paper presents a comparison of resonance stabilization energies of 29 aromatic hydrocarbons calculated by valence bond (VB), HMO, and SCF-MO methods. The results show that resonance energies per π-electron from VB methods when limited to Kekulé structures agree remarkably well with SCF-MO calculations.

KINETICS AND ENERGETICS

I. ACTIVATION PARAMETERS AND KINETICS OF SELECTED REACTIONS

This section is designed to serve as a guide to the rates and energetics of some common chemical reactions; all reactions discussed are thermal.

A. Useful Equations for Determining Activation Parameters from Rate Data

1. General Rate Expressions

The table below summarizes the most common rate expressions, using the following symbols:

x (or y) = Concentration (usually in moles/liter) of reactant (x or y) at time, t (usually in seconds)
x_o (or y_o) = Initial concentration of x (or y) (at time zero)
x' (or y') = Decrease in concentration of reactant ($x = x_o - x'$), that is the amount consumed up to time t
k = Rate constant (specific rate) in units of $x^{(1-n)}/t$, where n = order of reaction

Order	Differential Form	Integrated Form	Half-Life, $t_{1/2}$
0	$-dx/dt = k$	$kt = x_o - x$	$x_o/(2k)$
1	$-dx/dt = kx$	$kt = \ln (x_o/x)$	$0.693/k$
1	$-dx'/dt = k(x_o - x')$	$kt = \ln [x_o/(x_o - x')]$	$0.693/k$
2	$-dx/dt = kx^2$	$kt = (1/x) - (1/x_o)$	$(kx_o)^{-1}$
2	$-dx'/dt = k(x_o - x')^2$	$kt = x'/[x_o(x_o - x')]$	$(kx_o)^{-1}$
2	$-dx/dt = kxy$	$kt = (y_o - x_o)^{-1} \ln (x_o y/y_o x)$	
2	$-dx'/dt = k(x_o - x')(y_o - y')$	$kt = (x_o - y_o)^{-1} \ln \dfrac{y_o(x_o - x')}{x_o(y_o - x')}$	

2. Symbols and Constants in the Arrhenius and Eyring Equations

(For other values and conversion factors, see pp. 458, 462, and 166.)

Gas constant: $R = 1.987$ cal/deg-mole

$\ln = \log_e = 2.303 \log_{10}$

Boltzmann constant: $\kappa = 1.380 \times 10^{-16}$ erg/deg

Planck constant: $h = 6.626 \times 10^{-27}$ erg-sec

Temperature: T, in °K (0°C ≡ 273.16°K)

Rate constant: k (first order in sec^{-1}; second order in l/mole-sec)

Refs. p. 143.

3. Arrhenius Equation

$$k = A \exp \frac{-E}{RT}$$

A = Preexponential factor in units of k; typical values for unimolecular reactions are about 10^{10} to 10^{12} sec^{-1}

E = Arrhenius (empirical) energy of activation (cal/mole); from a plot of log k versus 1/T:

$E = -4.576 \times \text{slope}$ $\qquad$ $\log A = E/4.576T + \log k$

[For discussion of deviations from the Arrhenius equation, see J. R. Hulett, Quart. Rev., 18, 227 (1964).]

4. Eyring (Transition State Theory) Equation

$$k = K \frac{\kappa T}{h} \exp \frac{-\Delta G^{\dagger}}{RT}$$

K = Transmission coefficient (fraction of molecules that, having reaching the transition state, proceed to product); usually assumed = 1

$\Delta G^{\dagger}$ = Free energy of activation = $\Delta H^{\dagger} - T\,\Delta S^{\dagger}$ ($\Delta G^{\dagger}$, and $\Delta H^{\dagger}$ the enthalpy of activation, are in cal/mole; entropy of activation, $\Delta S^{\dagger}$, is in cal/deg, or simply eu (entropy units))

$\Delta G^{\dagger}$ = $4.576T(10.319 + \log T - \log k)$ for k at a specific temperature; from a plot of log (k/T) versus 1/T:

$\Delta H^{\dagger} = -4.576 \times \text{slope}$ $\qquad$ $\Delta S^{\dagger} = 4.576 \log (k/T) + (\Delta H^{\dagger}/T) - 47.22$

Alternatively (for solution phase reactions):

$\Delta H^{\dagger} = E - RT$ (T taken as mean value over range studied)

$\Delta S^{\dagger} = 4.576 \log (k/T) + (E/T) - 49.21$

$\Delta S^{\dagger}$ may also be determined from log A (Arrhenius equation), where T is taken as the mean value for the range studied (alternatively a value of $\Delta S^{\dagger}$ at a certain T may be reported):

$$\Delta S^{\dagger} = 4.576 \log (A/T) - 49.21$$

5. Volume of Activation

The change in volume of a reacting species in going from an initial state to an activated complex is indicated by:

$$\Delta V^{\dagger} = V - V_o$$

$\Delta V^{\dagger}$ may be (+) or (-), and values are best reported as those extrapolated to zero pressure:

$$\left[\frac{d(\ln k)}{dp}\right]_T = -\frac{\Delta V^{\dagger}}{RT}$$

Refs. p. 143.

Typical values are ± 5 to 20 cm^3/mole. For example, in solution:

Reaction	$\sim\Delta V^{\dagger}$	Reaction	$\sim\Delta V^{\dagger}$
$R\text{-}R \rightarrow 2\ R\cdot$	5-15	$C_2H_5I + C_2H_5O^-$	-5
Diels-Alder reactions	-25	$Py + RBr \rightarrow PyR^+Br^-$	-26
$R\cdot + CCl_4 \rightarrow RCl + \dot{C}Cl_3$	-20	S_N1 reactions	-15

For details, see E. Whalley, Advan. Phys. Org. Chem., 2, 93 (1964); W. J. le Noble, J. Chem. Educ., 44, 729 (1967) and Prog. Phys. Org. Chem., 5, 207 (1967).

6. Isotope Effects

For determining directly the difference in Eyring activation parameters for reactions of systems differing by the presence of isotopes A and B (or, in fact, for reactions of any two systems) the following are useful (assumes no isotope effect on transmission coefficient):

$$\log \frac{k_A}{k_B} = \frac{1}{4.576T}\left[\Delta G_B^{\dagger} - \Delta G_A^{\dagger}\right] = \frac{1}{4.576}\left[\frac{(\Delta H_B^{\dagger} - \Delta H_A^{\dagger})}{T} - (\Delta S_B^{\dagger} - \Delta S_A^{\dagger})\right]$$

From plot of $\log(k_A/k_B)$ versus $1/T$:

$$\Delta\Delta H^{\dagger} = \Delta H_B^{\dagger} - \Delta H_A^{\dagger} = 4.576 \times \text{slope}$$

$$\Delta\Delta S^{\dagger} = \Delta S_B^{\dagger} - \Delta S_A^{\dagger} = (1/T)(\Delta H_B^{\dagger} - \Delta H_A^{\dagger}) - 4.576 \log (k_A/k_B)$$

It is common to refer to an isotope effect in terms of free energy difference per isotopic atom (n = number of such atoms involved in the effect).

$$\Delta\Delta G^{\dagger} = \Delta G_B^{\dagger} - \Delta G_A^{\dagger} = -\frac{4.576\ T}{n} \log \frac{k_B}{k_A}$$

7. Diffusion Controlled Reactions

Some bimolecular reactions are successful at every encounter --- radical combination and disproportionation, excited state quenching, energy transfer, and some ion reactions in solution; their rates are dependent mainly on the viscosity of the solvent:

$$k = \frac{8RT}{3000\eta}$$

where k is in liter/mole-sec, $R = 8.31 \times 10^7$ ergs/mole-deg K, and η = viscosity in poise (g/sec-cm). The equation assumes spherical species of equal diameter; if diameters d_1 and d_2 are unequal, multiply the equation by $(1/4)[2 + (d_1/d_2) + (d_2/d_1)]$.

Refs. p. 143.

For typical organic solvents at 25°, k ∿ 10^9 - 10^{10} and for typical diffusion controlled reactions, apparent activation energies are 2 to 3 kcal/mole. For reactions with activation energies > 10 kcal/mole, viscosity usually will have no effect until a glassy state is reached. Typical rates, assuming $d_1 = d_2$, are as follows:

Solvent	10^{-10}k(20°C)	Solvent	10^{-10}k(20°C)
n-C_6H_{14}	2.1	Cyclo-C_6H_{12}	0.66
$CHCl_3$	1.2	H_2O	0.64
CH_3OH	1.1	C_2H_5OH	0.55
C_6H_6	1.0	$HOCH_2CH_2OH$	0.038
CCl_4	0.67	$HOCH_2CH(OH)CH_2OH$	0.00061

For details, see J. G. Calvert and J. N. Pitts, Jr., Photochemistry, Wiley, New York, 1966, p. 626 (from which the data above were taken), and J. Leffler and E. Grunwald, Rates and Equilibria of Organic Reactions, Wiley, New York, 1963, p. 59.

B. Substitution and Solvolysis

In this section and others below, we summarize activation parameters for typical reactions; activation energies or enthalpies are in kcal/mole, A is in units of k, and $\Delta S^\dagger$ is in eu.

1. S_N2 Reactions (R = primary carbon)

Reaction	Solvent	E	log A	$\Delta H^\dagger$	$\Delta S^\dagger$	Ref.
RBr + LiCl	Acetone	∿18	∿8.5			1
RBr + Et_4NCl	DMF	∿20	∿12			1
RCl + I^-	Acetone	18.5	8			3a
EtI + Et_3N	Acetone			11.5	-39	2
EtI + Et_3N	Hexane			15.3	-42	2
RX + $R'O^-$(X = Cl,Br,I)	EtOH	19.5-21.5	10-11.5			3a
CH_3I + Et_2S	ROH			17	-25	3b

2. S_N1 Reactions (3f)

Reaction	Solvent	$\Delta H^\dagger$	$\Delta S^\dagger$	Ref.
CH_3X (X=Cl,Br,I)	H_2O	24	-10	4
CH_3F	H_2O	21.4	-26.2	4
$CH_3OSO_2C_6H_5$	H_2O	20.9	-10.9	4
$CH_3OSO_2CH_3$	H_2O	21.2	-11.7	4
CH_2CHCH_2X(X=Br,I)	H_2O	21	- 7	4
$C_6H_5CH_2Cl$	H_2O	20.8	-11.0	4
t-BuCl	H_2O	23.8	14.4	3b
t-BuCl	ROH	26	- 3	3b
t-BuCl	90% aq. diox. or acetone	22	-17	3b

Refs. p. 143.

(Continued)

Reaction	Solvent	ΔH†	ΔS†	Ref.
$C_6H_5CH(Cl)CH_3$	EtOH	25.6[a]	11.97[b]	5
$C_6H_5CH(Cl)CH_3$	80% aq. EtOH	20.4[a]	10.00[b]	5
$(CH_3)_3CCH_2OTS$	HOAc	31.5	- 1.0	6
$(C_6H_5)_3CCH_2OTS$	HOAc	25.2	- 2.5	6
Cycloalkyl tosylates, $C_nH_{2n-1}OTs$				
n = 5	HOAc	24.1	- 4.2	7
5(OBs)	HOAc	22.7	- 6.3	7
6	HOAc	27.3	- 0.5	7
6(OBs)	HOAc	26.8	0.4	8
7	HOAc	23.3	- 5.7	7
Bicyclo[2.2.1]heptyl (norbornyl) tosylates and brosylates				
2-exo (OBs)	HOAc	23.3	7.7	10
2-endo (OBs)	HOAc	26.0	- 1.5	10
7 (OTs)	HOAc	35.7	- 3.5	9
2-ene-7-anti(OTs)	HOAc	23.3	5.7	9
2,3-benzo-5-exo(OBs)	HOAc	24.1	- 1.2	11
2,3-benzo-5-endo(OBs)	HOAc	27.8	- 7.0	11
Bicyclo [2.2.2]octyl (OBs)	HOAc	24.6	1.2	12

[a]Arrhenius E. [b]Arrhenius log A.

C. Addition

Reaction	Solvent	E	log A	ΔH†	ΔS†	Ref.
$C_2H_4 + H_2$	Vapor	43.2	10.6			3c
$(CH_3)_2CCH_2 + HBr$	Vapor	22.5	7.2			3c
Cyclopentadiene (Diels-Alder)	Vapor, neat or EtOH			15.5	-34	3d
$2\ C_2H_5\cdot \rightarrow C_4H_{10}$	Vapor	2.0	11.2			13
Cyclic ketones, $C_nH_{2n-2}O$ + $NaBH_4$/i-PrOH						
n = 4				8.1	-36.4	14
5				9.3	-38.8	14
6				5.1	-48.1	14

Refs. p. 143.

D. Unimolecular Elimination and Decomposition in the Vapor Phase

Reaction	E	log A	Ref.	Reaction	E	log A	Ref.
$C_2H_5X \rightarrow C_2H_4$				$CH_3CO_2R \rightarrow$ olefin + CH_3CO_2H			
X = Cl	56.9	13.6	15	R = C_2H_5	47.8	12.5	15
Br	53.9	13.4	15	$(CH_3)_3C$	40.5	13.3	15
I	50.0	13.7	15	$CH_3CHC_6H_5$	43.7	12.8	15
$(CH_3)_3CX \rightarrow$				Dicyclopentadiene	35	13	15
$(CH_3)_2C{=}CH_2$					$\Delta H^{\dagger}$	$\Delta S^{\dagger}$	Ref.
X = Cl	45.0	13.9	15	$CH_3C(=O)O C(CH_3)_3$	39	20	17a
Br	45.5	13.0	15	$C_6H_5CH_2C(=O)OC(CH_3)_3$	28	2	17a
$CH_3N{=}NCH_3$	52.5	16.54	3e				
$C_2H_5N_3$	39.7	14.30	3e				
$(n\text{-}Pr)_2O_2$	36.5	15.36	3e				

E. Unimolecular Rearrangement (Isomerization)

Reaction	Solvent	E	log A	$\Delta H^{\dagger}$	$\Delta S^{\dagger}$	Ref.
Azobenzenes (cis→trans)	Methylcyclohexane	23-24	12-13			18
Stilbene (cis→trans)	Vapor	42.8	12.78		-1	3e
Dimethyl maleate (cis→trans)	Vapor	23.5	5.11		-36	3e
Cyclopropane→propene	Vapor	65.0	15.18		11	3e
Allyl phenyl ether (o-Claisen)	$(C_6H_5)_2O$	32			-12	19
7-Phenyl to 3-phenyl-cycloheptatriene (1,5 H-migration)	Neat	27.6	10.8	26.9	-11.7	17b
cis-1,2-Divinylcyclobutane to 1,5-cyclooctadiene (Cope rearr.)	Neat			23.1	-11.7	17b

F. Redox in Aqueous Solution (20)

Reaction	$\Delta H^{\dagger}$	$\Delta S^{\dagger}$
$Co(L)^{+2}$ + $Co(L)^{+3}$ (Various ligands, L)	∿13	-26 to -38
Fe^{+2} + FeX^{+2} (X = Cl_2, Br_2, SCN)	7-9	-21 to -27
Fe^{+3} + HO_2^-	27	50
Co^{+3} + HO_2^-	18	55

Refs. p. 143.

G. Isotope Effects

For details, the following recent literature is recommended: M. J. Goldstein "Kinetic Isotope Effects and Organic Reaction Mechanisms," Science, 154, 1616 (1966); R. E. Weston, Jr., "Transition-State Models and Hydrogen-Isotope Effects," Science, 158, 332 (1967); E. A. Halevi, "Secondary Deuterium Isotope Effects," in Progress in Physical Organic Chemistry, Vol. 1, Interscience, New York, 1963, pp. 109-122; W. H. Saunders, "Kinetic Isotope Effects," in Survey of Progress in Chemistry, Vol. 3, Academic Press, 1966; Isotope Effects in Chemical Reactions, C. J. Collins and N. S. Bowman, Eds., Van Nostrand Reinhold, New York, 1971 (a modern guide to kinetic isotope effects).

1. Useful Equations

a. Maximum Possible Primary CH versus CD Isotope Effect (25). If the Eyring theory is valid, then the maximum effect is given as

$$\frac{k_H}{k_D} = \exp \frac{h\Delta\nu_o}{2RT}$$

where h = Planck's constant and $\Delta\nu_o$ = bond stretching frequency difference in the ground state. This equation is true only if the following conditions are met: only zero-point energy differences contribute to the isotope effect; bending and other modes cancel each other between reactants and transition state; and quantum mechanical effects (tunneling, e.g.) are negligible. For example:

T (°C):	0	25	100	200	300	500
$(k_H/k_D)_{max}$:	8.3	6.9	4.7	3.4	2.7	2.1

At very high temperature, assuming that the conditions mentioned are met, k_H/k_D approaches $\sqrt{2}$ for reactions with similar transition states. For more details on these points, see R. A. More, et al., J. Amer. Chem. Soc., 93, 9 (1971).

b. Semiempirical Calculation of α-Secondary Deuterium Effects. (A. Streitwieser et al., J. Amer. Chem. Soc., 80, 2376 (1957); Ann. N. Y. Acad. Sci., 84, 580 (1960)).

$$\frac{k_H}{k_D} = \Sigma(\nu_H - \nu_H^{\dagger}) \exp \frac{0.187}{T}$$

where sums are taken over stretching and bending fundamental vibrations for ground state and activated state C-H bonds.

c. Relation between Tritium and Deuterium Isotope Effects. [For summary see E. S. Lewis and J. K. Robinson, J. Amer. Chem. Soc., 90, 4337 (1968)].

$$k_H/k_T = (k_H/k_D)^r \qquad 1.58 \geq r \geq 1.33$$

The lower value of r applies to small effects ($k_H/k_D < 2$), the higher value to temperature-independent isotope effects. In the commonly used "Swain equations" r = 1.442, determined with the neglect of a tunneling contribution.

Refs. p. 143.

2. Kinetic Deuterium Isotope Effects

Reaction	T(°C)	k_H/k_D	Ref.
Primary Effects			
t-BuOK + $C_6H_5CD_2CH_2Br \rightarrow C_6H_5CDCH_2$	30	7.9	21
$(CH_3)_2CDOH + HCrO_4^- \rightarrow (CH_3)_2CO$	25	6.7	22
$C_6H_5CDO + MnO_4^- \rightarrow C_6H_5CO_2^-$ (pH 7)	25	7.5	23
$CF_3CD(C_6H_5)OH + MnO_4^- \rightarrow$ ketone	25	16	27
$CF_3CT(C_6H_5)OH + MnO_4^- \rightarrow$ ketone	25	k_H/k_T = 57	27
$(CH_3)_2CTNO_2$ + collidine → anion	32	k_H/k_T = 155	27
1-Methylcyclohexyl-2-d acetate → olefin	500	1.7	24
$CD_3COCD_3 + Br_2/H^+ \rightarrow BrCD_2COCD_3$	25	7.7	25
$ND_3 + h\nu$ (Hg sens.) $\rightarrow N_2 + D_2$	20	10	25
$CH_2CH_2 + D_2 \rightarrow DCH_2CH_2D$ (gas phase)	550	2.5	25
Phenol-2,4,6-d_3 + I_2/polar solv. → p-Iodophenol	25	3.97	26
$CDCl_3 + Cl_2 \rightarrow CCl_4$ (gas phase)	23	5.2	25
Secondary Effects			
$C_6H_5CH_2CD_2Br + EtO^-$ (E_2)	60	1.17	28
Cyclo-C_5H_9OTs-1-d + HOAc(S_N1)	50	1.15	30
Cyclo-C_5H_9OTs-2,2,5,5-d_4 + HOAc(S_N1)	50	2.06	30
$(CH_3)_2CDBr + H_2O$ (S_N1)	60	1.07	29
+ EtO^- (E_2)	25	1.05	31
+ EtO^- (S_N2)	25	1.00	31
[structure: D_3C, CD_3] ⇄ (racemization in benzene)	42	1.13	32

3. Equilibrium, Solvent, and Other Isotope Effects

System	T(°C)	Effect	Ref.
$2\ H_2O \rightleftarrows H_2O^+ + HO^-$	25	$K = 1.00 \times 10^{-14}$	25
$2\ D_2O \rightleftarrows D_3O^+ + DO^-$	25	$K = 1.54 \times 10^{-15}$	25
$^{15}NH_3 + {}^{14}NH_4^+ \rightleftarrows {}^{15}NH_4^+ + {}^{14}NH_3$	25	K = 1.034	33
$(D_3C)_3BN(CH_3)_3 \rightleftarrows (D_3C)_3B + N(CH_3)_3$	25	K_H/K_D = 1.25	34
S_N2 reactions in D_2O	25	$k_H/k_D \sim 1.2$	35
$CH_3COCH_3 + DO^-/D_2O$ (exchange)	25	k_H/k_D = 0.77	35
t-BuCl + 90% D_2O/10% dioxane	25	k_H/k_D = 1.37	35
Specific catalysis, rapid pre-equilibrium			
HO^- (DO^-)		$k_H/k_D \sim$ 0.5-0.7	25,35
H_3O^+ (D_3O^+)		$k_H/k_D \sim$ 0.3-0.4	25,35
Biphenyl-d_{10}, uv spectrum (cyclohexane)		$\lambda_H = \lambda_D$; $\varepsilon_H/\varepsilon_D$ = 0.98	36

Refs. p. 143.

H. References

1. A. J. Parker, Advan. Phys. Org. Chem., 5, 224 (1967).
2. K. B. Wiberg, Physical Organic Chemistry, Wiley, New York, 1964, p. 380.
3. A. A. Frost and R. G. Pearson, Kinetics and Mechanism, 2nd ed., Wiley, New York, 1965. (a) p. 148; (b) p. 138; (c) p. 104; (d) p. 376; (e) p. 110; (f) for considerable data on tertiary chlorides and tosylates, see H. Tanida and T. Tsushima, J. Amer. Chem. Soc., 92, 3397 (1970).
4. R. E. Robertson, Prog. Phys. Org. Chem., 4, 231 (1967).
5. E. Eliel, in M. S. Newman, Ed., Steric Effects in Organic Chemistry, Wiley, New York, 1956, p. 88.
6. S. Winstein et al., J. Amer. Chem. Soc., 74, 1113 (1952).
7. H. C. Brown and G. Ham, ibid., 78, 2735 (1956).
8. S. Winstein and D. Trifan, ibid., 74, 1147 (1952).
9. S. Winstein, et al., ibid., 77, 4183 (1955).
10. J. C. Martin and P. D. Bartlett, ibid., 79, 2533 (1957).
11. P. D. Bartlett and W. P. Giddings, ibid., 82, 1240 (1960).
12. H. M. Walborsky et al., ibid., 83, 988 (1961).
13. W. A. Pryor, Free Radicals, McGraw-Hill, New York, 1966, p. 314.
14. H. C. Brown and I. Ichikawa, Tetrahedron, 1, 221 (1957).
15. A. Maccoll, Advan. Phys. Org. Chem., 3, 99 (1965).
16. A. Maccoll, in The Chemistry of Alkenes, S. Patai, Ed., Interscience, New York, 1964, pp. 215-220.
17. E. Kosower, An Introduction to Physical Organic Chemistry, Wiley, New York, 1968. (a) p. 163; (b) pp. 242, 245.
18. D. Gegion, K. Muszkat, and E. Fischer, J. Amer. Chem. Soc., 90, 3907 (1968). Log A values were calculated from their data.
19. S. J. Rhoads, in Molecular Rearrangements, P. deMayo, Ed., Interscience, New York, 1963, p. 655.
20. F. Basolo and R. G. Pearson, Mechanism of Inorganic Reactions, 2nd ed., Wiley, New York, 1967, p. 475.
21. W. H. Saunders, Jr., and D. Edison, J. Amer. Chem. Soc., 82, 138 (1960).
22. F. Westheimer, Chem. Rev., 45, 419 (1949).
23. K. Wiberg and J. Steward, J. Amer. Chem. Soc., 77, 1786 (1955).
24. C. H. DePuy and R. W. King, Chem. Rev., 60, 431 (1960).
25. K. Wiberg, ibid., 55, 713 (1955).
26. E. Grovenstein and D. Kilby, J. Amer. Chem. Soc., 79, 2972 (1957).
27. E. S. Lewis and J. K. Robinson, ibid., 90, 4337 (1968).
28. S. Asperger, N. Ilakovac, and D. Pavlovic, Croat. Chem. Acta, 34, 7 (1962).
29. K. T. Leffek, J. Llewellyn, and R. Robertson, Can. J. Chem., 38, 1505 (1960).

30. A. Streitwieser, Jr., R. Jagow, R. Fahey, and S. Suzuki, J. Amer. Chem. Soc., 80, 2376 (1957).

31. V. J. Shiner, Jr., ibid., 74, 5285 (1952).

32. K. Mislow, R. Graeve, A. J. Gordon, and G. H. Wahl, Jr., ibid., 86, 1733 (1964).

33. J. Bigeleisen, Science, 147, 463 (1965). This paper provides a survey of equilibrium isotope effects.

34. P. Love, R. W. Taft, Jr., and T. Wartik, Tetrahedron, 5, 116 (1959).

35. R. E. Weston, Jr., Ann. Rev. Nucl. Sci., 11, 439 (1961).

36. K. T. Leffek, Can. J. Chem., 45, 2115 (1967).

II. LINEAR FREE ENERGY RELATIONSHIPS

A. Introduction (1,5,6)

The linear free energy principle is based on linear correlations between the logarithm of the rate (or equilibrium constant) for one reaction and that for a second reaction run with the same variations in reactant structure or reaction conditions. As Wells has reviewed the subject (6), such relationships may correlate changes in the reaction medium (such as the Winstein-Grunwald equation; see p. 22 of this book), changes in the structure of substrate (Hammett and Taft equations), or changes in the structure of the reagent (Bronsted catalysis law; Swain-Scott and Edwards equations).

All such correlations take the following form:

$$\log k_A = m \log k_B + C$$

where k might be a rate or equilibrium constant, m is the slope, and C is the intercept of the straight line. Obviously, the relationships can be expressed in terms of standard free energy change (for an equilibrium situation) or free energy of activation (for a kinetic situation).

One of the oldest applications of this principle is the Bronsted catalysis laws:

$$\text{(acid catalysis)} \quad \log k_a = \alpha \log K_a + \log G_a$$

$$\text{(base catalysis)} \quad \log k_b = \beta \log K_b + \log G_b$$

The rate constants k_a and k_b are known as catalytic constants for general acid and general base catalyzed reactions, using acid or base of ionization constant K_a or K_b, respectively. The Bronsted coefficients, α and β, are nearly always less than 1.0, and like G_a and G_b are constants characteristic of the reaction series and conditions. These relationships hold only when the acid (or base) catalysts are of the same type (such as a series of aliphatic carboxylic acids), and break down when the structures are altered drastically.

The value of α is taken as a measure of the position of the transition state along the reaction coordinate (in a proton transfer reaction); values are expected to lie between 0 and 1. However, deviant α-values (negative or >1) may occur when some new effect enters the transition state that is not present in the reference proton transfer reaction. For recent data, see F. Bordwell, et al., J. Amer. Chem. Soc., 91, 4002 (1969).

B. The Hammett Equation

This application of the linear free energy principle involves the examination of the behavior of molecules whose structures are arbitrarily divided into a reacting center, X, and a nonreacting residue containing structural elements that influence a rate or equilibrium reaction at X. Changes in the nature of substituents on an aromatic ring or aliphatic chain, for example, will be linearly correlated with a change in rate or equilibrium constant at X.

The Hammett equation is the best known and most widely studied linear free energy relationship:

$$\log \frac{k}{k_o} = \sigma\rho$$

where k is the rate (or equilibrium) constant of the reaction of a substituted compound, k_o is the value for the unsubstituted (or standard) substance, σ is a constant characteristic of the substituent, and ρ is a constant for the particular reaction and is a measure of the sensitivity of the reaction to substituent changes. The equation originally was based on the ionization of substituted benzoic acids in water at 25°, for which $\rho = 1.00$ and k_o is the equilibrium constant for the unsubstituted acid. Separate σ-values are defined by this reaction for meta and para substituents and provide a measure of the total electronic influence (polar, inductive, field effects) in the absence of conjugation effects; in addition, they are not valid for substituents ortho to the reaction center due to anomalous (mainly steric) effects. Some recent studies on ortho-substituent effects have been reported (25). The following σ-values are taken from references 1, 2, and 3.

1. Ordinary Hammett Substituent Constants

Substituent	σ-meta	Uncertainty Limit	σ-para	Uncertainty Limit
H	0		0	
CH_3	-0.069	0.02	-0.170	0.02
CH_2CH_3	-0.07	0.1	-0.151	0.02
$CH_2CH_2CH_3$	-0.05		-0.151	0.02
$CH(CH_3)_2$	-0.10	0.03	-0.197	0.02
$CH_2CH(CH_3)_2$			-0.12	
$C(CH_3)_3$	-0.10	0.03	-0.197	0.02
$CH_2CH_2CH(CH_3)_2$			-0.23	
$C(CH_3)_2CH_2CH_3$			-0.12	
C_6H_5	0.06	0.05	-0.01	0.05
C_6F_5	-0.12		-0.03	

(Continued)

Refs. p. 154.

Substituent	σ-meta	Uncertainty Limit	σ-para	Uncertainty Limit
$CH=CHC_6H_5$	0.14			
$CH=CHNO_2$	0.34	0.03	0.26	0.03
$CH_2C_6H_5$	0.34		0.46	
CN	0.56	0.05	0.660	0.02
CH_2CN			0.01	
CHO	0.36		0.22	
$COCH_3$	0.376	0.02	0.502	0.02
$COCF_3$	0.65			
$CONH_2$	0.28		0.36	
CO_2H	0.37	0.1	0.45	0.1
CO_2^-	-0.1	0.1	0.0	0.1
CO_2CH_3	0.32		0.39	
$CO_2CH_2CH_3$	0.37	0.1	0.45	0.1
CF_3	0.43	0.1	0.54	0.1
CH_2Cl			0.18	
$CH_2Si(CH_3)_3$	-0.19		-0.22	
N_2^+	1.76	0.2	1.91	0.2
NH_2	-0.16	0.1	-0.66	0.1
$NHCH_3$	-0.30		-0.84	0.1
$NHCH_2CH_3$	-0.24		-0.61	
$N(CH_3)_2$	-0.05		-0.83	0.1
N^+H_3	1.13		1.70	
$N^+H_2CH_3$	0.96			
$N^+(CH_3)_3$	0.88	>0.2	0.82	>0.2
$NHCOCH_3$	0.21	0.1	0.00	0.1
$NHNH_2$	-0.02		-0.55	
NHOH	-0.04		-0.34	
$N=N-C_6H_5$			0.64	
NO			0.12	
NO_2	0.710	0.02	0.778	0.02
O^-	-0.71		-0.52	
OH	0.121	0.02	-0.37	0.04
OCH_3	0.115	0.02	-0.268	0.02
OCH_2CH_3	0.1	0.1	-0.24	0.1
$OCH_2C_6H_5$			-0.42	
OC_6H_5	0.252	0.02	-0.320	0.02
$OCOCH_3$	0.39	0.1	0.31	0.1
OCF_3	0.40		0.35	
F	0.337	0.02	0.062	0.02
Cl	0.373	0.02	0.227	0.02
Br	0.391	0.02	0.232	0.02
I	0.352	0.02	0.18	0.1
IO_2	0.70	0.1	0.76	0.1
AsO_3H^-			-0.02	>0.1
$B(OH)_2$	0.01		0.45	
$Ge(CH_3)_3$			0.0	0.1
$Ge(CH_2CH_3)_3$			0.0	0.1
PO_3H^-	0.2	>0.1	0.26	>0.1
SH	0.25	0.1	0.15	0.1
SCH_3	0.15	0.1	0.00	0.1

Refs. p. 154.

Substituent	σ-meta	Uncertainty Limit	σ-para	Uncertainty Limit
$S^{+}(CH_3)_2$	1.0	>0.1	0.9	>0.1
SCN			0.52	0.1
$SCOCH_3$	0.39	0.1	0.44	0.1
$SOCH_3$	0.52	0.1	0.49	0.1
SCF_3	0.40		0.50	
SO_2CH_3	0.56		0.68	
SO_2CF_3	0.79		0.93	
SO_2NH_2	0.55	0.1	0.62	0.1
SO_3^-	0.05	>0.1	0.09	>0.1
$SeCH_3$	0.1	0.1	0.0	0.1
$Si(CH_3)_3$	-0.04	0.1	-0.07	0.1
$Si(CH_2CH_3)_3$			0.0	0.1
$Sn(CH_3)_3$			0.0	0.1
$Sn(CH_2CH_3)_3$			0.0	0.1

2. ρ-Values for Selected Reactions

A few selected examples of ρ-values from the hundreds published (4-6) are listed below. From these examples it is obvious that extrapolation from one reaction to another is risky. For (+) ρ, the rate or equilibrium constant increases with increasing σ, reflecting a reaction that is enhanced by withdrawal of electron density from the reaction site; the opposite holds for a (-) ρ-value. A ρ-value of zero means that the reaction is totally unaffected by the nature of the substituent. There are many examples of nonlinear Hammett plots, which arise mainly in deviations from an otherwise constant mechanism for the series of compounds; such deviations most often develop by a change in rate-determining step or by a change in transition state (4,6,26).

It should be noted that the substituent constants and ρ-values discussed here apply to ionic reactions. Results with free radical reactions are often poor. Alternative equations have been proposed containing the usual σ-values plus a correction term that accounts for special resonance effects (21).

Reaction	Solvent	Temperature	ρ
Equilibria			
Standard Reaction by Definition			
$ArCO_2H + H_2O \rightleftarrows ArCO_2^- + H_3O^+$	H_2O	25°	1.000
Other Acid Ionization Reactions			
2-Methylbenzoic acids	H_2O	25°	1.430
2-Nitrobenzoic acids	H_2O	25°	0.905
2-Hydroxybenzoic acids	H_2O	25°	1.103
2-Chlorobenzoic acids	H_2O	25°	0.855

(Continued)

Refs. p. 154.

Reaction	Solvent	Temp.	ρ
$ArCH_2CO_2H$	H_2O	25°	0.489
$ArCH=CHCO_2H$	H_2O	25°	0.466
$ArCH_2CH_2CO_2H$	H_2O	25°	0.212
ArOH	H_2O	25°	2.113
ArSH	48% EtOH	20-22°	2.236
$ArNH_3^+$	H_2O	25°	2.767
$ArCH_2NH_3^+$	H_2O	25°	0.723
Other Equilibria			
$ArNH_2 + HCO_2H \rightleftarrows HCONHAr + H_2O$	67% C_5H_5N	100°	-1.429
$ArCHO + HCN \rightleftarrows ArCHOHCN$	95% EtOH	20°	-1.492
$Ar_3CCl \rightleftarrows Ar_3C^+ + Cl^-$	SO_2	0°	-3.974
Reaction Rates			
$ArCO_2H + CH_3OH + H^+ \rightarrow ArCO_2CH_3 + H_3O^+$	CH_3OH	25°	-0.229
$ArCO_2H + C_5H_{11}OH + H^+ \rightarrow ArCO_2C_5H_{11} + H_3O^+$	Cyclohexanol	55°	0.555
$ArCO_2C_2H_5 + OH^- \rightarrow ArCO_2^- + CH_3OH$	60% $(CH_3)_2CO$	25°	2.265
$ArCO_2C_2H_5 + H_3O^+ \rightarrow ArCO_2H + C_2H_5OH_2^+$	60% $(CH_3)_2CO$	25°	2.146
$ArCOCl + H_2O \rightarrow ArCO_2H + HCl$	50% $(CH_3)_2CO$	0°	0.797
$ArCOCl + CH_3OH \rightarrow ArCO_2CH_3 + HCl$	CH_3OH	0°	1.469
$ArCONH_2 + OH^- \rightarrow ArCO_2^- + NH_3$	H_2O	100°	1.068
$ArCONH_2 + H_3O^+ \rightarrow ArCO_2H + NH_4^+$	H_2O	100°	0.119
$(ArCO)_2O + H_2O \rightarrow ArCO_2H$	75% dioxane	58°	1.568
$ArO^- + C_2H_5I \rightarrow ArOC_2H_5 + I^-$	EtOH	42.5°	-0.994
$ArO^- + (CH_2)_2O \rightarrow ArOCH_2CH_2OH$	98% EtOH	70.4°	-0.947
$ArCH_2Cl + OH^-(0.0506N) \rightarrow ArCH_2OH + Cl^-$	H_2O	30°	-0.333
$ArCH_2Cl + H_2O \rightarrow ArCH_2OH + HCl$	50% $(CH_3)_2CO$	30.4°	-1.816
$ArCH_2Cl + I^- \rightarrow ArCH_2I + Cl^-$	$(CH_3)_2CO$	20°	0.786
$Ar_2CHCHCl_2 + NaOH \rightarrow Ar_2C=CHCl$	92.5% EtOH	20.1°	2.456
<u>cis</u> - $ArN=N-C_6H_5 \rightarrow$ <u>trans</u>	C_6H_6	25°	-0.610
$ArNO_2 + H_2 + Rh \rightarrow ArNH_2$	75% EtOH	22°	0.247
$2\ ArCHO \rightarrow ArCH_2OH + ArCO_2H$	50% MeOH	40°	3.633
$(ArCO_2)_2 \rightarrow 2\ ArCO_2\cdot$	$CH_3COC_6H_5$	80°	-0.201
$ArCH_3$ + N-bromosuccinimide $\rightarrow ArCH_2Br$	CCl_4	80°	-1.806
$ArCH_3 + Br_2 \xrightarrow{h\nu} ArCH_2Br$	CCl_4	80°	-1.369
$ArH + C_6H_5\cdot \rightarrow ArC_6H_5$	ArH	80°	0.675
$ArH + NO_2^+ \rightarrow ArNO_2$	$(CH_3CO)_2O$	18°	-5.926

C. Modified Hammett and Taft Equations

There are several situations in which ordinary Hammett σ-values do not correlate well with rates or equilibria. Principally, these occur when a substituent is capable of direct conjugation ("through-resonance") with the reaction center (from para position particularly); when the system in question is nonbenzenoid, say aliphatic; and when substituents interact sterically with each other or with the reaction center. For example, the Okamoto-Brown equation [$\log(k/k_o) = \rho\sigma^+$] is used for reactions involving generation of a positive charge in a position capable of direct

Refs. p. 154.

conjugation with a substituent; the standard reaction was solvolysis of cumyl chlorides [$XC_6H_4C(Cl)(CH_3)_2$] at 25° in 90% acetone (aq.). The σ^+ values were established by first using several ordinary σ-meta values to obtain ρ (since resonance interactions are unlikely from the meta position). There are several such substituents constant scales, the most important of which are summarized in the following table.

1. Definitions of Various σ-Value Scales

Symbol	Definition
σ	Ordinary Hammett substituent constant (see above).
σ^n	"Normal" values derived by van Bekkum, Verkade, and Wepster (19) from series of reactions where substituent constants were "well behaved" (correlated very well) and where direct conjugation was considered very unlikely. These values are compared to $\sigma°$, below.
$\sigma°$	Derived from reactions where the aromatic ring is shielded from the reaction center (e.g., phenylacetic acids) or where substituent is in meta position only. Taft (20) proposed that these values are more representative of non-conjugation effects than σ^n values.
σ^+	Substituent participates in direct conjugation with reaction center in an electron demanding transition state (13). See the text above.
σ^-	Analogous to σ^+, but for an electron-rich transition state reaction (14). For example, basicity of phenoxide ions.
σ'	Values derived for substituents in the 4-position (bridgehead carbon) of bicyclo[2.2.2]octane-1-carboxylic acids and their ethyl esters [$\sigma' = (1/1.464)\log(K/K_o)$], where K_o is ionization of acid in 50% aq. ethanol at 25°; the divisor (1.464) is the value of ρ for ionization of benzoic acids under the same conditions, and is included to facilitate comparison of σ and σ' values.
σ''	Values derived for trans-4-substituted cyclohexanecarboxylic acids and esters.
F and R	Swain and Lupton (9) proposed separate field and resonance values, based on the proposition that any σ is a linear combination of these two contributions: $\sigma = fF + rR$. The values of f and r are characteristic of the reaction type; for example $\sigma_p^+ = 0.51\,F + 1.00R$.

Refs. p. 154.

2. The Taft Equation

For situations involving direct steric interaction of substituent and reaction site, Taft (15) has shown that polar and steric factors can be separated. Using the hydrolysis of orthosubstituted benzoate esters as a model reaction, and assuming the relative $\Delta G^{\dagger}$ for the reactions involves the sum of independent factors

$$\log \frac{k}{k_o} = \rho^*\sigma^* - SE_s$$

where σ^* is known as the polar substituent parameter, ρ^* is a measure of sensitivity to polar effects, E_s is the steric effect parameter, and S is a measure of sensitivity to steric effects. Since the steric effect is expected to be about the same for both acid- and base-catalyzed ester hydrolyses, and the polar effect of the acid-catalyzed reaction is very small ($\rho_A^* \approx 0$), then using $\rho_B^* = 2.48$ from the usual Hammett expression (para and meta substituents only), the σ^* values should be on about the same scale as Hammett's σ-values. By definition, $\sigma^* = (1/2.48)[\log(k/k_o)_{Base} - \log(k/k_o)_{Acid}]$. From the acid-catalyzed reaction, $SE_s = \log(k/k_o)$; setting S = 1 for the standard reaction allows determination of E_s. The σ^* and E_s values are also applicable to aliphatic systems, using CH_3CO_2R as standard.

Furthermore, the close correlation found between σ^* and σ' led to another scale, σ_I, a measure of polar effects exclusive of resonance:

$$\sigma_I = 0.45\ \sigma^* \simeq \sigma'$$

Chemical shifts in the nmr spectra of m-substituted fluorobenzenes correlate with σ_I better than with σ (27). For limitations of E_s parameters in groups exhibiting conformational preferences see ref. 29.

3. Values of Various Substituent Parameters

The comprehensive table on pages 152 and 153 summarizes data taken from many sources, especially references 6-12. Where small discrepancies were noted, the most recently reported value was recorded. In the case of large discrepancies, the value was omitted unless two or more recent reports agreed.

D. Nucleophilic Reactivity (23)

1. The Swain-Scott Equation

Swain and Scott (16,17) have correlated the reactivities of several nucleophiles using the equation

$$\log k/k_o = sn$$

where s = 1.00 for reactions of methyl bromide (in any given solvent)

Refs. p. 154.

and the value of n = 0.00 for water. Values of n for a number of nucleophiles are given below (6,7,23,24). Correlations are good for reactions for which the methyl bromide reaction is a good model. Very recently, new scales of solvent nucleophilicity have been developed for use with Swain-Scott type relationships (28).

Reagent	n	Reagent	n	Reagent	n
ClO_3^-, ClO_4^-	<0	$CH_3CO_2^-$	2.7	HO^-	4.2
BrO_3^-, IO_3^-	<0	Cl^-	2.7	SCN^-	4.4
H_2O	0.00	HCO_2^-	2.75	$C_6H_5NH_2$	4.5
p-$CH_3C_6H_4SO_3^-$	<1.0	$C_6H_5O^-$	3.5	I^-	5.0
NO_3^-	1.03	Br^-	3.5	HS^-	5.1
Picrate anion	1.9	C_5H_5N	3.6	SO_3^{-2}	5.1
F^-	2.0	HCO_3^-	3.8	CN^-	5.1
$ClCH_2CO_2^-$	2.2	HPO_4^{-2}	3.8	$S_2O_3^{-2}$	5.35
$HOCH_2CO_2^-$	2.5	N_3^-	4.00	$HPSO_3^{-2}$	6.6
SO_4^{-2}	2.5	$(NH_2)_2CS$	4.1		

2. The Edwards Equation (18,22)

Edwards developed a more general equation for nucleophilic reactivity based on the standard electrode potential, E , of the reagent, in the reaction

$$2X^- \rightarrow X_2 + 2e^-$$

and on the pK_a of the conjugate acid of the reagent in water.

$$\log \frac{k}{k_o} = \alpha E_n + \beta H$$

where $E_n = E° + 2.60$, $H = pK_a + 1.74$ and α and β are constants for the particular reaction. A modification of this equation takes into account the polarizability of the reagent:

$$\log \frac{k}{k_o} = aP + bH$$

where P is a polarizability factor defined by $P \equiv \log (R/R_{H_2O})$ in which R is the molar refraction. This equation may be transformed into the other by means of

$$a = 3.60\alpha$$

$$b = \beta + 0.62\alpha$$

$$E = 3.60P + 0.062H$$

Refs. p. 154.

Group	σ_p^n	σ_m^n	σ_p^{+a}	σ_m^+	σ_p^-	σ_p°
CH_3	-0.129	-0.069	-0.256	-0.065		-0.15
CH_2CH_3	-0.117	-0.072	-0.218	-0.064		
$CH_2CH_2CH_3$						
$CH(CH_3)_2$	-0.098	-0.068	-0.275	-0.060		
$C(CH_3)_3$	-0.136	-0.067	-0.275	-0.059		
C_6H_5			-0.085	0.109		
$CH_2C_6H_5$						
$C\equiv N$	0.674	0.613	0.674	0.562	0.89	0.69
CHO					1.13	
$COCH_3$	0.502	0.376	0.567		0.85	0.46
$CONH_2$					0.62	
CO_2H	0.406	0.348	0.472	0.322	0.728	
CO_2^-	0.120	0.069	0.109	-0.028		
CO_2CH_3	0.463	0.321	(0.489)	0.368	0.68	0.46
$CO_2CH_2CH_3$			0.472	0.366	0.68	
CF_3	0.532	0.467	0.582	0.520		
CH_2Cl			(-0.01)	-0.14		
$CH_2Si(CH_3)_3$			-0.234			
N_2^+			1.797		3.2	
NH_2	-0.172	-0.038	-1.111	-0.016		-0.38
$N(CH_3)_2$	-0.172	-0.049	(-1.7)			-0.44
NH_3^+						
$N(CH_3)_3^+$	0.800	0.855	0.636	0.359		
$NHCOCH_3$			-0.249			
NO_2	0.778	0.710	0.740	0.674	1.25	0.82
O^-					-0.81	
OH	-0.178	-0.095	-0.853			
OCH_3	-0.111	-0.076	-0.648	0.047	-0.2	-0.12
OCH_2CH_3			-0.577			
OC_6H_5			-0.899			
$OCOCH_3$			0.178			
F	0.056	0.337	-0.247	0.352	0.02	0.17
Cl	0.238	0.373	0.035	0.399		0.27
Br	0.265	0.391	0.025	0.405	0.26	0.26
I	0.299	0.352	-0.034	0.359		0.27
IO_2			0.716			
PO_3H^-			0.288			
SH			0.019			
SCH_3	0.220	0.225	-0.164	0.158		
$S(CH_3)_2^+$	1.199	1.032	0.660			
$SCOCH_3$			0.431			
$SOCH_3$			0.386			
SO_2CH_3	0.686	0.678	0.747		1.05	
SO_2NH_2			0.608			
SO_2^-			0.121			
$SeCH_3$			-0.109			
$Si(CH_3)_3$	0.011	-0.053	0.093			

[a]Except for the values in parentheses, these data are from reference 9 and are the result of least-squares analysis. [b]For aliphatic substrates. [c]For o-substituted benzoates. [d]In aqueous solution.

Refs. p. 154.

Substituent Parameters (see p. 150)

σ°_{m}	σ^{*}	σ_{I}	F	R	E_{s}	Group
-0.07	0.00	-0.05	-0.052	-0.141	0.000	CH_3
	-0.10	-0.05	-0.065	-0.114	-0.07[b]	CH_2CH_3
	-0.115				-0.36[b]	$CH_2CH_2CH_3$
	-0.19	-0.07			-0.47[b]	$CH(CH_3)_2$
	-0.30		-0.104	-0.138	-1.54[b]	$C(CH_3)_3$
	0.60	0.10	0.139	-0.088	-0.90[c]	C_6H_5
	0.215				-0.38[b]	$CH_2C_6H_5$
0.62		0.56	0.847	0.184		$C\equiv N$
		0.31[d]				CHO
0.34	1.65	0.28[d]	0.534	0.202		$COCH_3$
		0.21				$CONH_2$
0.36			0.552	0.140		CO_2H
		0.05	-0.221	0.124		CO_2^-
0.36	2.00	0.30				CO_2CH_3
		0.30[d]	0.552	0.140		$CO_2CH_2CH_3$
0.42		0.42	0.631	0.186	-1.16[b]	CF_3
	1.05	0.17				CH_2Cl
			-0.229	-0.081		$CH_2Si(CH_3)_3$
			2.760	0.360		N_2^+
-0.14		0.10	0.037	-0.681		NH_2
-0.15		0.10[d]				$N(CH_3)_2$
		0.60[d]				NH_3^+
		0.92[d]	1.460	0.000		$N(CH_3)_3^+$
		0.28[d]	0.470	-0.274		$NHCOCH_3$
0.70		0.63	1.109	0.155	-0.71[b]	NO_2
		0.12[d]				O^-
		0.25	0.487	-0.643		OH
0.13		0.26	0.413	-0.500	0.97[b]	OCH_3
			0.363	-0.444	0.86[b]	OCH_2CH_3
		0.38	0.747	-0.740		OC_6H_5
		0.39[d]		0.679	-0.071[b]	$OCOCH_3$
0.35		0.52	0.708	-0.336	0.49[b]	F
0.35		0.47	0.690	-0.161	0.18[b]	Cl
0.38		0.45	0.727	-0.176	0.01[b]	Br
		0.40	0.672	-0.197	-0.20[b]	I
			1.098	0.144		IO_2
			0.288	0.098		PO_3H^-
		0.23	0.464	-0.111		SH
0.13		0.19	0.332	-0.186		SCH_3
			1.687	-0.042		$S(CH_3)_2^+$
			0.602	0.102		$SCOCH_3$
		0.56[d]	0.860	0.007		$SOCH_3$
0.66		0.62	0.900	0.215		SO_2CH_3
			0.679	0.188		SO_2NH_2
			0.057	0.058		SO_2^-
			0.221	-0.124		$SeCH_3$
		-0.11	-0.047	-0.044		$Si(CH_3)_3$

Refs. p. 154.

The following table lists constants for use in the Edwards equations. Values were taken from references 6 and 23; those in parentheses are in some doubt. See page 151 for discussion.

Group	E_n	H	P	R
$(NH_2)_2CS$	2.18	0.08		
CN^-	2.79	10.88	0.373	8.66
CNS^-	1.83	(1.00)		
$ClCH_2CO_2^-$	0.79	4.54		
$C_2H_5SO_2S^-$	2.06	(-5.00)		
$H_2N(CH_2)_2NH_2$		11.7	0.696	-18.22
$HOCH_2CH_2S^-$	(2.9)	11.2		
$(CH_3)_2NOH$		6.9	0.647	16.3
C_5H_5N	1.20	7.04		
$C_6H_5O^-$	1.46	11.74		
$C_6H_5S^-$	(2.9)	8.1		
$C_6H_5NH_2$	1.78	6.3	0.921	30.6
$(C_2H_5)_3N$		12.46	0.955	33.1
$C_6H_5N(CH_3)_2$		6.95	1.046	40.8
$S^=$	3.08	14.66	0.611	15.0
HS^-	2.60	8.70		
ClO_4^-	<0	(-9.0)	-0.046	3.3
F^-	-0.27	4.9	-0.150	2.6
Cl^-	1.24	(-3.00)	0.389	9.0
Br^-	1.51	(-6.0)	0.539	12.7
I^-	2.06	(-9.0)	0.718	19.2
HO^-	1.65	17.48	0.143	5.1
H_2O	0.0	0.0	0.0	3.67
NH_3	1.36	11.22	0.184	5.61
NH_2OH		7.7	0.292	7.19
NO_2^-	1.73	5.1		
NO_3^-	0.29	(0.40)	0.0	3.67
NH_2NH_2		9.8	0.381	8.82
N_3^-	1.58	6.46	0.524	12.27
$SO_3^=$	2.57	9.0		
$SO_4^=$	0.59	3.74	0.0035	3.70
$S_2O_3^=$	2.52	3.60		
$CO_3^=$	1.1	12.1	0.043	4.05
$CH_3CO_2^-$	0.95	6.46		

E. References

1. L. P. Hammett, Physical Organic Chemistry, 2nd ed., McGraw-Hill, New York, 1970.
2. D. H. McDaniel and H. C. Brown, J. Org. Chem., 23, 420 (1958).
3. (a) G. B. Parlin and D. D. Perrin, Quart. Rev., 1966, 75. (b) J. Clark and D. D. Perrin, Quart. Rev. 1964, 295.
4. J. E. Leffler and E. Grunwald, Rates and Equilibria of Organic Reactions, Wiley, New York, 1963.
5. H. H. Jaffé, Chem. Rev., 53, 191 (1953).

6. P. R. Wells, Chem. Rev., 63, 171 (1963), and Linear Free Energy Relationships, Academic Press, New York, 1968.

7. R. D. Gilliom, Introduction to Physical Organic Chemistry, Addison-Wesley, Reading, Mass., 1970.

8. E. M. Kosower, Physical Organic Chemistry, Wiley, New York, 1968.

9. C. G. Swain and E. C. Lupton, Jr., J. Amer. Chem. Soc., 90, 4328 (1968).

10. J. Shorter, Chem. Brit., 5, 269 (1969).

11. S. Ehrenson, Progr. Phys. Org. Chem., 2, 195 (1964).

12. C. D. Ritchie and W. F. Sager, Progr. Phys. Org. Chem., 2, 323 (1964).

13. H. C. Brown and Y. Okamoto, J. Amer. Chem. Soc., 80, 4979 (1958).

14. L. P. Hammett, Trans. Faraday Soc., 34, 156 (1938).

15. R. W. Taft, Jr., J. Amer. Chem. Soc., 74, 3120 (1952); 75, 4231 (1953).

16. C. G. Swain and C. B. Scott, J. Amer. Chem. Soc., 75, 141 (1953).

17. C. G. Swain, R. B. Mosely, and D. E. Brown, J. Amer. Chem. Soc., 77, 3731 (1955).

18. J. O. Edwards, J. Amer. Chem. Soc., 76, 1541 (1941).

19. H. van Bekkum, P. E. Verkade, and B. M. Wepster, Rec. Trav. Chim., 78, 815 (1959).

20. R. W. Taft, Jr., J. Phys. Chem., 64, 1805 (1960).

21. T. Yamamoto and T. Otsu, Chem and Ind., #19, 787 (1967).

22. J. O. Edwards, J. Amer. Chem. Soc., 76, 1540 (1954), and 78, 1819 (1956).

23. For a concise review, see K. M. Ibne-Rasa, J. Chem. Educ., 44, 89 (1967).

24. J. Hine, Physical Organic Chemistry, McGraw-Hill, New York, 1962.

25. M. Charton, J. Amer. Chem. Soc., 91, 6649 (1969), and other papers in the series.

26. A concise and recent review is offered by J. O. Schneck, J. Chem. Educ., 48, 103 (1971).

27. R. W. Taft, et al., J. Amer. Chem. Soc., 85, 709 (1963).

28. P. E. Peterson and F. J. Waller, ibid., 94, 991 (1972) and T. W. Bentley, F. L. Schadt, and P. v. R. Schleyer, ibid., 94, 993 (1972).

29. A. Babadjamian, M. Chanon, R. Gallo, and J. Metzger, ibid, 95, 3807 (1973).

III. CONFORMATIONAL FREE ENERGY VALUES

The "conformational free energy" (also called A-value) of a substituent in a molecule is defined as the energy difference between a given conformation and the conformation of lowest energy. The most common values are those for axial-equatorial equilibria in cyclohexanes, for which the conformational free energy of X is given in terms of the equilibrium shown:

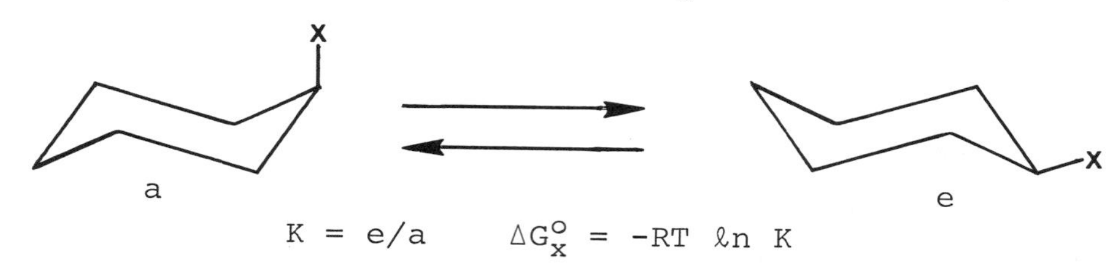

$$K = e/a \qquad \Delta G^{\circ}_{X} = -RT \ln K$$

The energy values are expressed as $-\Delta G°$; since it is determined from concentrations rather than activities, $-\Delta G°$ is solvent-dependent. For a discussion of the effect of solvent "internal pressure" on conformational equilibria, see R. J. Ouellette and S. H. Williams, J. Amer. Chem. Soc., 93, 466 (1971). Values are usually temperature-dependent except in those cases where $\Delta S° \sim 0$. The following key references are the most recent:

"Tabulation and *a priori* Calculations of Conformational Energies," Chapter 7 in Conformational Analysis, E. L. Eliel, N. L. Allinger, S. J. Angyal, and G. A. Morrison, Eds., Interscience, New York, 1966.

"Table of Conformational Energies -- 1967," J. A. Hirsch, Topics in Stereochemistry, 1, 199 (1967).

"Quantitative Conformational Analysis of Cyclohexane Systems," J. Reisse, in *Conformational Analysis. Scope and Present Limitations*, G. Chiurdoglu, Ed., Academic Press, New York, 1971.

"NMR Spectroscopy as an Aid in Conformational Analysis," W. A. Thomas, in *Annual Reports on NMR Spectroscopy*, Vol. 3, E. F. Mooney, Ed., Academic Press, New York, 1970, pp. 113-131.

A similar parameter, known as the R-value, has recently been introduced as a quantitative measure of actual conformation of six-membered rings, both carbocyclic and heterocyclic. The R-value provides a quantitative description of the dihedral arrangements within a fragment XCH_2CH_2Y in X⬡Y, independent of electronegativity of X and Y:

$$R = \frac{J_{aa} + J_{ee}}{J_{ae} + J_{ea}}$$

J = nmr coupling constant between the various axial and equatorial protons in the $-CH_2CH_2-$ fragment. For details, see J. B. Lambert, Accts. Chem. Res., 4, 87 (1971).

Unless otherwise indicated, the $-\Delta G°$ data in the following table are the "best values" recommended by Hirsch (op. cit.) or are the only values available. When two values are given, the first refers to aprotic and the second to protic solvents. Most of the values were determined by nmr spectroscopy and refer to room temperature. Using the nmr method, values are usually derived from the equation, $K = (\delta_a - \delta)/(\delta - \delta_e)$, where δ = nmr chemical shift of some nucleus or nuclei in a rapidly exchanging mixture (e.g., the proton geminal to group X) while δ_a and δ_e are the values for the stereochemically homogeneous axial-X and equatorial-X compounds, respectively. The latter values are usually determined with the stereochemically rigid 4-t-butyl compounds, or by measurements at very low temperature (nonexchange) conditions. Equilibrium constants, K, were calculated from the given $-\Delta G°$ values. Direct conversion of $-\Delta G°$ into percent equatorial isomer is possible with the use of the table on page 158.

Group	$-\Delta G°$ (kcal/mole)	K
F	0.15	1.31
Cl	0.43	2.15
Br	0.38	1.96
I	0.43	2.15
CH_3	1.70[a]	20.5
CF_3	2.1	41.8
C_2H_5	1.75	22.4
$CH(CH_3)_2$	2.15	45.7
$C(CH_3)_3$	>5	>7240
$CH_2C(CH_3)_3$	2.0	35.1
cyclo-C_6H_{11}	2.15[a]	45.7
C_6H_5 (phenyl)	3.0	209
$CH{=}CH_2$	1.35[a]	11.0
$C{\equiv}CH$	0.18[a] (0.41)[f,g]	1.38
CN	0.17	1.35
CO_2H	1.35	11.0
CO_2^-	1.92	30.3
CO_2CH_3	1.27	9.55
$CO_2C_2H_5$	1.20	8.43
$COCH_3$	1.17[b]	8.02
COCl	1.25	17.8
OH[e]	0.52[c], 0.87	2.52, 470
OCH_3	0.60	2.90
OCD_3	0.56	2.71

Group	$-\Delta G°$ (kcal/mole)	K
$OCOCH_3$	0.60	2.90
OTs	0.50	2.43
$OSiMe_3$	0.89[d]	4.86
OSO_2CH_3	0.97[i]	5.06
NH_2	1.20, 1.60	8.43, 17.2
$NHCH_3$	1.0[a]	5.92
$N(CH_3)_2$	2.1[b]	41.8
NH_3^+	1.9[b]	29.3
$NH(CH_3)_2^+$	2.4[b]	71.1
$NHSiMe_3$	1.21[d]	8.59
NO_2	1.10	7.08
-NC	0.21[f,g]	1.74
NCO	0.51[f,g]	3.78
NCS	0.28[f,g]	2.07
$NCNC_6H_{11}$	1.00[g,h]	13.3
SH	0.9	4.96
S^-	1.3[a]	10.1
SCH_3	0.7[a]	3.47
SC_6H_5	0.8[a]	4.14
SOC_6H_5	1.9 (0°)[b]	29.3
$SO_2C_6H_5$	2.5[a]	85.1
HgBr	0 (95°)[a]	1.0
HgCl	0.3[b]	1.71

[a]Measured in aprotic solvent. [b]Measured in protic solvent. [c]Value may be too low. See E. L. Eliel et al., J. Chem. Soc. (D), 360 (1968). [d]Value in CH_2Cl_2. J. P. Hardy and W. D. Cumming, J. Amer. Chem. Soc., 93, 928 (1971). [e]For recent work on the OH group in several solvents, see C. H. Bushweller et al., J. Org. Chem., 35, 2086 (1970). [f]In 90% CS_2 - 10% TMS at -80°. [g]C. H. Bushweller and J. W. O'Neil, J. Org. Chem., 35, 276 (1970). [h]The cyclohexylcarbodiimido group; in CS_2 at -80°. [i]See reference in footnote e.

IV. FREE ENERGY - COMPOSITION CHART

The following table gives calculated values of $\Delta G°$ (= -RT ln K) in kcal/mole for a reaction $A \rightleftharpoons B$; the percent and equilibrium constant (K) headings are in terms of the more stable isomer. For example, a 65/35 mixture at 240°K has $\Delta G°$ = 0.295.

Percent:	55.00	60.00	65.00	70.00	75.00	80.00	85.00
K :	1.22	1.50	1.86	2.33	3.00	4.00	5.67
Temp. °K							
25	0.010	0.020	0.031	0.042	0.055	0.069	0.086
50	0.020	0.040	0.061	0.084	0.109	0.138	0.172
75	0.030	0.060	0.092	0.126	0.164	0.206	0.258
100	0.040	0.081	0.123	0.168	0.218	0.275	0.344
110	0.044	0.089	0.135	0.185	0.240	0.303	0.379
120	0.048	0.097	0.148	0.202	0.262	0.330	0.413
130	0.052	0.105	0.160	0.219	0.284	0.358	0.448
140	0.056	0.113	0.172	0.236	0.305	0.385	0.482
150	0.060	0.121	0.184	0.252	0.327	0.413	0.517
160	0.064	0.129	0.197	0.269	0.349	0.441	0.551
170	0.068	0.137	0.209	0.286	0.371	0.468	0.586
180	0.072	0.145	0.221	0.303	0.393	0.496	0.620
190	0.076	0.153	0.234	0.320	0.415	0.523	0.655
200	0.080	0.161	0.246	0.337	0.436	0.551	0.689
210	0.084	0.169	0.258	0.353	0.458	0.578	0.723
220	0.088	0.177	0.270	0.370	0.480	0.606	0.758
230	0.092	0.185	0.283	0.387	0.502	0.633	0.792
240	0.096	0.193	0.295	0.404	0.524	0.661	0.827
250	0.100	0.201	0.307	0.421	0.545	0.688	0.861
260	0.104	0.209	0.320	0.438	0.567	0.716	0.896
270	0.108	0.217	0.332	0.454	0.589	0.743	0.930
280	0.112	0.225	0.344	0.471	0.611	0.771	0.965
290	0.116	0.234	0.357	0.488	0.633	0.798	0.999
300	0.120	0.242	0.369	0.505	0.655	0.826	1.033
310	0.124	0.250	0.381	0.522	0.676	0.853	1.068
320	0.128	0.258	0.393	0.538	0.698	0.881	1.102
330	0.132	0.266	0.406	0.555	0.720	0.909	1.137
340	0.136	0.274	0.418	0.572	0.742	0.936	1.171
350	0.139	0.282	0.430	0.589	0.764	0.964	1.206
360	0.143	0.290	0.443	0.606	0.785	0.991	1.240
370	0.147	0.298	0.455	0.623	0.807	1.019	1.275
380	0.151	0.306	0.467	0.639	0.829	1.046	1.309
390	0.155	0.314	0.479	0.656	0.851	1.074	1.344
400	0.159	0.322	0.492	0.673	0.873	1.101	1.378
410	0.163	0.330	0.504	0.690	0.895	1.129	1.412
420	0.167	0.338	0.516	0.707	0.916	1.156	1.447
430	0.171	0.346	0.529	0.724	0.938	1.184	1.481
440	0.175	0.354	0.541	0.740	0.960	1.211	1.516
450	0.179	0.362	0.553	0.757	0.982	1.239	1.550
460	0.183	0.370	0.566	0.774	1.004	1.266	1.585
470	0.187	0.378	0.578	0.791	1.025	1.294	1.619
480	0.191	0.387	0.590	0.808	1.047	1.322	1.654

90.00 9.00	95.00 19.00	99.00 99.00	99.50 199.00	99.90 999.00	99.95 1999.00	99.99 9999.00
0.109	0.146	0.228	0.263	0.343	0.377	0.457
0.218	0.292	0.456	0.526	0.686	0.755	0.915
0.327	0.439	0.684	0.788	1.029	1.132	1.372
0.436	0.585	0.913	1.051	1.372	1.509	1.829
0.480	0.643	1.004	1.156	1.509	1.660	2.012
0.524	0.702	1.095	1.262	1.646	1.811	2.195
0.567	0.760	1.186	1.367	1.783	1.962	2.378
0.611	0.819	1.278	1.472	1.920	2.113	2.561
0.655	0.877	1.369	1.577	2.058	2.264	2.744
0.698	0.936	1.460	1.682	2.195	2.415	2.927
0.742	0.994	1.551	1.787	2.332	2.566	3.110
0.785	1.053	1.643	1.892	2.469	2.717	3.292
0.829	1.111	1.734	1.997	2.606	2.868	3.475
0.873	1.170	1.825	2.103	2.743	3.019	3.658
0.916	1.228	1.916	2.208	2.881	3.170	3.841
0.960	1.286	2.008	2.313	3.018	3.321	4.024
1.004	1.345	2.099	2.418	3.155	3.472	4.207
1.047	1.403	2.190	2.523	3.292	3.623	4.390
1.091	1.462	2.281	2.628	3.429	3.774	4.573
1.135	1.520	2.373	2.733	3.566	3.925	4.756
1.178	1.579	2.464	2.838	3.704	4.075	4.939
1.222	1.637	2.555	2.944	3.841	4.226	5.122
1.265	1.696	2.647	3.049	3.978	4.377	5.305
1.309	1.754	2.738	3.154	4.115	4.528	5.487
1.353	1.813	2.829	3.259	4.252	4.679	5.670
1.396	1.871	2.920	3.364	4.389	4.830	5.853
1.440	1.930	3.012	3.469	4.527	4.981	6.036
1.484	1.988	3.103	3.574	4.664	5.132	6.219
1.527	2.047	3.194	3.679	4.801	5.283	6.402
1.571	2.105	3.285	3.785	4.938	5.434	6.585
1.615	2.164	3.377	3.890	5.075	5.585	6.768
1.658	2.222	3.468	3.995	5.212	5.736	6.951
1.702	2.281	3.559	4.100	5.350	5.887	7.134
1.745	2.339	3.650	4.205	5.487	6.038	7.317
1.789	2.398	3.742	4.310	5.624	6.189	7.500
1.833	2.456	3.833	4.415	5.761	6.340	7.682
1.876	2.514	3.924	4.520	5.898	6.491	7.865
1.920	2.573	4.015	4.626	6.035	6.642	8.048
1.964	2.631	4.107	4.731	6.173	6.792	8.231
2.007	2.690	4.198	4.836	6.310	6.943	8.414
2.051	2.748	4.289	4.941	6.447	7.094	8.597
2.095	2.807	4.380	5.046	6.584	7.245	8.780

(Continued)

Percent: K : Temp. °K	55.00 1.22	60.00 1.50	65.00 1.86	70.00 2.33	75.00 3.00	80.00 4.00	85.00 5.67
490	0.195	0.395	0.602	0.825	1.069	1.349	1.688
500	0.199	0.403	0.615	0.841	1.091	1.377	1.722
510	0.203	0.411	0.627	0.858	1.113	1.404	1.757
520	0.207	0.419	0.639	0.875	1.135	1.432	1.791
530	0.211	0.427	0.652	0.892	1.156	1.459	1.826
540	0.215	0.435	0.664	0.909	1.178	1.487	1.860
550	0.219	0.443	0.676	0.926	1.200	1.514	1.895
560	0.223	0.451	0.688	0.942	1.222	1.542	1.929
570	0.227	0.459	0.701	0.959	1.244	1.569	1.964
580	0.231	0.467	0.713	0.976	1.265	1.597	1.998
590	0.235	0.475	0.725	0.993	1.287	1.624	2.033

90.00	95.00	99.00	99.50	99.90	99.95	99.99
9.00	19.00	99.00	199.00	999.00	1999.00	9999.00
2.138	2.865	4.472	5.151	6.721	7.396	8.963
2.182	2.924	4.563	5.256	6.858	7.547	9.146
2.225	2.982	4.654	5.361	6.996	7.698	9.329
2.269	3.041	4.745	5.467	7.133	7.849	9.512
2.313	3.099	4.837	5.572	7.270	8.000	9.695
2.356	3.158	4.928	5.677	7.407	8.151	9.877
2.400	3.216	5.019	5.782	7.544	8.302	10.060
2.444	3.275	5.111	5.887	7.681	8.453	10.243
2.487	3.333	5.202	5.992	7.819	8.604	10.426
2.531	3.392	5.293	6.097	7.956	8.755	10.609
2.575	3.450	5.384	6.202	8.093	8.906	10.792

SPECTROSCOPY

I. THE ELECTROMAGNETIC SPECTRUM

The various regions depicted below are somewhat arbitrary and not very sharply defined; each overlaps at both ends the adjacent regions. The units defining the regions are in frequency (Hz, cm^{-1}), energy (kcal/mole, eV) and wavelength (m and λ, the "official" or most common unit used for a given region). Some useful relationships and constants are the following: 1 nanometer (nm) $\equiv$ 1 millimicron (mμ) = 10^{-9} m; 1 μm (formerly micron) = 10^{-6} m = 10^{-3} mm; 1 angstrom (Å) = 0.1 nm = 10^{-8} cm; 1 eV = 23.06 kcal/mole = 8063 cm^{-1}. Energy, $E = h\nu$ [h = Planck's constant = 9.534×10^{-14} kcal-sec/mole = 1.583×10^{-34} cal-sec/molecule; ν in Hz (cycles/sec)]. E in kcal/mole = $2.8635/\lambda$ (λ in μm); E in eV = $12,345/\lambda$ (λ in Å).

Region	Spectral Use	Hz	cm^{-1}	kcal/mole	eV	m	λ
Cosmic rays							
		10^{22}	$3x10^{11}$	$9.5x10^{8}$	$4.1x10^{7}$	$3x10^{-14}$	$3x10^{-4}$ Å
Gamma rays	Nuclear transitions						
		10^{19}	$3x10^{7}$	$9.5x10^{5}$	$4.1x10^{4}$	$3x10^{-10}$	3 Å
Soft X-rays	Inner electron transitions						
		10^{17}	$3x10^{5}$	$9.5x10^{3}$	$4.1x10^{2}$	$3x10^{-8}$	300 Å
Vacuum uv	Valence electron transitions (electronic spectra)						
		$1.5x10^{15}$	$5x10^{4}$	143	6.2	$2x10^{-7}$	200 nm
Quartz uv							
		$7.5x10^{14}$	$2.5x10^{4}$	71.5	3.1	$4x10^{-7}$	400 nm
Visible							
		$3.8x10^{14}$	$1.2x10^{4}$	36.2	1.6	$8x10^{-7}$	800 nm
Near ir	Vibrational ir overtone and combination bands						
		$1.2x10^{14}$	$4x10^{3}$	12	0.52	$2.5x10^{-6}$	2.5 μm
Vibrational ir	Fundamental region						
		$2.5x10^{13}$	$8x10^{2}$	2.4	0.10	$1.2x10^{-5}$	12.5 μm
Far ir	Skeletal, ring torsional, solid state (lattice) deformations						
		10^{12}	33	$9.5x10^{-2}$	$4.1x10^{-3}$	$3x10^{-4}$	300 μm
Microwave (radar)	Molecular rotations; bond torsions						
		10^{9}	$3.3x10^{-2}$	$9.5x10^{-5}$	$4.1x10^{-6}$	$3x10^{-1}$	300 mm
Short Radio	Spin orientations (nmr, esr, nqr)						
		$1.5x10^{6}$	$5x10^{-5}$	$1.4x10^{-7}$	$6.1x10^{-9}$	$2x10^{2}$	200 m
Broadcast	Radio, tv, etc.						
		$5.5x10^{5}$	$1.8x10^{-5}$	$5.2x10^{-8}$	$2.2x10^{-9}$	$5.6x10^{2}$	550 m
Long Radio	Induction heating; long-wave communication						
		$3x10^{3}$	10^{-7}	$2.9x10^{-10}$	$1.3x10^{-11}$	10^{5}	10^{5} m
Electric Power	Commercial power and light						
		$3x10^{-1}$	10^{-11}	$2.9x10^{-14}$	$1.3x10^{-15}$	10^{9}	10^{9} m

II. SOLVENTS AND OTHER MEDIA FOR SPECTRAL MEASUREMENTS

This section covers most of the common, as well as some uncommon, solvents and media for spectral measurements in the uv-vis and ir regions. Similar information for nmr spectroscopy is found on pp. 249 and 450 and that for esr spectroscopy (use of frozen glasses, e.g.) on page 336. Many actual ir spectra are reproduced in this book, beginning on page 169. There are also available, at no cost, several handy booklets containing spectra of common solvents:

Spectro-Solv Solvents -- uv, near ir, mid ir; Beckman Instruments, Fullerton, Calif. 92634 (Bulletin 810).
IR Spectra of Solvents in the KBr Region; Beckman Instruments (Publication IR-93-MI).
Spectrophotometric Solvents -- uv, mid ir; Eastman Kodak, Distillation Products, Rochester, N. Y. 16403.
Spectroquality Solvents, MC & B, 2909 Highland Avenue, Norwood, Ohio 45212.
26 Frequently Used Prism Spectra for the IR Spectroscopist and Most Useful Standard IR Grating Spectra, Sadtler Research Labs., 3316 Spring Garden Street, Philadelphia, Pa. 19104.

A. Ultraviolet-Visible Absorption and Fluorescence Spectra

The "cut-off" point" in the uv is the wavelength at which the absorbance approaches 1.0 for a thickness (usually 1 or 10 mm) of the noncompensated solvent. To avoid possible faulty spectrum interpretation and poor resolution, a solvent or other medium should not be used for measurements below the cut-off point, even though a compensating (reference) cell is employed. Most of the solvents listed below are available in highly purified "spectrograde;" cut-off values are given for 1 and 10 mm path lengths.

Some solvents for fluorescence measurements often must be of greater purity than spectrograde. Techniques are published for the purification of solvents to "fluorescence grade" [e.g., Brodie et al., J. Biol. Chem., 168, 299 (1947); S. Udenfriend, Fluorescence Assay in Biology and Medicine, Academic Press, New York, 1964, Ch. 4]. Such solvents are also available commercially from Harleco, 60th and Woodland Avenues, Philadelphia, Pa. 19143.

		Cut-off(nm)				Cut-off (nm)	
Substance	Bp	1 mm	10 mm	Substance	Bp	1 mm	10 mm
2-Methylbutane	28		210	Decalin	≃190		200
Pentane	36		200				
Hexane	69	190	200	H_2O	100	187	191
Heptane	96		195	D_2O	101	190	195
i-Octane	98	190	205	Methanol	65	200	205
Cyclopentane	49		210	Ethanol	78	200	205
Methylcyclopentane	72		200	2-Propanol	81	200	210
Cyclohexane	81	190	195	95% Ethanol	78	200	205
Methylcyclohexane	101		205	Glycerol	290d	200	205
Benzene	80	275	280	96% H_2SO_4	≃300		210
Toluene	110	280	285				
m-Xylene	139	285	290	Et_2O	35	205	215

(Continued)

Substance	Bp	Cut-off (nm) 1 mm	10 mm	Substance	Bp	Cut-off (nm) 1 mm	10 mm
THF	67		220	N,N-DMF	152		270
1,4-Dioxan	102	210	220	DMSO	189d	250	265
$(C_4H_9)_2O$	142		210				
CS_2	46		380	CH_3COCH_3	56	320	330
$CHCl_3$	60	230	245	C_2H_5OAc	76	245	255
CCl_4	76	245	260	C_2H_5OCHO	55		260
CH_2Cl_2	40	220	230	CH_3OCHO	32	250	260
$ClCH_2CH_2Cl$	84	220	230	$n-C_4H_9OAc$	125	245	255
$Cl_2C{=}CCl_2$	120	280	290	$C_2H_5CO_2C_2H_5$	99		255
$CHBr_3$	150	315	330				
$BrCCl_3$	105	320	340	KBr[a]	-		200
				KCl[a]	-		200
CH_3CN	81	190	195	AgCl[a]	-		400
CH_3NO_2	100	360	380	TlBr[a]	-		450
Pyridine	112	300	305				

[a]Values for solid, as used in a pellet for example.

B. Infrared Regions

Different solvents and suspension media (for pellets or mulls) are often required for the different regions of the spectrum. These regions are usually defined as follows:

Near ir: 1 to 3 μm(10,000 to 3333 cm^{-1})
Mid ir (NaCl region): 2 to 15 μm (5000 to 667 cm^{-1})
Far ir (KBr region): 12.5 to 25 μm(800 to 400 cm^{-1})
(CsBr region): 15 to 35 μm(667 to 286 cm^{-1})

References to details of sample handling may be found in the sections on vibrational spectra (p. 184) and optical materials (pp. 178-184).
The data summarized below were taken from a large variety of published and unpublished sources. In addition, there are several useful, but relatively little known techniques for sample handling, for which the following references are provided:

Water as an IR Solvent, H. Sternglanz, Appl. Spec., 10, 77 (1956).
Pressed AgCl Discs, L. Pytlewski and V. Marchesani, Anal. Chem., 37, 618 (1965).
Lamination or Impregnation of Samples into Sheets of Mica, Polyethylene or AgCl, J. Sands and G. Turner, Anal. Chem., 24, 791 (1952).
Polyethylene Powder or Sheet for Sample Preparation, K. Schwing and L. May, Anal. Chem., 38, 523 (1966).
Molten Sulfur as a Solvent, T. Wiewiorowski, et al., Anal. Chem., 37, 1080 (1965).

1. Pelletizing and Mulling Media

The usual pellets or disks are prepared under pressure using 1 to 5% sample in the dry, finely ground (<400 mesh) pellet medium; liquid mulls require higher concentrations. The table below shows the transparent regions ($\geq$80% T) for the substances common to these techniques [for a brief review, see A. Nicholson, Anal. Chem., 31, 519 (1959)]. For additional data, see pages 179-182.

	Usable Regions of the Spectrum: cm^{-1} (μm)		
Substance	Near IR[a]	Mid IR	Far IR
KBr	10000-3333(1-3)	5000-667(2-15)	800-250(12.5-40)
KCl	10000-3333(1-3)	5000-667(2-15)	800-526(12.5-19)
CsBr	10000-3333(1-3)	5000-667(2-15)	800-250(12.5-40)
CsI	10000-3333(1-3)	5000-667(2-15)	800-130(12.5-77)
AgCl	10000-3333(1-3)	5000-667(2-15)	800-530(12.5-19)
TlCl	10000-3333(1-3)	5000-667(2-15)	800-530(12.5-19)
Polyethylene	10000-3333(1-3)[b]	2500-1540(4-6.5 1250-741(8-13.5)	625-278(16-36)
Polystyrene		-	400-278(25-36)[c]
Teflon		5000-1333(2-7.5)[d] 1111-690(9-14.5)	
Kel-F		5000-1333(2-7.5) 870-690(11.5-14.5)	
Nujol[e]		5000-3333(2-3) 2500-1540(4-6.5) 1250-667(8-15)	667-286(15-35)
Fluorolube[f]		5000-1430(2-7)	
C_4Cl_6[g]		5000-1667(2-6) 1430-1250(7-8)	

[a]Very little work has been reported for near ir spectra of pellets or mulls; most measurements are made in solution (see below for solvents). [b]For a 0.1 mm sheet when compensated; otherwise, interfering bands occur at 1.9, 2.3, and 2.6-2.8 μm. [c]For 0.025 mm sheet. [d]For 0.01 mm thick specimen; Teflon and Kel-F powders are very good for observing details in the OH stretch region. [e]Common brand of heavy mineral oil; for its actual spectrum, see below. [f]A saturated chlorofluorocarbon oil (Hooker Chemical Co., 2600 47th Street, Niagara Falls, N.Y. 14302). For its actual spectrum, see below. [g]Hexachloro-1,3-butadiene; see spectrum below.

2. Infrared Spectra of Common Solvents and Other Substances

The spectra on the following pages are included not only to assist in the choice of a desired ir medium, but also to provide often needed ir information on commonly used general solvents. The spectra are reproduced, with generous permission, from various collections of Sadtler Research Laboratories. The mid-IR spectra of methylene chloride, tetrahydrofuran, dimethylformamide, and hexachloro-1,3-butadiene were measured on a prism instrument, while the remainder (chloroform, carbon tetrachloride, tetrachloroethylene, carbon disulfide, cyclohexane, benzene, pyridine, diethyl ether, dioxane, acetone, ethyl methyl ketone, ethyl acetate, methanol, dimethyl sulfoxide, nitromethane, Nujol, Fluorolube, silicone stopcock grease) are grating spectra and cover the mid- and far-IR regions.

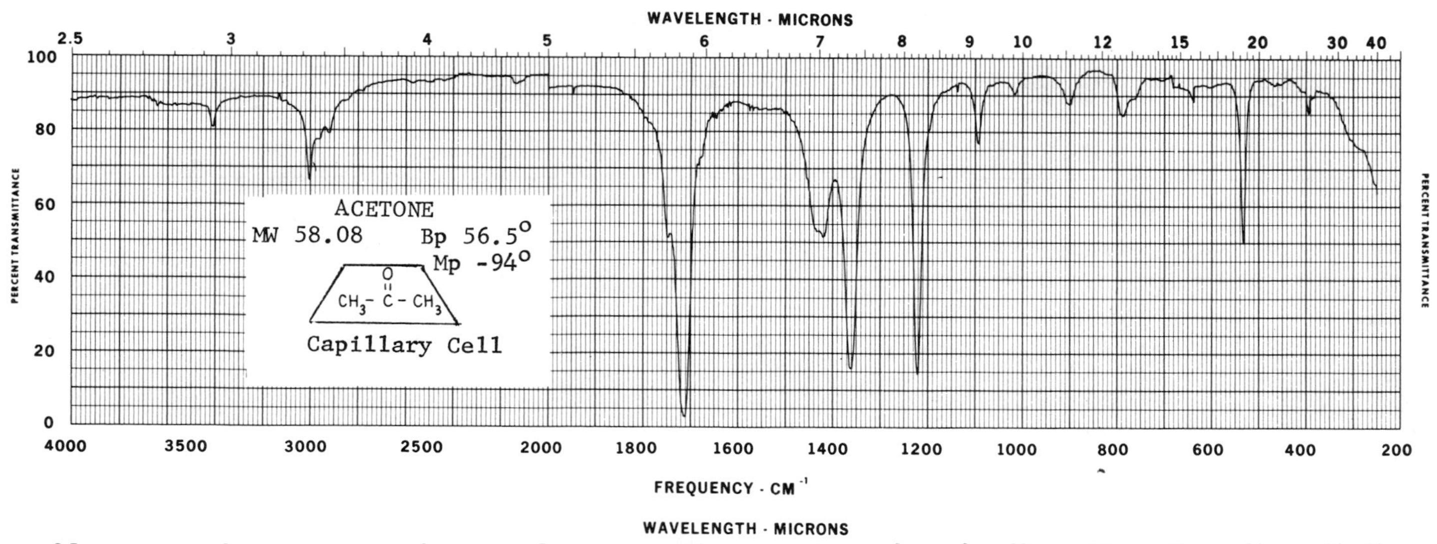

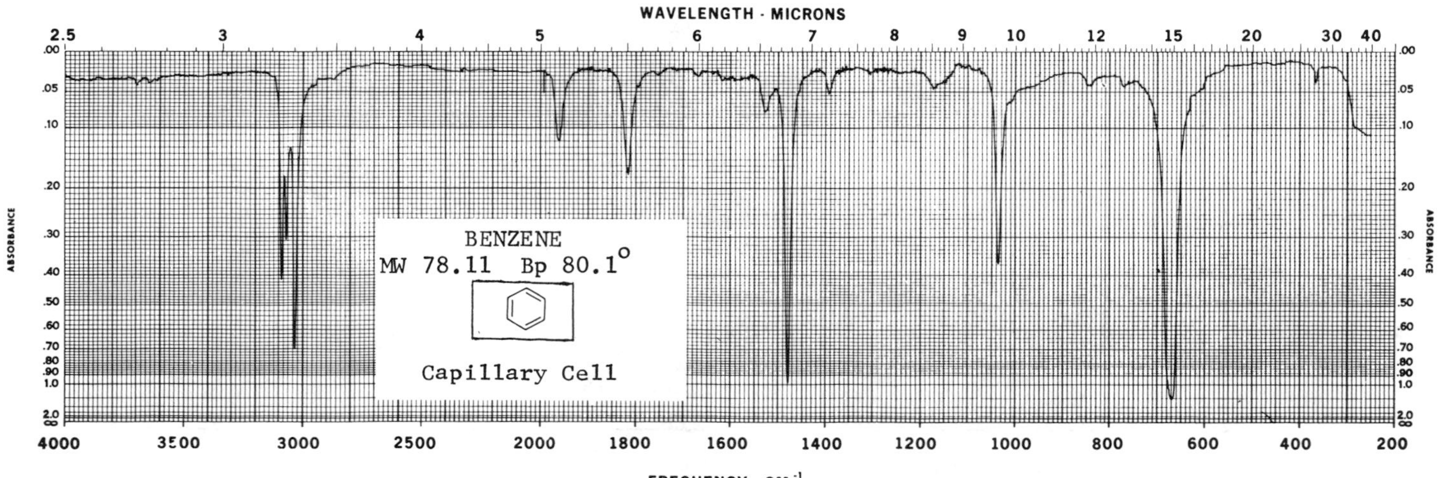

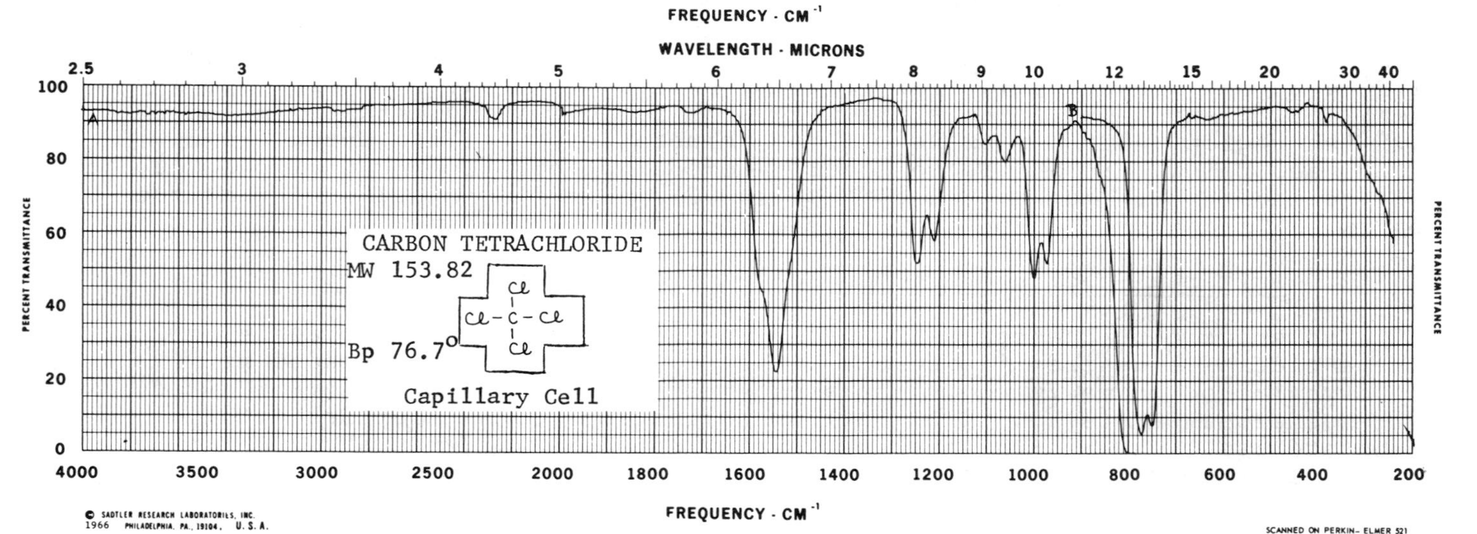

SCANNED ON PERKIN- ELMER 521

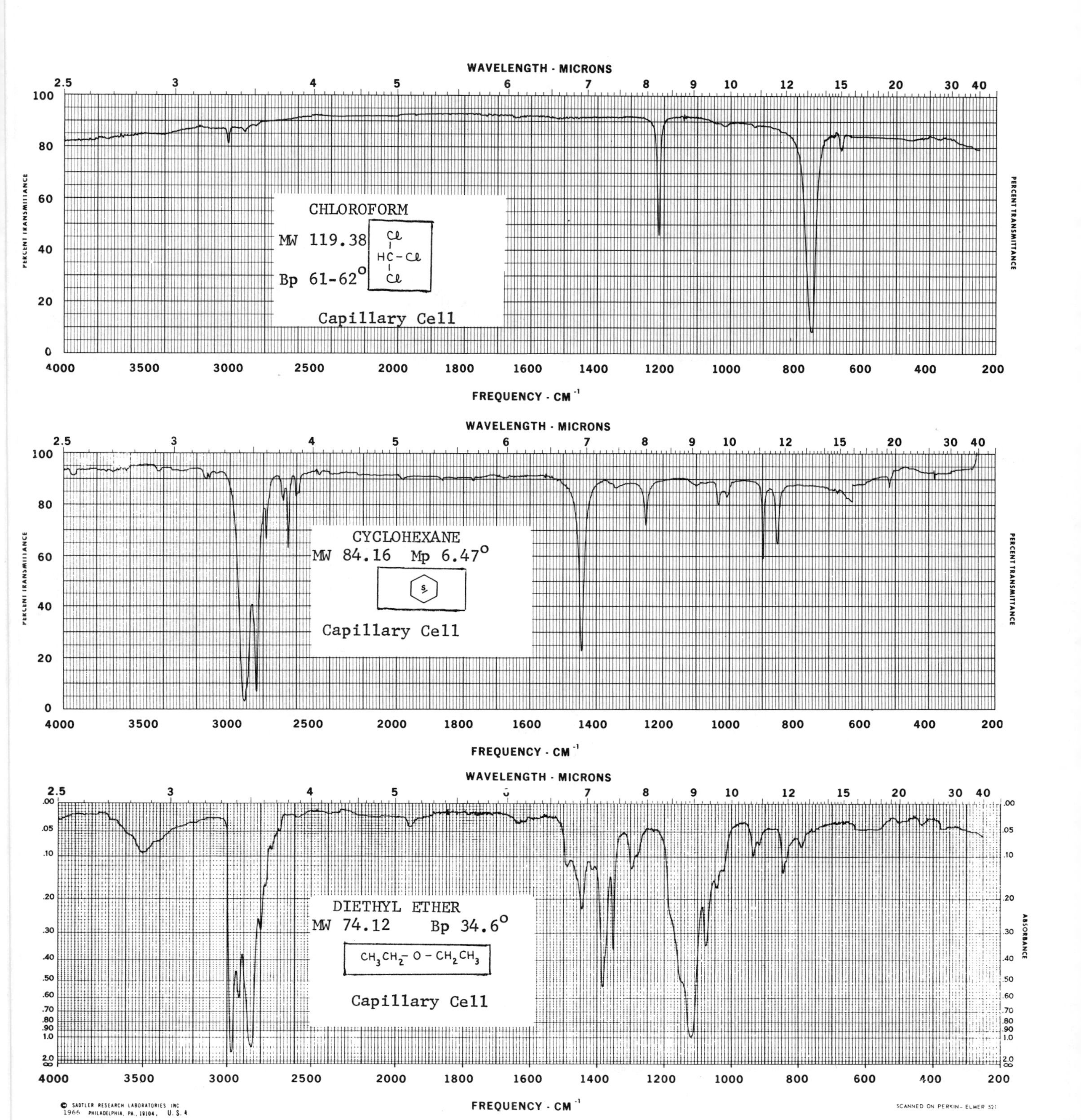

WAVELENGTH - MICRONS
CHLOROFORM
MW 119.38
Bp 61-62°
Capillary Cell
FREQUENCY - CM⁻¹
PERCENT TRANSMITTANCE
WAVELENGTH - MICRONS
CYCLOHEXANE
MW 84.16 Mp 6.47°
Capillary Cell
FREQUENCY - CM⁻¹
PERCENT TRANSMITTANCE
WAVELENGTH - MICRONS
DIETHYL ETHER
MW 74.12 Bp 34.6°
CH3CH2-O-CH2CH3
Capillary Cell
FREQUENCY - CM⁻¹
ABSORBANCE
SCANNED ON PERKIN-ELMER 521

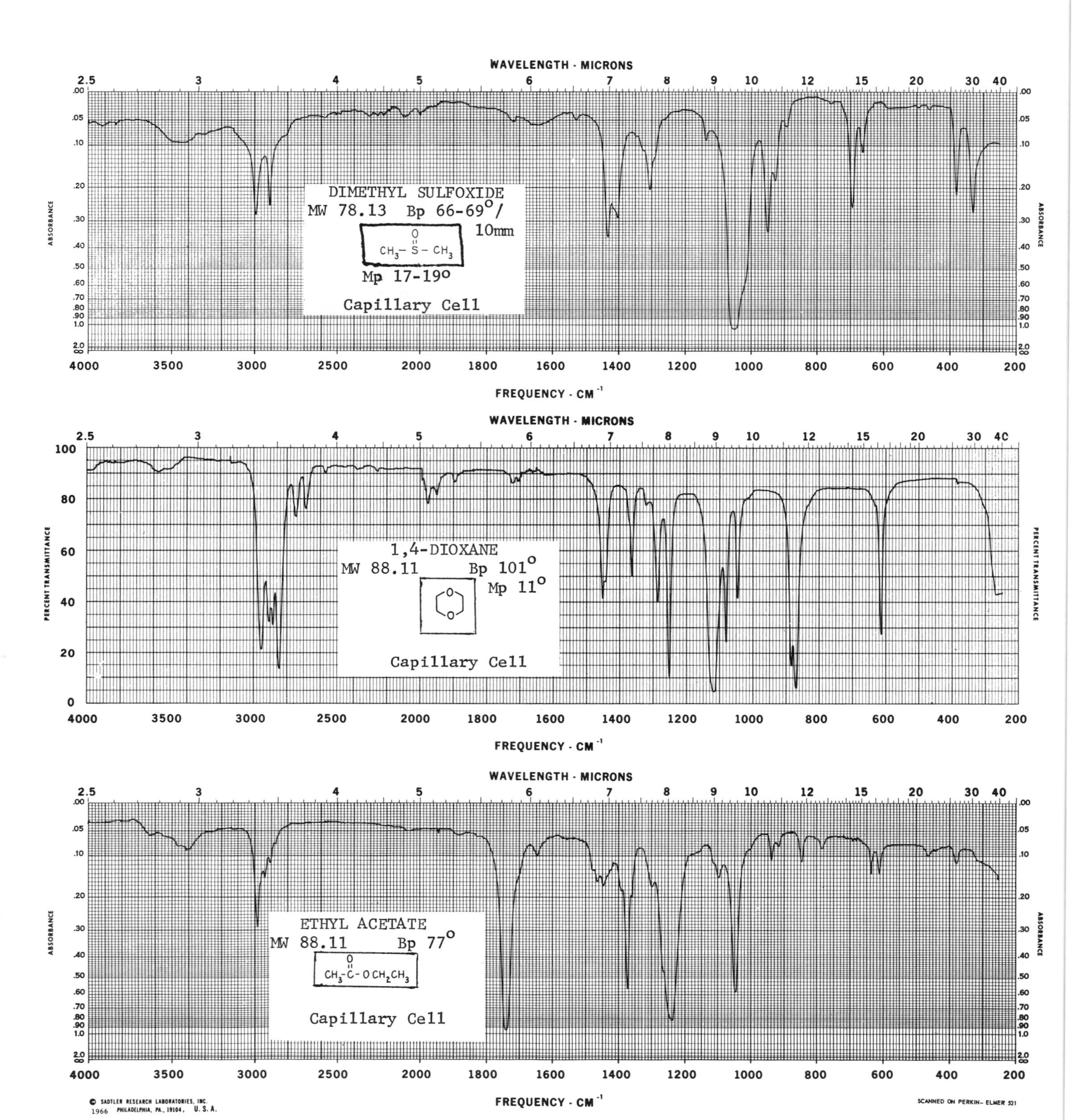
WAVELENGTH - MICRONS
DIMETHYL SULFOXIDE
MW 78.13 Bp 66-69°/10mm
Mp 17-19°
Capillary Cell
FREQUENCY - CM⁻¹
ABSORBANCE
1,4-DIOXANE
MW 88.11 Bp 101°
Mp 11°
Capillary Cell
PERCENT TRANSMITTANCE
ETHYL ACETATE
MW 88.11 Bp 77°
Capillary Cell
© SADTLER RESEARCH LABORATORIES, INC. 1966 PHILADELPHIA, PA., 19104, U. S. A.
SCANNED ON PERKIN-ELMER 521

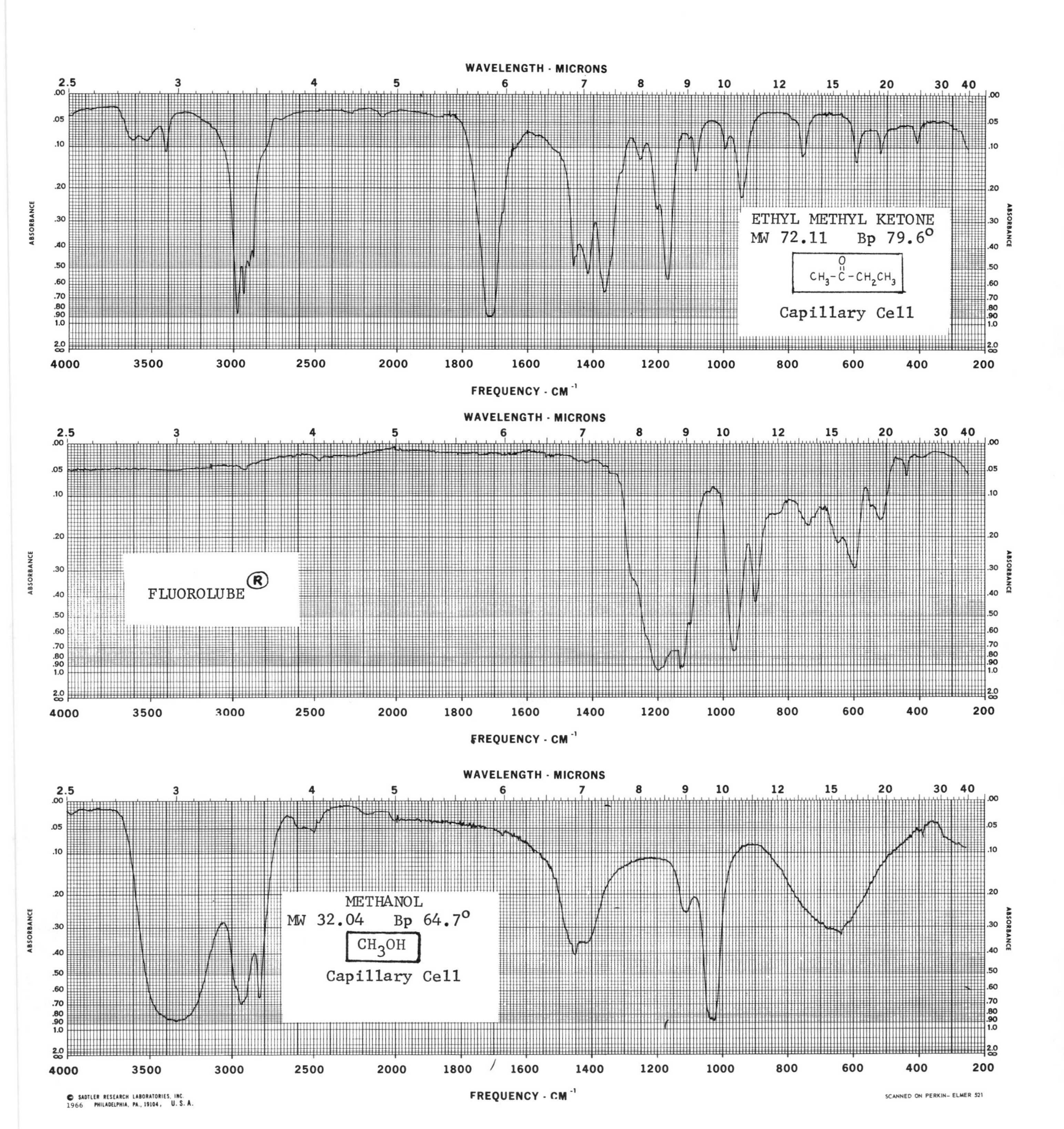
WAVELENGTH - MICRONS
FREQUENCY - CM⁻¹
ABSORBANCE
ETHYL METHYL KETONE
MW 72.11 Bp 79.6°
CH3-C(=O)-CH2CH3
Capillary Cell
FLUOROLUBE®
METHANOL
MW 32.04 Bp 64.7°
CH3OH
Capillary Cell
© SADTLER RESEARCH LABORATORIES, INC.
1966 PHILADELPHIA, PA., 19104, U. S. A.
SCANNED ON PERKIN-ELMER 521

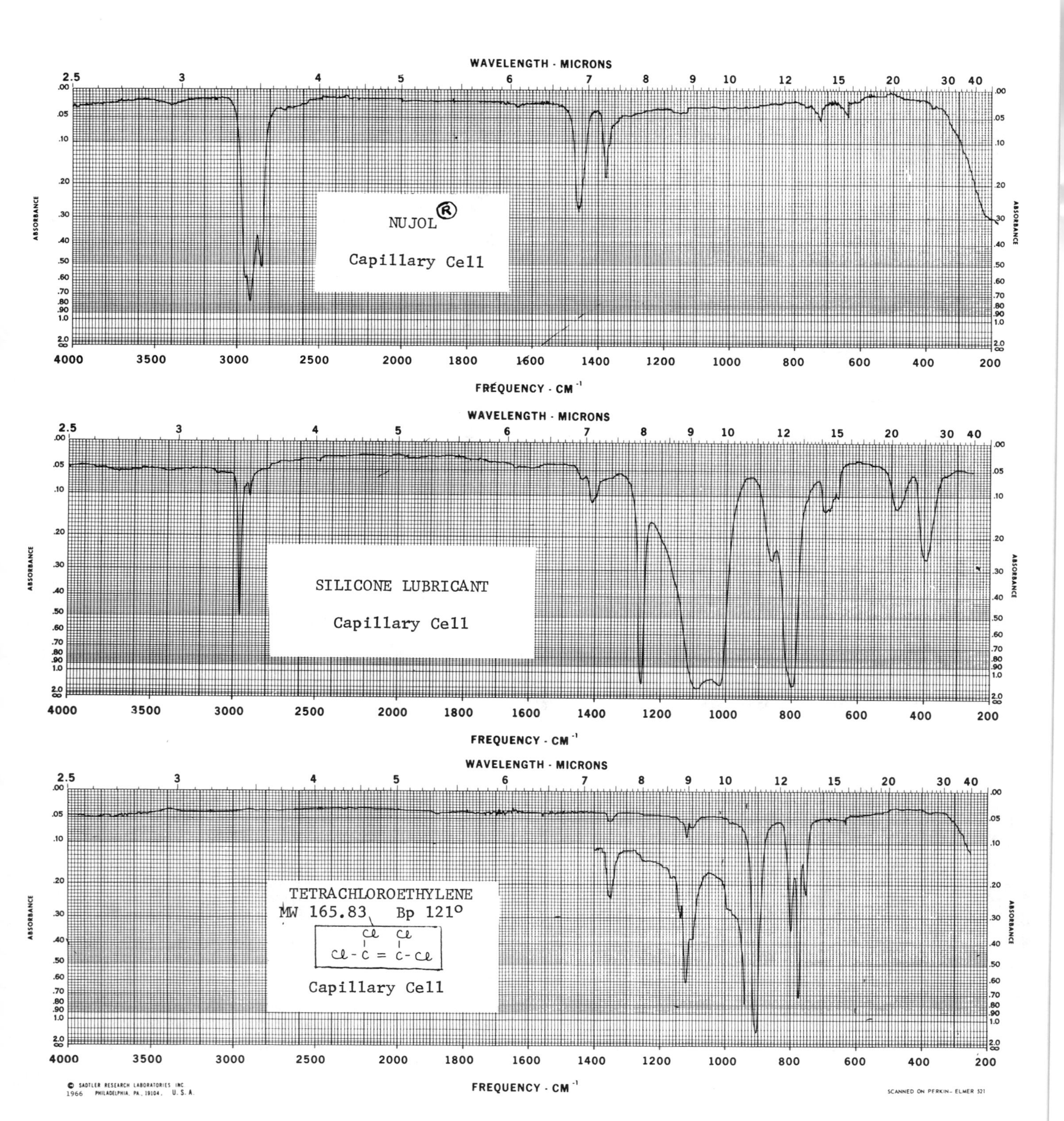
WAVELENGTH - MICRONS
FREQUENCY - CM⁻¹
ABSORBANCE
NUJOL®
Capillary Cell
SILICONE LUBRICANT
Capillary Cell
TETRACHLOROETHYLENE
MW 165.83 Bp 121°
Cl-C = C-Cl
Capillary Cell
SCANNED ON PERKIN-ELMER 521

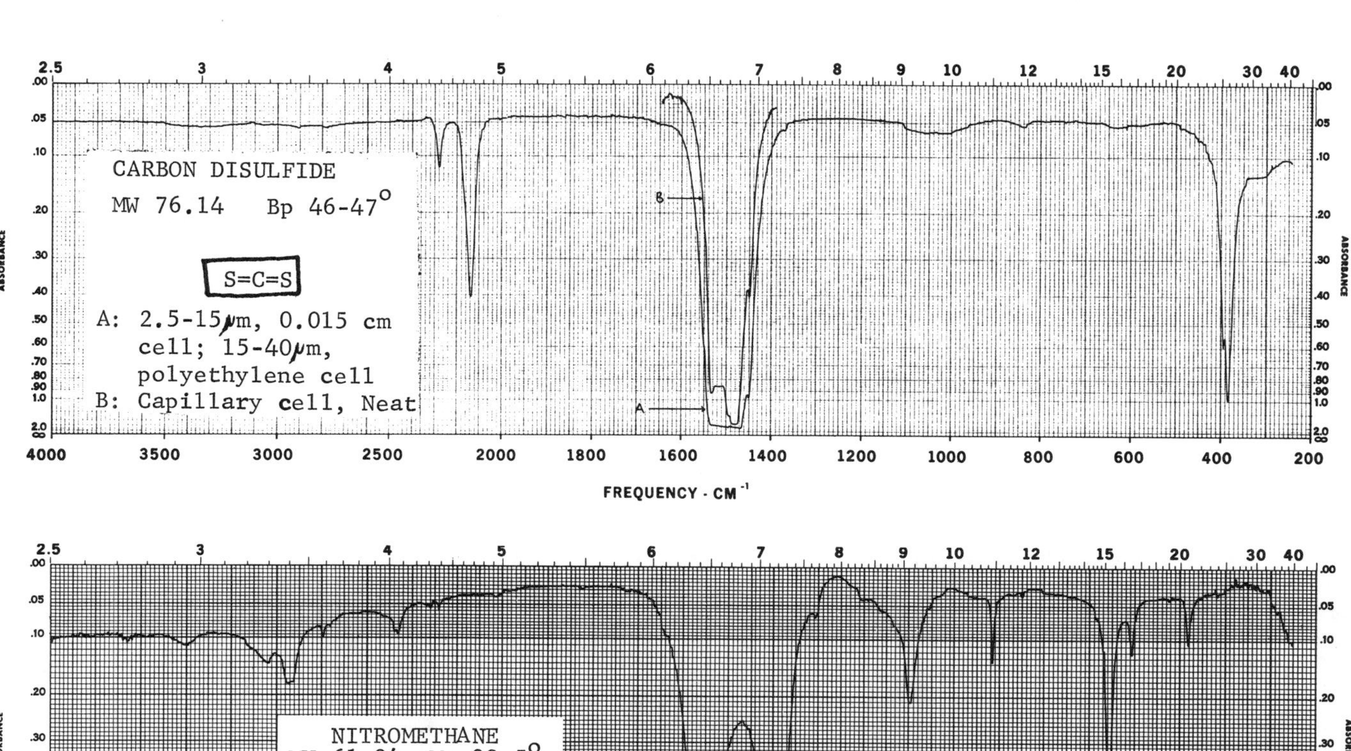
CARBON DISULFIDE
MW 76.14 Bp 46-47°
S=C=S
A: 2.5-15μm, 0.015 cm cell; 15-40μm, polyethylene cell
B: Capillary cell, Neat
FREQUENCY - CM⁻¹
ABSORBANCE

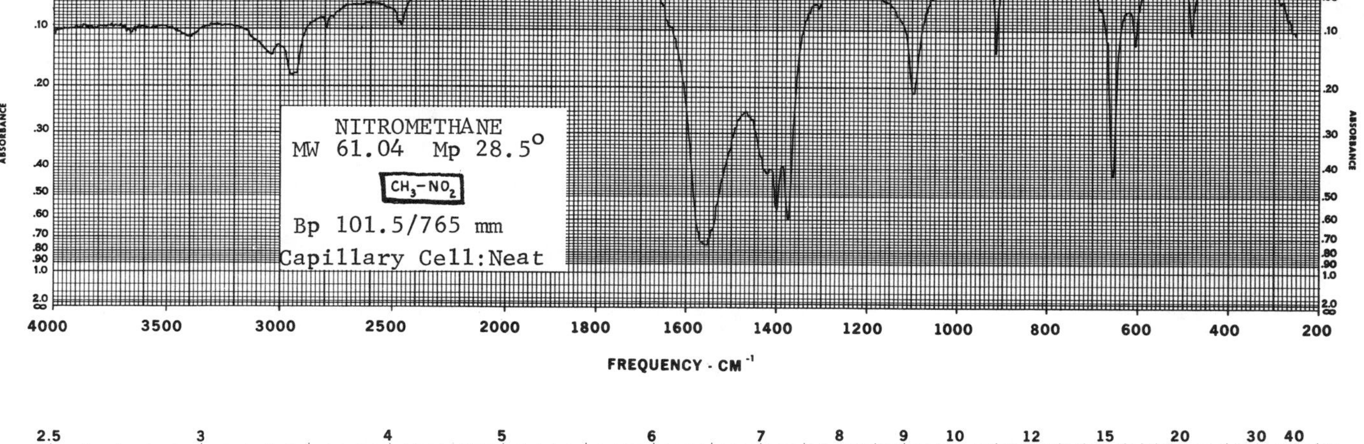
NITROMETHANE
MW 61.04 Mp 28.5°
CH3-NO2
Bp 101.5/765 mm
Capillary Cell:Neat
FREQUENCY - CM⁻¹
ABSORBANCE

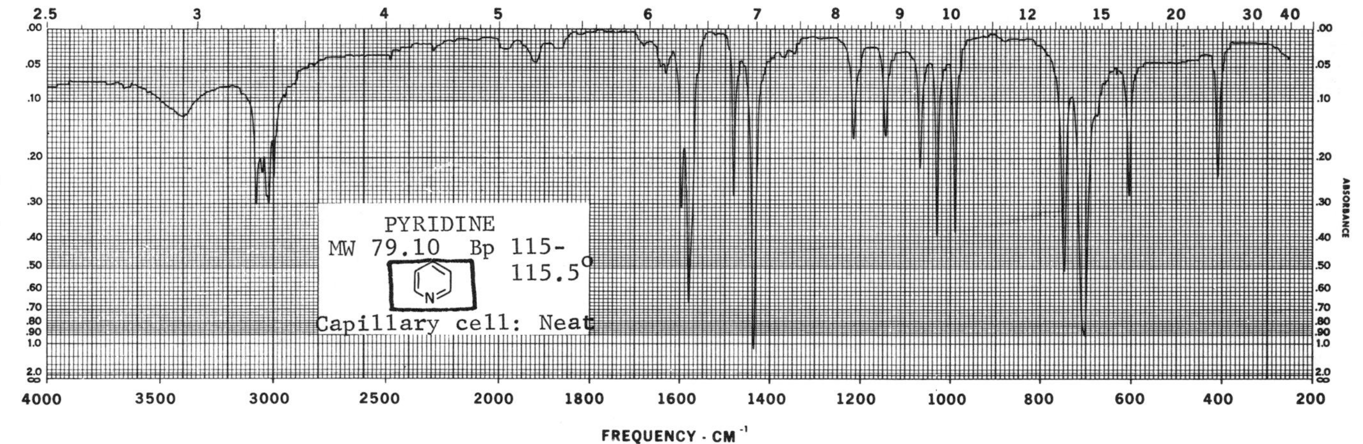
PYRIDINE
MW 79.10 Bp 115-115.5°
Capillary cell: Neat
FREQUENCY - CM⁻¹
ABSORBANCE

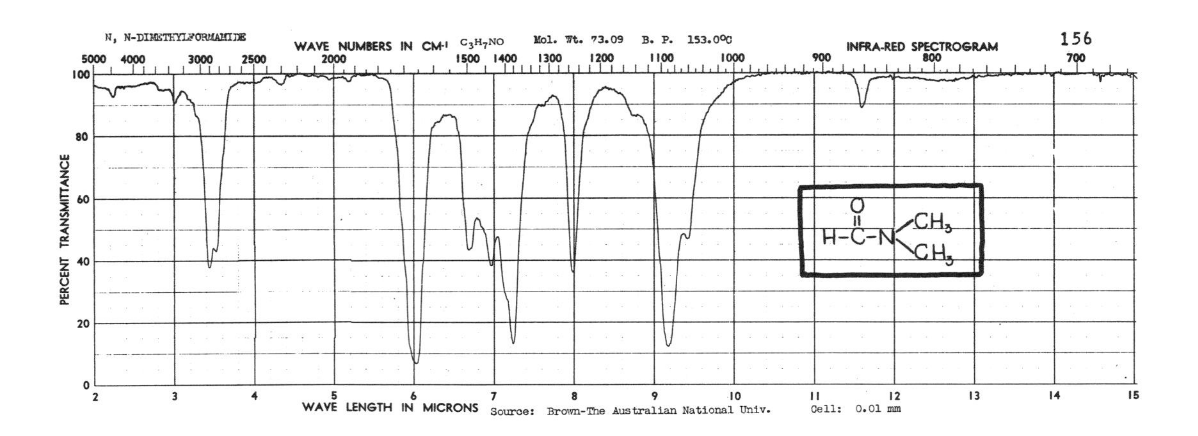
N, N-DIMETHYLFORMAMIDE
WAVE NUMBERS IN CM-1
C_3H_7NO
Mol. Wt. 73.09
B. P. 153.0°C
INFRA-RED SPECTROGRAM
156
PERCENT TRANSMITTANCE
WAVE LENGTH IN MICRONS
Source: Brown-The Australian National Univ.
Cell: 0.01 mm

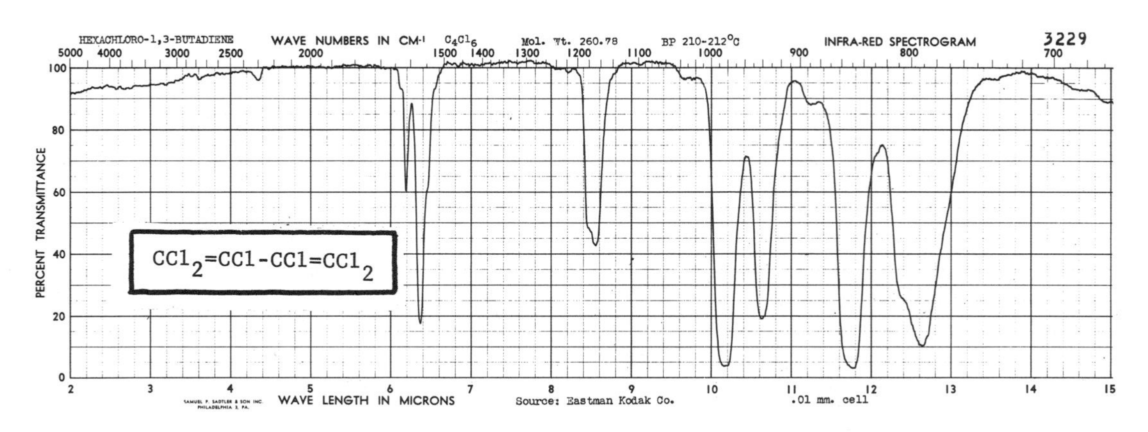
HEXACHLORO-1,3-BUTADIENE
WAVE NUMBERS IN CM-1
C_4Cl_6
Mol. Wt. 260.78
BP 210-212°C
INFRA-RED SPECTROGRAM
3229
PERCENT TRANSMITTANCE
$CCl_2=CCl-CCl=CCl_2$
WAVE LENGTH IN MICRONS
Source: Eastman Kodak Co.
.01 mm. cell

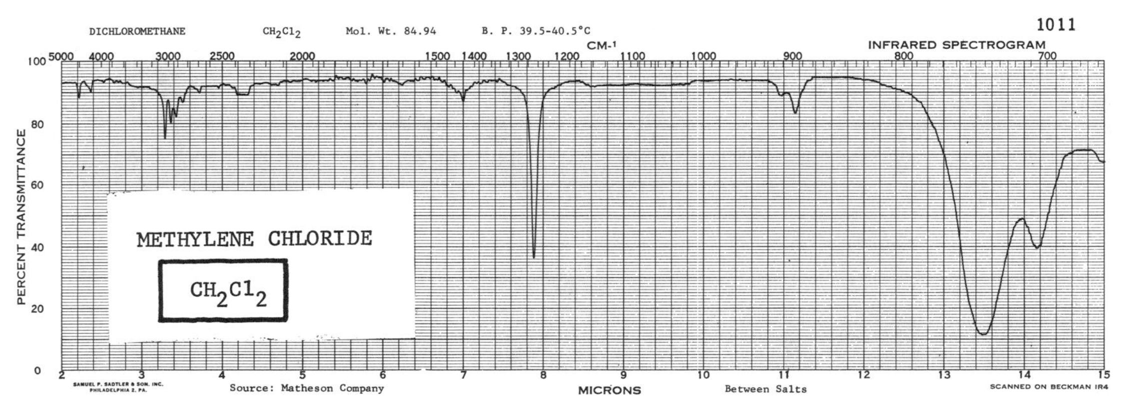
DICHLOROMETHANE
CH_2Cl_2
Mol. Wt. 84.94
B. P. 39.5-40.5°C
1011
INFRARED SPECTROGRAM
PERCENT TRANSMITTANCE
METHYLENE CHLORIDE
CH_2Cl_2
Source: Matheson Company
MICRONS
Between Salts
SCANNED ON BECKMAN IR4

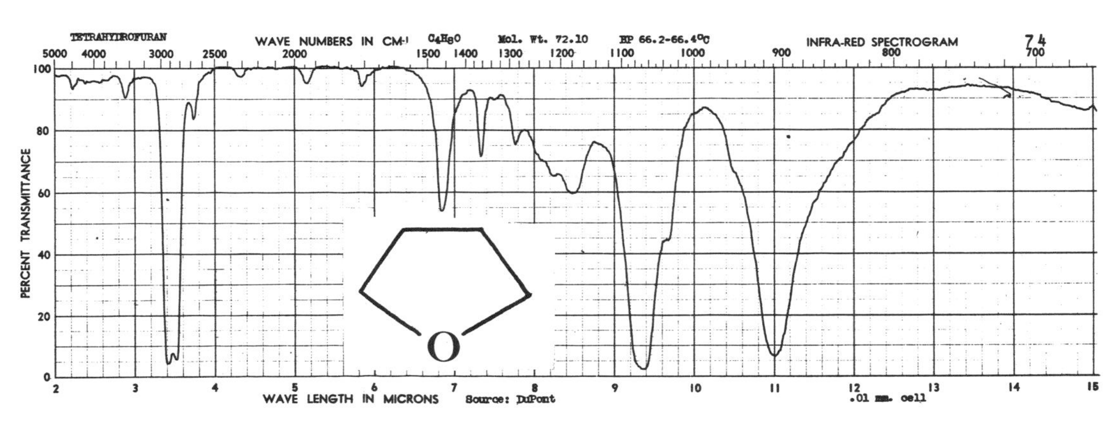
TETRAHYDROFURAN
WAVE NUMBERS IN CM-1
C_4H_8O
Mol. Wt. 72.10
BP 66.2-66.4°C
INFRA-RED SPECTROGRAM
74
PERCENT TRANSMITTANCE
WAVE LENGTH IN MICRONS
Source: DuPont
.01 mm. cell

3. Near-Infrared Solvents

The following is reproduced from R. F. Goddu and D. A. Delker, Anal. Chem., 32, 140 (1960) with permission of the ACS. Solid lines indicate usable regions of the spectrum; numbers above the lines are the maximum desirable path lengths in cm. In addition to the solvents in the table, tetrachloroethylene ($Cl_2C{=}CCl_2$) is an excellent one, since it is transparent over the whole region (10-cm path length). For a slightly larger table write for the spectrosolvent chart of Matheson, Coleman, and Bell (2909 Highland Avenue, Norwood, Ohio 45212) which also covers uv and mid ir.

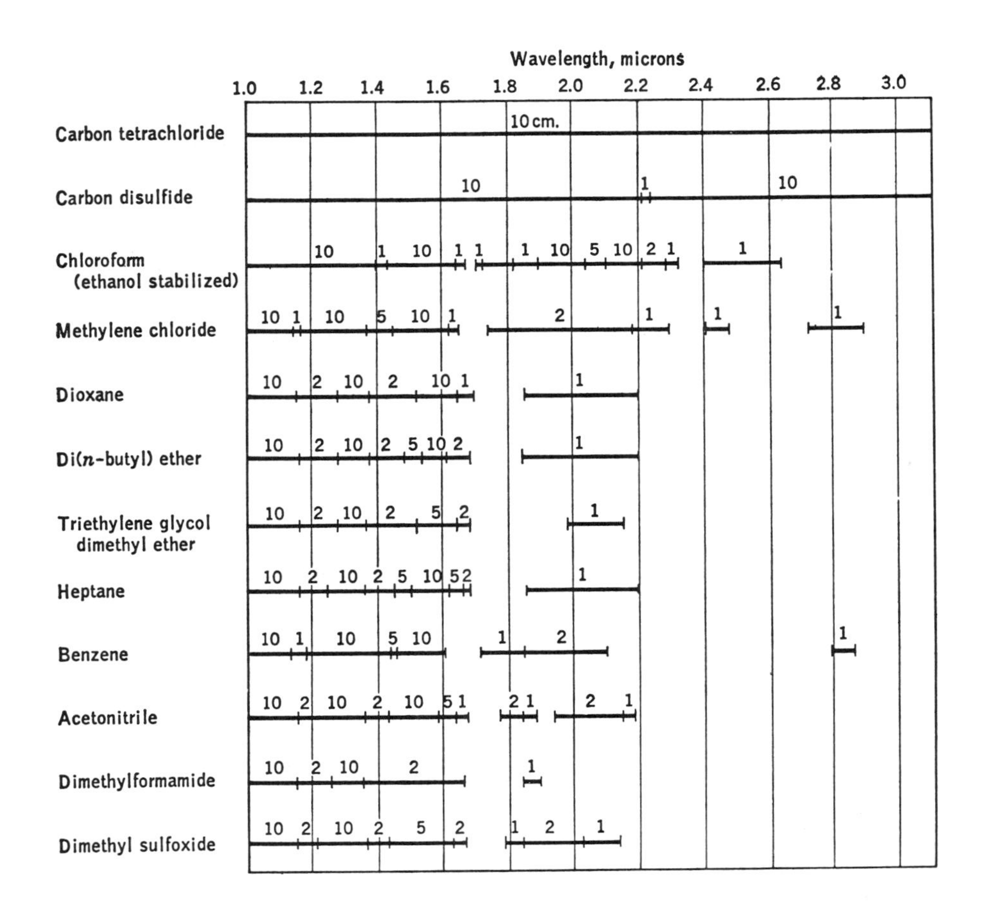

4. Far-Infrared Solvents

The following table is reproduced from F. Bentley et al., IR Spectra and Characteristic Frequencies 700-300 cm^{-1}, Interscience, New York, 1968, p. 12; the solid lines indicate useful regions.

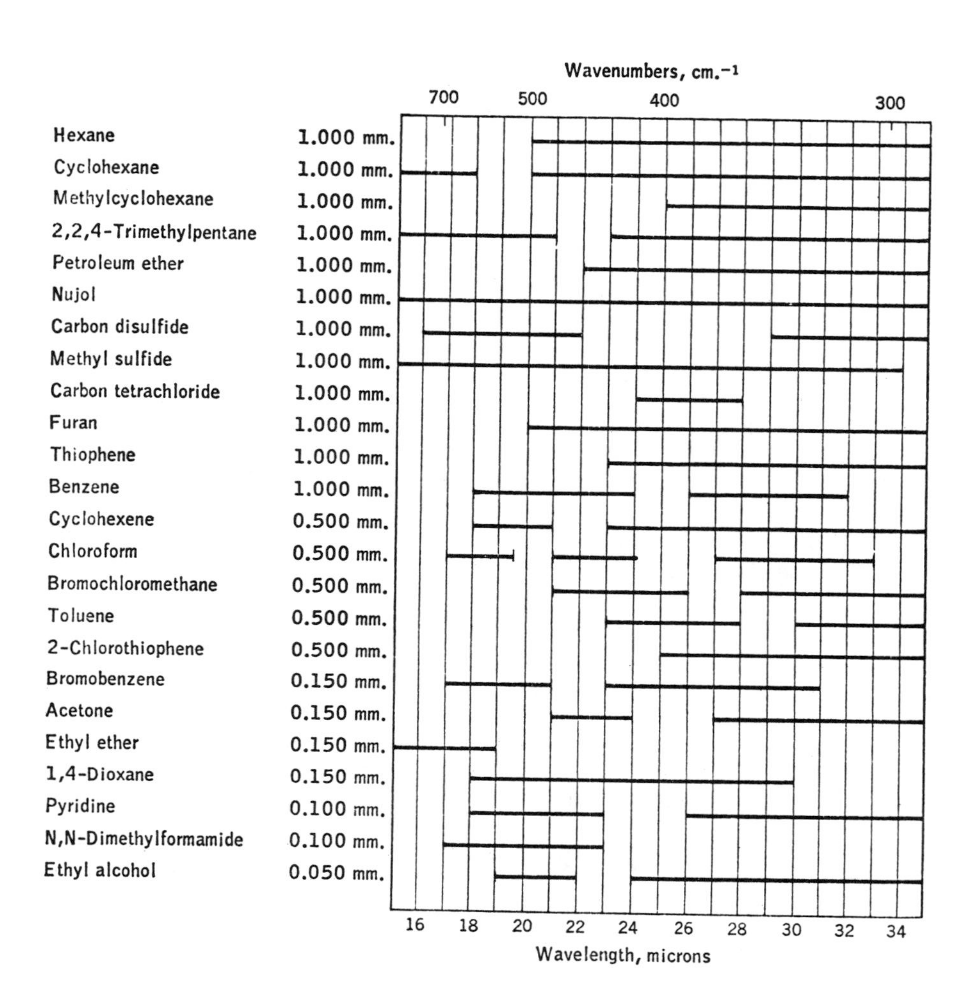

III. OPTICAL MATERIALS FOR SPECTROSCOPY AND PHOTOCHEMISTRY

The exact specifications of the optical properties of different materials, especially glasses, will vary somewhat depending on the manufacturer. The data below serve as a general guide to the proper choice of optical material. Three informative booklets, all entitled Optical Crystals, are available from Optovac, Inc., Isomet Corp., and Harshaw Chemical Co. (see addresses below); in addition to much useful technical data, the booklets include

Suppliers p. 182.

interesting discussions on the handling and preparation of optical crystals. One section below lists representative companies where cells and window materials can be obtained.

Specialized cells for variable high- and low-temperature work are described in G. Konn and D. Avery, Infrared Methods, Academic Press, New York, 1960, pp. 47-48, and in N. Colthup, L. Daly, and S. Wiberley, Introduction to IR and Raman Spectroscopy, Academic Press, New York, 1964, pp. 69-71. Some commercial units are available, such as the Spectrodewar (Rho Scientific, Box 295, Commack, N. Y. 11725) and a system from Andonian Associates (26 Thayer Road, Waltham 54, Mass.). Both of the abovementioned books have useful sections on sample-handling techniques for ir measurements. For a review of low-temperature uv measurements, see B. Meyer, Science, 168, 783 (1970), and B. Meyer, Low Temperature Spectroscopy. Optical Properties of Molecules in Matrices, Mixed Crystals, and Frozen Solutions, Elsevier, New York, 1971.

A. Prism Materials for Infrared

	Optimum Operating Range		
Material	cm^{-1}	μm	Comments
Glass	33333-5000	0.3-2	
Quartz	33333-2857	0.3-3.5	
LiF	4000-1695	2.5-5.9	High resolution for OH,
CaF_2 (fluorite)	4200-1300	2.4-7.7	NH, and CH stretch
NaCl (rock salt, halite)	5000-650	2-15.4	Most common material
KCl (sylvite)	>500	<20	
AgCl	5000-500	2-20	
KBr	1100-385	9-26	C-Br stretch region
KI	<323	>31	
KRS-5 (TlBr/TlI, 42/58 mole %)	10000-250	1-40	
CsBr	1100-250	9-40	C-Br stretch
CsI	1000-200	10-50	C-Br stretch

B. Transmission Characteristics of Glasses

Most applications in the electronic spectral region (uv-vis) rely on pyrex and various types of quartz or fused silica. The following table provides a "cut-off point" -- the wavelength below which transmission is effectively zero for any reasonable glass thickness -- as well as additional transmittance data. Note that many of these materials are useful in the near-ir region (nir). For vacuum uv applications, see the next section. Most of the data below were obtained from manufacturers' catalogs. For related information on filters, see page 360.

Suppliers p. 182.

Material	Cut-off (nm)	Useful Range (nm)	Other Transmittance Data: Thickness (mm)	≃%T	λ (nm)
Optical glass	280	360-2500	5	75	360
Pyrex[b]	275	320-2500	4	50[a]	330
Corex D[c]	260	320-2500	4	50[a]	304
Vycor 7905[d]	220	269-3250	5	40	250
Vycor 7910[d]	210	250-2250	4	50[a]	236
Quartz(crystal)			5	50[a]	185
Far-uv Silica[e]	160	170-2600 2850-3600	10	50	170
NIR silica[f]	190	200-3600	5	80	220
Standard silica	170	220-2500	5	70	220

[a]Taken from Calvert and Pitts, Photochemistry, Wiley, New York, 1966, p. 748. [b]Corning glass 7740. [c]Corning glass 9700. [d]Corning glass numbers. [e]Forms of fused silica, often under different cell manufacturer's names, such as Suprasil, Quarasil, Spectrosil, Silica Q; all have essentially the same characteristics. [f]Often sold under names such as WF Spectrosil, Infrasil, Silica I.

C. Transmission Characteristics of Optical Crystals and Related Materials

Materials listed below are useful in both the ir and electronic spectral regions, including vacuum uv. Note that for most optical materials, absorption of short wavelength radiation increases substantially with increasing temperature. The data are taken mainly from booklets provided by Harshaw, Optovac, Isomet, Barnes Engineering, and Wilks. See also D. W. McCarthy, "The Reflection and Transmission of Infrared Materials", Appl. Opt., 2, 591-603 (1963) (available as Reprint 6197 from Beckman Instruments, Fullerton, Calif. 92634).

"Useful range" is defined as that over which there is >40% transmission for moderate thickness (≃5 to 10 mm). The symbol ω besides a substance indicates its usefulness as cell material for aqueous solutions (see L. May, in H. A. Szymanski, Ed., Progress in Infrared Spectroscopy, Vol. 2, Plenum Press, New York, 1964, p. 217). The symbol p indicates that the material in finely powdered form -- usually <400 mesh (37 μm particles) -- is useful as a medium for making compressed pellet sample preparations [for a review, see A. Nicholson, Anal. Chem., 31, 519 (1959)].

According to Wilks Scientific (see address below), for internal reflection work (ATR), the best surfaces are provided by: KRS-5, Si, Ge, AgCl, AgBr, Al_2O_3 (sapphire), Irtran-4 (ZnSe), and Irtran-6 (CdTe). Arsenic selenide (As_2Se_3) is also useful (see table entry below).

Material	Useful Range: μm	cm^{-1}	Solubility in Water, g/100 g(rt)	Mp (°C)
LiF(ω)[a]	0.11-7 200-425	90910-1429 50-23.5	0.27	870
NaF	0.15-14	66667-714	4.2	997

Suppliers p. 182. (Continued)

Material	Useful Range		Solubility in Water, g/100 g(rt)	Mp (°C)
	μm	cm^{-1}		
NaCl(ω)[n]	0.20-16	50000-625	36	801
	300-425	33.3-23.5		
NaBr	0.22-23	45450-435	91	755
NaI	0.15-12	66667-833	185	651
$NaNO_3$	0.35-2	28570-5000	88	307
KF	0.20-15	50000-667	92	846
KCl(p)	0.20-20	50000-500	35	776
KBr(p)	0.21-25	47620-400	65	730
	300-425	33.3-23.5		
KI	0.25-35	40000-286	144	723
CsF	0.30-15	33333-667	367	682
CsCl	0.20-30	50000-333	186	646
CsBr	0.25-37	40000-270	124	630
CsI	0.25-50	40000-200	160	621
α-Al_2O_3[b,o]	0.20-6.5	50000-1538	Insol.	2015
MgF_2	0.13-7	76920-1429	0.008	1310
CaF_2[a](ω)	0.12-9	83333-1111	0.001	1360
$CaWO_4$	0.28-5	35710-2000	Insol.	1650
$CaCO_3$[c]	0.25-2	40000-5000	0.001	1339(1000 mm)
SrF_2	0.13-10	76920-1000	0.012	1400
BaF_2[d](ω)	0.14-12	71430-833	0.12	1280
TlCl[e]	0.40-30	25000-333	0.32	430
TlBr[e](p)	0.45-45	22222-222	0.048	480
KRS-5[e,f](ω)	0.55-35	18180-286	0.02(0°)	414
KRS-6[e,g]	0.40-32	25000-312	0.1 (0°)	
β-PbF_2	0.25-10	40000-1000	0.064	822
As_2S_2[h](ω)	1-12	10000-833	2.3	(195)
As_2Se_3[p](ω)	0.8-16.7	12500-600	Insol.	
MnF_2	0.12-8	83333-1250	0.008	1255
AgCl(ω)(p)	0.42-24	23810-417	0.0002	458
AgBr	0.50-35	20000-286	0.00001	432
CdF_2	0.25-9	40000-1111	4.3	1100
LaF_3	0.40-9	25000-1111	Insol.	1493
Si	1.5-8	6667-1250	Insol.	1410
Ge(ω)	2-18	5000-556	Insol.	937
Fused silica(ω)	0.2-3	50000-3333	Insol.	(1710)
	80-400	125-25		
Sapphire[i,o](ω)	0.2-6	50000-1667	Insol.	2000
T-12[j]	1-10	10000-1000	0.16	
Irtran 1[k]	0.5-9	20000-1111	Insol.	>900
Irtran 2(ω)[k]	0.7-14.5	14290-690	Insol.	>800
Irtran 3[k]	<0.4-11.5	>25000-870	Insol.	>1000
Irtran 4[k]	0.55-22	18182-455	Insol.	>300
Irtran 5[k]	<0.4-9.5	>25000-1053	Insol.	>800
Irtran 6[k]	1.5-31	6667-323	Insol.	>300
$NH_4H_2PO_4$(ADP)	<0.2-1.3	>50000-7692	23(0°)	Dec.
KH_2PO_4(KDP)	0.2-1.5	50000-6667	33	253
$Ni_2SO_4 \cdot 6H_2O$	0.2-0.3	50000-33333	63(0°)	(53.3)
Polyethylene (ω)(p)[l]	16-300	625-33	Insol.	
Plexiglas[m]	0.31-0.6	32260-16667	Insol.	
Teflon(ω)(p)	50-400	200-25	Insol.	
Polystyrene	30-400	333-25	Insol.	
Mica	200-425	50-23.5	Insol.	

Suppliers p. 182.

[a]The transmission properties of LiF in the short λ range are less uniform than those of CaF_2 (manufacturing aberrations); also, LiF is less hard, more easily scratched, and less readily ground than CaF_2. For photochemical work at $\lambda < 110$ nm, very thin LiF windows or windowless cells are required (Calvert and Pitts, Photochemistry, Wiley, New York, 1966, p. 749). CaF_2 will not fog, and is especially useful in high-pressure cells. [For more details, including proper cements for such windows, see McNesby and Okabe, Adv. Photochem., 3, 157 (1964).] [b]Natural corundum. [c]Calcite. [d]Shock sensitive; handle with care. Resistant to F_2 and fluorides. [e]All thallium compounds are TOXIC; use extreme care in handling, especially grinding or polishing. [f]42 TlBr/58 TlI (mole %); one of best materials for ATR work. [g]40 TlBr/60 TlCl (mole %). [h]Also known as "Servofrax" (Servo Corporation of America). [i]Hexagonal alumina (corundum) with traces of Fe and Ti, imparting blue color. [j]An optical window material developed by Harshaw Chemical Co. [k]The "Irtrans" (1 through 6) are compressed, polycrystalline materials, respectively of MgF_2, ZnS, CaF_2, ZnSe, MgO, and CdTe. Under mp, the numbers refer to upper use limit in air. For details, write for pamphlet U-71 or 2-67 from Eastman Kodak, Special Products, Rochester, N.Y. 14650. [l]High density. Excellent for far ir. [m]Poly(methyl methacrylate). [n]Use saturated NaCl solution for aqueous solution spectra. [o]Only ATR material with good uv transmission (always use with a 60° angle of incidence). [p]A new material (Unimetrics Universal Corp.) insoluble in all organic solvents and in water and acids up to pH 11. Almost completely opaque to 12 μm. Has n_D 2.8 and is useful for ATR spectra.

D. Transmission Region Graph: Windows, Sources, and Detectors (10 to 1000 μm)

The figure that appears on page 183 is reproduced with permission, from L. Meites, Ed., Handbook of Analytical Chemistry, McGraw-Hill, New York, 1963, p. 6-156.

E. Addresses of Suppliers

Most spectroscopic instrument companies carry some cell materials. The following, however, are the addresses and telephone numbers of representative companies specializing in optical materials and cells. A key publication by the company is mentioned in some cases.

Barnes Engineering Company , 30 Commerce Road, Stamford, Conn., 06902 (203-348-5381) (ir materials).
Bausch and Lomb, Inc., 11825 Linden Ave., Rochester, N. Y. 14623 (716-385-1000) (uv).
Bolab, Inc., 359 Main Street, Reading, Mass. 01867 (uv).
Fisher Scientific Co., consult local office (uv; Bulletin 14-385-902).
Harshaw Chemical Co., 1945 East 97th Street, Cleveland, Ohio 44106 (216-721-8300) (ir, uv; Harshaw Optical Crystals).
Hellma Cells, Inc., Box 544, Borough Hall Station, Jamaica, N. Y. 11424 (212-544-9534) (uv).
International Crystal Laboratories (ir) and Luminon Inc. (uv), 120 Coit Street, Irvington, N. J. 07111 (201-373-4242).
Isomet Corp., 433 Commercial Avenue, Palisades Park, N. J. 07650 (201-944-4100) (uv, ir; Optical Crystals).
Opticell, 10792 Tucker Street, Beltsville, Md. 20705 (301-345-2343) (uv).
Optovac, Inc., North Brookfield, Mass. 01535 (617-867-3767) (uv, ir; Optical Crystals).

Transmission Region Graph: Windows, Sources, and Detectors (10 to 1000 μm) (see page 182)

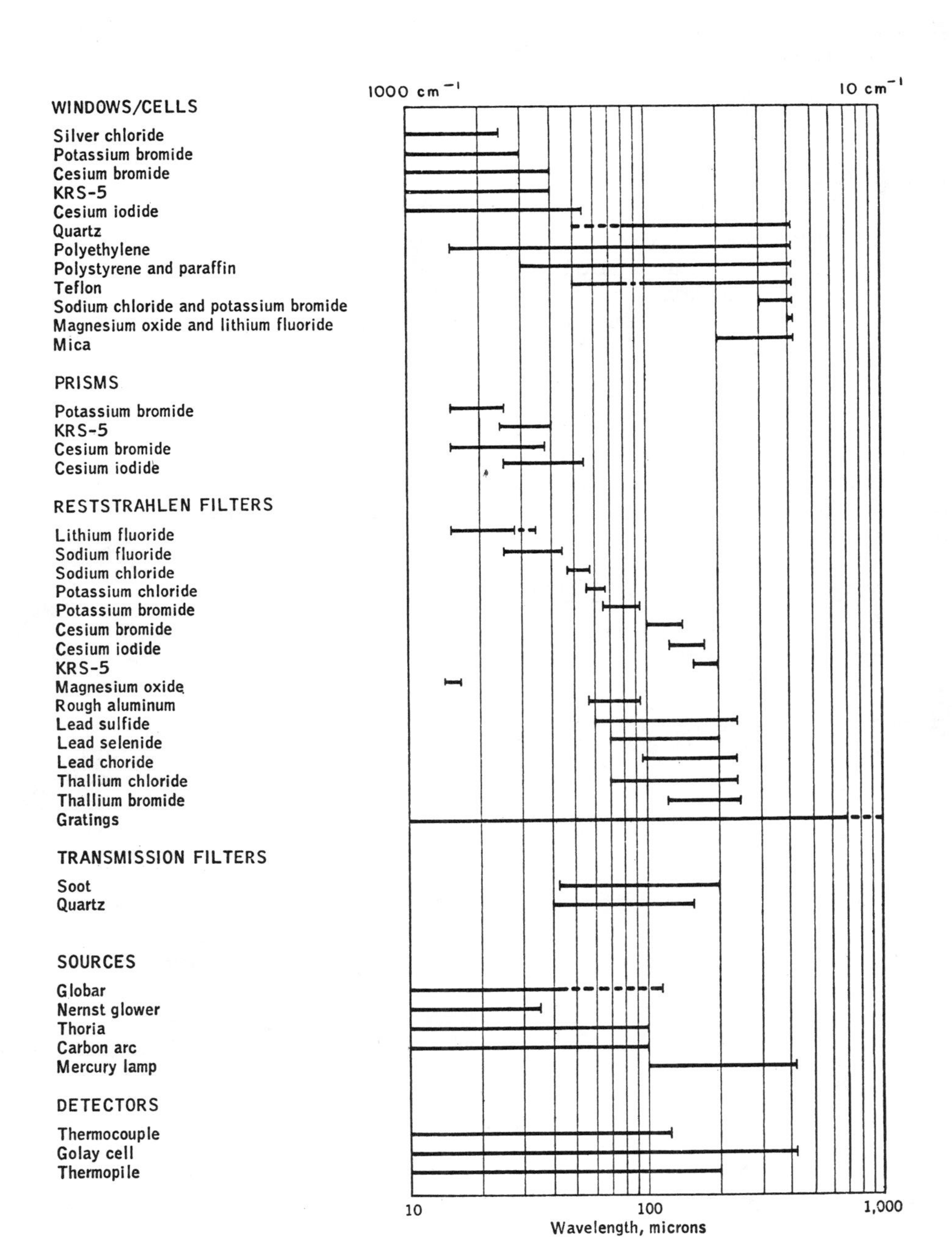

Precision Cells, Inc., 401 Broadway, New York, N. Y. 10013 (212-226-1373) (uv, vis).
Pyrocell Manufacturing Co., 91 Carver Avenue, Westwood, N. J. 07675 (201-664-6660) (uv, vis).
Servo Corporation of America, 111 New South Road, Hicksville, N. J. (516-938-9700) (uv, ir).
Wilks Scientific Corp., 140 Water Street, Norwalk, Conn. 06856 (203-838-4537) (ir; Wilks Infrared Accessories).

IV. VIBRATION SPECTRA

A. Introduction (1)

This section summarizes the most often called-for data associated with the various ir regions of the electromagnetic spectrum. The bulk of information is in the form of updated correlation tables of absorption data for functional groups, especially for the "vibrational" (mid) ir region. All data are given in wavenumbers (cm^{-1}; also called Kaisers) followed by the value in wavelength (μm) in parentheses.

Wavenumber is the much preferred unit for ir data; conversion to wavelength is accomplished with the equation, $\lambda\bar{\nu} = 10^4$ (λ in μm, $\bar{\nu}$ in cm^{-1}), or with the conversion table on p. 464.

Several general books on various aspects of ir are recommended in reference 1. Appropriate cell materials and solvents for spectral measurements are covered separately on pp. 168 and 180.

1. Spectral Regions and Applications

Region	cm^{-1}	μm	Comments
Near ir	12500-4000	0.8-2.5	Qualitative and quantitative analysis for H-containing functional groups; many overtone bands of fundamentals in mid-ir region
Mid ir	4000-600	2.5-16.7	Also called vibrational or fundamental ir region. Typical functional group absorptions, mainly in 1400-4000 cm^{-1} (6.7-2.5 μm) region; "fingerprint" region from 600-1400 cm^{-1} (16.7-6.7 μm)
Far ir	600-30	16.7-3.30	For low-energy vibrations, especially for inorganic and organometallic compounds. Good for skeletal and ring deformations (torsions) as well as solid state (lattice) vibrations

2. Liquid Cell Path-Length Determination

From the fringe pattern obtained with an empty cell count the number (n) of fringes (peak to peak, e.g.) between two wavenumbers, $\bar{\nu}_1$ and $\bar{\nu}_2$ (wavelengths, λ_1 and λ_2):

$$\text{path length} = \frac{n}{2(\bar{\nu}_1 - \bar{\nu}_2)} = \frac{n\lambda_1\lambda_2}{2(\lambda_2 - \lambda_1)}$$

3. Infrared Absorption versus Raman Scattering Spectra (2)

Raman scattering ir spectra are complementary to the absorption spectra; their measurement is rapidly becoming routine through recent instrumentation developments (laser excitation-source). However, frequency correlation charts are not that readily available. In general, Raman excitation depends on bond polarizability changes, while ir absorption depends on bond dipole moment changes. Weak ir absorption lines (e.g., S-H, $R_2C{=}CR_2$, $-C{\equiv}N$ stretch) become strong Raman lines, and vice versa. For a review on "A State of the Art Comparison to IR," see H. J. Sloane, Appl. Spec., 25, 430 (1971).

B. Vibrational (Mid) Infrared Spectral Data

1. Abbreviations

Vibration Modes	Intensity	Other
as = antisymmetric	S = strong	br = broad
s = symmetric	M = medium	pr = primary
ν = stretch	W = weak	sec = secondary
δ = bend or deformation	V = very	t = tertiary
		var = variable

2. General Regions of Interest

The following table pinpoints the fairly narrow regions in which characteristic and usually intense bands appear for specific functional groups. Only the most useful and consistently reliable assignments are mentioned. More details and additional data are found in the more extensive tables below.

Region	Spectral Information
3600-2500 (2.78-4.00)	Stretching vibrations of H attached to C, O, N, and S; all fairly strong except S-H (weak). Weak bands may also be overtones of 1600 (6.6) region bands.
2400-2100 (4.16-4.76)	Triple bond stretch ($C{\equiv}C$, $-C{\equiv}N$, $-\overset{+}{N}{\equiv}\overset{-}{C}$, $N{\equiv}\overset{+}{N}-$); also C-D, O-D, P-H and Si-H stretch.
2200-1900 (4.55-5.27)	Cumulated double-bond stretch (C=C=C, C=C=O, -N=C=C, -N=C=S, -N=C=N-, $R_2C{=}\overset{+}{N}{=}\overset{-}{N}$, CO_2, N_3^-); $-O-C{\equiv}N$, $-S-C{\equiv}N$. Isocyanates (-N=C=O) absorb at somewhat higher wavenumber (2240-2275).
1850-1750 (5.41-5.71)	Carbonyl stretch for carboxylic acid anhydrides, acid halides, acyl peroxides, and carbonates (alkyl; aryl).

Refs. p. 210. (Continued)

1780-1600 (5.61-6.22)	Carbonyl stretch for ketones, aldehydes, carboxylic acids, esters (lactones), and amides (lactams)(amide-I band).
1700-1500 (5.89-6.68)	Double-bond stretch: C=C (olefinic, aromatic), C=N, -N=0, $-NO_2$(as), $C=NR_2$.
1650-1450 (6.06-6.89)	N-H bending (amines, ammonium ions, amides (amide-II band); N=N stretch (trans 1440-1410; cis 1510).
1450-1300 (6.89-7.69)	C-H bending for sp^3-C; $-NO_2$ stretch (sym)(1320-1380).
1400-1300 (7.14-7.69)	S=O stretch (as) for sulfones, sulfonamides, sulfonates, sulfonyl halides; symmetric S=O stretch 1000-1200 (10.00-8.50).
1400-1250 (7.69-8.00)	O-H bending; C-N stretch in amides (amide-III band).
1300-1000 (7.69-10.00)	C-O bend and/or stretch (alcohols, ethers, acetals, ketals, esters, acids); C-N stretch (amines); C=S stretch (thiones); P=O stretch; C-F stretch.
1100-1000 (9.09-10.00)	Sulfoxide S=O stretch; Si-O- stretch; P-O-C and P-OH stretch (or bend). Symmetric S=O stretch for sulfones, sulfonamides, sulfonates; sulfonyl halides at 1000-1200 (10.00-8.50).
1000-800 (10.00-12.50)	C-H bending (deformations) in alkenes (cis HC=CH often near 690-700, but variable and uncertain).
900-690 (11.11-14.49)	Aromatic substitution pattern (C-H bending); S-O stretch; N-H bending (RNH_2, $R_2\underset{+}{N}H_2$).
750-695 (13.33-14.39)	C-Cl stretch; R_2CCl_2 850-795 (11.76-12.58). C-Br 600-500 (16.67-20.00); C-I 500-200 (20.00-50.00).

3. Index to Vibrational Infrared Correlation Data

Refs. p. 210.

4. Vibrational Infrared Correlation Data

The data are for relatively dilute solutions in nonpolar solvent (CCl_4 especially) unless mentioned otherwise; only data of practical use are included. If a single position is cited rather than a range, the value should be considered generally within ± 5 to 10 cm^{-1}.

a. Saturated C-H

Group	Position or Range	Mode and Intensity	Remarks
CH_3	2962 (3.38)	ν-as(S)	δ-s for CH_3 on heteroelement:
	2872 (3.48)	ν-s (S)	B 1300 · C 1375 · N 1426 · O 1455
	1460 (6.85)	δ-as(S)	Si 1253 · P 1299 · S 1325
	1375 (7.27)	δ-s (M)	Ge 1242 · As 1249 · Se 1282
	(CH_3 as acetyl or acetoxyl enhances 1375 intensity greatly)		Sn 1190
CH_2	2925 (3.42)	ν-as(S)	Intensities of ν-s are additive (proportional to the number of CH_2's in molecule); heteroatoms (CH_2-X) weaken intensity.
	2850 (3.51)	ν-s (S)	
	1465 (6.83)	δ-as(M)	
	(Active methylenes, $-CH_2CO-$, e.g., have VS bands, 1400-1440)		
$(CH_2)_x$	720-740 (13.89-13.50)	δ (S)	$x > 4$
▷	3050 (3.28)	ν (W)	Also for epoxides, aziridines and $-CH_2$-halogen
	1010 (9.9)	(M)	Skeletal vibration
$>C(CH_3)_2$	1385(7.22),1370(7.30)	δ (M)	Characteristic doublet
	1145(8.73),1170(8.55)	δ (M)	
$-C(CH_3)_3$	1397(7.16),1370(7.30)	δ (M,S)	Characteristic doublet
	1250(8.00),1210(8.26)	(S,S)	Skeletal; 1210 more constant than 1250

Note: C-D bands ($\bar{\nu}$) appear at ∼ (1/1.37) times those for C-H.

Refs. p. 210.

b. Olefins

ν (C-H) (W)	ν (C=C)	δ-Out of plane (C-H) (S)	Other
$CH{=}CH_2$			
3085 (3.23) 3015 (3.32)	1638-1648 (M) (6.11-6.07)	990 (10.10) 910 (10.99)	Overtone of δ at ∿1825 (5.48). Vinyl halides $CH_2{=}CHCl$, Br, I· $\bar{\nu}$(C=C) 1601, 1593, 1581
$C{=}CH_2$			
3080 (3.24) 2975 (3.36)	1648-1658 (M) (6.07-6.03)	890 (11.24)	Overtone of δ at ∿1800 (5.55)
C=CH			
3020 (3.31)	1665-1675 (M) (6.01-5.97)	790-840 (12.66-11.90)	
CH=CH (cis)			
3020 (3.31)	1652-1662 (M) (6.05-6.02)	∿690 (14.49) (Variable; ambiguous)	
CH=CH (trans)			
3020 (3.31)	1668-1678 (W) (5.99-5.96)	965-980 (10.36-10.20)	Conjugation $(-CH{=}CH-)_n$ ($n \geq 2$) introduces ≥2 bands (C=C) (S) from 1600-1650; δ-bands relatively unaffected in such cases
$R_2C{=}CR_2$			
---	1665-1675 (VW) (6.01-5.97)	---	

c. Cyclic Olefins: ν (C=C)

n	 C_n	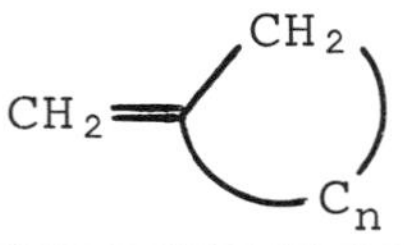 CH_2 CH_2 C_n
1	1640	1730
2	1566	1678
3	1611	1657
4	1646	1651
5	1650	

Refs. p. 210.

d. Aromatic Rings

ν C-H (W)	3070-3030 (3.26-3.30)	
ν C=C (M)	1580-1600 (6.33-6.25)	Ring "breathing" modes (very sharp); 1500 usually stronger than 1600. Both stronger and at lower range (cm^{-1}) if ring
ν C=C (M)	1450-1500 (6.90-6.67)	

conjugated. These values (C-H and C=C) apply as well to condensed and heteroaromatic rings.

Benzene Substitution Patterns. 1667-2000 (6.0-5.0) (overtone and combination bands--usually VW, requiring concentrated sample and no obscuring bands) and 675-890 (14.81-11.24) (δC-H, out-of-plane, a function of number of adjacent H's); fairly intense, but often obscured if chlorinated solvent used. Band positions for substituted naphthalenes and other polycondensed aromatics are found in Colthup (2), while those for pyridines, furans, thiophenes, and pyrroles are in reference 3.

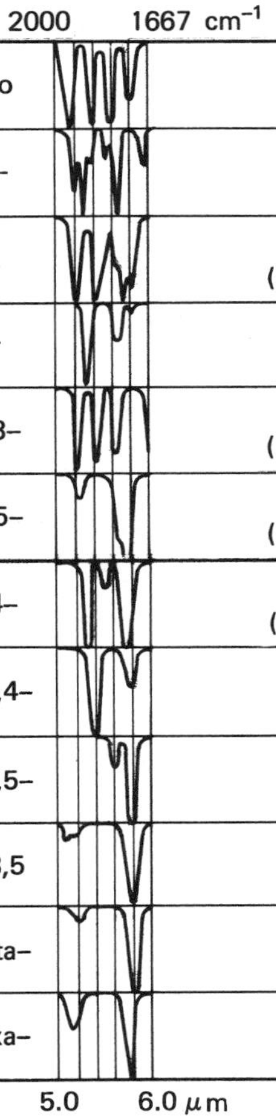

	2000–1667 cm^{-1} (5.0–6.0 μm)	675–890 Region		
Mono		690 (14.49)	and	770–730 (12.99–13.70)
1,2–		770–730 (12.99–13.70)		
1,3		810–750 (12.35–13.33)	and	710–690 (14.08–14.49)
1,4		840–810 (11.90–12.35)		
1,2,3–		780–760 (12.82–13.16)	and	745–705 (13.42–14.18)
1,3,5–		865–810 (11.56–12.35)	and	730–675 (13.70–14.82)
1,2,4–		825–805 (12.12–12.42)	and	885–870 (11.30–11.49)
1,2,3,4–		810–800 (12.35–12.50)		
1,2,4,5–		870–855 (11.49–11.70)		
1,2,3,5		850–840 (11.76–11.90)		
Penta–		870 (11.49)		
Hexa–				

Refs. p. 210.

e. Cumulated Double Bonds and Triple Bonds

Group	Position or Range	Mode and Intensity	Remarks
C=C=C (allenes)	1950 (5.13)	ν(S)	Doublet (1930-1970) if coupling groups attached (-COR, etc.)
C=C=CH_2	850 (11.76)	δ C-H(S)	
C=C=O (ketenes)	2150 (4.65)	ν(S)	
	1120 (8.93)	(S)	
O=C=O (CO_2)	2349.3 (4.256)	ν-as(VS)	Used for calibration with 720.5 and 667.3
C=C=N (ketenimines)	2000 (5.00)	ν(S)	
N=C=N (carbodiimides)	2140-2130 (4.67-4.69)	ν(VS)	Aliphatic
	2145 (4.66); 2115 (4.73)	ν(VS)	Aryl; 2145 stronger band
N=C=O (isocyanates)	2275-2260 (4.39-4.42)	ν-as(S)	Unaffected by conjugation
	1390-1350 (7.19-7.41)	ν-s(W)	
N=C=S (isothiocyanate)	2140-1990 (4.67-5.26)	ν(S)	Aliphatic
	2130-2140 (4.70-4.90)	ν(S)	Aromatic [also 930(S)]
N=S=O (a nonlinear group)	1300-1200 (7.69-8.33), 1180-1100 (8.47-9.09)		
N_3 (azides)	2160-2120 (4.63-4.72)	ν-as(S)	
	1340-1180 (7.46-8.47)	ν-s(S)	
$RCH=\overset{+}{N}=\overset{-}{N}$ (diazo)	2050-2035 (4.88-4.92)	ν(S)	If R = carbonyl group: 2100-2080
$RR'C=\overset{+}{N}=\overset{-}{N}$	2030-2000 (4.93-5.00)	ν(S)	If R = carbonyl group: 2075-2050
R-C≡C-H (alkynes)	2140-2100 (4.67-4.76)	ν(M)	
R-C≡C-R'	2260-2190 (4.42-4.57)	ν(VW)	Absent if R = R'
≡C-H	3310-3200 (3.02-3.12)	ν(M)	Sharp
R-C≡N (nitriles)	2260-2240 (4.42-4.46)	ν(M)	R = aliphatic, nonconjugated
	2230-2220 (4.48-4.50)	ν(M)	R = aliphatic, conjugated
	2240-2200 (4.46-4.50)	ν(M)	R = aryl
$R-\overset{+}{N}\equiv\overset{-}{C}$ (isonitriles)	2200-2100 (4.55-4.76)	ν(S)	
R-S-C≡N (thiocyanates)	2140 (4.90)	ν(S)	Aliphatic
	2175-2160 (4.60-4.63)	ν(S)	Aromatic
$R-\overset{+}{N}\equiv N$ (diazonium)	2280-2240 (4.39-4.46)	ν(S)	Insensitive to anion type

Refs. p. 210.

f. Alcohols, Phenols, and Other O-H

Hydroxyl	ν O-H(W)	ν C-O(S)	δ O-H(M)	Remarks
H_2O	3710 (2.695)	---	1640-1615 (6.10-6.19)	
pr	3635 (2.749)	1045 (9.57)	1350-1250	All these values are for the
sec	3625 (2.758)	1100 (9.09)	(7.41-8.00)	free or "monomeric" (non-H-
t	3615 (2.765)	1150 (8.69)	1410-1310	bonded) in dil. CCl_4 Sol.
ArOH	3600 (2.778)	1200 (8.33)	(7.09-7.63)	
ROOH	3550 (2.816)			
RCO_2H(monomer)	3550 (2.816)		920 (10.87)(br)(dimer)	

For pr, sec, and t-alcohols of structure:

$C-\overset{\alpha}{C}-CH_2OH$ $\qquad$ $(C-C^{\alpha})(C-C_{\alpha'})CH-OH$ $\qquad$ $(C-C^{\alpha})(C-C^{\alpha'})(C-C_{\alpha''})C-OH$

the νC-O absorptions can be crudely predicted by subtracting the following increments for branching from the standard (unbranched) values (1045, 1100, 1150): α-branching, 15; α-unsaturation, 30; ring formation between α and α', 50; α-unsaturation and α'-branching, 90; α- and α'-unsaturation, 90; α, α', and α"-unsaturation, 140. From Nakanishi (4). Bending bands (δ OH) may not be useful with complex molecules. Solvent corrections for νO-H, applied to CCl_4 values, are as follows: hydrocarbons (not Nujol mull), +10; $CHCl_3$ and $CHBr_3$, -15; diethyl ether, ∿-300. From Schwarz (1e).

Intermolecular H-Bonding (νOH)

Dimeric. 3600-3500 (2.76-2.86), W but sharp; usually hidden or overlapped by polymeric band. Also in this region is H-bonding with polar solvents (ether, amines, ketones, etc.).

Polymeric. 3400-3200 (2.94-3.13), S and br, the only band observed for concentrated sample. N-H bands (amines, amides) also in these regions (see below).

Intramolecular H-Bonding (νOH)

Polyhydroxy Compounds. 3600-3500 (2.78-2.86), fairly sharp, M; critical H·····O distance for such bonding is ∿3.3 Å.

Chelation. 3200-2500 (2.86-4.00), br, diffuse band, may be overlooked since often W; characteristic of enol systems (C(OH)=C-CO-) and OH bonded to NO_2, etc.

π-Cloud. 3600-3500 (2.78-2.86)-olefins or aromatic rings.

For details on characterizing H-bonding by $\Delta\bar{\nu}$ (shifts from free OH), see Bellamy, (1b, 5); for H-bonding to cyclopropane rings, see reference 6.

Refs. p. 210.

Refs. p. 210.

g. Ethers, Acetals, Ketals, Peroxides, and Ozonides

Group	Position or Range	Mode and Intensity	Remarks
R-O-R'			
R = R' = alkyl	1150-1100 (8.69-9.09)	ν-as(VS)	
R = alkyl, R' = vinyl, phenyl	1275-1200 (7.85-8.33)	ν-as(S)	Stronger than ν-s; vinyl C=C at 1610 and 1620-1650, both S
	1075-1020 (9.30-9.80)	ν-s(S)	
R = CH_3, R' = alkyl	2830-2815 (3.54-3.55)	ν-s(CH_3)	Not found if R=CH_3CH_2
R = CH_3, R' = aryl	2850 (3.51)	ν-s(CH_3)	
R = $CH_2\phi$, R' = alkyl	1090 (9.18)	ν-s(S)	
	950-810 (10.53-12.35)	ν-as	Ring stretch
	1250 (8.00)	ν-s	Ring stretch
	840-750 (11.90-13.33)	δ C-H	ν C-H 3040-3000

For terminal epoxy (CH_2–epoxide ring), ν CH for CH_2 at 3050 (3.28). For oxetane rings (oxetane ring), ν at ∿980 (10.20); THF rings, 1070 (9.35).

R-O-C-O-R'

Group of four bands: 1190-1160 (8.40-8.62); 1195-1125 (8.37-8.89); 1098-1063 (9.11-9.41); 1055-1035 (9.48-9.66). For acetals only, a fifth at 1116-1103 (8.96-902).

Group	Position or Range	Mode and Intensity	Remarks
R-O-O-R'			
R = R' = alkyl	890-820 (11.24-12.19)	W	Difficult to detect
R = R' = aryl	∿1000 (10.00)	W	
Ozonides	1060-1040 (9.43-9.62)	See Helv. Chim. Acta, 36, 1900 (1953).	

h. Amines, Imines, and N-Onium

Amine (Non-H-Bonded)	ν N-H (W)	ν C-N*	δ N-H
RNH_2	∿3500 (2.86) (as) ∿3400 (2.94) (s) (For N-H in pr amines, $\bar{\nu}_s = 345.5 + 0.876\ \bar{\nu}_{as}$)	1230-1030 (M) (8.13-9.71)	1640-1560(6.10-6.41)(S) 900-650(11.11-15.38)(br,M)

R_2NH	3350-3310 (2.99-3.02)	See RNH_2	1580-1490 (6.33-6.71) (W)
R_3N	---	Two bands in 1230-1030 region for ν C-N	

[*For aromatic or vinyl amines (pr, sec, or t) two bands for ν C-N: 1360-1250 (7.38-8.00) (S) and 1280-1180 (7.81-8.48) (M)].

$R_2C=N-H$	3400-3300 (2.94-3.33)	1670 (5.99) (R alkyl) 1640 (6.10) (R aryl) 1618 (6.18) (extended conjugation)	1590-1500 (6.29-6.67)

<u>H-Bonding</u>: shifts less than those for OH. In general, for pr and sec amines <u>inter</u> appears at 3300-3000 (3.03-3.33), often complex, always br; <u>intra</u> appears at 3500-3200 (2.86-3.13), also complex.

Ammonium Salts[a]	ν N-H	δ N-H
NH_4^+	3300-3030 (3.03-3.30)	1430-1390 (7.00-7.20) Both br, S
RNH_3^+	3000 (3.33) (br, S)	1600-1575 (6.25-6.35) (as)
Several bands (Sometimes absent)	2500 (4.00) (M) 2000 (5.00) (M)	1500 (6.67) (as)
$R_2NH_2^+$	2700-2250 (3.70-4.44) (S)[b] 2000 (5.00) (M) (usually absent)	1600-1575 (6.25-6.35) (M)
R_3NH^+	2700-2250 (3.70-4.44) (S)[b]	
R_4N^+	No characteristic bands	
$R_2C=\overset{+}{N}(H)-$	2500-2300 (4.00-4.34)[b] 2200-1800 (4.55-5.56) (M)[b]	[ν C=N at ∿1680 (5.95)] "immonium band"
$(>N)_3C^+$ (guanidinium)	∿3300 (3.03) (br) νC=N: neutral guanidine 1660 (6.02); for hydrochlorides: monosubst. 1660, 1630 (6.02, 6.14); disubst. 1680, 1595 (5.95, 6.27); trisubst. 1635 (6.12).	

[a]KBr wafer spectra of hydrochlorides. [b]Usually over br range; often group of relatively sharp bands.

Refs. p. 210.

i. N-O Compounds: Oximes, Nitro, N-Oxides, Nitrates

Group	Position or Range	Mode and Intensity	Other
R_2C=N-OH (oximes)	3650-3500(2.74-2.86) 1685-1650(5.94-6.06) 960-930 (10.42-10.75)	ν OH (W) ν C=N (W) ν N-O (S)	At lower $\bar{\nu}$ and S if H-bonded. C=N is S when conjugated; subject to same ring size effects as carbonyl compounds (see below).
	1075-975(9.30-10.26)	ν N-O (S) for quinone monoximes	
R-NO_2 (nitro)	ν-as (VS)(N-O)	ν-s (VS)(N-O)	ν C-N (M)
Alkyl Olefinic Aryl	1570-1550(6.37-6.45) 1505-1500(6.65-6.67) 1560-1500(6.41-6.67)	1360-1320(7.36-7.58) 1360-1330(7.36-7.52) 1356-1340(7.37-7.46)	All 3 types at 870-830(11.49-12.05); often obscured in aryl
N-NO_2 (nitramine)	1600-1530(6.25-6.54) 1300-1260(7.69-7.94)	ν-as (N-O) ν-s (N-O)	
R-N=O (nitroso)	1600-1500(6.25-6.67)	ν (N=O)(S)	For monomer
Aliphatic dimer Aryl dimer	cis: 1330(7.52) and 1410(7.09) cis: 1390(7.20) and 1410(7.09)	trans: 1290-1176(7.75-8.50) trans: 1300-1250(7.69-8.00)	
N-N=O (nitrosamine)	1500-1450(6.67-6.90)	ν (N=O)	
N-Oxides	970-950(10.31-10.53) 1300-1200(7.69-8.33)	ν N-O (VS) ν N-O (VS)	Aliphatic Aromatic (pyridine)
	H-bonding (protonic solvent, e.g.) lowers $\bar{\nu}$ by 20-40 cm^{-1}		
N=N(-)-O (azoxy)	1310-1250(7.65-8.00)	ν N-O (S)	
R-O-N=O (nitrites)	1680-1650(5.95-6.06) (trans) and 1625-1610(6.16-6.21)(cis) ν N=O (S) 850-750(11.76-13.33)	 ν O-N (S)	
R-O-NO_2 (nitrates)	1650-1620(6.06-6.17) 1300-1250(7.69-8.00) 870-830(11.50-12.05)	ν N=O (as)(S) ν N=O (s) (S) ν O-N (S)	Also, δ NO_2 band near 700(14.29)

Refs. p. 210.

j. Miscellaneous Unsaturated N-Compounds (Azo, Hydrazones, etc.): ν C=N

$R_1R_2C{=}NR_3$	1660-1640(6.02-6.10)	R_1, R_2 = alkyl or aryl; R_3 = aryl, alkyl, H
$R_1R_2C{=}NNHR_3$	1635-1625(6.12-6.15)	R_1, R_2 = alkyl or aryl; R_3 = alkyl
$R_1R_2C{=}NN(R_3)_2$	1650-1625(6.06-6.15)	R_1, R_2 = alkyl or aryl; R_3 = alkyl
$CH_3O(R_1)C{=}NAr$	1665 (6.01)	R_1 = aryl
$CH_3S(R_1)C{=}NAr$	1611 (6.21)	R_1 = aryl
ArN=NAr	trans 1440-1410(6.94-7.09)	cis 1510(6.62) (ν N=N; not very useful)

k. C-Halogen Functions: ν C-X

C-F	1000-1100(10.00-9.09)(S)	C-Cl	850-800(11.76-20.0)(M)
CF_2, CF_3	1350-1200(7.41-8.33)(VS)(as)	$C{-}Cl_x$	∿800-850 (x = 2-4)
	1200-1100(8.33-9.09)(VS)(s)	ArCl	∿1100(9.09)(M)
=CF, ArF	1250-1100(8.00-9.09)(VS)	C-Br	680-500(14.71-20.0)(M)
		C-I	500-200(20.0-50.0)(M)

l. Carbonyl Compounds. In general, ν C=O values are sensitive to phase and solvent and are in the order (cm^{-1}): vapor > hexane > $CCl_4(CS_2)$ > $CHCl_3$ (e.g., $CH_3COC_6H_5$ 1709, 1697, 1692, 1683, respectively); values in $CHCl_3$ can be ∿10-20 cm^{-1} lower than in CCl_4. Values for KBr and other matrix phases are sensitive to crystal structure (especially amides), as well as to H-bonding (acids, pr and sec amides, imides, ureas, etc.) in which case absorptions are often shifted to lower $\bar{\nu}$ by 30 to 40 cm^{-1} or more compared to diluted $CHCl_3$ values (values for more concentrated solutions fall in between). Conjugation lowers ν C=O by ∿30 cm^{-1} for the first C=C (or aryl ring) and by another ∿15 cm^{-1} for the second; further conjugation effects are small. In such cases, ν C=C is intensified and ν C=O broadens compared to nonconjugated species. H-bonding also lowers ν C=O, as in hydroxyketones, β-diketo (enol) structures, o-amino or o-hydroxyaryl ketones, etc.; $\Delta\bar{\nu}$ for weak, medium and strong H-bonding is, respectively, <10, 10-50, and 50-150 cm^{-1}. Alkyl substitution at the α-position to a carbonyl effects a decrease of ∿4 to 5 cm^{-1} per group (e.g., in diluted CCl_4: $(CH_3CH_2)_2CO$ 1720; $[(CH_3)_2CH]_2CO$ 1712; $CH_3COC(CH_3)_3$ 1711; $[(CH_3)_3C]_2CO$ 1688. Note, however, cyclohexanone 1717; 2,6-dimethylcyclohexanone 1717; but 2,2,6- and 2,2,6,6-tetramethylcyclohexanone 1708 and 1698, respectively). Note also that such α-effects operate for the ether oxygen in esters and lactones (i.e., $C_6H_5CO_2CH_2CH_3$ 1724, $C_6H_5CO_2CH(CH_3)_2$ 1720, $C_6H_5CO_2C(CH_3)_3$ 1717 as well as for the carbonyl α-position. In addition, an α-halo substituent (Cl or Br) in the carbonyl nodal plane (e.g., equatorial 2-chlorocyclohexanone) causes an increase of 15-20 cm^{-1} (∿10 for I); an axial-halo (parallel to π-bond) has almost no effect. (For effects in the uv, see p. 225)

Refs. p. 210.

The following summary table for simple systems was constructed with data from many sources, especially references 1b, 1e, 4, 7, and 8.

Carbonyl Stretch for Aliphatic Cyclic Functions (Dilute CCl_4)

Function	Acyclic Reference[a]	Ring Size			
		3	4	5	6
Ketone[b]	1719 (5.82)	1820 (5.49)	1780 (5.62)	1745 (5.73)	1717 (5.82)
Lactone	1750 (5.71)		1841 (5.43)	1775 (5.63)	1750 (5.71)
Anhydride[c]	1832, 1762 (5.46, 5.68)		1824, 1760[d] (5.48, 5.68)	1871, 1795[e] (5.34, 5.57)	1821, 1774 (5.49, 5.64)
Lactam (N-H)	1700 (5.88)		1772 (5.64)	1716 (5.83)	1690 (5.92)
Lactam ($N-CH_3$)	1653 (6.05)	1835[f] (5.45)	1750[g] (5.71)	1698 (5.89)	1752[h] (5.71)
Lactam ($N-COCH_3$)	1706 (5.86)			1745, 1702[i] (5.73, 5.88)	1701[j] (5.88)
Imide (N-H)[c]	1748, 1721[k] (5.72, 5.81)		1761[l] (5.68)	1753, 1727[m] (5.70, 5.79)	1710, 1700 (5.85, 5.88)
Imide ($N-CH_3$)[c]	1706 (5.86)		1735-1750[n] (5.76-5.71)	1721, 1705 (5.81, 5.87)	1729, 1686 (5.78, 5.93)
Carbonate	1740 (5.75)			1819 (5.50)	1771 (5.65)
Urea (N-H)	1717(W), 1695 (5.82, 5.90)	1880, 1862[o] (5.32, 5.37)		1735(W), 1718[p] (5.76, 5.82)	1718[q] (5.82)
Urethane (N-H) (carbamate)	1735 (5.76)			1783[r] (5.61)	1743[s] (5.74)

[a]In the order of functions listed: CH_3COCH_3, $CH_3CO_2CH_3$, $(CH_3CO)_2O$, $CH_3CONHCH_3$, $CH_3CON(CH_3)_2$, $(CH_3CO)_2NCH_3$, $(CH_3CO)_2NH$, $(CH_3CO)_2NCH_3$, $(RO)_2CO$, $(RNH)_2CO$, $RNHCO_2C_2H_5$. [b]For larger rings: 7 (1706 cm^{-1}), 8 (1704), 9 (1700), then gradually increasing again approaching 6-ring value. [c]High-frequency band usually the more intense one in acyclic compound; the opposite for cyclic C-disubstituted. [d]No

Refs. p. 210.

solvent given [A. Duckworth, J. Org. Chem., 27, 3146 (1962)]. [e]Maleic anhydride 1855, 1784; phthalic anhydride 1854, 1779. [f]N-t-Bu, neat. [g]N-t-Bu in $CHCl_3$. [h]For 7-ring (caprolactam). [i]1735, 1685 neat. [j]1685 neat. [k]Third peak at 1698. [l]Band at 1824 in Raman. [m]Also 1780 (VW). Maleimide 1790, 1710; phthalimide 1775, 1735. [n]In CH_2Cl_2; value depends on C-substituent. Weak bands also at 1810 and 1890. A. Ebnöther, et al., Helv. Chim. Acta, 42, 918 (1959). In Nujol, 1740-1720 [E. Testa, Ann. Chem., 660, 118 (1962)]. For more on malomidides, see C. Fayat, Comp. Rend., 264 (C), 2009 (1967). [o]Di-t-Bu compound. [p]1661 in KBr wafer. For N-alkyl, 1667 neat or KBr. [q]1692 in KBr wafer. [r]1750 neat (1741, KBr wafer). For 1,3-thiazolidin-2-one 1645 (CCl_4). [s]1667 in Nujol mull (br, with several peaks).

Aldehydes: ν C=O (S)

Alkyl	1725 (5.80)	ν C-H characteristic for aldehydes: 2720 (3.67) (M) and a combination band at 2820 (3.55) (M). H-Bonding has strong effect on C=O [e.g., salicaldehyde, 1666 (6.00)]. CCl_3CHO at 1768 (5.65).
α,β-unsaturated	1685 (5.94)	
α,β,γ,δ-unsaturated	1675 (5.97)	
Aryl	1700 (5.88)	

Ketones (RCOR'): ν C=O (S)

R=R'=alkyl 1720-1710 (5.81-5.85) R=alkyl, R'=cyclopropyl 1695 (5.90)
R=alkyl, R'=aryl 1690 (5.93) R=alkyl, R'=α,β-unsaturated 1675 (5.97)
R=R'=aryl; R=R'=α,β-unsaturated; R=alkyl, R'=α,β,γ,δ-unsaturated 1665 (6.01)

[For s-cis ( ) ν C=C may appear <1600 cm^{-1}; its intensity is enhanced, and is comparable to that of ν C=O, which is itself less intense than in s-trans for ( ); no intensity enhancement for ν C=C in s-trans].

All types have C-CO-C bending and C-CO stretch: aliphatic 1200-1100 (8.33-9.09), aromatic 1300-1200 (7.70-8.33). Both are M and may appear as one to several bands. Bicyclic ketones [e.g., (2.2.1)-7-one, etc.] are strained and absorb at >1750 cm^{-1}.

α-Diketones (-CO-CO-)

s-trans (acyclic) 1720-1705 (5.81-5.87) (as)
s-cis (cyclic) 6 ring: 1760, 1730 (5.68, 5.78) 5-ring: 1775, 1760 (5.63, 5.68)
For enol [C=C(OH)-CO-]: 1675 (5.97) with ν C=C at 1650 (6.06) (S). Tropolone has ν C=O 1600 (6.25) [tropone 1650 (6.06)].

1,3-Diketones [keto-form ~1720 (5.81)]

"Free"-enol 1650 (6.06) Intra H-bond 1615-1600 (6.19-6.25)
Both enol forms have ν C=C ∿ 1600 (6.25), intense as ν C=O.

Refs. p. 210

Quinones (1,2 and 1,4)

6-Ring 1690-1660 (5.92-6.02) 5-ring (acenaphthene) 1720-1700 (5.81-5.88)
1,4-Quinones may show >1 band (Fermi resonance). If carbonyls are in different rings (extended conjugation), ν 1660-1635 (6.02-6.12). Intra H-bonding from peri-OH (e.g., 1-hydroxy-9,10-anthraquinone), ν ∿ 1630 (6.13).

Carboxylic Acids (Dimeric State Mostly)

Aliphatic	1765-1750 (5.67-5.71)	ν C=O (VS)	Monomer
	1720-1710 (5.81-5.85)	ν C=O (VS)	H-Bond dimer
	3000-2500 (3.33-4.00)	Group (br) of small bands for dimer (ν O-H)	
	3550 (2.82)	ν O-H (M)	Monomer
	1420 (7.04) and 1300-1200 (7.69-8.33)		Combination bands (M) for dimer
	920-860 (10.87-11.63)	δ O-H (M)(br)	for dimer
Aromatic, vinyl	1720 (5.81)	ν C=O (VS)	Monomer
	1690 (5.92)	ν C=O (VS)	Dimer
α-Halo	Shifts of +10-20 cm^{-1} (Br, Cl); +50 (CF_3)		
Carboxylate (RCO_2^-)	1610-1550 (6.06-6.45),	ν-as (VS),	
	1400 (7.15)	ν-s (S)	

Carboxylic Acid Halides: RCOCl [ν C=O (S)]

R=alkyl 1810-1790 (5.53-5.59)
R=aryl or α,β-unsatd. 1780-1750 (5.62-5.71) Also 1750-1700 (5.71-5.88)(W)(1 or 2 bands).
RCOBr slightly > RCOCl; RCOI < RCOCl; RCOF (vapor) ∿ 1870 (5.35)

Carboxylic Acid Anhydrides (and Diacylperoxides)

For cyclic, see table on page 196; in all cases below, the low frequency (s) is less intense than the high (as).

Aliphatic 1830-1810 (5.46-5.53) and 1770-1750 (5.65-5.71)
Aromatic, vinyl 1795-1775 (5.57-5.63) and 1735-1715 (5.76-5.83)
[ν C-O-C (S) from 1300-1050 (7.69-9.52); if conjugated, the higher $\bar{\nu}$ range]

RC(=O)-O-O-C(=O)R

R=aliphatic 1820-1810 (5.49-5.52) and 1795-1785 (5.57-5.60)
R=aryl 1805-1780 (5.54-5.62) and 1780-1760 (5.62-5.68)

Refs. p. 210.

Esters and Lactones (See Table on Page 196 for More Data)

RCO_2R'			
R	R'	ν C=O (S)	ν C-O-C (VS) (br)
Alkyl	Alkyl	1745 (5.73)	1275-1050 (7.85-8.70) (2 bands)
H	Alkyl	1735 (5.76)	1185 (8.44), 1160 (8.62)
CH_3	Alkyl	1745 (5.73)	1245 (8.03)
Aryl, vinyl	Alkyl	1725-1715 (5.80-5.83)	1300-1250 (7.69-8.00) and 1200-1100 (8.33-9.09)
Phthalates	Alkyl	1780 (5.62)	1120 (8.93), 1070 (9.35)
Alkyl	Aryl, vinyl	1770-1755 (5.65-5.70)	1210 (8.26); ν C=C (vinyl) 1690-1650 (5.92-6.06)
Aryl, vinyl	Aryl, vinyl	1735 (5.76)	1260 (7.94), 1200 (8.33) (benzoates)
-C(=O)-	Alkyl	1745 (5.73)	
-C(=O)-C-	Alkyl	1650 (6.06) (enol) (VS) (keto unaffected)	ν C=C 1630 (6.14) (VS, br)
α-Pyrones		1740-1720 (5.75-5.81) (split; higher band stronger)	ν C=C 1650-1620 (6.06-6.17), 1570-1540 (6.37-6.49)
γ-Pyrones		1680-1650 (5.95-6.06) (infrequently split)	ν C=C 1650-1600 (6.06-6.25), 1590-1560 (6.29-6.41)
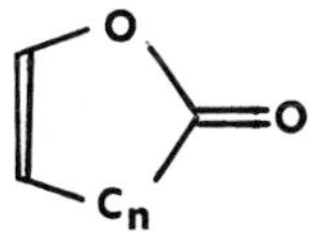 Unsaturated lactone ring (O, C=O, C_n), n=1		1800 (5.56)	ν C=C 1660 (6.02)
n=2		1760 (5.68)	ν C=C 1685 (5.93)

Refs. p. 210.

Refs. p. 210.

Amides and Lactams (See Table on Page 196 for More Data)

Type	ν N-H (M)		Amide I ≡ ν C=O (S)		Amide II ≡ δ N-H (S)	
	Free	Associated	Free	Associated	Free	Associated
$-CONH_2$ (aliphatic)	3520, 3400 (2.84, 2.94)	3200-3050 (3.12-3.28) (several)	1690 (5.92)	1650 (6.06)	1600 (6.25)	1640 (6.10)
			Amide III (ν C-N) at 1420-1405 (7.04-7.12)			
-CONHR (aliphatic)	3440 (2.91)	3300 (3.03)	1680 (5.95)	1640 (6.10)	1530 (6.54) 1260 (7.94)	1570 (6.37) 1300 (7.69)
	Amide II and III are mixed (two bands)					
$-CONR_2$ (aliphatic)	---	---	1650 (6.06)		---	---
	Amide II and III absent; Amide I not very sensitive to phase or conc.					
Lactams (N-H)	3440 (2.91)	3175 (3.15) (dimer)	See table on page 196 for Amide I.			
	If highly associated, band at 3070 (3.26).					
α-Pyridones	3400 (2.94)	3200-2400 (3.12-4.17) (several)	1690-1650 (5.92-6.06)	Slightly lower	Similarly for α-quinolones	
γ-Pyridones	3400 (2.94)	3200-2400 (3.12-4.17) (several)	1650-1630 (6.06-6.13)	1550 (6.45) (KBr)	Similarly for γ-quinolones	

For electron withdrawing or unsaturated groups on N (halo, vinyl, aryl), and for C=C-CON, Amide I rises ∿15 cm^{-1}; for C=C-CON-C=C, band decreases by ∿15 cm^{-1}. Vinylogous amides (C-CO-C=C-N) have ν C=O ∿30 to 40 cm^{-1} lower than ordinary. Amide solid state (matrix) spectra are very sensitive to crystal structure; association is always observed if N-H present.

Imides (7,10) (See Table on Page 196 for Some Cyclic)

RCO-NX-COR'			ν C=O	Solvent
R	R'	X		
CH_3	Alkyl	H[a]	1745,1715(5.73,5.83)(S,M)	KBr
CH_3	Alkyl	Alkyl	1705,1690(5.87,5.92)(S,W)	CCl_4
Aryl,vinyl	Alkyl	H[a]	1740,1700(5.75,5.88)(W,S)	KBr (1730,1700 $CHCl_3$)

Aryl	Aryl	Aryl'	1670(5.99)(S)(2 bands if O_2N-aryl)	$CHCl_3$ (1680,1645 Nujol)
Aryl	Aryl	ArCO	1730-1690(5.78-5.92)(S)	$CHCl_3$
$-(CH_2)_2-$		Alkyl	1760,1690(5.68,5.92)(M,S)	KBr
$-(CH_2)_2-$		ArCO	1780,1720,1680(5.62,5.81,5.95)(W,M,S)	$CHCl_3$
$-(CH_2)_2-$		Vinyl,aryl	1780,1720(5.62,5.81)(W,S)	$CHCl_3$
(cyclic imide: C_n ring, N–C(=O)R)		R=alkyl	1730,1690(5.78,5.92)(W,S)	CCl_4 } n = 3,4
		R=aryl	1690(5.92) (S)	CCl_4 }

[a]For X=H, ν N-H at 3400 (2.94) (W)--very characteristic (solvent insensitive).

Miscellaneous (ν C=O) (for Cyclic, See Table on Page 196)

RNHCONHR (ureas) 1660(6.02). In solid state, 1640(6.00)(br) for N,N'-diaryl
RCONHNHCOR 1740-1700(5.75-5.89) and 1707-1683(5.86-5.94)
RCONHOH (Hydroxamic acids) Alkyl: 1640(6.00) in solid state

ROCON< (carbamates)	N-Aliphatic	N-Aromatic	
pr (NH_2):	1728-1722(5.79-5.81)	---	These are all in $CHCl_3$ solution
sec:	1722-1705(5.81-5.86)	1739-1719(5.75-5.82)	
t:	1691-1683(5.91-5.94)		

(pr and sec urethanes show typical amide ν N-H and Amide II bands)
ROCOCl (chloroformates) Alkyl: 1780-1775(5.62-5.63) Aryl: 1785(5.60)
R_2NCOCl (carbamoyl chlorides) Alkyl-aryl: 1745-1739(5.73-5.75)
$(RO)_2CO$ (carbonates) Dialkyl: 1740(5.75) Diaryl: 1815-1775(5.51-5.63)
Alkyl-aryl: 1787-1757(5.60-5.69) All have ν C-O 1250(8.00)(S)

m. Silicon, Germanium, and Tin Compounds

Si-H	2250-2150(4.44-4.65)	ν(S)	900-800(11.11-12.50)	δ(S)
SiH_2	980-930(10.20-10.75)	δ(S)		
SiH_3	960-900(10.42-11.11)	δ(S)	Two bands	

(Electronegative groups on Si raise ν Si-H ≥ 2225 range)

(Continued)

Refs. p. 210.

Group	Data
$Si-CH_3$	800(12.50) ν(S) 1260(7.93) δ(VS, sharp)
$Si(CH_3)_2$	820-800(12.19-12.50) ν(S) 1260(7.93) δ(VS, br)
$Si(CH_3)_3$	840 and 755(11.90 and 13.25) ν(S) 1260(7.93) δ(S)
$R_3Si-COR'$	1620(6.17) ν C=O
$Si-C_6H_5$	1630(6.13), 1430(7.00) ν C-C 1125(8.89)(VS) [doublet for $(C_6H_5)_2Si$)]
Si-OH	3700-3650(2.70-2.74) ν OH (M, free) [Associated 3400-3200(2.94-3.13)]
	870-820(11.49-12.20) δ OH
Si-O-R	ν Si-O(S): R=H 910-830(11.00-12.05); R=alkyl 1100-1050(9.09-9.52) (VS doublet for Et); R=aryl 970-920(10.31-10.87); R=Si 1100-1000(9.09-10.00)
Si-Cl	<670(14.92)
Si-F	1000-800(10.00-12.50)(S); SiF_2 943-910(10.60-11.00)(S) and 910-870(11.00-11.50)(M); SiF_3 980-945(10.20-10.58)(S) and 910-860(11.00-11.63)(M)
Ge-H	2080-2050(4.81-4.88)
Sn-H	1850-1800(5.40-5.56)

n. Phosphorus Compounds

Group	Data
P-H	2450-2300(4.08-4.35) ν(M) [δ VW ∿1250-1000(8.00-10.00); for PH_2 1090-1080(9.19-9.26)(M)]
P-OR	R=H 2700-2600(3.70-3.85) νOH (associated, br, S)
	R=alkyl 1050-1000(9.52-10.00) ν P-O(S)
	R=aryl 1260-1160(7.94-8.62), 950-875(10.53-11.42) ν P-O, C-O
	R=P 970-900(9.52-11.11)
P=O	(ν, S) (data mostly for condensed phases - neat films, matrix)
$(RO)_3PO$	R=alkyl 1286-1258(7.78-7.95) R=aryl 1314-1290(7.61-7.75)
	(With OH on α-carbon, 1240-1180(8.06-8.47))
$X(RO)_2PO$	X = amino, halo, hydroxy, etc. 1300-1250(7.69-8.00)
R_3PO	R = alkyl 1180-1100(8.48-9.09) R=aryl 1145-1100(8.73-9.09)
R_2XPO (R=alkyl)	X = O-alkyl, OH, Cl, S-alkyl 1220-1140(8.20-8.77)
P=S	800-650(12.50-15.38) (W)
P-N	960(10.42) ν(M)

o. Sulfur and Selenium Compounds

Group	Data
S-H	2600-2550(3.85-3.92) ν(W) H-bonding shifts small
Se-H	2300-2280(4.35-4.39) ν(neat)
S-R	R = aryl, vinyl ∿600(16.67)(M) (too weak for use if R = alkyl or S)
$S-CH_3$	1325(7.55) δ CH(M)

R	R'	$\bar{\nu}$	
RR'C=S		C=S (M)	
Aryl	Aryl	1225-1200(8.16-8.33)	(Alkyl exist as trimeric sulfides)
Alkyl, aryl	S-R"	1225-1190(8.16-8.40)	Thioesters
S-R"	S-R"	1100-1060(9.09-9.36)	Thiocarbonates
S-Alkyl	O-Alkyl	1070-1020(9.35-9.80) [Also ∿ 1125(8.89) and 1225 (8.16)]	
Alkyl, aryl (includes thiolactams)	N<	1570-1400(6.37-7.15), 1420-1260(7.04-7.93), and 1140-940(8.77-10.64)	Similar to amide I, II, III bands
Alkyl, aryl	Cl	1235-1225(8.10-8.16)	
R-SO-R'		S=O (S)	
Alkyl (sulfoxides)	Alkyl	1060-1040(9.43-9.62) Conjugation lowers $\bar{\nu}$ 10-20 cm^{-1} (aryl or vinyl)	Dimethyl 1070; diethyl 1066
Alkyl, aryl	S-Alkyl, aryl	1108-1095(9.02-9.12)	Thiosulfoxides
Alkyl, aryl	OH	1090(9.17)	Sulfinic acid
Alkyl	O-Alkyl	1130(8.85)	Sulfinates
O-Alkyl	O-Alkyl	1200(8.33)	Sulfites
O-Aryl	O-Aryl	1245(8.03)	Sulfites
$R-SO_2-R'$		S=O (S)(as,s)	
Alkyl	Alkyl	1340-1300(7.46-7.69), 1150-1135(8.70-8.81)	
Alkyl	Aryl	1335-1325(7.49-7.55), 1160-1150(8.62-8.70)	
Aryl	Aryl	1360-1335(7.35-7.49), 1170-1160(8.55-8.62)	
Alkyl, aryl	OH	1350-1340(7.41-7.46), 1165-1150(8.59-8.70)	(anhydrous)
(Note. Sulfonic acid hydrates have		1250-1150, 1100-1000, as do salts, RSO_3^-)	
Alkyl, aryl	O-Alkyl, aryl	1380-1340(7.24-7.46), 1195-1170(8.37-8.55)	
O-Alkyl, aryl	O-Alkyl, aryl	1420-1380(7.04-7.24), 1200-1150(8.33-8.70)	
Alkyl, aryl	N<	1360-1330(7.35-7.52), 1180-1160(8.47-8.62)	
NR_2	NR_2	1320(7.58), 1145(8.73)	
Alkyl, aryl	Cl	1390-1360(7.20-7.35), 1185-1170(8.44-8.55)	

o. Inorganic Ions (1d,4) (Solid Phase Spectra)[a]

Ion	$\bar{\nu}$	Ion	$\bar{\nu}$
NH_4^+	3300-3030(3.03-3.30)(VS)	$SO_3^=$	990-930(10.10-10.75)(M)
	1430-1390(7.00-7.20)(S)		680-620(14.70-16.13)(VS)
CN^-, SCN^-, NCO^-	2200-2000(4.55-5.00)(S)	$SO_4^=$	1130-1080(8.85-9.26)(VS)
$CO_3^=$	1450-1410(6.90-7.09)(S)		680-610(14.70-16.39)(S)
	880-860(11.36-11.63)(M)	HSO_4^-	1200-1050(8.33-9.52)(2M)
NO_2^-	1250-1230(8.00-8.13)(VS)		870-850(11.49-11.76)(M)
	840-800(11.90-12.50)(W)		600-570(16.67-17.53)(S)
NO_3^-	1380-1350(7.25-7.42)(VS)	PO_4^{-3}	1025-1000(9.76-10.00)(2S)
	840-815(11.90-12.27)(M)	$HPO_4^=$	1100-1000(9.09-10.00)(S) and
Silicates	1100-900(9.09-11.11)(S)	$H_2PO_4^-$	1000-800(10.00-12.50)(M)

[a]A recent monograph on inorganic ir spectra is available (1g).

Refs. p. 210.

C. Far-Infrared Spectral Data

For details, see reference 9. Several charts are presented here that summarize far-infrared absorptions. The following table is reprinted by permission of Beckman Instruments, Inc., Fullerton, Calif.

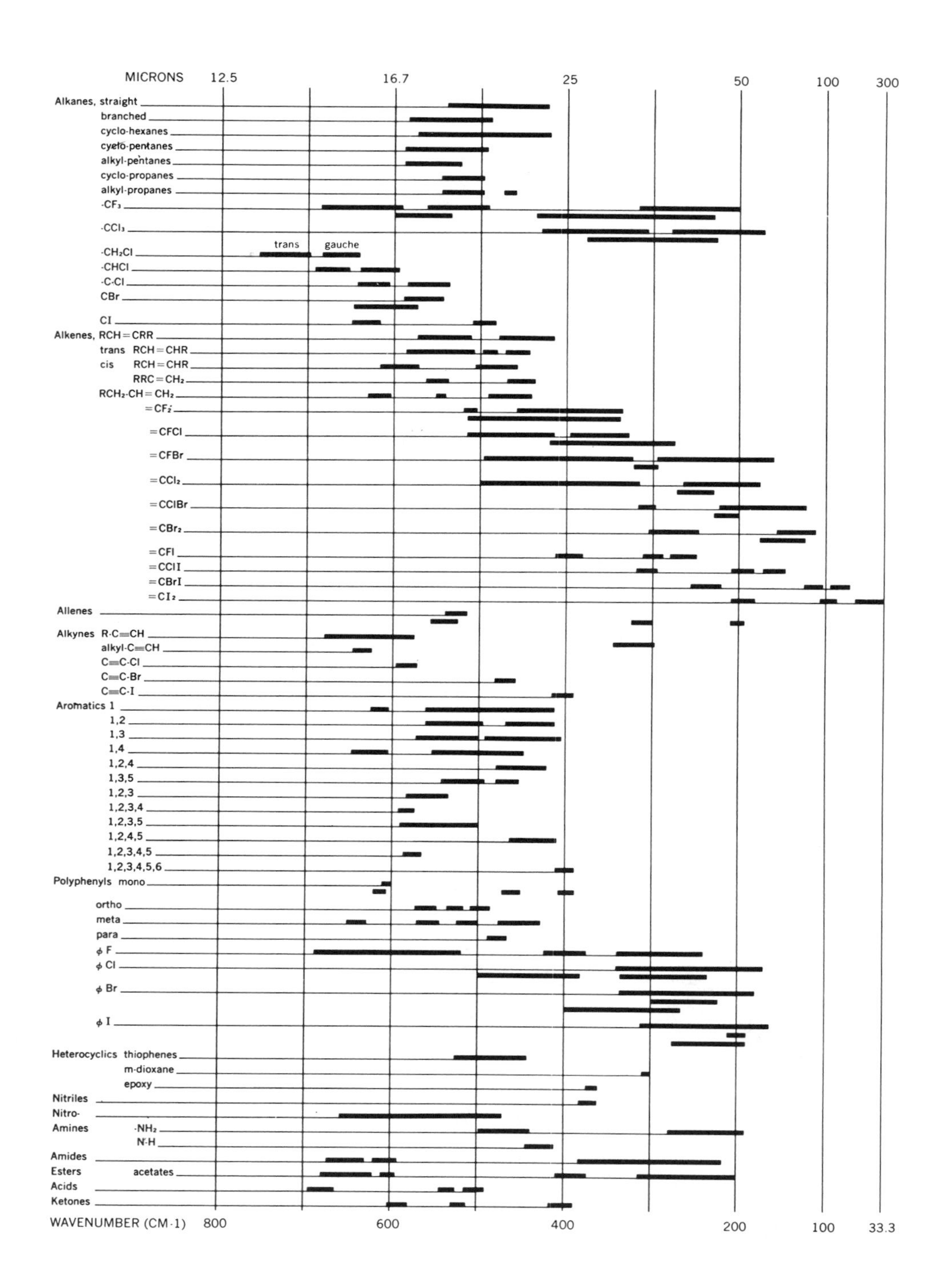

Refs. p. 210.

Beckman Far Infrared Chart (Continued)

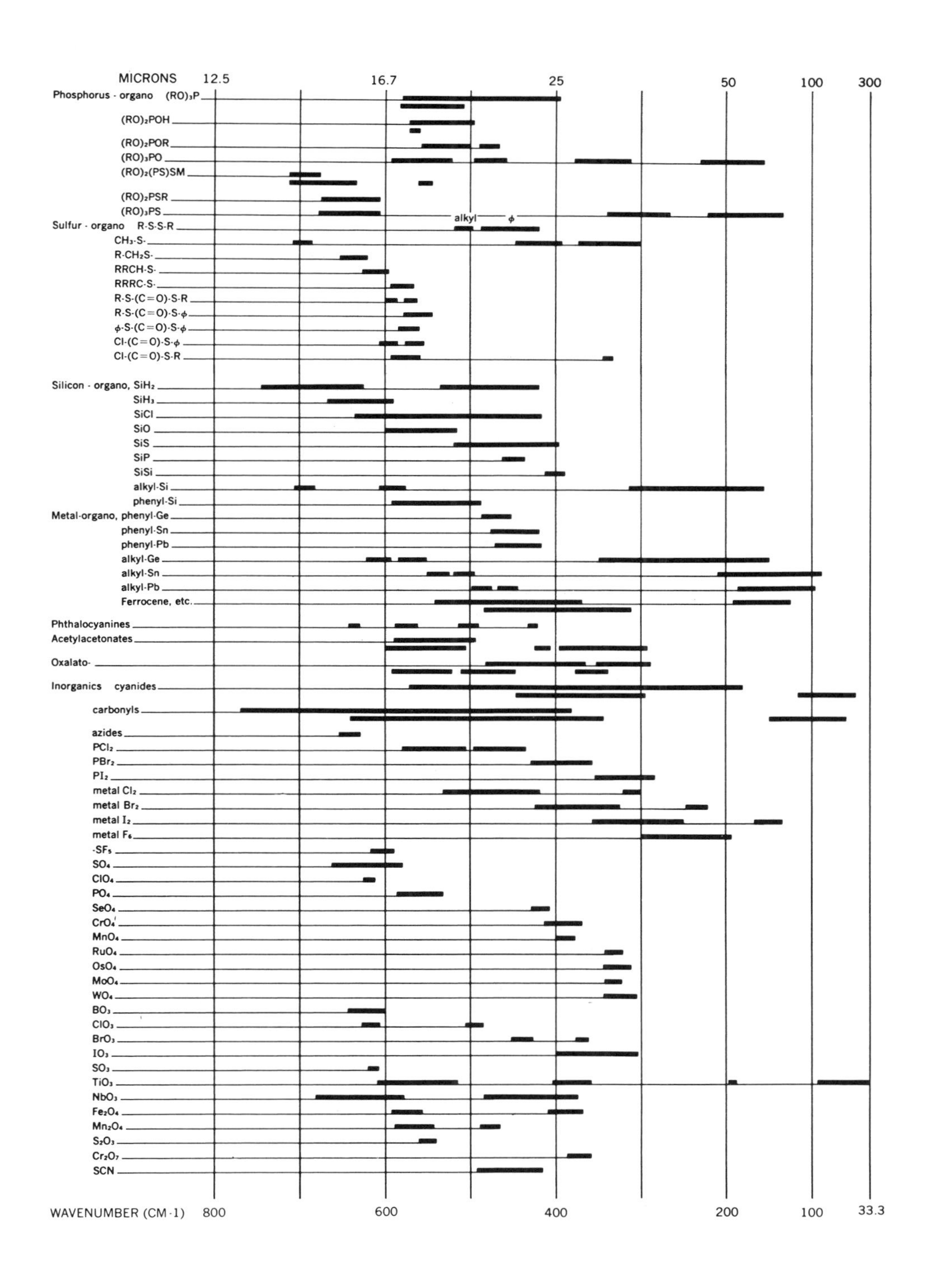

MICRONS 12.5 16.7 25 50 100 300
Phosphorus - organo (RO)₃P
(RO)₂POH
(RO)₂POR
(RO)₃PO
(RO)₂(PS)SM
(RO)₂PSR
(RO)₃PS
alkyl
ϕ
Sulfur - organo R-S-S-R
CH₃-S-
R-CH₂S-
RRCH-S-
RRRC-S-
R-S-(C=O)-S-R
R-S-(C=O)-S-ϕ
ϕ-S-(C=O)-S-ϕ
Cl-(C=O)-S-ϕ
Cl-(C=O)-S-R
Silicon - organo, SiH₂
SiH₃
SiCl
SiO
SiS
SiP
SiSi
alkyl-Si
phenyl-Si
Metal-organo, phenyl-Ge
phenyl-Sn
phenyl-Pb
alkyl-Ge
alkyl-Sn
alkyl-Pb
Ferrocene, etc.
Phthalocyanines
Acetylacetonates
Oxalato-
Inorganics cyanides
carbonyls
azides
PCl₃
PBr₃
PI₃
metal Cl₂
metal Br₂
metal I₂
metal F₆
-SF₅
SO₄
ClO₄
PO₄
SeO₄
CrO₄
MnO₄
RuO₄
OsO₄
MoO₄
WO₄
BO₃
ClO₃
BrO₃
IO₃
SO₃
TiO₃
NbO₃
Fe₂O₄
Mn₂O₄
S₂O₃
Cr₂O₇
SCN
WAVENUMBER (CM-1) 800 600 400 200 100 33.3

Refs. p. 210.

The following two charts are reproduced with permission from Handbook of Analytical Chemistry, McGraw-Hill, New York, 1963, pp. 6-160 to 6-163. The first one is a spectra-structure correlation chart from 15 to 35 μm for alkanes, alkenes, cycloalkanes, and aromatic hydrocarbons, while the second chart (p. 207) is for heterocyclic and organometallic compounds, and aliphatic derivatives. V = variable, W = weak, M = medium, S = strong, M-S = medium to strong. The bands indicated near 300 cm^{-1} by dotted lines are often not real.

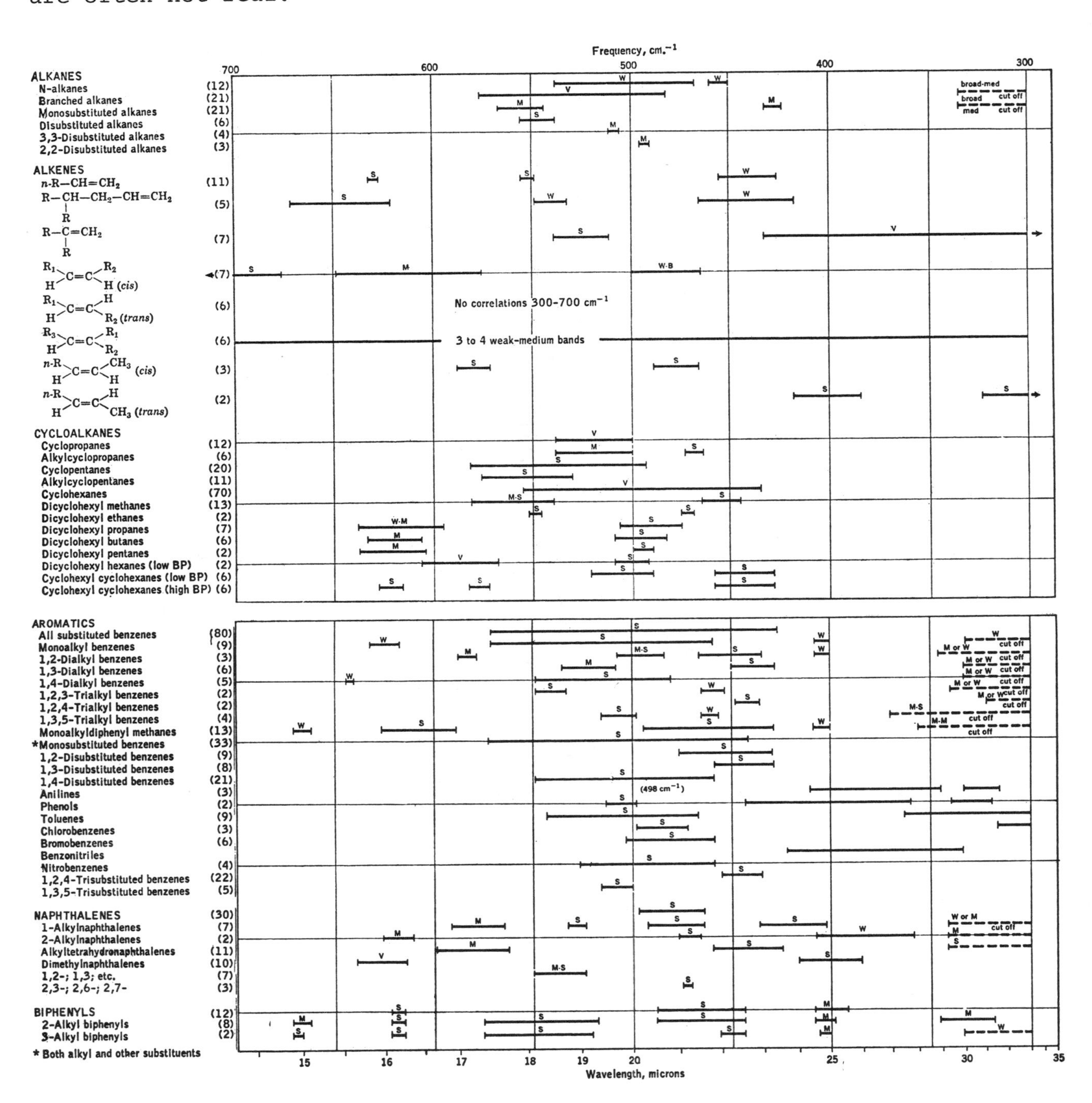

Refs. p. 210.

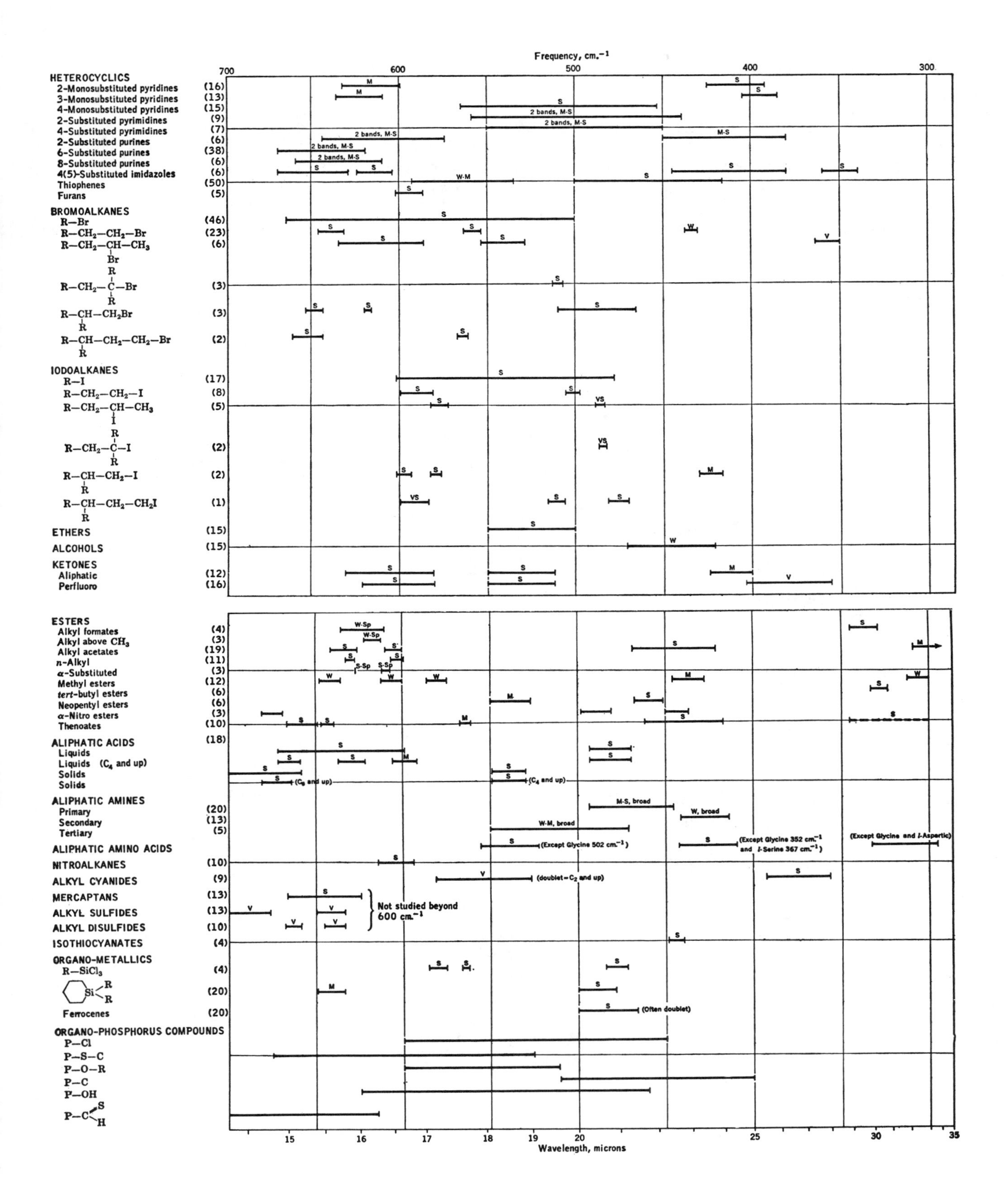

Frequency, cm.$^{-1}$
700
600
500
400
300
HETEROCYCLICS
2-Monosubstituted pyridines (16)
3-Monosubstituted pyridines (13)
4-Monosubstituted pyridines (15)
2-Substituted pyrimidines (9)
4-Substituted pyrimidines (7)
2-Substituted purines (6)
6-Substituted purines (38)
8-Substituted purines (6)
4(5)-Substituted imidazoles (6)
Thiophenes (50)
Furans (5)
2 bands, M-S
W-M
M-S
BROMOALKANES
R—Br (46)
R—CH$_2$—CH$_2$—Br (23)
R—CH$_2$—CH—CH$_3$ (6)
Br
R—CH$_2$—C(R)(R)—Br (3)
R—CH(R)—CH$_2$Br (3)
R—CH(R)—CH$_2$—CH$_2$—Br (2)
IODOALKANES
R—I (17)
R—CH$_2$—CH$_2$—I (8)
R—CH$_2$—CH(I)—CH$_3$ (5)
R—CH$_2$—C(R)(R)—I (2)
R—CH(R)—CH$_2$—I (2)
R—CH(R)—CH$_2$—CH$_2$I (1)
ETHERS (15)
ALCOHOLS (15)
KETONES
Aliphatic (12)
Perfluoro (16)
ESTERS
Alkyl formates (4)
Alkyl above CH$_3$ (3)
Alkyl acetates (19)
n-Alkyl (11)
α-Substituted (3)
Methyl esters (12)
tert-butyl esters (6)
Neopentyl esters (6)
α-Nitro esters (3)
Thenoates (10)
W-Sp
S-Sp
ALIPHATIC ACIDS (18)
Liquids
Liquids (C$_4$ and up)
Solids
Solids
(C$_6$ and up)
(C$_4$ and up)
ALIPHATIC AMINES
Primary (20)
Secondary (13)
Tertiary (5)
M-S, broad
W, broad
W-M, broad
ALIPHATIC AMINO ACIDS
(Except Glycine 502 cm.$^{-1}$)
(Except Glycine 352 cm.$^{-1}$ and l-Serine 367 cm.$^{-1}$)
(Except Glycine and l-Aspartic)
NITROALKANES (10)
ALKYL CYANIDES (9)
(doublet—C$_2$ and up)
MERCAPTANS (13)
ALKYL SULFIDES (13)
ALKYL DISULFIDES (10)
Not studied beyond 600 cm.$^{-1}$
ISOTHIOCYANATES (4)
ORGANO-METALLICS
R—SiCl$_3$ (4)
Si R R (20)
Ferrocenes (20)
(Often doublet)
ORGANO-PHOSPHORUS COMPOUNDS
P—Cl
P—S—C
P—O—R
P—C
P—OH
P—C(=S)H
15
16
17
18
19
20
25
30
35
Wavelength, microns

Refs. p. 210.

The chart that follows is a spectra-structure correlation chart from 15 to 35 μm for inorganic ions. It is reproduced from Handbook of Analytical Chemistry, McGraw-Hill, New York, 1963, p. 6-164. It is a revised version of a chart appearing in F. A. Miller et al., Spectrochim. Acta, 16, 135 (1960) and is included with permission of the copyright owner, Maxwell Scientific Int., Inc., Elmsford, N. Y.

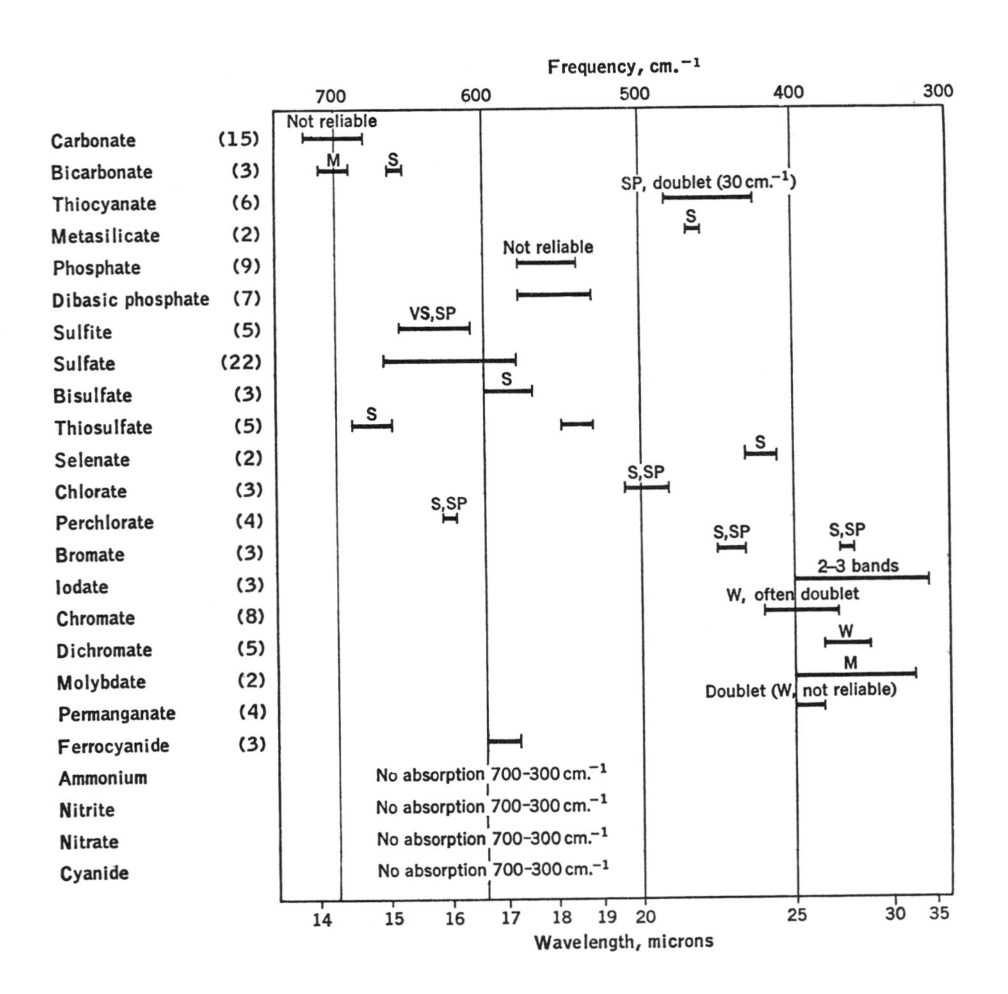

Refs. p. 210.

D. Near-Infrared Spectral Data

The following table summarizes correlations and average molar absorptivity data (l/mole-cm), mostly in CCl_4 [reprinted from R. Goddu and D. Delker, Anal. Chem., 32, 140 (1960) by permission of the American Chemical Society]. For reviews of the subject, see reference 11.

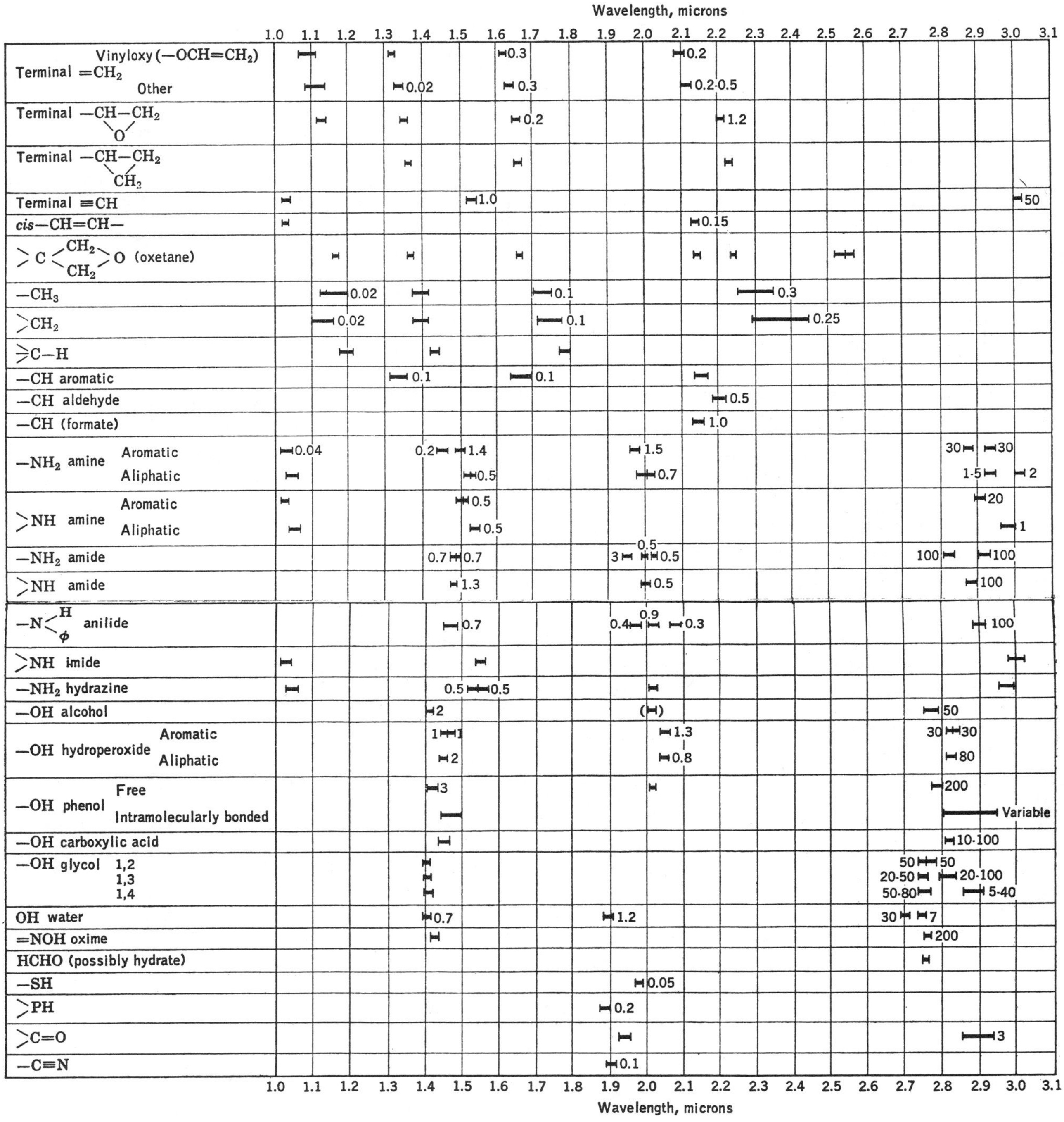

Wavelength, microns

Group	1.0	1.1	1.2	1.3	1.4	1.5	1.6	1.7	1.8	1.9	2.0	2.1	2.2	2.3	2.4	2.5	2.6	2.7	2.8	2.9	3.0
Terminal $=CH_2$, Vinyloxy ($-OCH=CH_2$)	⊢⊣			⊢⊣			⊢⊣ 0.3				⊢⊣ 0.2										
Terminal $=CH_2$, Other		⊢⊣		⊢⊣ 0.02			⊢⊣ 0.3					⊢⊣ 0.2-0.5									
Terminal $-CH-CH_2$ (epoxide, $-O-$)		⊢⊣		⊢⊣			⊢⊣ 0.2						⊢⊣ 1.2								
Terminal $-CH-CH_2$ (cyclopropane, $-CH_2-$)				⊢⊣			⊢⊣						⊢⊣								
Terminal $\equiv CH$	⊢⊣					⊢⊣ 1.0															⊢⊣ 50
cis—$CH=CH$—	⊢⊣											⊢⊣ 0.15									
$>C<(CH_2)_2>O$ (oxetane)		⊢⊣		⊢⊣			⊢⊣					⊢⊣	⊢⊣			⊢⊣					
$-CH_3$		⊢⊣ 0.02		⊢⊣				⊢⊣ 0.1					⊢⊣ 0.3								
$>CH_2$		⊢⊣ 0.02		⊢⊣				⊢⊣ 0.1						⊢⊣ 0.25							
$\geq C-H$		⊢⊣			⊢⊣			⊢⊣													
$-CH$ aromatic				⊢⊣ 0.1			⊢⊣ 0.1					⊢⊣									
$-CH$ aldehyde													⊢⊣ 0.5								
$-CH$ (formate)												⊢⊣ 1.0									
$-NH_2$ amine, Aromatic	⊢⊣ 0.04				0.2 ⊢⊣	⊢⊣ 1.4				⊢⊣ 1.5									30 ⊢⊣	⊢⊣ 30	
$-NH_2$ amine, Aliphatic	⊢⊣					⊢⊣ 0.5				⊢⊣ 0.7										1-5 ⊢⊣	⊢⊣ 2
$>NH$ amine, Aromatic	⊢⊣					⊢⊣ 0.5														⊢⊣ 20	
$>NH$ amine, Aliphatic	⊢⊣					⊢⊣ 0.5														⊢⊣ 1	
$-NH_2$ amide					0.7 ⊢⊣⊢⊣ 0.7					3 ⊢⊣	⊢⊣ 0.5 ⊢⊣ 0.5								100 ⊢⊣	⊢⊣ 100	
$>NH$ amide					⊢⊣ 1.3						⊢⊣ 0.5								⊢⊣ 100		
$-N<(H)(\phi)$ anilide					⊢⊣ 0.7					0.4 ⊢⊣	⊢⊣ 0.9 ⊢⊣ 0.3									⊢⊣ 100	
$>NH$ imide	⊢⊣					⊢⊣															⊢⊣
$-NH_2$ hydrazine	⊢⊣					0.5 ⊢⊣⊢⊣ 0.5					⊢⊣									⊢⊣	
$-OH$ alcohol					⊢⊣ 2						(⊢⊣)							⊢⊣ 50			
$-OH$ hydroperoxide, Aromatic					1 ⊢⊣⊢⊣ 1						⊢⊣ 1.3								30 ⊢⊣⊢⊣ 30		
$-OH$ hydroperoxide, Aliphatic					⊢⊣ 2						⊢⊣ 0.8								⊢⊣ 80		
$-OH$ phenol, Free					⊢⊣ 3						⊢⊣							⊢⊣ 200			
$-OH$ phenol, Intramolecularly bonded					⊢⊣														⊢⊣ Variable		
$-OH$ carboxylic acid					⊢⊣														⊢⊣ 10-100		
$-OH$ glycol 1,2					⊢⊣													50 ⊢⊣⊢⊣ 50			
$-OH$ glycol 1,3					⊢⊣													20-50 ⊢⊣	⊢⊣ 20-100		
$-OH$ glycol 1,4					⊢⊣													50-80 ⊢⊣	⊢⊣ 5-40		
OH water					⊢⊣ 0.7					⊢⊣ 1.2								30 ⊢⊣ ⊢⊣ 7			
$=NOH$ oxime					⊢⊣													⊢⊣ 200			
HCHO (possibly hydrate)																		⊢⊣			
$-SH$										⊢⊣ 0.05											
$>PH$									⊢⊣ 0.2												
$>C=O$										⊢⊣										⊢⊣ 3	
$-C\equiv N$										⊢⊣ 0.1											

Wavelength, microns

Refs. p. 210.

E. References

1. There are innumerable books that discuss ir spectroscopy. The following are particularly recommended. (a) R. T. Conley, Infrared Spectroscopy, Allyn and Bacon, Boston, 1966--an excellent, thorough, and easy-to-use text. (b) L. J. Bellamy, The IR Spectra of Complex Molecules, 2nd ed., and Advances in IR Group Frequencies, Methuen, London, 1966 and 1968. (c) For gases, see R. Pierson, A. Fletcher, and E. Gantz, "Catalog of IR Spectra for Qualitative Analysis of Gases," Anal. Chem., 28, 1218 (1956). (d) For inorganic compounds, see reference 1a as well as F. Miller and C. Wilkins, Anal. Chem., 24, 1253 (1952), F. Miller et al., Spectrochim. Acta, 16, 135 (1960), and K. Nakamoto, IR Spectra of Inorganic and Coordination Compounds, 2nd ed., Wiley, New York, 1970. Cf. K. Nakamoto and P. McCarthy, Spectroscopy and Structure of Metal Chelate Compounds, Wiley, New York, 1968, and D. M. Adams, Metal-Ligand and Related Vibrations, Arnold, London, 1967. (e) A particularly good chapter with many useful tables and discussions is by G. Eglinton in Physical Methods in Organic Chemistry, J. C. P. Schwarz, Ed., Holden-Day, San Francisco, and Oliver and Boyd Ltd., London, 1965, Ch. 3. (f) For attenuated total reflectance spectra (ATR) see J. Fahrenfort, Spectrochim. Acta, 17, 698 (1961); H. Hecht, Internal Reflectance Spectroscopy, Interscience, New York, 1967; W. Wendlandt, Modern Aspects of Reflectance Spectroscopy, Plenum Press, New York, 1968; and W. Wendlandt and H. Hecht, Reflectance Spectroscopy, Interscience, New York, 1966. (g) R. A. Nyquist and R. O. Kagel, Infrared Spectra of Inorganic Compounds (3800-45 cm^{-1}), Academic Press, New York, 1971.
2. For example, see reference 1e, p. 116; N. B. Colthup et al., Introduction to IR and Raman Spectroscopy, Academic Press, New York, 1964; J. Brandmüller and H. Moser, Einführung in die Ramanspektroskopie, Steinkopff, Darmstadt, 1962; R. N. Jones et al., J. Org. Chem., 30, 1822 (1965); and H. Szymanski, Ed., Raman Spectroscopy. Theory and Practice, Plenum Press, New York, 1967.
3. A. R. Katritzky, Physical Methods in Heterocyclic Chemistry, Vol. II, Academic Press, New York and London, 1963.
4. K. Nakanishi, Infrared Absorption Spectroscopy, Holden-Day, San Francisco, and Nankodo, Tokyo, 1962.
5. E. M. Arnett et al., J. Amer. Chem. Soc., 92, 2365 (1970)--a discussion of H-bond complex formation.
6. L. Joris et al., ibid., 90, 327 (1968).
7. A. J. Gordon, R. L. E. Ehrenkaufer, and G. C. Ting, unpublished data on imides and amides.
8. H. K. Hall and R. Zbinden, J. Amer. Chem. Soc., 80, 6428 (1958).
9. For a very complete discussion and extensive data, see F. Bentley et al., IR Spectra and Characteristic Frequencies 700-300 cm^{-1}, Interscience, New York, 1968, and A. Finch et al., Chemical Applications of Far Infrared Spectroscopy, Academic Press, New York and London, 1970. See also J. Brasch, Y. Mikawa and R. Jakobsen, Appl. Spectrosc. Rev., 1, 187 (1968); for a bibliography, see A. Coste, "Far IR Spectrophotometry," Commis. Energ. At. (France) Serv. Dom., Ser., "Biblio.," CEA-BIB-126 (1968); for inorganic substances, J. Ferraro, Anal. Chem., 40, 24A (1968). A recent survey of the field is K. D. Moller and W. G. Rothschild, Far-Infrared Spectroscopy, Wiley-Interscience, New York, 1971.
10. Aryl imides: S. Shalaby and E. McCaffery, Anal. Chem., 40, 823 (1968). N-Acyl lactams, thioimides, and others: P. Bassignana et al., Spectrochim. Acta, 21, 677 (1965).

11. J. D. McCallum, in Progress in Infra Red Spectroscopy, Vol. 2, H. A. Szymanski, Ed., Plenum Press, New York, 1964, and R. F. Goddu, in Advances in Analytical Chemistry and Instrumentation, Vol. 1, C. N. Reilley, Ed., Interscience, New York, 1960.

V. ELECTRONIC ABSORPTION AND EMISSION SPECTRA: ULTRAVIOLET AND VISIBLE

Absorption in the vacuum uv (100-200 nm; 10^5-50,000 cm^{-1}), quartz uv (200-350 nm; 50,000-28,570 cm^{-1}), and visible (350-800 nm; 28,570-12,500 cm^{-1}) regions is described quantitatively by the Beer-Lambert-Bouguet law:

$$A = -\log T = \varepsilon bc$$

where A = absorbance or optical density; T = transmittance, that is the ratio of transmitted to incident light intensity, I/I_o; b = cell thickness in cm; c = concentration, usually in mole/liter; ε = molar absorptivity, often called extinction coefficient, in liter/mole-cm. Theoretical details are covered in reference 1. For compilations of organic chromophore types and their uv-vis properties, see reference 2; other useful treatments are references 3 and 4. Theory and extensive absorption data for inorganic ions and complexes are found in references 5 and 6. Appropriate solvents and cell materials are discussed on pages 167 and 179. An article and a book on low temperature uv measurement techniques are also available (7).

Emission spectra (fluorescence and phosphorescence) are a much more sensitive and specific measurement than absorption spectra and are particularly useful in trace analysis and in biology and medicine (8, 9). A fluorescence spectrum in many cases is the mirror image of the lowest energy absorption band, and is usually located adjacent to that band on the long wavelength side. After removal of an excitation source, fluorescence intensity shows a first-order decay process:

$$F = F_o \exp(-t/\tau)$$

where F_o = initial intensity (at some arbitrary time zero), F = intensity at time t, and τ is a decay constant characteristic of the substance and is the time required for F to fall to (1/e)th of its initial value (τ is also called the average lifetime of the excited state). Theoretical details and much useful data are covered in reference 10. Related data are included in the section on photochemistry (p. 350). For very dilute solutions, observed fluorescence intensity is directly proportional to concentration:

$$F = 2.3(I_o \varepsilon bc)(\phi)$$

where ϕ = quantum yield of fluorescence (see p. 362 for explanation of quantum yield). In practice, for this equation to hold, the solution should absorb $\leq 2\%$ of the exciting radiation and εbc should be < 0.05 (8,9). If $\phi = 1$ (thus every excited state molecule decays by fluorescence), τ becomes τ_o, the natural or intrinsic lifetime; however, ordinarily $\phi < 1$ because of processes competing with fluorescence ($\phi = \tau/\tau_o$). A useful bibliography of fluorescence literature is published as a series (12).

Refs. p. 220.

A. Nomenclature and Notations for Spectral Data

Item	Description
Red shift or bathochromic shift	To longer wavelength.
Blue shift or hypsochromic shift	To shorter wavelength.
Hyperchomic shift	Leading to increased ε.
Hypochromic shift	Leading to decreased ε.
Chromophore	Atom or group of atoms and/or electrons within a molecule that are mainly responsible for the particular light absorption.
Auxochrome	A functional group, which alone does not absorb in the uv, but when conjugated to a chromophore, a red shift and hyperchromic effect (e.g., $-NH_2$ on a double bond) result.
Isosbestic point	That wavelength at which two compounds that can be converted into each other have equal absorptivity. For example, two compounds may be in equilibrium ($ArOH \rightleftarrows ArO^-$); the family of absorption curves generated by changes in pH will have one point at which absorbance is constant, independent of $ArOH/ArO^-$ ratio. This "isosbestic" (or isobestic) point depends only on the total number of "equivalents" of the two equilibrating species.
Oscillator strength, f	A theoretical measure of intensity of a transition: $f = 4.315 \times 10^{-9} \int \varepsilon \, d\bar{\nu}$ (the integral represents the area under an absorption band, where $\bar{\nu}$ is in cm^{-1}). For highly allowed transitions, $f \rightarrow 1$.
Aromatic band nomenclature	Principal transitions in spectra of polycyclic aromatics are described by three notation systems, all referring to the comparable transitions in benzene as the "parent" [256 nm(ε220), 203(6900), 183 (46,000), —]:

Log ε_{max} range:	2.3-3.2	3.6-4.1	4.5-5.2	≃ 4.3-4.5
Platt	^1L_b	^1L_a	^1B_b	^1B_a
Clar	α	Para	β	
Moffitt	V	U	X	Y

Refs. p. 220.

Item	Description
Burawoy classification	Empirical designation of bands in substituted aromatic systems: E and B bands (also called local excitation or LE bands) for $\pi \rightarrow \pi^*$ transitions; K band for $\pi \rightarrow \pi^*$ transitions involving a conjugating group (also called electron transfer or ET band); R band for $n \rightarrow \pi^*$ transition when group on aromatic ring has nonbonded electron pair. Ranges for λ_{max} are as follows: E (180-220 nm, $\varepsilon \simeq 5000$), B (250-290, $\varepsilon \simeq 100\text{-}1000$), K (220-250, $\varepsilon \simeq 20{,}000$), R (275-300, $\varepsilon \simeq 10\text{-}100$).
Inorganic-complex transitions	Designated as ligand field bands (d-d transition of central atom), charge-transfer bands (intramolecular; higher oscillator strengths than d-d); internal ligand bands (transitions of ligands themselves).
Visible colors	The colors observed in the visible spectrum correspond to the following absorption regions:

Color Observed	Absorption Range, cm^{-1} (nm)
Red	14,000-16,000 (714-625)
Yellow	18,000 (556)
Green	20,000 (500)
Blue	21,000-25,000 (476-400)
Violet	>25,000 (<400)

B. Standard Spectral Data for Instrument Calibration

To check the photometric scale (A or T readings) of a spectrophotometer, record the spectrum of the following solution: 0.0400 g/liter K_2CrO_4 in 0.05M aq. KOH in a 1.0 cm cell at 25°C. The following selection of T and A values were taken from Haupt, J. Res. Nat. Bur. Stds., 48, 414 (1952). At 375 nm, the temperature coefficients of T and A are $\Delta T/\Delta t = 0.0022$/deg and $\Delta A/\Delta t = -0.00093$/deg.

Refs. p. 220.

λ (nm)	%T	A	λ (nm)	%T	A
220	35.8	0.446	350	27.6	0.559
230	67.4	0.171	360	14.8	0.830
240	50.7	0.295	370	10.3	0.987
250	31.9	0.496	375	10.2	0.991
260	23.3	0.633	390	20.2	0.695
275	17.5	0.757	400	40.2	0.396
290	37.3	0.428	420	75.1	0.124
300	70.9	0.149	440	88.2	0.054
315	90.0	0.046	460	96.0	0.018
330	71.0	0.149	480	99.1	0.004
340	48.3	0.316	500	100.0	0.000

Wavelength accuracy is conveniently checked by recording the spectrum of a didymium or a holmium oxide glass filter. Such filters are available to fit the sample compartments of various spectrophotomers; the glasses are also available in various thicknesses from Corning as C.S. No. 1-60 (Glass No. 5120) for didymium and C.S. No. 3-138 (Glass No. 3130) for holmium oxide. The following calibration peaks (absorption maxima) are used for holmium oxide: 279.3, 287.6, 333.8, 360.8, 385.8, 418.5, 446.0, 453.4, 536.4, and 637.5 nm.

These filters are also used to detect stray light in the instrument.* Holmium oxide is completely opaque (T = 0) below 225 nm and didymium below 340 nm. Any measurable transmittance below these wavelengths indicates stray light. In addition, the holmium oxide filter can be used to test resolution; a well focused instrument with good resolution will display three distinct and intense bands within the broad absorption between 460 and 440 nm.

Very recently, the National Bureau of Standards has made available standard reference materials for spectrophotometry. SRM 930 is a set of three solid glass filters calibrated from 440 to 635 nm. It is used to check photometric scale accuracy at three T levels and at four wavelengths. SRM 931 is a set of four ampoules used as liquid absorbance standards for uv and vis spectrophotometry. Three ampoules contain empirical mixtures of cobalt, nickel and nitrate ions in 0.1 N perchloric acid; the fourth contains a blank. NBS reference materials are discussed on page 496.

C. Organic Spectral Data

1. Absorption Characteristics for Typical Unconjugated Chromophores

The following values are found in nonpolar, usually hydrocarbon solvents. Data are taken from several sources, especially reference 2, Ch. 1 . For interesting discussions on the cyclopropane ring as a chromophore, see reference 13.

*See R. B. Cook and R. Jankow, J. Chem. Educ., 49, 405 (1972) for a recent review on effects of stray light in spectroscopy.

Refs. p. 220.

Compound	Transition	λ_{max} (nm)	ε_{max} (approx.)
C_2H_6	$\sigma \rightarrow \sigma^*$	135	
H_2O	$n \rightarrow \sigma^*$	167	7000
ROH	$n \rightarrow \sigma^*$	180-185	500
RSH	$n \rightarrow \sigma^*$	190-200; 225-230(sh)	1500; 150
RCl	$n \rightarrow \sigma^*$	170-175	300
RBr	$n \rightarrow \sigma^*$	200-210	400
RI	$n \rightarrow \sigma^*$	255-260	500
R_2O	$n \rightarrow \sigma^*$	180-185	3000
R_2S	$n \rightarrow \sigma^*$	210-215; 235-240(sh)	1250; 100
RSSR	$n \rightarrow \sigma^*$	250	400
Amines (1°, 2°, 3°)	$n \rightarrow \sigma^*$	190-200	2500-4000
C_2H_4	$\pi \rightarrow \pi^*$	163; 174	15,000; 5500
C_2H_2	$\pi \rightarrow \pi^*$	173	6000 (vapor)
R-C≡CH	$\pi \rightarrow \pi^*$	185; 223	2200; 120
R-C≡C-R	$\pi \rightarrow \pi^*$	178; 196; 223	10,000; 2000; 160
C=C=C	$\pi \rightarrow \pi^*$	170-185; 230	5000-10^4; 600
C=C=O	$n \rightarrow \pi^*$	380	20
	$\pi \rightarrow \pi^*$	225	400
RCOCl	$n \rightarrow \pi^*$	280	10-15
RCO_2H, RCO_2R'	$n \rightarrow \pi^*$	195-210	40-100
R-$CONH_2$	$n \rightarrow \pi^*$	175	7000
R-CN		<170	
RCONHCOR	$n \rightarrow \pi^*$	230-240	80-100
	$\pi \rightarrow \pi^*$	190-200	10,000-15,000
RNO_2	$n \rightarrow \pi^*$	270-280	20-30
	$\pi \rightarrow \pi^*$	200-210	15,000
RONO	$n \rightarrow \pi^*$	350	150
	$\pi \rightarrow \pi^*$	220	1000
RNO (monomer)	$n \rightarrow \pi^*$	600-650; 300	20; 100
RN=NR	$n \rightarrow \pi^*$	350-370	10-15
	$\pi \rightarrow \pi^*$	<200	
RCHO	$n \rightarrow \pi^*$	290	15
	$\pi \rightarrow \pi^*$	185-195	
R_2CO	$n \rightarrow \pi^*$	270-290	10-20
	$\pi \rightarrow \pi^*$	180-190	2000-10,000
RCOCOR	$n \rightarrow \pi^*$	420-460; 280-285	10; 20
RSOR	$n \rightarrow \pi^*$	210-230	1500-2500
RSO_2R		<190	

2. Absorption Maxima of Aromatic Compounds

The transitions of the bands cited are not assigned here; for the polycyclic aromatic series (first nine entries), bands for the different compounds arising from the same transition type are so arranged. For fine structure band groups, only λ_{max} of the band center is given. For details and more data, see references 1 to 4.

Refs. p. 220.

Compound	Solvent[a]	Band Maxima: λ (log ε)				
Benzene	C			183(4.66)	204(3.90)	256(2.30)
Naphthalene	E	167(4.48)	190(4.00)	220(5.12)	286(3.97)	312(2.46)
Anthracene	C	186(4.51)	221(4.16)	256(5.26)	375(3.95)	(buried)
Naphthacene	B	187(4.20); 211(4.64)	230(3.23)	272(5.26)	474(4.10)	
Azulene	C	[700(2.48)]	193(4.26)	236(4.34)	269(4.67)	357(3.60)
Phenanthrene	C		222(4.38)	252(4.82)	292(4.20);	345(2.32)
Quinoline	C			228(4.60)	270(3.50)	315(3.40)
Isoquinoline	C			218(4.80)	265(3.62)	313(3.26)
Acridine	E			250(5.30)	358(4.00)	
Pyridine	H	195(3.88)	251(3.30)		270(2.65)	
Pyrimidine	C		243(3.31)		298(2.48)	
Pyrazine	C		260(3.80)		327(2.00)	
Pyridazine	C		246(3.11)		340(2.50)	
Purine	W	<220(3.48)	263(3.90)			
Pyrrole	H	210(3.71)			240(2.48)	
Furan	H	205(3.81)				
Thiophene	H	231(3.85)				
Imidazole	E	207(3.70)				
Pyrazole	E	210(3.50)				
Isoxazole	E	211(3.60)				
Thiazole	E	240(3.60)				
C_6H_5R:						
R = OH	W	210.5(3.78)	270(3.16)			
O^-	W	235(3.97)	287(3.42)			
OCH_3	W	217(3.81)	269(3.17)			
SH	H	236(4.00)	269(2.85)			
NH_2	W	230(3.93)	280(3.16)			
NH_3^+	W	203(3.88)	254(2.20)			
NO_2	H	252(4.00)	280(3.00)		330(2.10)	
CHO	E	244(4.18)	280(3.18)		328(1.30)	
$COCH_3$	E	240(3.11)	278(3.04)		319(1.70)	
CO_2H	W	230(4.00)	270(2.90)			
CO_2^-	W	224(3.94)	268(2.75)			
CN	W	224(4.11)	271(3.00)			
F	E	204(3.80)	254(3.00)			
Cl	W	209.5(3.87)	263.5(2.28)			
Br	W	210(3.90)	261(2.28)			
I	W	207(3.85)	257(2.85)			
CH_3	W	207(3.85)	261(2.35)			
$CH{=}CH_2$	E	244(4.08)	282(2.65)			
$C{\equiv}C{-}C_6H_5$	H	236(4.10)	278(2.81)			
C_6H_5	E	246(4.30)				

[a]C = cyclohexane, E = ethanol, B = benzene, H = hexane, W = water.

Refs. p. 220.

3. Empirical Rules for Calculating $\pi \rightarrow \pi^*$ Maxima in Conjugated Systems

For discussion and references, see reference 1, Ch. 10, reference 2, Ch. 3, reference 3, Ch. 3, and reference 4, Ch. 5.

a. Dienes (Fieser-Woodward Rules)

Parent Systems:[a]

C = C - C = C

Acyclic diene: 214 nm

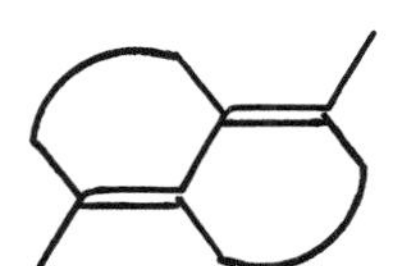

Heteroannular diene: 217 nm

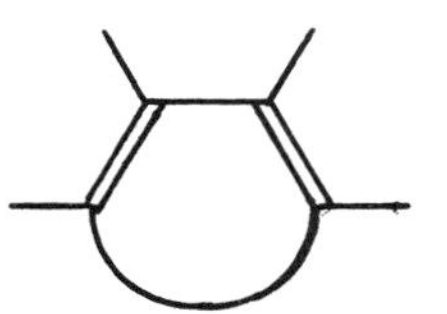

Homoannular diene: 253 nm

Structural Feature		Increment (nm)[c]
Extended conjugation (per double bond)		30
Alkyl substituent or ring residue		5
Exocyclic double bond		5
Polar groups:[b]	RCO_2	0
	RO	6
	RS	30
	Cl, Br	17
	NR_2	60

[a]Values same in all solvents. [b]R = alkyl only. [c]All are +.

b. Conjugated Carbonyl Compounds (Woodward-Fieser Rules)

Parent Systems:[a]

β, β >C = C(α) - C(=O) -   (ring: C_n)

Acyclic or ring, n ≥ 3
215 nm

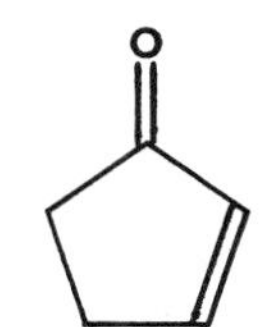

Cyclopentenone
205 nm

β, β >C = C(α) - C(=O) - H

Aldehydes
210 nm

Structural Feature	Increment (nm)[b] α	β	γ and higher
Extended conjugation (per double bond)	30	--	--
Homodiene component	39	--	--
Exocyclic double bond	5	--	--
Alkyl groups	10	12	18
OH	35	30	50

(Continued)

Refs. p. 220.

Structural Feature	Increment (nm)[b] α	β	γ and higher
OR	35	30	17 (δ, + 31)
SR		85	
CH_3CO_2	6	6	6 (δ)
Cl	15	12	
Br	25	30	
NR_2		95	

Solvent corrections:

Solvent	Correction	Solvent	Correction
Water	-8	Diethyl ether	+7
Methanol	0	Hexane	+11
Dioxane	+5	Cyclohexane	+11
Chloroform	+1		

[a]Data for spectra in ethanol; apply corrections for other solvents.
[b]All are +.

c. Substituted Benzene Derivatives: K-Band[a]

Parent Systems[b] $C_6H_5\overset{O}{\overset{\|}{C}}R$: R = alkyl, 246 nm; R = H, 250 nm; R = OH and O-alkyl, 230 nm — C_6H_5CN: 224 nm

Structural Feature	Increment per Substituent[c] Ortho	Meta	Para
Alkyl (or ring)	3	3	10
HO, Alkyl-O	7	7	25
O^-	11	20	78
Cl	0	0	10
Br	2	2	15
NH_2	13	13	58
NHAc	20	20	45
NHR			73
NR_2	20	20	85

[a]Also called principal electron transfer (ET) band. [b]Values apply for ethanol as solvent. [c]All are +.

D. Inorganic Spectral Data

1. Spectrochemical Series (5,6)

The type of ligand will influence the position of a ligand-field band; the following series gives rise to absorption at progressively shorter wavelength, and is virtually independent of the central atom or ion. The first listed atom is the one attached directly to the central metal; for the abbreviated ligands, the atoms attached are given in parentheses:

Refs. p. 220.

I^-, Br^-, $OCrO_3^-$, Cl^-, SCN^-, N_3^-, dtp (diethyldithiophosphate; 2-S), F^-, SSO_3^{-2}, urea (O), OCO_2^{-2}, OCO_2R^-, ONO^-, OH^-, OSO_3^{-2}, ONO_2^-, ox^{-2} (oxalate; 2-O), OH_2, mal^{-2} (malonate; 2-O), NCS^-, gly^- (glycinate; 1-O, 1-N), $enta^{-4}$ (ethylenediaminetetraacetate; 4-O, 2-N), py (pyridine; N), NH_3, en(ethylenediamine; 2N), dien (diethylenediamine; 3-N), tren [tris(aminoethyl)amine; 4-N], SO_3^{-2}, dip(α,α'-dipyridyl; 2-N), phen(o-phenanthroline; 2-N), NO_2^-, $C_5H_6^-$(cyclopentadienide), CN^-.

Similarly, for a constant ligand group, the metal ions belong to a series of progressively shorter wavelength of absorption:

Mn(II), Ni(II), Co(II), Fe(III), Cr(III), V(III), Co(III), Mn(IV), Mo(III), Rh(III), Ru(III), Pd(IV), Ir(III), Re(IV), Pt(IV).

2. Spectral Data for Free Ions

Many anions (ligands) absorb as free-ions in the near uv; when coordinated to a metal, the internal transitions of some of these ions show a red shift (6).

Ion	cm^{-1}	nm	Ion	cm^{-1}	nm
$(COS)_2^{-2}$	39,000	256	$S_2O_3^{-2}$	46,000	217
ox^{-2}	40,000	250	SCN^-	47,000	213
$SeCN^-$	42,500	235	NO_2^-	47,600	210
N_3^-	42,500	235	NO_3^-	49,000	204
I^-	44,200	226	SO_3^{-2}	50,000	200

The following table presents the principal bands of metal ions in aqueous solutions. The data refer to the hexaaquo species, $M(H_2O)_6^{+n}$, except for $Pd(H_2O)_4^{+2}$ and $Ag(H_2O)_2^{+1}$ (5, 11).

Ion	Color Observed	k cm^{-1} (ε)	Ion	Color Observed	k cm^{-1} (ε)
$Ti^{+3}(3d^1)$	Purple	20.3(4)	$Co^{+2}(3d^7)$	Pink	8.0(1.3)
$V^{+3}(3d^2)$	Green	17.2(6)			20.0(5.0)
$Cr^{+3}(3d^3)$	Blue	17.0(14)	$Ni^{+2}(3d^8)$	Light green	8.7(1.6)
		24.0(15)			14.5(2.0)
$Mn^{+3}(3d^4)$	Violet	21.0	$Cu^{+2}(3d^9)$	Blue	12.0(11)
$Cr^{+2}(3d^4)$	Pale blue	14.1(4.2)	$Ru^{+3}(4d^5)$		25.0(sh)
$Mn^{+2}(3d^5)$	Very pale pink	Weak, narrow			44.5(2300)
$Fe^{+3}(3d^5)$	(Yellow)[a]	14.0	$Pd^{+2}(4d^8)$		26.4(86)
$Fe^{+2}(3d^6)$	Pale green	10.0(1.1)	$Ag^{+1}(4d^{10})$		44.7(400)
					47.5(900)
					51.9(1500)

[a]When pure, Fe^{+3} is colorless, but solutions are usually yellow due to hydrolysis.

Refs. p. 220.

E. References

1. H. H. Jaffé and M. Orchin, Theory and Applications of Ultraviolet Spectroscopy, Wiley, New York, 1962.
2. A. I. Scott, Interpretation of the UV Spectra of Natural Products, Pergamon Press, New York, 1962.
3. D. J. Pasto and C. Johnson, Organic Structure Determination, Prentice-Hall, New York, 1969.
4. R. M. Silvestein and G. C. Bassler, Spectrometric Identification of Organic Compounds, Wiley, New York, 1967.
5. C. K. Jorgensen, Absorption Spectra and Chemical Bonding in Complexes, Pergamon Press, London, 1962, Ch. 15.
6. H. H. Schmidtke, Ch. 4 in Physical Methods in Advanced Inorganic Chemistry, H. Hill and P. Day, Eds., Interscience, New York, 1968.
7. B. Meyer, Science, 168, 783 (1970). See also B. Meyer, Low Temperature Spectroscopy, Elsevier, New York, 1971.
8. S. Udenfriend, Fluorescence Assay in Biology and Medicine, Academic Press, New York, 1962, and S. V. Konev, Fluorescence and Phosphorescence of Proteins and Nucleic Acids, Plenum Press, New York, 1967.
9. Handbook of Analytical Chemistry, McGraw-Hill, New York, 1963, pp. 6-176 to 6-196.
10. R. S. Becker, Theory and Interpretation of Fluorescence and Phosphorescence, Wiley-Interscience, New York, 1969.
11. B. N. Figgis, Introduction to Ligand Fields, Interscience, New York, 1966.
12. Guide to Fluorescence Literature, Vol. 2, R. A. Passwater, Ed., IFI/Plenum Data Corp., New York, 1970.
13. A. J. Gordon, J. Chem. Educ., 44, 461 (1967); R. C. Hahn, P. H. Howard, and G. A. Lorenzo, J. Amer. Chem. Soc., 93, 5816 (1971).

VI. OPTICAL ACTIVITY AND OPTICAL ROTATION

A. Nomenclature and Definitions

1. Monochromatic Optical Rotation Measurements

Rotation at a single wavelength is reported, along with solvent and concentration, as either of the following:

(a) Specific rotation $[\alpha]_{\lambda}^{t} = \alpha/lc$ where α = observed rotation in degrees, c = concentration (or density for a neat liquid) in g/cc, t = °C, l = pathlength in dm, and λ = wavelength of measurement.
(b) Molecular rotation $[\Phi]_{\lambda}^{t} = M[\alpha]/100$ where M = molecular weight.

2. Optical Rotatory Dispersion (ORD) and Circular Dichroism (CD)

ORD is the measurement of optical rotation as a function of wavelength using plane-polarized light; CD is the unequal absorption of right and left circularly polarized light as a function of wavelength. Nearly all measurements are made in the uv-vis spectral regions. CD measurements in the ir region using the Raman effect have been reported [B. Bosnich, et al., J. Amer. Chem. Soc., 94, 4750 (1972)] but the observations may arise through a spurious mechanism [L. D. Barron, et al., ibid., 95, 603 (1973)]. Liquid crystal induced CD (LCICD) spectra for achiral solutes in cholesteric phases have recently been reported (37).

Refs. p. 234.

a. Normal and Plain Dispersion Curves. Plain dispersion curves (A and B) are those exhibiting no inflections or extrema, and are obtained in λ regions where the substance does not absorb. Curve A is a plain positive

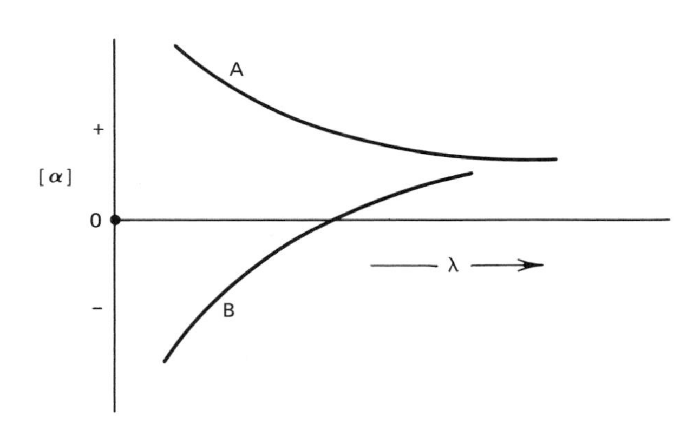

dispersion curve (rotation increases as λ decreases); curve B is a plain negative dispersion curve. Alternatively, A may be referred to as a normal positive curve (no change in sign, i.e., crossing of the zero line).

b. Drude equation (1,2,3). Outside the region of absorption (plain curve region), rotation as a function of λ can be approximated (λ_o = wavelength in nm of closest absorption maximum, $\lambda_o < \lambda$; A = empirical constant for the chromophore):

$$[\alpha] = \frac{A}{\lambda^2 - \lambda_o^2} \qquad [\Phi] = \frac{K}{\lambda^2 - \lambda_o^2} \qquad K = \frac{MA}{100}$$

For j-absorption maxima, the multiple-term Drude equation is:

$$[\Phi] = \frac{\sum_{j=o}^{j} K_j}{\lambda^2 - \lambda_j^2}$$

in practice, more than two terms are rarely workable; see reference 3.

Refs. p. 234.

c. Anomalous Dispersion (ORD) and Absorption (CD)

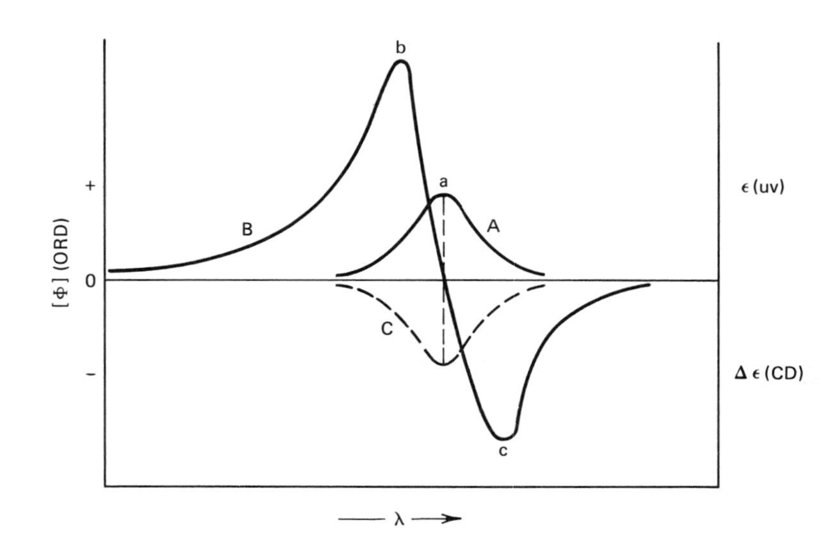

The figure above represents a theoretical (idealized) plot for a single optically active absorption band, giving rise to a Cotton effect (1,3,4): unequal absorption (CD) and unequal transmission velocity (ORD) of left and right circularly polarized light.

Curve A: uv absorption spectrum (ε = molar absorptivity).

Curve B: a negative Cotton effect (peak = point b; at lower λ than the trough = point c; b and c are the extrema of the Cotton effect). The molecular amplitude is defined in hundreds of degrees as $([\Phi]_c - [\Phi]_b)/100$; the breadth of the Cotton effect is $\lambda_c - \lambda_b$ in nm.

Curve C: circular dichroism curve, a plot of λ verses differential dichroic absorption, $\Delta\varepsilon = \varepsilon_L - \varepsilon_R$ (ε_L and ε_R = molar absorptivities for left and right circularly polarized light). By analogy with $[\alpha]$, CD can be reported as specific ellipticity (emerging light in CD experiment is elliptically polarized) $[\Psi] = \Psi/\ell c$ (ℓ in dm, c in g/cc) or molecular ellipticity $[\theta] = M\ [\Psi]/100 = 3300\ \Delta\varepsilon$. The sign of $\Delta\varepsilon$ (or $[\theta]$) is the same as the sign of the Cotton effect ORD curve. Curve C shows a negative maximum for the CD extremum. The λ for a CD extremum (max, min, infl) coincides almost exactly with that of the uv absorption curve (max, min, or infl).

For $n \rightarrow \pi^*$ transitions in saturated, optically active carbonyl compounds (1): ORD molecular amplitude = $40.28\ \Delta\varepsilon = 0.0122\ [\theta]$, where $\Delta\varepsilon$ and $[\theta]$ refer to the values for the CD maximum.

3. (R,S)-Nomenclature

A scheme for unambiguously designating absolute configuration of any optically active molecule, it is based on a priority rule (assigning each ligand a priority based principally on atomic number) and a sequence rule (ordering the ligands according to priority). There is no relationship between the R,S system and the (+) or (-) sign of rotation. For complete details of this so-called Cahn-Ingold-Prelog system, see Angew. Chem.

Refs. p. 234.

Intern. Ed. Engl., 5, 385 (1966), and K. R. Hanson, J. Chem. Educ., 88, 2731 (1966). For recent development in stereochemical nomenclature, see E. L. Eliel, J. Chem. Educ., 48, 163 (1971).

B. Methods for Predicting Absolute Configuration and Rotation

For extensive discussion and data on experimentally determined absolute configuration, see J. Mills and W. Klyne, Progress in Stereochemistry, 1, 177 (1954), and F. Allen and D. Rogers, J. Chem. Soc. (D), 838 (1966), 308 (1968), 452 (1969); the latter provide a summary of all reported (through 1968) X-ray determinations. For inorganic compounds, see C. L. Hawkins, Absolute Configuration of Metal Complexes, Wiley-Interscience, New York, 1971.

1. Brewster's Rules (2,5)

A method for predicting the sign and in some cases the magnitude of rotation at 589 nm (Na D-line) for acyclic molecules. For the absolute configuration shown below, rotation is (+) if the polarizabilities (R_D, molar refractivities) of the substituent atoms or groups directly attached to the asymmetric C are: a > b > c > d.

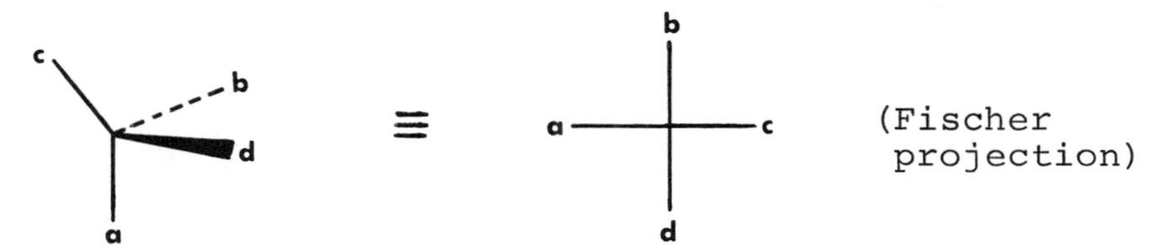

For example, in the molecule $CH_3CH(Br)CN$, the dextro enantiomer will have a = Br, b = CN, c = CH_3, d = H (the R-isomer) as predicted from R_D values in the table below. This rule strictly applies only when a, b, c, and d are conformationally symmetric (halogen, methyl, etc.) and when there are no intramolecular H-bonding situations; under these criteria, the system is said to have only atomic asymmetry. For more complicated groups, conformational asymmetry contributions must be used instead. The latter are determined empirically or are calculated, and are a function of R_D values of the attached atoms. The following assumptions are applied when using conformational rotatory values (listed in table below);

(a) Only staggered (energy minima) conformations are considered.

(b) Doubly skewed conformations X Y Z and 1,3-diaxial interactions X Y are excluded.

(c) All allowed conformations contribute equally to the total rotation.

The rotating power (M_D) of each allowed conformer is the sum of individual skew (i.e. 1, 2) interactions whose signs are defined as follows:

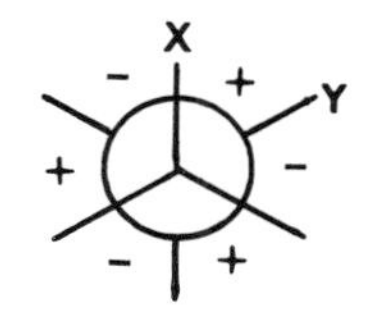

Refs. p. 234.

For example, consider 2-chlorobutane (reference 2, p. 405). Only conformations A and B are allowed. C in the equations refers to the C on CH_3; k is a constant, and C,H,Cl refer to functions of R_D for each attached atom; a term such as C•H or H•Cl refers (in structure A) to CH_3-H and H-Cl interactions of the spatially adjacent groups:

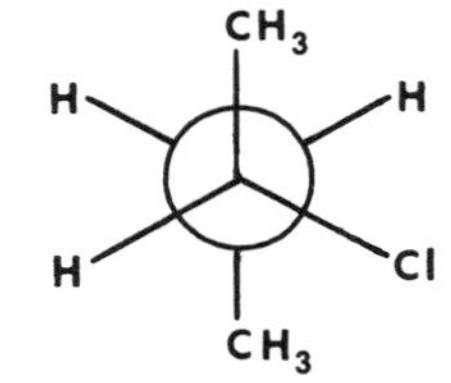

A

$$M_A = k(C\cdot H - H\cdot Cl + Cl\cdot C - C\cdot H + H\cdot H - H\cdot C)$$
$$= k(Cl - H)(C - H) \quad \text{(as written symbolically)}$$

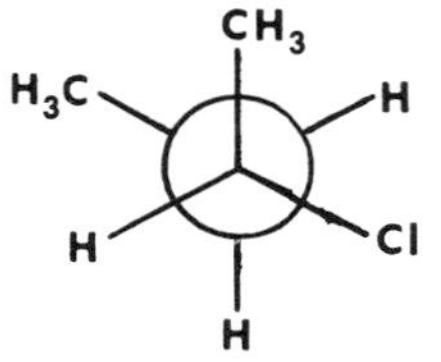

B

$$M_B = k(C\cdot H - H\cdot Cl + Cl\cdot H - H\cdot H + H\cdot C - C\cdot C)$$
$$= -k(C-H)(C-H)$$

Using the values from the table below and since A and B contribute equally, $M_D = (M_A + M_B)/2 = (139 - 60)/2 = +39.5°$ (highest observed value is +36°); note that all H interactions do not contribute since its rotatory contribution is zero.

R_D - Values and Conformational Rotatory Contributions (2,5)

The following conformational asymmetry contributions (calculated values) refer only to the case where one of the substituents (say X or Y above) is an alkyl group; thus k(X - H)(C - H).

X	R_D[a]	Rotatory Contribution[a]
I	13.954	268
Br	8.741	192
SH	7.729	174
Cl	5.844	139
CN	3.580 (5.459)	87 (131)
C_6H_5	3.379 (6.757)	82 (158)
CO_2H	3.379 (4.680)	82 (114)
CH_3	2.591	60
NH_2[b]	2.382	53

Refs. p. 234.

X	R_D[a]	Rotatory Contribution[a]
OH[b]	1.518	23
H	1.028	0
D	1.004	
F	0.81	-10

[a]For atom directly attached to central carbon; numbers in parentheses refer to the total 2-atom unsaturated unit (CN, CC, CO). [b]When OH and NH_2 are attached to benzylic carbon, they rank <u>ahead</u> of carbon substituents.

More details and applications to allenes, hexahelicene, and proteins are given in reference 2; for recent application to cyclopropanes, see R. G. Bergman, J. Amer. Chem. Soc., 91, 7405 (1969). For the "benzoate rule" of cyclic, secondary alcohols, see J. Brewster, Tetrahedron, 13, 106 (1961).

2. The Axial α-Haloketone Rule (2,3)

(a) The sign of the Cotton effect and the $n \rightarrow \pi^*$ wavelength maximum for cyclohexanones are relatively insensitive to an equatorial α-halogen. See section 3b below.

(b) An axial chlorine, bromine, or iodine (fluorine is anomalous) will cause a bathochromic uv shift (up to 25 nm) and may change the sign of the Cotton effect of the parent ketone; the following rule applies:

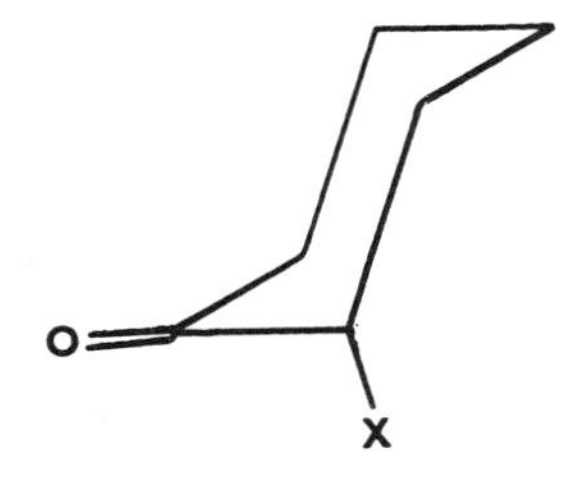

(+) Cotton effect

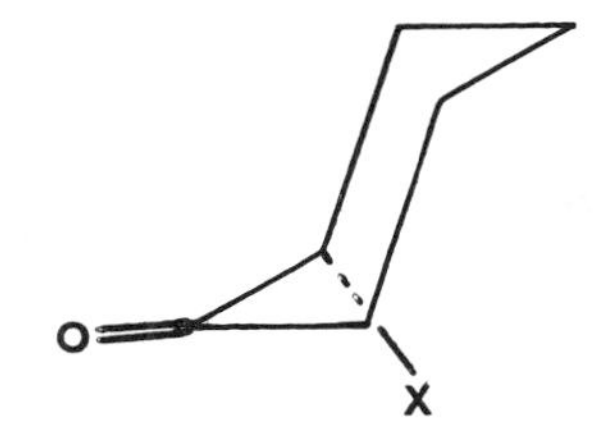

(-) Cotton effect

Refs. p. 234.

3. The Octant Rule (1-5)

Based on orbital symmetry principles, it relates the sign of a Cotton effect exhibited by an optically active, nonconjugated cyclohexanone to the positions of groups relative to the carbonyl group.

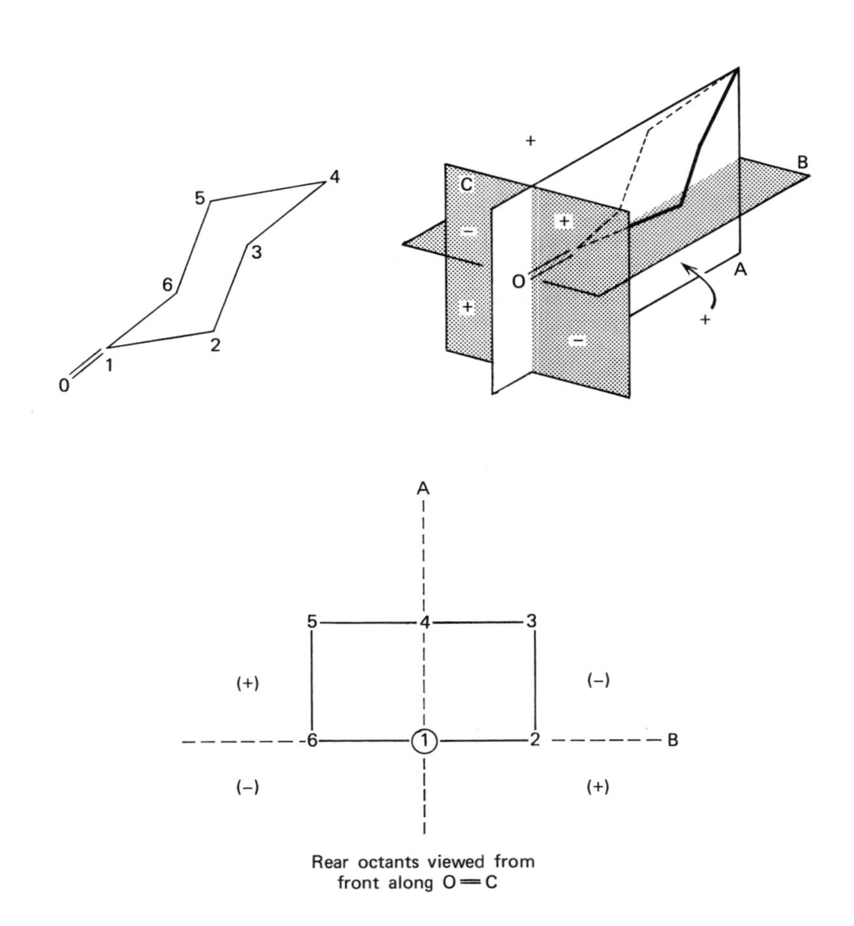

Rear octants viewed from front along O═C

The mutually orthogonal planes A, B, and C are defined as follows. Plane A bisects angles 6-1-2 and 5-4-3, and contains the carbonyl group; plane B contains O, C_1, C_2, and C_6; plane C intersects the carbonyl bond (approximately bisecting it). Octants behind plane C are the rear octants, those in front are the front octants.

(a) Substituents lying exactly in a plane (axial and equatorial on C-4; equatorial on C-2 and C-6) make no contribution to the Cotton effect. In practice (1), equatorial alkyl groups on C-2 and C-6 may not be exactly in the plane; rotatory contributions to the molecular amplitude in such a case are as follows: methyl, <9°; i-propyl, 15-21°; t-butyl, 33-39° (positive when on C-2, negative when on C-6).

(b) Positive contributions are made by axial C-2 and axial or equatorial groups on C-5; the contributions are negative for the same situation on C-6 and C-3; the axial haloketone rule is thus a specific application of the octant rule.

Refs. p. 234.

(c) The front octants (rarely "occupied") have signs opposite to those of the rear.

For example:

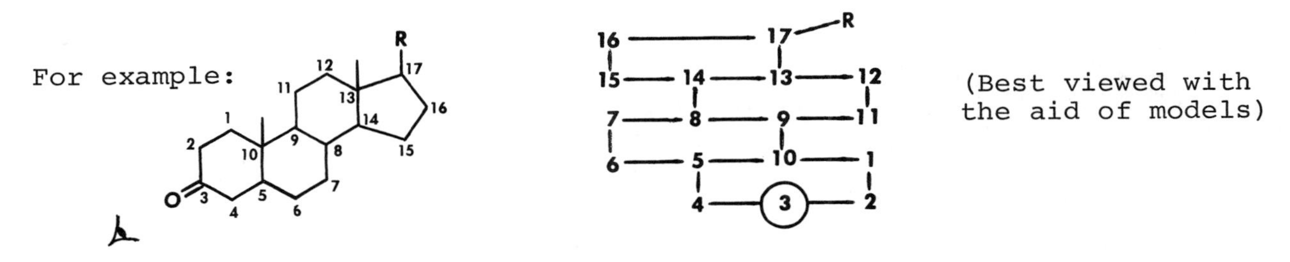

(Best viewed with the aid of models)

Net positive contributions: 5, 6, 7, 8, 14, 15, 16; net negative contributions: 1, 11, 12, R.

(d) Extensions of the octant rule; quadrant rules (1,1a).

(i) Skewed α,β-conjugated and homoconjugated (1,4) olefins show a positive Cotton effect for a right-hand helical conformation and a negative Cotton effect for a left-hand one (helicity rule for skewed dienes) (1,5):

Right-hand helix (X = C or O)

An olefin octant rule has also been proposed for chiral olefins including trans-cyclooctene and twistene (36).

(ii) Application to cyclobutanones, cyclopentanones, and cycloheptanones (reference 1, p. 83).

(iii) Chirality of glycols: N. Harada and K. Nakanishi, J. Amer. Chem. Soc., 91, 3989 (1969).

(iv) The benzene ring as a chromophore: J. H. Brewster and J. G. Buta, J. Amer. Chem. Soc., 88, 2233 (1966).

(v) Hudson's lactone sector rule (reference 1, p. 265).

(vi) Quadrant rule for peptide linkage ($n \rightarrow \pi^*$ transition): B. J. Litman and J. A. Schellman, J. Phys. Chem., 69, 978 (1965); for an exception to the rule with a lactam, see N. J. Greenfield and G. D. Fasman, J. Amer. Chem. Soc., 92, 177 (1970). See also the quadrant rule for $n \rightarrow \pi^*$ Cotton effects in dithiourethanes of amines, H. Ripperger, Angew. Chem., 79, 721 (1967), and a similar rule for α- and β-phenylalkylamine hydrochlorides, H. E. Smith and T. C. Willis, J. Amer. Chem. Soc., 93, 2282 (1971).

(vii) Benzoate sector rule for cyclic, secondary alcohols: N. Harada, M. Ohashi, and K. Nakanishi, J. Amer. Chem. Soc., 90, 7349 (1968). For recent extensions, see N. Harada, et al., ibid., 93, 5577 (1971).

4. Miscellaneous Methods

Absolute and relative configuration have been determined by asymmetric synthesis, chemical transformations, Fredga's method of quasiracemates, and optical comparison (Freudenberg's "rule of shift" and Van't Hoff's "rule of superposition"). For details, see reference 2, pages 95-123. Recently, nmr spectroscopy has also been used; for example, W. Pirkle, et al., J. Amer. Chem. Soc., 93, 2817 (1971).

Refs. p. 234.

Refs. p. 234.

C. Characteristic Cotton Effects

The table below summarizes ORD and CD parameters for representative Cotton effects. All data refer to measurements at 25° or room temperature and, for the most part, were calculated or read from graphical representations in the cited references. For more details and references to many types of optically active chromophores, see references 1, 1a, and 3. Also, compilations of rotation data for natural products (steroids, terpenes, amino acids, alkaloids) are available from Pergamon Press in the series, Tables de Constantes Selectionées, Vols. 6, 9, 10, 11, and 14 (Pouvoir Rotatoire Naturel). Other collections of optical rotatory data are: T. M. Lowry, in International Critical Tables, Vol. VII, McGraw-Hill, New York, 1930, pages 355-488; W. Kuhn and H. Martin, in Landolt-Bornstein. Zahlenwerte und Functionen, Vol. II, Part 8, Springer-Verlag, Berlin, 1962, pages 5-676 to 5-814; and L. Meites, Ed., Handbook of Analytical Chemistry, McGraw-Hill, New York, 1963, pages 6-240 to 6-270. For recent discussions on compounds whose chirality is due to the presence of deuterium, see reference 32.

Considerable data on the ORD and CD spectra of inorganic complexes are available in R. D. Gillard, Ch. 5 in Physical Methods in Advanced Inorganic Chemistry, H. Hill and P. Day, Eds., Interscience, New York, 1968.

Compound	Absolute Configuration	ORD[a]			CD[a,b]		Ref.[d]
		Molecular Amplitude	Peak[c]	Trough[c]	$[\theta]_{max} \times 10^{-2}$	λ_{max}	
Cyclohexanones							
(+)-3-methyl	3R	+26(D)	320	270			3-103
(+)-3,6,6-trimethyl	3R	+81(D)					1-76
(-)-trans-3-methyl-6-i-propyl [(-)-menthone]	3R, 6R	+11(M)	320	280	-4.2(DC)	280	1-78,82
(+)-cis-3-methyl-6-i-propyl [(+)-isomenthane]	3R, 6S	+92(M)	320	270			3-105
(+)-trans-6-chloro-3-methyl:							
axial-Cl	3R, 6R	-33(I)	300	330	-24(I)	318	1-79,80
equatorial-Cl		+22(M)	305	270			1-79
(+)-Camphor	1S				+52.8(D)	300	4-211
(+)-Norcamphor	1S	-28(I)	280	320			7
(+)-Dehydronorcamphor	1S	+780(I)	320	280	+150(I)	305	7,8

(+)- [bicyclic ketone structure]		+520(I)	325	275	+130(I)	303	7,8
(+)-trans-9-Methyl-1,4,9,10-tetrahydronaphthalene	9S, 10S	+440(C)	275	245			1-245
Cholesta-2,4-diene		+520(I)	280	245			1-246
Camphorquinone[e]					-15 +15	475 290	1-173 1-173
Hexahelicene	R	∿+10^{6}(CH)	358	319	($[\alpha]_D^{25}$+3700(ref 2-176))		6
(+)-p-$CH_3C_6H_4\overset{O}{S}C_2H_5$	R	+1280(E)	257	222			9
Twisted Biphenyls							
$X=CH_3$; $Y=NO_2$	S	+60(D) -170(D)	335 270	320 320	-16(D) +90(D) +25(D)	251 298 350	1-182
$X=CH_3$; $Y,Y=CH_2COCH_2$	S	-1720(I)	290	325	-950(I) -105(I)	255 300	8, 10
$X=CH_3$; $Y,Y=(CH_2)_3$	S	-1430(I)	230	253	-104(I) +60(I)	250 275	8, 10

[a]Solvent in parentheses: I = isooctane, M = methanol, D = dioxane, DC = decalin, C = cyclohexane, CH = chloroform, E = 95% ethanol. [b]All CD data refer to maxima at λ_{max} (nm). [c]Wavelengths at extrema (nm). [d]Page number of reference given after hyphen. [e]Center of complex CD curves.

Refs. p. 234.

D. Resolution Methods and Agents

This section summarizes most of the common optical resolving methods or agents for organic systems; for general discussion, the reader is referred to references 24 and 25 and to reference 2, pages 47-83. For very recent and fairly thorough reviews see Methods of Optical Resolution, by P. H. Boyle, Quart. Rev., 25, 323 (1971) (132 references), and Resolving Agents and Resolutions in Organic Chemistry, by S. Wilen, Topics in Stereochemistry, 6, 107 (1971) (384 references). Some of the agents given below may be available as the enantiomer, and many are sold as their salts (alkaloids and amines, especially). At least one company (Norse Laboratories, Inc., Santa Barbara, Calif. 93103) offers a custom resolution service as well as a large variety of resolving agents. Hoffman-La Roche, Inc. (Kingsland Street, Nutley, N. J.) sells a test kit containing 5 mM each of (+) and (-)-α-methylbenzylamine and the corresponding N-methyl and N,N-dimethyl derivatives, all as the free bases.

Resolution of inorganic complexes, especially electrolytes, is readily accomplished with many of the agents listed below. Alkaloids are used as the protonated species and acids (camphorsulfonic, tartaric, etc.) as the anions. Resolved metal complexes themselves may be used, such as $Co(en)_3^{+3}$, $Ni(phen)_3^{+2}$, cis-$Co(en)_2(NO_2)_2^{+}$, and $Co(edta)^{-}$. In addition, other methods are employed such as chromatography, zone melting, and diffusion. For details, see reference 34.

The preparation of optically active molecules by asymmetric synthesis has been reviewed recently: J. D. Morrison and H. S. Mosher, Asymmetric Organic Reactions, Prentice-Hall, Englewood Cliffs, N. J., 1971, and H. Brunner, Optical Activity from Asymmetric Transition Metal Atoms, Angew. Chem. Internat. Ed. Engl., 10, 249 (1971). Also of interest is the "spontaneous generation" of optical activity by chance development, in the absence of dissymmetric forces; for a recent discussion, see R. E. Pincock et al., Science, 174, 1018 (1971).

1. Chemical

Agent	MW	Absolute Configuration	$[\alpha]_D$ (Solvent)[b]	Supplier[c]	Ref.[d]
		For Acids			
2-Amino-1-butanol	89.14		-10.1(n)	A	
Amphetamine	135.21	S	+38(B)	A,K	F-838(1)
Brucine	394.45	a	-127(Ch)	A,K,FL,KL	
Cinchonidine	294.38	a	-109(Et)	A,C,KL,K	
Cinchonine	294.38	a	+229(Et)	FL,K,KL	
Dehydroabietylamine	285.46	a	+46(M)	A,H	F-97(2)
Deoxyephedrine	149.23	S	+17.9(W)[f]		
Ephedrine	165.23	αR,βR	-41(W/HCl)	FL,K,KL	
Menthol	156.26	1R,2R,5R	-50(W/Et)	A,F,K	

Refs. p. 234.

Menthyl amine	155.18	1R,2R,5R	-32(Et)		2-54
Morphine	285.33	a	-132(M)[e]	A	
1-(1-Naphthyl)ethylamine	171.24		+81.35(n)	A	
1-(2-Naphthyl)ethylamine	171.24				
α-Phenethylamine	121.18	R	+39(M)	A,FL	2-51
Quinidine	324.41	a	+230(Ch)	A,FL,K	
Quinine	324.41	a	-117(Ch)	A,C,K	
Stilbenediamine	212.3				
Strychnine	334.4	a	-139(Ch)	A,K	
		For Amines and Alcohols			
3β-Acetoxy-Δ[5]-etienic acid	360.48	a		K	F-9(1)
3-Bromocamphor-8-sulfonic acid	311.21	a	-92(W)	A,K	F-40(2)
Camphoric acid	200.23	1S,3S	+48(Et)	A,F,FL,K	
Camphor-10-sulfonic acid	232.30	1R	+43(Et)	A,B,F,	F-108(1)
Camphor-10-sulfonyl chloride	250.74	1R	+32(Ch)		F-109(1)
O,O'-Diacetyltartaric acid	234.2				
O,O'-Dibenzoyltartaric acid	258.31	2R,3R	-118(Et)	A,F(Kodak)	
6,6'-Dinitro-2,2'-diphenic acid	332.2				
O,O'-Di-p-toluyltartaric acid	386.36	2S,3S	+132(Et)	A	F-213(2)
Glutamic acid	147.13	2S	+31(6N HCl)	A,C,F,KL	2, p. 53
Hydratropic acid	150.18				
Malic acid	134.09	2S	-2(W)	A,F,K,KL	
Mandelic acid	152.15		-156(W)	A	
Menthoxyacetic acid	214.3	1R,2R,5R	-93(M)	A,K	2, p. 57
Menthoxyacetyl chloride	232.8	1R,2R,5R			29
Menthyl isocyanate	181.3	1R,2R,5R			2, p. 57
α-Phenethyl isocyanate	147.18			FL	
L-Pyrrolidone-5-carboxylic acid	129.1	5S	-12(W)	A,FL,K,KL	F-977(1)
Tartaric acid	150.09	2S,3S	-12(W)	A,F,K,KL	2, p. 53
Tartranilic acid	225.2	2R,3R	+115(M)		26
		Miscellaneous			
Complexing and Related Agents (27)					
Cyclodextrin		a			14
Desoxycholic acid	392.6	a		A,C,FL,K	2, p. 58
Digitonin	1229.30	a	-54(M)	FL,K,KL	30
2-(2,4,5,7-Tetranitro-9-fluorenylideneaminooxy)propanoic acid (TAPA) (resolution of hexahelicene; methylcholanthenes)					11

(Continued)

Refs. p. 234.

Agent	MW	Absolute Configuration	$[\alpha]_D$ (Solvent)[b]	Supplier[c]	Ref.[d]
For Aldehydes and Ketones					
(+)-Butane-2,3-dithiol	122.25	1S,2S	-13(n)		12
Via mandelic acid derivative of hydrazone					28
Via ketal of D-(-)-butane-2,3-diol	90.42		-12.9(n)	K	F-81(1)
4-(4-Carboxyphenyl) semicarbazide derivative (resolved via the carboxy group)					F-118(1)
Via enamines using camphor-10-sulfonic acid					F-58(2)
General techniques					2, p. 56
For Skewed Olefins					
$PtCl_2$/olefin/optically active amine complex (trans-cyclooctene)					13; F-272(2)
For Resolution of Salts					
Via silver dibenzoylhydrogen tartrate					31
Asymmetric Induction, Destruction, or Kinetic Resolution Techniques					20; 25
Biochemical Resolution (Use of Enzymes)					20; 25
For recent method of resolving α,β-unsaturated steroidal ketones and determining their absolute configuration by reaction with the steroid dehydrogenase Arthrobacter simplex, see J. Fried, M. J. Green, and G. V. Nair, J. Amer. Chem. Soc., 92, 4136 (1970).					

[a]See Merck Index (7th or 8th ed.) or Heilbron's Dictionary of Organic Compounds (1st or 2nd ed.) among others, for structures of complex molecules. [b]Rotation at or near 20° for presumably optically pure substance; the values have been obtained from several sources, including issues of the CRC Handbook of Chemistry and Physics (45th ed., 1964, p. E130; 50th ed., 1969, p. E255) and the manufacturers. Please note that general agreement is lacking on correct values for optically pure material, and those cited here are the highest found and are to be considered tentative in most cases (except perhaps for the natural products--alkaloids, etc.). Solvents: B = benzene, Ch = chloroform, Et = ethanol, M = methanol, W = water, n = neat. [c]Representative commercial sources of the optically active forms: A = Aldrich, B = J. T Baker, C = Columbia Organic, F = Fischer Scientific, FL = Fluka (Buchs SG, Switzerland), H = Hercules Powder (compound available as mixture "Amine D"), K = K and K, KL = Koch-Light (Colburn, Bucks, England); for complete addresses, see Fieser and Fieser, Reagents for Organic Synthesis, Wiley, New York, (Vol. 1, 1967; Vol. 2, 1969). [d]F-n(1 or 2) refers to page n in Vol. 1 or 2, Fieser and Fieser (previous footnote); only representative references given. In many cases, no general reference is available; see reference 25 instead. [e]As monohydrate. [f]As hydrochloride. J. Jacobus, personal communication.

Refs. p. 234.

2. Chromatographic

These methods in principal might be particularly useful for compounds without convenient functional groups (handles); however, the technique is still rather undeveloped. A review has appeared (15) (197 references through 1966) of methods using adsorption, glc, and ion exchange with dissymmetric sorbents; see also reference 2, p. 61. The following additional examples are typical of recent techniques.

Chromatographic Method	Description	Ref.
Paper	Amino acids with a benzene ring; aromatic rings with at least two polar functions; and alkaloids resolved on natural paper and enzyme-sprayed paper.	16
Thin layer	Optically active silica gel sorbent mixture for general use.	17
Column	Variety of compounds resolved with acetyl cellulose; available commercially as "Acetyl cellulose Woelm" (see chromatography section for addresses of Woelm suppliers).	18
Column	Partial resolution of amino acids on polystyrene resin containing chemically bound (N-carboxymethyl-L-valine) copper (II).	35
Glc	Resolution of 2-n-alkanols on 50 to 100 m columns of N-trifluoroacetyl α-amino acid esters. For references to more examples of glc, see reference 20, page 211.	19

E. Optical Purity

Many techniques for optical purity determination are intimately related to those of resolution; for details see the recent review by Raban and Mislow (20), from which most of the following discussion is taken.

1. Definitions

Optical purity: $p = [\alpha]/[A]$ (polarimetric purity) where $[\alpha]$ = specific rotation of a substance and $[A]$ = specific rotation of the pure enantiomer ($[A]$ also called "absolute rotation").

Enantiomeric purity: measure of the excess of one enantiomer over the other, namely,

$$\left|\frac{R - S}{R + S}\right| = 2R - 1$$

where R and S are mole fractions of the enantiomers ($R + S = 1$). Optical purity usually equals enantiomeric purity; in practice, the term optical purity is used for both. However, in at least one case (2-ethyl-2-methylsuccinic acid)(22), $[\alpha]$ is <u>not</u> a linear function of enantiomer concentration and thus enantiomeric <u>and</u> optical purity are <u>unequal</u>; it is not known how general this phenomenon is.

Refs. p. 234.

2. Methods Involving Actual Separation of Enantiomers or Their Diastereomeric Derivatives

	Method	Ref.
a.	Kinetic resolution. (Differences in rates of enantiomers with chiral reagents.)	20; 2, p. 65
b.	Vapor phase chromatography. (Separation and quantitative analysis of diastereomeric derivatives.)	20

3. Methods Not Involving Separation

a. Isotopic Dilution (20). Typically, (m) g of labeled (^{14}C, ^{2}H, e.g.) racemic material of activity S_o are added to (n) g of unlabeled material, rotation $[\alpha]$; reresolution (or, in general, optical fractionation) gives sample of new rotation, $[\alpha_i]$ and new activity, S_i:

$$[A]^2 = \frac{S_i n^2 [\alpha]^2 - S_o mn[\alpha][\alpha_i]}{S_i(m + n)^2 - S_o m(m + n)}$$

(whence $p = [\alpha]/[A]$). A very sensitive radioassay technique for racemic content in peptides (concentration range 0.001-1%) has recently been developed (23).

b. Differential Microcalorimetry (20). For enantiomers R and S that form racemic compounds (racemates) with heat of fusion ΔH_r and melting point T_r, an unknown R,S mixture of melting point T has optical purity given by (R_g = gas constant; R and S = mole fractions of enantiomers):

$$\ln 4RS = \ln 4(1-R^2) = [\Delta H_r/R_g][(1/T_r) - (1/T)]$$

c. NMR Spectroscopy (20, 21). Integrated intensities of chemically shifted probe nuclei of diastereomeric derivatives in an achiral (optically inactive) solvent; or of the mixture of enantiomers in a chiral solvent (e.g., α-phenethylamine; 2,2,2-trifluoro-1-phenylethanol; various carbinols). Clever techniques for enhancing this method have been reported (33), using optically active nmr shift reagents.

F. References

1. P. Crabbé, ORD and CD in Organic Chemistry, Holden-Day, San Francisco, 1965.
1a. P. Crabbé, Topics in Stereochemistry, 1, 93 (1967).
2. E. Eliel, Stereochemistry of Carbon Compounds, McGraw-Hill, New York, 1962.
3. C. Djerassi, Optical Rotatory Dispersion, McGraw-Hill, New York, 1960.
4. L. Velluz, M. Legrand, and M. Grosjean, Optical Circular Dichroism, Academic Press, New York, 1965.
5. J. H. Brewster, Topics in Stereochemistry, 2, 1 (1967) and references therein.

6. A. Moscowitz, Tetrahedron, 13, 48 (1961).
7. K. Mislow and J. Berger, J. Amer. Chem. Soc., 84, 1956 (1962).
8. E. Bunnenberg et al., ibid., 84, 2823 (1962).
9. K. Mislow et al., ibid., 85, 2329 (1963).
10. K. Mislow, et al., ibid., 84, 1455 (1962).
11. M. S. Newman and O. Lednicer, ibid., 78, 4765 (1956); M. S. Newman et al., J. Org. Chem., 31, 4293 (1966). For the agent, see Org. Syn., 48 (1968).
12. E. J. Corey and R. Mitra, J. Amer. Chem. Soc., 84, 2938 (1962).
13. A. C. Cope et al., ibid., 85, 3276 (1963); 92, 1243 (1970).
14. F. Cramer and W. Dietsche, Chem. Ber., 92, 378 (1959).
15. S. Rogozhin and V. Davankov, Usp. Khim., 37, 1327 (1968). [In Russian; English translation available in Russ. Chem. Rev., 37 (1968)]. See also D. Buss and T. Vermeulen, Ind. Eng. Chem., 60, 12 (1968).
16. I. M. Hais and K. Macek, Paper Chromatography, Academic Press, New York, 1963, pp. 74, 450, 462, 580, 586, 848.
17. G. Marini-Bettòlo, This Layer Chromatography, Elsevier, New York, 1964, p. 9.
18. For example, A. Lüttringhaus, V. Hess, and H. Rosenbaum, Z. Naturforsch., 22b, 1296 (1967).
19. E. Gil-Av, B. Feibush, and R. Charles-Singer, Tetrahedron Letters, 1009 (1966); R. Charles-Singer and E. Gil-Av, ibid., 4231 (1966).
20. M. Raban and K. Mislow, Topics in Stereochemistry, 2, 199 (1967).
21. W. H. Pirkle and S. D. Beare, J. Amer. Chem. Soc., 91, 5150 (1969), and references cited; 93, 2817 (1971).
22. A. Horeau, Tetrahedron Letters, 3121 (1969).
23. D. Kemp et al., J. Amer. Chem. Soc., 92, 1043 (1970).
24. L. Velluz, Substances Naturelles de Syntheses, Vol. IX, Masson et Cie, Paris, 1954.
25. W. Theilaker in Houben-Weyl's Methoden der Organischen Chemie, Vol. 4, Part 2, Ch. 7, Georg Thieme Verlag, Stuttgart, 1955.
26. T. Montzka, T. Pindell, and J. Matiskella, J. Org. Chem., 33, 3993 (1968), describe several substituted tartranilic acids.
27. For general discussion, see M. S. Newman, Steric Effects in Organic Chemistry, J. Wiley, New York, 1956, Ch. 10, pp. 472-478.
28. K. Bott, Tetrahedron Letters, 4569 (1965).
29. W. S. Johnson et al., J. Amer. Chem. Soc., 74, 2832 (1952); W. Bachmann, W. Cole, and A. Wilds, ibid., 62, 824 (1940); J. W. Cook, et al., J. Chem. Soc., 911 (1950). For the resolving agent itself, see Org. Syn., Coll. Vol. III, pp. 544, 547 (1955).
30. R. B. Woodward, J. Amer. Chem. Soc., 74, 4223 (1952).
31. B. Bosnick and S. Wild, ibid., 92, 459 (1970).
32. D. Arigoni and E. L. Eliel, Topics in Stereochemistry, 4 (1969), and L. Verbit, in Progress in Physical Organic Chemistry, Vol. 7, A. Streitwieser, Jr., and R. W. Taft, Eds., Wiley-Interscience, New York, 1970.
33. G. M. Whitesides and D. W. Lewis, J. Amer. Chem. Soc., 92, 6979 (1970); 93, 5914 (1971). H. L. Goering, J. Eikenberry, and G. Koermer, ibid., 93, 5913 (1971).
34. A. M. Sargeson, in Chelating Agents and Metal Chelates, F. P. Dwyer and D. P. Mellor, Eds., Academic Press, New York, 1964, pp. 193-198.
35. R. V. Snyder, R. J. Angelici, and R. B. Meck, J. Amer. Chem. Soc., 94, 2660 (1972).
36. M. Fétizon, I. Hanna, A. I. Scott, A. D. Wrixon, and T. K. Devon, J. Chem. Soc. D, 545 (1971) and references cited. Cf. C. Levin and R. Hoffmann, J. Amer. Chem. Soc., 94, 3446 (1972).
37. J. Eric Nordlander and T. McCrary, Jr., ibid., 94, 5135 (1972).

VII. MASS SPECTROMETRY

This section contains common data useful for interpretation of spectra and calculation of ion mass. For detailed values of natural abundances and exact isotopic masses, see page 88. Several books are available that provide extensive detail beyond that given here; the following bibliography is recommended:

J. H. Beynon, Mass Spectrometry and Its Applications to Organic Chemistry, Evsevier, Amsterdam, 1960 (640 pp.).
J. H. Beynon, R. A. Saunders, and A. E. Williams, The Mass Spectra of Organic Molecules, Elsevier, New York, 1968 (519 pp.).
J. H. Beynon et al., Table of Ion Energies for Metastable Transitions in Mass Spectrometry, Elsevier, New York, 1970.
J. H. Beynon and A. E. Williams, Mass and Abundance Tables for Use in Mass Spectrometry, Elsevier, New York, 1963.
K. Biemann, Mass Spectrometry. Organic Chemical Applications, McGraw-Hill, New York, 1962 (384 pp.).
H. Budzikiewicz, C. Djerassi, and D. Williams, Mass Spectrometry of Organic Compounds, Holden-Day, San Francisco, 1967 (690 pp.).
A. L. Burlingame, Topics in Organic Mass Spectrometry, Wiley-Interscience, New York, 1970 (472 pp.). This book covers current areas of active research.
R. Kiser, Introduction to Mass Spectrometry and Its Applications, Prentice-Hall, New York, 1965.
J. Lederberg, Computation of Molecular Formulas for Mass Spectrometry, Holden-Day, San Francisco, 1964 (78 pp.).
F. W. McLafferty, Ed., Mass Spectrometry of Organic Ions, Academic Press, New York, 1963 (730 pp.).
Q. N. Porter and J. Baldas, Mass Spectrometry of Heterocyclic Compounds, Wiley-Interscience, New York, 1971 (564 pp.).
J. Roboz, Introduction to Mass Spectrometry Instrumentation and Techniques, Interscience, New York, 1968.

A. Selected Ionization Potentials of Neutral Substances

The data below represent the Franck-Condon (vertical) ionization potentials (IP) ($R + e^- \rightarrow R^{+\bullet} + 2e^-$). Unless indicated otherwise, the values cited have been determined by photoionization and are among the best available. All data have been extracted from "Ionization Potentials, Appearance Potentials, and Heats of Formation of Gaseous Positive Ions", NSRDS-NBS 26, June, 1969 (available from Superintendent of Documents, Government Printing Office, Washington, D. C. 20402); values are in eV (1 eV = 23.0609 kcal/mole).

A letter in parentheses following the IP represents a method other than photoionization: S = optical spectroscopy, SL = semilog plot, VC = vanishing current, EVD = extropolated voltage difference, RPD = retarding potential difference. In some cases, the electronic state of the product ion is given after the formula, for example, $Ar(^2P_{3/2})$.

R	IP (eV)
1. Inorganic	
H	12.6 (S)
D	13.6 (S)
H_2	15.42
D_2	15.46
He	24.6
Ne	21.8
Ar ($^2P_{3/2}$)	15.75
Kr ($^2P_{3/2}$)	14.0
Xe ($^2P_{3/2}$)	12.13
F_2	15.7 (S)
Cl_2 ($^2\Pi_{3/2g}$)	11.48
Br_2 ($^2\Pi_{3/2g}$)	10.53
I_2 ($^2\Pi_{3/2g}$)	9.28
HF ($^2\Pi$)	15.77 (RPD)
HCl ($^2\Pi_{3/2}$)	12.74
HBr	11.62
HI ($^2\Pi_{3/2}$)	10.38
2. Hydrocarbons	
C	11.3 (S)
$CH_3\cdot$	9.82
CH_4	12.70
CD_4	12.88
C_2H_2	11.40
C_2D_2	11.42
C_2H_4	10.51
C_2H_6	11.52
C_3H_4	
Allene	10.2 (SL)
Propyne	10.36
Cyclopropene	9.95 (SL)
C_3H_6	
Propene	9.73
Cyclopropane	10.1
C_3H_8	11.14
C_4H_2 (1,3-butadiyne)	10.2 (VC)
C_4H_6	
1,2-Butadiene	9.57 (SL)
1,3-Butadiene	9.08
1-Butyne	10.18
2-Butyne	9.9 (VC)
C_4H_8	
1-Butene	9.58
cis-2-Butene	9.13
trans-2-Butene	9.13
iso-Butene	9.23
Cyclobutane	10.58 (SL)
C_4H_{10}	
Butane	10.63
Isobutane	10.57
C_5H_6 (Cyclopentadiene)	8.9 (SL)
C_5H_8	
1,2-Pentadiene	9.42 (SL)
1,3-Pentadiene	8.68 (SL)
2,3-Pentadiene	9.26 (SL)
1,4-Pentadiene	9.58 (SL)
Cyclopentene	9.01
C_5H_{10}	
1-Pentene	9.50
cis-2-Pentene	9.11 (SL)
trans-2-Pentene	9.06 (SL)
2-Methyl-1-butene	9.12
2-Methyl-2-butene	8.67
Cyclopentane	10.9 (SL)
C_5H_{12}	
Pentane	10.35
Isopentane	10.32
Neopentane	10.35
C_6H_4 (benzyne)	9.75
C_6H_6 (benzene)	9.25
C_6D_6 (benzene-d_6)	9.25
C_6H_{10} (cyclohexene)	8.94
C_6H_{12}	
1-Hexene	9.45
2,3-Dimethyl-2-butene	8.30
Cyclohexane	9.88
C_6H_{14}	
Hexane	10.18
Isohexane	10.12
3-Ethylbutane	10.08
C_7H_8	
Toluene	8.82
Cycloheptatriene	8.55 (SL)
C_7H_{10} (norbornene)	9.0 (SL)
C_7H_{14}	
Methylcyclohexane	9.85
C_7H_{16} (heptane)	9.90
C_8H_6 (phenylacetylene)	8.82
C_8H_8	
Styrene	8.47
Cyclooctatetraene	7.99
Cubane	8.74 (EVD)
C_8H_{10}	
Ethylbenzene	8.76
o-Xylene	8.56
m-Xylene	8.56
p-Xylene	8.44
C_8H_{18} (isooctane)	9.86
C_9H_{12} (trimethylbenzene)	
1,2,3	8.48
1,2,4	8.40
1,3,5	8.39

(Continued)

R	IP(eV)
$C_{10}H_8$	
Naphthalene	8.12
Azulene	7.43(S)
$C_{10}H_{18}$(decalins)	
cis	9.61(EVD)
trans	9.61(EVD)
$C_{12}H_{10}$(biphenyl)	8.27
$C_{14}H_{10}$	
Diphenylacetylene	8.85(SL)
Anthracene	7.55(SL)
Phenanthrene	8.10(SL)

3. Heteroelement Compounds

R	IP(eV)
$N_2(X^2\Sigma_g^+)$	15.56
NH_3	10.15
$NO(X^1\Sigma^+_{3/2})$	9.25
$N_2O(X^2\Pi_{3/2})$	12.89(S)
NO_2	9.78
CH_3NH_2	8.97
CH_2N_2(diazomethane)	8.99(S)
CH_3CN	12.21
$C_2H_5NH_2$	8.86
$(CH_3)_2NH$	8.24
CH_2CHCN	10.91
C_2H_5CN	11.84
$C_2H_6N_2$(azomethane)	8.65(SL)
$(CH_3)_3N$	7.82
C_4H_5N(pyrrole)	8.20
C_5H_5N(pyridine)	9.23
$C_6H_5NH_2$(aniline)	7.69
C_6H_5CN(benzonitrile)	9.71
$C_6H_5CH_2NH_2$	7.56
$(C_6H_5)_2NH$	7.25
$(C_6H_5)_3N$	6.86
$O_2(X^2\Pi g)$	12.06
O_3	11.7
H_2O	12.61
D_2O	12.64
$CO(X^2\Sigma^+)$	14.01(S)
$CO_2(X^2\Pi g)$	13.79(S)
CH_2O	10.90
CH_3OH	10.85
HCOOH	11.05
CH_3CHO	10.21
CH_3OCH_3	9.96(S)
CH_3CH_2OH	10.48
CH_3COOH	10.35
CH_3COCH_3	9.69
CH_3COOCH_3	10.27
CH_3CH_2CHO	9.98
CH_2CHCH_2OH	9.67
C_4H_4O(furan)	8.89
$(C_2H_5)_2O$	9.53
C_4H_8O(tetrahydrofuran)	9.42(S)
$C_4H_8O_2$(p-dioxan)	9.13
C_6H_5OH	8.50
$C_6H_4O_2$(p-benzoquinone)	9.67
$C_6H_5OCH_3$	8.20
C_6H_5CHO	9.51
$C_6H_5COCH_3$	9.27
$(C_6H_5)_2O$	8.82(SL)
$(C_6H_5)_2CO$	9.46(SL)
$HCONH_2$	10.25
CH_3CONH_2	9.77
$C_6H_5CONH_2$	9.4(SL)
$C_6H_5NHCOCH_3$	8.39(SL)
CH_3NO_2	11.08
$C_6H_5NO_2$	9.92
$C_6H_6N_2O_2$(p-nitroaniline)	8.85(SL)
$C_6H_5NO_3$(p-nitrophenol)	9.52(SL)
$CH_3Cl(^2E_{1/2})$	11.26
$CH_3Br(^2E_{1/2})$	10.53
$CH_3I(^2E_{1/2})$	9.54
C_6H_5F	9.19
C_6H_5Cl	9.07
C_6H_5Br	8.98
C_6H_5I	8.62
$(CH_3)_4Si$	9.5(SL)
H_2S	10.42
$CS_2(^2\Pi_{3/2}g)$	10.08
SO_2	12.34
CH_3SH	9.44
$(CH_3)_2S$	8.68
C_4H_4S(thiophene)	8.86
C_6H_5SH	8.32
$C_6H_5SCH_3$	8.9(SL)

B. Common Doublets and Their Mass Differences

m/e	Doublet	$10^3\Delta M$(amu)	m/e	Doublet	$10^3\Delta M$(amu)
2	H_2-D	1.549	30	COD-$CO^{17}H$	2.060
12	C-H_{12}	93.900		N_2H_2-CO^{18}	22.637
13	CH-C^{13}	4.468		N_2D-CO^{18}	21.088
	1/2C_2D-C^{13}	3.694		$N^{15}NH$-CO^{18}	11.847
14	CD-N	11.030		$N^{15}NH$-COD	1.991
	N-$C^{13}H$	8.109		N_2D-$C^{13}OH$	14.153
	CH_2-N	12.579		N_2D-$CO^{17}H$	13.294
	1/2CO-$C^{13}H$	13.726		C_2H_6-NO	48.962
15	NH-N^{15}	10.789	31	C_2H_7-FC	56.373
	CH_3- N^{15}	23.365		C_2H_7-$N^{15}O$	59.748
	CHD-N^{15}	21.817		C_2H_7-NO^{17}	52.570
16	CH_4- O	36.384		$C^{13}CH_6$-$N^{15}O$	55.284
	NH_2-O	23.809		C_2H_5D-$N^{15}O$	58.203
	$C^{13}H_3$-O	31.919		C_2H_5D-NO^{17}	51.020
	$N^{15}H$-O	13.021	32	O_2-S	17.754
	ND-O	22.262		C_2H_8-S	90.524
	CH_2D-O	34.838		C_2H_8-NO^{18}	60.363
	CH_4-ND	14.124		$C^{13}CH_7$-NO^{18}	55.897
	$C^{13}H_3$-ND	9.653		C_2H_6D-NO^{18}	58.815
17	NH_3-O^{17}	27.415	36	C_3-HCl^{35}	23.320
	$C^{13}H_4$-O^{17}	35.527	37	C_3H-Cl^{37}	41.920
	$N^{15}H_2$-O^{17}	16.629		$Cl^{35}D$-Cl^{37}	17.051
	NHD-O^{17}	25.870	40	C_3H_4-Ar	68.918
	CH_3D-O^{17}	38.444		C_2NH_2-Ar	56.340
18	$N^{15}H_3$-O^{18}	24.421		C_2O-Ar	32.531
	HO^{17}-O^{18}	7.793		C_3H_2D-Ar	67.371
	NH_2D-O^{18}	33.663		C_2ND-Ar	54.793
19	H_{19}-F	150.270		$C_2N^{15}H$-Ar	45.553
28	N_2-CO	11.234	44	C_3H_8-N_2O	61.535
	C_2H_2D-N_2	23.603		N_2O-CS	28.987
	$C^{13}CH_3$-N_2	20.685		C_3H_8-$Si^{28}O$	90.761
29	$N^{15}N$-COH	0.445	79	C_6H_7-Br^{79}	136.382
	$N^{15}N$-$C^{13}O$	4.912		C_5H_3O-Br^{79}	99.998
	$N^{15}N$-CO^{17}	4.053		C_5H_5N-Br^{79}	123.807
	N_2H-$C^{13}O$	15.701	81	C_6H_9-Br^{81}	154.066
	N_2H-CO^{17}	14.842		C_5H_5O-Br^{81}	117.685
	CO^{17}-$C^{13}O$	0.859		C_5H_7N-Br^{81}	141.500
30	COD-$C^{13}OH$	2.919		DBr^{79}-Br^{81}	16.141

C. Common Fragment Ions

The following are frequently occurring positive ions, coming either directly from the molecular ion or from subsequent fragments. For a glossary of fragment ion types, see F. McLafferty, Interpretation of Mass Spectra, Benjamin, New York, 1966, pages 214-221, and Beynon, Saunders and Williams, op. cit., pages 461-474.

m/e	Ions
14	CH_2
15	CH_3
16	O
17	OH
18	H_2O, NH_4
19	F, H_3O
26	CN, C_2H_2
27	C_2H_3
28	CO, N_2, C_2H_4, CHNH
29	C_2H_5, CHO
30	NO, CH_2NH_2
31	CH_3O, CH_2OH
32	O_2
33	SH, CH_2F
34	H_2S
35	Cl
36	HCl
39	C_3H_3
40	CH_2CN, Ar
41	C_3H_5, C_2H_2NH, CH_2CN+H
42	C_3H_6
43	CH_3CO, C_2H_5N, C_3H_7
44	CH_2CHO+H, CO_2, CH_3CHNH_2, NH_2CO, $(CH_3)_2N$
45	CH_3CHOH, CH_2CH_2OH, CH_2OCH_3, CO_2H
46	NO_2
47	CH_3S, CH_2SH
48	CH_3S+H
49	CH_2Cl
51	CHF_2
53	C_4H_5
54	CH_2CH_2CN
55	C_4H_7, CH_2CHCO
56	C_4H_8
57	C_4H_9, C_2H_5CO, CH_3COCH_2
58	CH_3COCH_2+H, $C_2H_5CHNH_2$, $(CH_3)_2NCH_2$, $C_2H_5NHCH_2$, C_2H_2S
59	CH_3OCO, $CH_2OC_2H_5$, $(CH_3)_2COH$, CH_3OCHCH_3, CH_3CHCH_2OH
60	CH_2CO_2H+H, CH_2ONO
61	CH_3OCO+2H, CH_2CH_2SH, CH_2SCH_3
65	Cyclopentadienyl cation
67	C_5H_7
68	$(CH_2)_3CN$
69	C_5H_9, CF_3, $CH_3CHCHCHO$, $CH_2C(CH_3)CO$
70	C_5H_{10}
71	C_5H_{11}, C_3H_7CO
72	$CH_2COC_2H_5$, $C_4H_{10}N$ isomers, $(CH_3)_2NCO$
73	Homologs (+CH_2 units) of 59
74	$CH_2CO_2CH_3$+H
75	C_2H_5OCO+2H, $CH_2SC_2H_5$, $CH(OCH_3)_2$, $(CH_3)_2CSH$
77	C_6H_5
78	C_6H_5+H
79	C_6H_5+2H, Br
80	Pyrrolyl-2-CH_2
81	Furanyl-2-CH_2, C_6H_9
82	CCl_2, C_6H_{10}, $(CH_2)_4CN$
83	C_6H_{11}, $CHCl_2$, thiophenyl
85	C_6H_{13}, C_4H_9CO, $CClF_2$
86	$C_3H_7COCH_2$+H, $C_5H_{12}N$ isomers
87	$C_3H_7CO_2$, homologs of 73, $CH_3OCOCH_2CH_2$
88	H+$CH_2CO_2C_2H_5$
89	C_3H_7OCO+2H, C_6H_5C
90	C_6H_5CH
91	$C_6H_5CH_2$ (C_7H_7), C_6H_5N, $(CH_2)_4Cl$
92	$C_5H_4NCH_2$ (pyridinyl CH_2)
93	CH_2Br, C_6H_5O (or C_6H_4OH), C_7H_9
94	2-Pyrroloyl (C_4H_4NCO), C_6H_5O+H
95	2-Furanoyl (C_4H_3OCO)
96	$(CH_2)_5CN$
97	C_7H_{13}, C_4H_3S-CH_2 (thiopheneyl-2-CH_2)
99	C_7H_{15}, $C_6H_{11}O$
100	$C_4H_9COCH_2$+H, $C_6H_{14}N$ isomers
101	C_4H_9OCO
102	$C_3H_7OCOCH_2$+H
103	C_4H_9OCO+2H, $C_5H_{11}S$, $(CH_3CH_2O)_2CH$
104	$C_2H_5CHONO_2$
105	C_6H_5CO, $C_6H_5C_2H_4$
106	$C_6H_5CHCH_2$
107	$C_6H_5CH_2O$, $HOC_6H_4CH_2$ (o- and p-)
108	N-methyl-2-pyrroloyl
111	2-Thiophenoyl
119	CF_3CF_2, $C_6H_5C_3H_6$, $CH_3C_6H_4C_2H_4$, $CH_3C_6H_4CO$
120	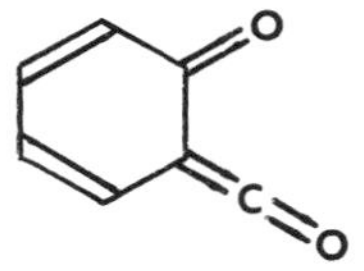

m/e	Ions	m/e	Ions
121	HOC_6H_4CO, $CH_2C_6H_4OCH_3$, C_9H_{13},	127	I
		131	C_3F_5, $C_6H_5CHCHCO$
		138	o-$HOC_6H_4CO_2$+H
		139	ClC_6H_4CO
		143	$(CH_2)_4Br$
		149	Phthalic anhydride + H
		154	Biphenyl
123	FC_6H_4CO		
125	C_6H_5SO		

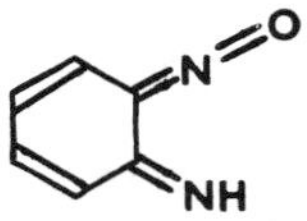

D. Common Neutral Fragments

These fragments are commonly lost from the molecular ion or daughter ions; for details, see F. McLafferty, op cit. (Section C above), and Beynon, Saunders, and Williams, op. cit., pages 461-474.

Formula Mass	Fragment Lost	Formula Mass	Fragment Lost
1	$H\cdot$	47	$CH_3S\cdot$
2	H_2	48	CH_3SH
15	$CH_3\cdot$	49	$ClCH_2\cdot$
16	$O\cdot$	52	$F_2CH\cdot$
17	$HO\cdot$	53	$C_4H_5\cdot$
18	$H_2O\cdot$	54	$CH_2CHCHCH_2$
19	$F\cdot$	55	$CH_2CH\underset{\bullet}{C}HCH_3$
20	HF	56	C_4H_8, CH_3CONH_2
26	C_2H_2, $NC\cdot$	57	$C_4H_9\cdot$, $C_2H_5CO\cdot$
27	$CH_2CH\cdot$, HCN	58	C_4H_{10}, $SCN\cdot$
28	C_2H_4, CO	59	$C_3H_7O\cdot$, $CH_3OCO\cdot$
29	$C_2H_5\cdot$, $\cdot CHO$	60	CH_3CO_2H, C_3H_7OH
30	NO, C_2H_6, CH_2O, $H_2NCH_2\cdot$	61	$C_2H_5S\cdot$
31	CH_3NH_2, $HOCH_2\cdot$, $CH_3O\cdot$	62	$H_2S+C_2H_4$
32	CH_3OH, $H_2O+C_2H_4$	63	$C_2H_4Cl\cdot$
33	$HS\cdot$	67	$C_5H_7\cdot$
34	H_2S	68	$CH_2C(CH_3)CHCH_2$
35	$Cl\cdot$	69	$F_3C\cdot$, $C_5H_9\cdot$
36	HCl	71	$C_5H_{11}\cdot$
37	H_2Cl	72	C_5H_{12}
39	$C_3H_3\cdot$	73	$C_2H_5CO_2\cdot$, $C_2H_5OCO\cdot$
40	CH_3CCH	74	$C_4H_{10}O$
41	$CH_2CHCH_2\cdot$	76	$H_2S+C_3H_6$
42	CH_2CO, cyclo-C_3H_6, CH_2CHCH_3	79	$Br\cdot$
43	$C_3H_7\cdot$, $CH_3CO\cdot$, $CH_2CHO\cdot$, HNCO	80	HBr
44	CO_2, C_3H_8, $H_2NCO\cdot$, C_2H_4O	88	$F_2ClC\cdot$
45	$C_4H_9\cdot$, $C_2H_5O\cdot$	122	$C_6H_5CO_2H$
46	C_2H_5OH, $H_2O+C_2H_4$, $O_2N\cdot$	127	$I\cdot$
		128	HI

E. Exact Mass Determination: Standards and Peak Matching

Certain substances that yield fragments of precisely known mass and composition are often added to an unknown as internal references for two purposes: (1) to fill in mass gaps (that might appear in the unknown's spectrum) to facilitate accurate counting; (2) for high-resolution peak matching to determine the "exact" mass and molecular formula for a peak (J. Lederberg, Computation of Molecular Formulas for Mass Spectrometry, Holden-Day, San Francisco, 1964, is useful in this regard). A computer program is also available for calculation of theoretical intensities of mass spectrum isotope peaks for a given formula; see B. Dombek, J. Lowther, and E. Carberry, J. Chem. Educ., 48, 729 (1971).

The most common broad range mass reference compounds are given below; the approximate mass range is that over which reasonably intense peaks are found.

Compound	Property[d]	Useful Mass Range	Suppliers[a,e]
Perfluorokersene	Low boiling (~90°)	19-300	1, 2, 3
Perfluorokerosene[b]	High boiling (~210°)	19-800	1, 2, 3
Perfluorotriheptyltriazine[c]		600-1600	2
Perfluorotributylamine	$(C_4F_9)_3N$ (bp 170°-180°)	65-650	1, 2, 3

[a]1. Pierce Chemical Company, P. O. Box 117, Rockford, Ill. 61105.
2. Peninsular Chemresearch, Inc., Box 14318, Gainesville, Fla. 32601.
3. Columbia Organic Chemicals Co., Inc., P. O. Box 5273, Columbia, S. C. 29209.

[b]For additional information, see R. S. Gohlke and L. H. Thompson, "Perfluorokerosene: A Useful Internal Reference for Negative Ion Mass Spectrometry," Anal. Chem., 40, 1004 (1968).

[c]For details, see T. Aczel, Anal. Chem., 40, 1917 (1968).

[d]For instructions on typical "Controlled Introduction of Mass Calibrant into a High-Resolution Mass Spectrometer," see L. A. Shadoff and L. B. Westover, Anal. Chem., 39, 1048 (1967).

[e]Varian Associates, 611 Hansen Way, Palo Alto, Calif. 94303, sells a complete "Mass Spectrometer Reference Compound Kit" consisting of the following compounds (useful mass range): perfluoroalkane 225 (60-750); bromoform (80-260); perfluorotributylamine (65-650); 1,2-dichlorooctafluoro-1-cyclohexene (100-300); 1,1,2-trifluoro-2-chloro-1, 4-dibromobutane (90-310); 1,2-dibromotetrafluorobenzene (90-310); perfluoroheptyl iodide (65-500); 1,1,2,2-tetrabromoethane (40-350); hexachlorocyclopentadiene (40-280); hexachlorobutadiene (50-265); 1,3-dibromopropane (40-190); 1,1-difluoro-1, 2,3,3-tetrachloro-2-propene (40-220). These compounds were chosen from among hundreds tested for their easily identified and intense peaks (R. Carr, personal communication); a booklet accompanies the kit, giving reproduced spectra and exact masses and compositions for the useful peaks. Individual components of the kit are available directly from supplier 1 in footnote a.

F. Mass Calibration Peaks of Perfluorokerosene

The following represent the peaks of most use for mass calibration found in the spectrum of high boiling perfluorokerosene. The data are taken from a standardized spectrum, run on an AEI MS-9 spectrometer.

992.9378	$C_{21}F_{39}$	492.9697	$C_{11}F_{19}$
980.9378	$C_{20}F_{39}$	480.9697	$C_{10}F_{19}$
966.941	$C_{22}F_{37}$	466.9729	$C_{12}F_{17}$
954.941	$C_{21}F_{37}$	454.9729	$C_{11}F_{17}$
942.941	$C_{20}F_{37}$	442.9729	$C_{10}F_{17}$
930.941	$C_{19}F_{37}$	430.9729	$C_{9}F_{17}$
916.9442	$C_{21}F_{35}$	418.9729	$C_{11}F_{15}$
904.9442	$C_{20}F_{35}$	404.9761	$C_{10}F_{15}$
892.9442	$C_{19}F_{35}$	392.9761	$C_{9}F_{15}$
880.9442	$C_{18}F_{35}$	380.9761	$C_{8}F_{15}$
866.9474	$C_{20}F_{33}$	368.9761	$C_{7}F_{15}$
854.9474	$C_{19}F_{33}$	354.9793	$C_{9}F_{13}$
842.9474	$C_{18}F_{33}$	342.9793	$C_{8}F_{13}$
830.9474	$C_{17}F_{33}$	330.9793	$C_{7}F_{13}$
816.9506	$C_{19}F_{31}$	318.9793	$C_{6}F_{13}$
804.9506	$C_{18}F_{31}$	304.9825	$C_{8}F_{11}$
792.9506	$C_{17}F_{31}$	292.9825	$C_{7}F_{11}$
780.9506	$C_{16}F_{31}$	280.9825	$C_{6}F_{11}$
766.9537	$C_{18}F_{29}$	268.9825	$C_{5}F_{11}$
754.9537	$C_{17}F_{29}$	254.9856	$C_{7}F_{9}$
742.9537	$C_{16}F_{29}$	242.9856	$C_{6}F_{9}$
730.9537	$C_{15}F_{29}$	230.9856	$C_{5}F_{9}$
716.9569	$C_{17}F_{27}$	218.9856	$C_{4}F_{9}$
704.9569	$C_{16}F_{27}$	204.9888	$C_{6}F_{7}$
692.9569	$C_{15}F_{27}$	192.9888	$C_{5}F_{7}$
680.9569	$C_{14}F_{27}$	180.9888	$C_{4}F_{7}$
666.9601	$C_{16}F_{25}$	168.9888	$C_{3}F_{7}$
654.9601	$C_{15}F_{25}$	161.9904	$C_{4}F_{6}$
642.9601	$C_{14}F_{25}$	149.9904	$C_{3}F_{6}$
630.9601	$C_{13}F_{25}$	142.992	$C_{4}F_{5}$
616.9633	$C_{15}F_{23}$	130.992	$C_{3}F_{5}$
604.9633	$C_{14}F_{23}$	118.992	$C_{2}F_{5}$
592.9633	$C_{13}F_{23}$	111.9936	$C_{3}F_{4}$
580.9633	$C_{12}F_{23}$	99.9936	$C_{2}F_{4}$
566.9665	$C_{14}F_{21}$	92.9952	$C_{3}F_{3}$
554.9665	$C_{13}F_{21}$	80.9952	$C_{2}F_{3}$
542.9665	$C_{12}F_{21}$	69.9986	$^{13}CF_{3}$
530.9665	$C_{11}F_{21}$	49.9968	CF_{2}
516.9697	$C_{13}F_{19}$	30.9984	CF
504.9697	$C_{12}F_{19}$	15.0235	CH_{3}

G. Metastable Transitions and Ions

Some of the ions formed in the ionization chamber of a mass spectrometer are not stable enough to pass to the ion collector unchanged. These ions are referred to as metastable; they have a half-life on the order of only 1 μsec. In a group of ions of original mass-to-charge ratio m_1/e, most will either proceed to the collector and be detected as m_1/e or will decompose to a new fragment, m_2/e, before leaving the ionization chamber; again m_2/e will be detected as a separate peak. However, some of the original ions (m_1/e) may be metastable and decompose (say to m_2/e) after leaving the ionization chamber and before reaching the collector. This "metastable transition" gives rise to a "metastable peak," m*, which is identified by its low intensity and diffuse shape and by its appearance usually at nonintegral mass. The actual peak will appear in the spectrum at an m/e value given by:

$$m^* = \frac{m_2^2}{m_1}$$

Such a peak can be very valuable in the analysis of a spectrum, since it is the result of a unique unimolecular decomposition thereby linking specific ions (fragments) within a molecule. If the decomposition is indeed unimolecular, the intensity of m* is linearly proportional to sample pressure. If the transition is collision-induced, however, intensity is a function of some higher power of pressure.

To determine the parent and daughter masses ($m_1$and m_2, respectively) that give rise to the observed m*, several methods are available. Beynon (op. cit., 1960) has published several nomograms; this trial-and-error method is somewhat cumbersome. McClafferty (op. cit., 1966) has described a trial-and-error method using a slide rule. Probably the easiest and most accurate determination is available in the computer-generated "Table of Meta-Stable Transitions for Use in Mass Spectrometry" (Beynon et al., op. cit.). Possible m_1 and m_2 combinations are included for m* values from 1.00 to 498.00 at intervals of 0.01 mass units.

A program in PDP-8 FOCAL is also available for "Laboratory Computer Analysis of Mass Spectrometric Metastable Ion Signals," from L. E. Brady, University of Toledo, Toledo, Ohio 43606 [see J. Chem. Educ., 48, 469 (1971).]

VIII. NUCLEAR MAGNETIC RESONANCE SPECTROSCOPY

A. Introduction

Many general references to theory and applications of nmr spectroscopy can be found at the end of this section (1-5). A recent review of instrumentation and commercially available spectrometers can be found in reference 6; other discussions of instrumentation are found in references 1a to 1e.

The basic condition for nmr is stated in the equation

$$h\nu = \frac{\mu H_o(1 - \sigma)}{I}$$

where ν is the radio frequency, H_o is the applied field (note that higher field strength requires higher frequency), σ is the shielding constant for the particular atom (resulting from the diamagnetic susceptibility), μ is the magnetic moment of the nucleus (see p. 313 for a table of magnetic properties of elements), I is the spin quantum number (±1/2 for protons), and h is Planck's constant.

Refs. p. 333.

B. ASTM Standard Conventions and Definitions

The following definitions and conventions for presentation of nmr data are those accepted by the American Society for Testing and Materials and are quoted with permission from the "Annual Book of ASTM Standards," 3rd ed., 1969. Added comments appear in brackets.

1. Definitions

Nuclear Magnetic Resonance (NMR) Spectroscopy--That form of spectroscopy concerned with radiofrequency induced transitions between magnetic energy levels of atomic nuclei. [Note that the American Chemical Society recommends small letters for the abreviation (nmr).]

NMR Absorption Band; NMR Band--A region of the spectrum in which a detectable signal exists and passes through one or more maxima. [A given spectrum usually consists of several absorption bands.]

NMR Band Width, W--The width, expressed in hertz (Hz) [cycles per second], of an observed NMR band measured at one half maximum height.

NMR Spectral Resolution--The magnitude of the minimum NMR band width that can be observed in a given spectrometer.

Chemical Shift--The relative difference in magnetic field (or frequency) [expressed in parts per million, ppm, as explained below] between an NMR signal under observation and a suitable reference signal. [A more exact definition and a general discussion of the chemical shift on a non-theoretical level has recently been given (7). Also see below for additional comments and for reference compounds. To convert shift in Hz (cycles per second) to ppm use the equation: δ = shift in Hz divided by the radiofrequency of the instrument in MHz.]

Spin-Spin Coupling Constant (NMR), J--A measure, expressed in hertz (Hz), of the indirect spin-spin interaction between different magnetic nuclei in a given molecule.

Spin-Spin Multiplicity--Multiplicity in a spectrum arising from indirect spin-spin interactions of different nuclei in a given molecule.

Sweep Rate--The rate, in hertz (Hz) per second, at which the magnetic field strength or frequency of applied radio-frequency field is varied to produce an NMR spectrum.

Spinning Sidebands--Bands, paired symmetrically or nearly symmetrically, about a principal band, arising from spinning of the sample. [Variation of the spin rate will vary the distance between the sidebands.]

^{13}C Satellites--Pair of bands, spaced nearly symmetrically about a principal band, arising from indirect spin-spin coupling of the observed nucleus with ^{13}C [natural abundance = 1.108%] present in the same molecule.

2. Conventions

a. The multiplying factor used with the dimensionless shift or shielding parameter should be 10^6 (ppm).

b. The unit used for "raw data" (peak positions) should be hertz.

c. The dimensionless and frequency scales should have a common origin.

d. The standard sweep direction should be from low to high applied field (or high to low radiofrequency).

e. The standard orientation of spectra should be with high applied field on the right.

f. Absorption mode peaks should point up.

Refs. p. 333.

3. Data Presentation

a. The following should be specified whenever NMR data are published:

(1) Name of solvent and concentration of solute.
(2) Name and concentration of internal reference.
(3) Name of external reference.
(4) Temperature of sample in probe.
(5) Procedure used for measuring peak positions.
(6) Radio frequency at which measurements were made.
(7) Mathematical operations used to analyze the spectra.
(8) Numbers on the frequency scale (if used). They should increase from high to low applied field.
(9) Dimensionless and frequency scales. They should run in the same directions.

b. The following additional information should be given when the spectra themselves are published:

(1) Sweep rate.
(2) Values of both r-f [radio frequency] fields when spin decoupling or double resonance is employed.

c. The shifts and couplings obtained from the spectra should be reported, the former in dimensionless units (ppm) and the latter in frequency units (hertz).

4. Additional Comments

Critical comments on some of the ASTM definitions given above have recently been made by F. H. A. Rummens (7). He made the following proposals.

For frequency swept spectra:

$$\delta_\nu = \frac{\nu_s - \nu_r}{\gamma\mu_g H_o} \times 10^6 = (\sigma_r - \sigma_s) \times 10^6$$

For field swept spectra:

$$\delta_H = \frac{\nu_o(H_r - H_s)}{\gamma\mu_g H_r H_s} \times 10^6 = (\sigma_r - \sigma_s) \times 10^6$$

where ν_s and ν_r are resonance frequencies of the sample and reference respectively, H_r and H_s are the external fields necessary for the reference and sample to reach resonance, γ is the magnetogyric ratio for the particular nucleus, σ_r and σ_s are the shielding constants for each nucleus (see p. 244), and μ_g is the permeability of the medium between the magnet and sample. (This includes air in the magnet gap, sample tube, and solvent.) Of course, the μ_g correction would not be necessary for most experiments.

C. Reference Compounds

As stated above, all chemical shift data should be reported in parts per million (ppm) relative to a reference signal of an appropriate standard. This standard may be dissolved in the sample solution (internal reference) or placed in a separate container such as a sealed capillary inside the sample tube (external reference). An internal reference is generally preferred if one can be found that is chemically inert to the solvent and the sample. If an external reference is used, correction for the difference in magnetic susceptibility of the standard and the solvent must be made (assuming both are in other than spherical sample tubes). For scales where the reference position is arbitrarily assigned zero and both sample and reference are in cylindrical tubes:

$$\delta_{corr} = \delta_{obs} - \frac{2\pi}{3}(K_R - K_S)(10^6)$$

where δ_{corr} is the corrected chemical shift, δ_{obs} is the observed chemical shift, K_R and K_S are the volume diamagnetic susceptibilities of the reference and the solvent respectively. An extensive list of magnetic susceptibilities can be found in reference 1c. (The volume susceptibility of tetramethylsilane is -0.543×10^{-6} cgs Gaussian units.)

In general a reference compound should have as many of the following properties as possible. Some of these requirements are obviously not necessary for external standards.

1. It should give one peak of as narrow a bandwidth as possible.
2. This peak should be easily recognizable and well separated from any absorption bands of the sample.
3. It should be chemically inert to both sample and solvent.
4. It should be magnetically isotropic.
5. It should be soluble in a wide range of solvents.
6. It should be volatile enough or possess some other property so as to allow easy removal from sample.

D. Proton Magnetic Resonance (PMR)

1. Reference Compounds

Tetramethylsilane (TMS) is the accepted standard for pmr (see also ^{13}C nmr) and should be used whenever possible. Other useful reference compounds for special cases are given below along with some of their properties and their absorption position (solvent given in parentheses) in ppm (δ) from TMS. These peak positions may be solvent dependent to a small extent. Suppliers (numbers refer to list on p. 332) are given for less common substances. In some of the older literature, water was used as a standard although its large temperature and solvent dependence make its use undesirable. For general reference, its approximate absorption position is 4.7 ppm downfield from TMS. Dow Corning silicone stopcock grease has recently been described for use as a quantitative internal standard for pmr [J. A. Kapecki, J. Chem. Educ., 48, 731 (1971)].

Refs. p. 333.

Refs. p. 333.

Name	Formula	δ^a	Comment
Tetramethylsilane (TMS)	$(CH_3)_4Si$	(0.00)	Most generally used standard; water-insoluble. Accepted by ASTM; bp 28°.
3-(Trimethylsilyl)-tetradeuteriopropionic acid sodium salt	$(CH_3)_3SiCD_2CD_2CO_2Na$	0.00 (H_2O)	Water-soluble; more soluble than Tier's salt(DSS); pH, 0.05N = 7.92; stable to 200°. Supplied by 4,6,10,16. Synthesis in reference 8b.
2,2-Dimethyl-2-silapentane-5-sulfonic acid (DSS) (Tier's salt)	$(CH_3)_3Si(CH_2)_3SO_3Na$	0.015 (H_2O)	Water-soluble; pH, 0.05N = 4.56. Broad bands of methylenes may show up at 0.6, 1.73, 2.93. Decomposes above 120°. Supplied by 8,16,18,19,22,23.
Hexamethyldisilane	$[(CH_3)_3Si]_2$	0.037 ($CDCl_3$)	Can be used at higher temperature than TMS; bp 112°, mp 13.0-13.5°. Insoluble in water. Supplied by 7,8,19,20.
Hexamethyldisiloxane	$[(CH_3)_3Si]_2O$	0.055 ($CDCl_3$)	Can be used at higher temperature than TMS, bp 100.5°, mp -69.8°. Insoluble in water. Supplied by 7,8,19,20.
Tetradecylmethylbicyclo[2,2,2]octasilane	Bicyclic structure: two CH_3–Si bridgeheads linked by three $-Si(CH_3)_2-Si(CH_3)_2-$ bridges	-0.205 and -0.125 (ratio 6/1) (CCl_4)	Soluble in ethyl ether, chloroform, and carbon tetrachloride. High melting point (360°) makes it useful to very high temperatures. Synthesis in reference 9.

Cyclosilane-d_{18}	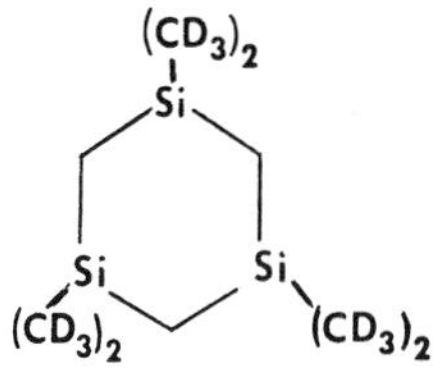 cyclic structure: three Si atoms, each bearing $(CD_3)_2$	-0.327 ($CDCl_3$)	Good for high temperature; bp 208°. Synthesis in reference 8a. Supplied by 4,6,10.
t-Butanol	$(CH_3)_3COH$	1.27 (H_2O)	Water-soluble.
Cyclohexane	C_6H_{12}	1.44 (CCl_4)	Broadens at low temperature.
Acetonitrile	CH_3CN	2.00 (H_2O)	Water-soluble.
Acetone	$(CH_3)_2CO$	2.08 (H_2O)	Water-soluble.
Dioxane	$C_4H_6O_2$	3.64 (H_2O)	Water-soluble.
Chloroform	$CHCl_3$	7.25 (CCl_4)	
Benzene	C_6H_6	7.266 (CCl_4)	

[a] A recent discussion of the effect of solvent on the resonance value of TMS can be found in reference 11c.

2. Solvents

The following is by no means a complete list of suitable solvents that could be used for pmr spectra. (For a more extensive list see reference 10). Of course, solvents with hydrogens could be used if the absorptions do not interfere with the region of interest. Three excellent solvents that contain no protons and are commonly used are carbon tetrachloride, carbon disulfide, and sulfur dioxide. The first of these is available with internal standard (1% TMS) added. Solvents listed

Refs. p. 333.

below are marked with (I) if available with internal standard. This is TMS in all cases except D_2O, in which DSS (see above) is used. All melting and boiling points are for hydrogen compounds except those that are starred (*). The approximate residual proton absorptions are given in ppm from TMS. It should be kept in mind that partial deuterium labeling might result in broadening or splitting of the residual peaks, due either to deuterium-hydrogen coupling or nonequivalence of the residual hydrogens. If the undeuterated solvent is used, the region obscured might be as wide as 2 or 3 ppm.

For solvent mixtures good for very low temperatures, see page 450. For two excellent reviews of solvent effects in nmr, see references 11a and b. The following suppliers (numbers refer to list on p. 332) sell a wide variety of deuterated solvents including most or all of those listed here: 5,9, 11,12,16,18,24. Suppliers that handle only the more common of these are 1,4,8,14,17,19,20,22.

Solvent	Formula	Bp	Mp	Residual Absorption (ppm from TMS)
Acetic acid-d_4 (I)	CD_3COOD	118	15.75*	2,06, 12.0[a]
Acetone-d_6 (I)	CD_3COCD_3	56.2	-95.4	2.07
Acetonitrile-d_3 (I)	CD_3CN	81.6	-45.7	1.96
Benzene-d_6 (I)	C_6D_6	80.1	5.5	7.24
Bromobenzene-d_5	C_6D_5Br	156	-30.8	7.1-7.5
Bromoform-d	$CDBr_3$	150	8.3	6.82
t-Butyl alcohol-d_{10}	$(CD_3)_3COD$	82	25.5	1.22, 1.35 [a]
Chloroform-d_1 (I)	$CDCl_3$	61.7	-63.5	7.25
Cyclohexane-d_{12}	C_6D_{12}	80.7	6.55	1.42
Deuterium chloride (38% in D_2O)	DCl	110		8.5 [a]
Deuterium oxide (I)	D_2O	101.42*	3.82*	4.61
1,2-Dichloroethane-d_4	$ClCD_2CD_2Cl$	83.5	-35	3.69
Diethyl ether-d_{10}	$C_2D_5OC_2D_5$	34.5	-116	1.2, 3.4
Dimethylformamide-d_7	$DCON(CD_3)_2$	152	-61	2.79, 2.94, 7.90
Dimethylsulfoxide-d_6 (I)	CD_3SOCD_3	189(d)	18.4	2.50
p-Dioxane-d_8	$C_4D_8O_2$	102	11.8	3.56

Refs. p. 333.

Ethanol-d_6 (95% in D_2O)	CD_3CD_2OD in D_2O	78.15	-120.0	1.17, 3.59, 4.1 [a]
Ethanol-d_6 (Anhydrous)	CD_3CD_2OD	78.3	-114.5	1.17, 3.59, 4.1 [a]
Formic acid-d_2	DCOOD	101	8.4	8.2, 10.8 [a]
Hexafluoroacetone deuterate	CF_3COCF_3 (1.6 D_2O)	106	40	---
Hexamethyl phosphoric triamide-d_{18}	$[(CD_3)_2N]_3PO$	232	4	2.64
Isopropanol-d_8	$CD_3CDODCD_3$	82.4	-89.5	1.2, 1.6 [a], 4.0
Methanol-d_4	CD_3OD	64.5	-97.5	3.34, 4.1 [a]
Methylcyclohexane-d_{14}	$C_6D_{11}CD_3$	100.9	-126.6	0.8-1.8
Methylene chloride-d_2	CD_2Cl_2	40	-95.1	5.28
Nitrobenzene-d_5	$C_6D_5NO_2$	211	5.8	7.4-8.3
Nitromethane-d_3	CD_3NO_2	101	-28.5	4.29
n-Octane-d_{18}	$CD_3(CD_2)_6CD_3$	125.7	-56.8	0.7-1.4
Pyridine-d_5 (I)	C_5D_5N	115.6	-41.8	7.0-7.8, 8.57
Sulfuric acid-d_2	D_2SO_4	>300	14.35*	10.9 [a]
Tetrahydrofuran-d_8	C_4D_8O	67	-65	1.6-2.0, 3.5-3.8
Tetramethylene sulfone-d_8	$C_4D_8SO_2$	283	28.9	2.0-2.5, 2.8-3.5
Tetramethylurea-d_{12}	$(CD_3)_2NCON(CD_3)_2$	167	-1	
Toluene-d_8	$C_6D_5CD_3$	110.6	-95	2.31, 7.10
Trifluoroacetic acid-d_1	CF_3COOD	72.4	-15	11.34
2,2,2-Trifluoroethanol-d_3	CF_3CD_2OD	73.5		6.1, 6.6 [a]

[a] These positions may vary considerably depending on temperature and solute.

Refs. p. 333.

3. General Regions of Chemical Shifts

The charts on these next four pages, compiled by E. Mohacsi, were taken from reference 12 and reproduced with permission of the copyright owner.

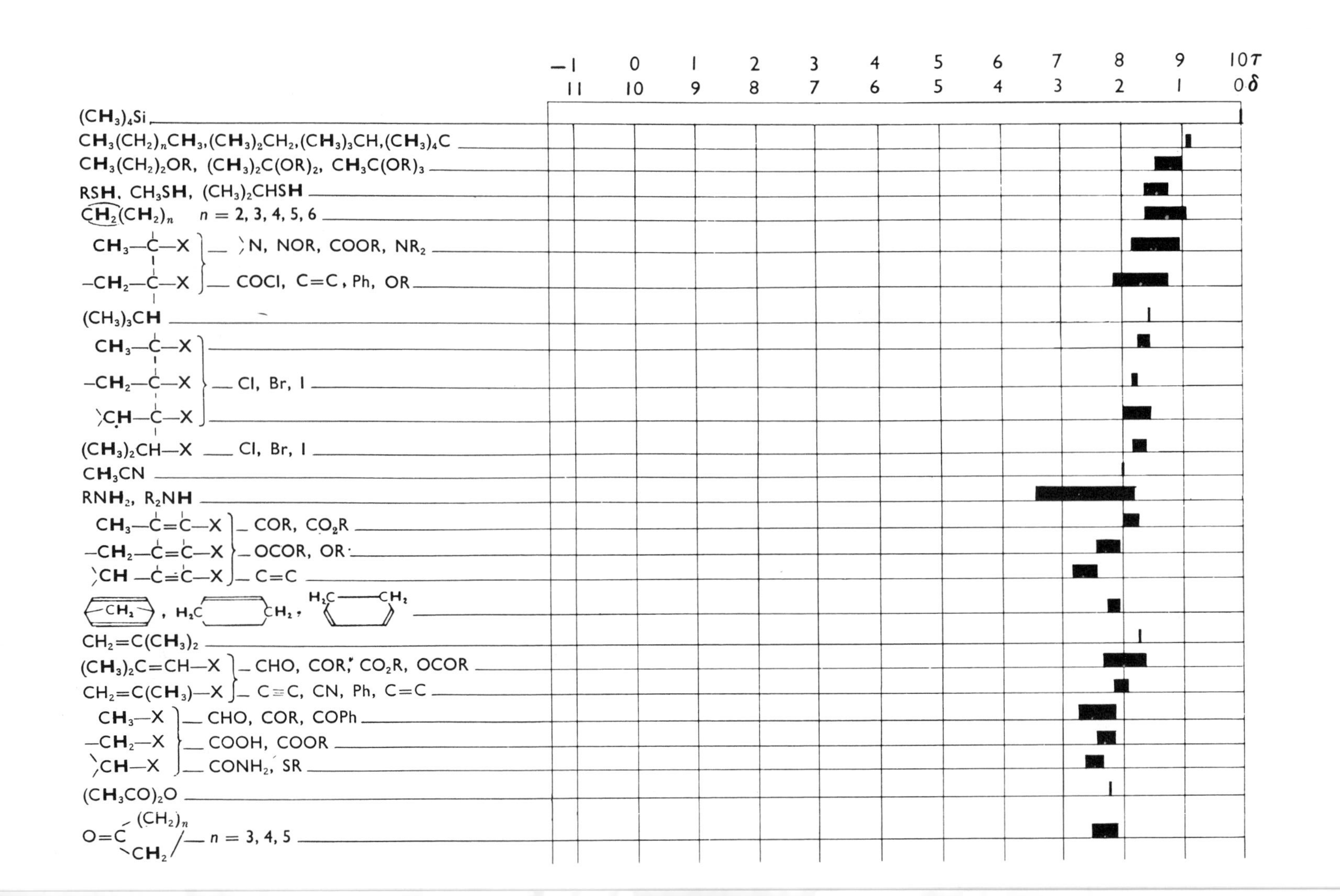

Refs. p. 333.

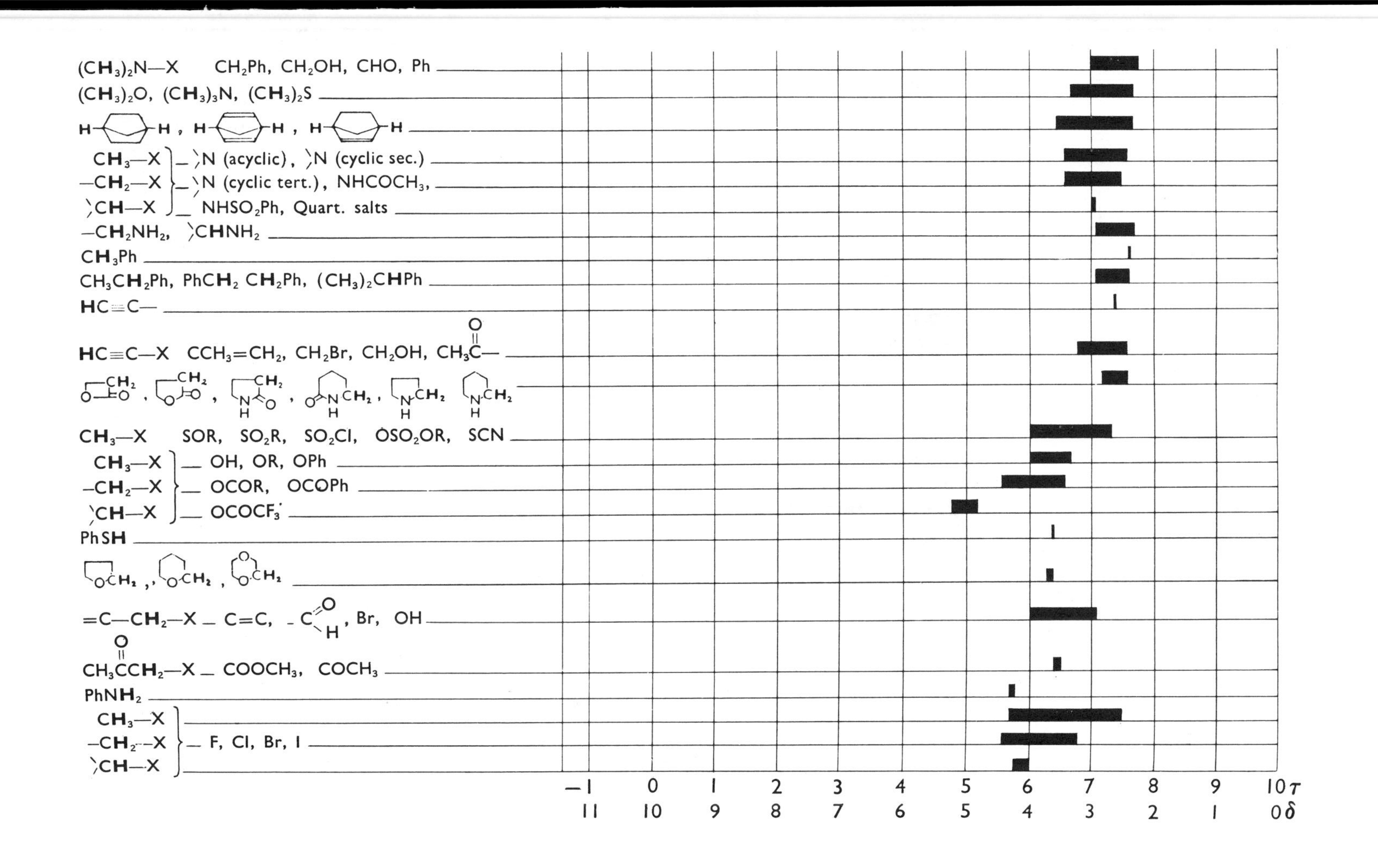

(CH3)2N—X CH2Ph, CH2OH, CHO, Ph
(CH3)2O, (CH3)3N, (CH3)2S
CH3—X — N (acyclic), N (cyclic sec.)
—CH2—X — N (cyclic tert.), NHCOCH3,
CH—X — NHSO2Ph, Quart. salts
—CH2NH2, CHNH2
CH3Ph
CH3CH2Ph, PhCH2 CH2Ph, (CH3)2CHPh
HC≡C—
HC≡C—X CCH3=CH2, CH2Br, CH2OH, CH3CO—
CH3—X SOR, SO2R, SO2Cl, OSO2OR, SCN
CH3—X — OH, OR, OPh
—CH2—X — OCOR, OCOPh
CH—X — OCOCF3
PhSH
=C—CH2—X — C=C, Br, OH
CH3COCH2—X — COOCH3, COCH3
PhNH2
CH3—X
—CH2—X — F, Cl, Br, I
CH—X
−1 0 1 2 3 4 5 6 7 8 9 10τ
11 10 9 8 7 6 5 4 3 2 1 0δ

Refs. p. 333.

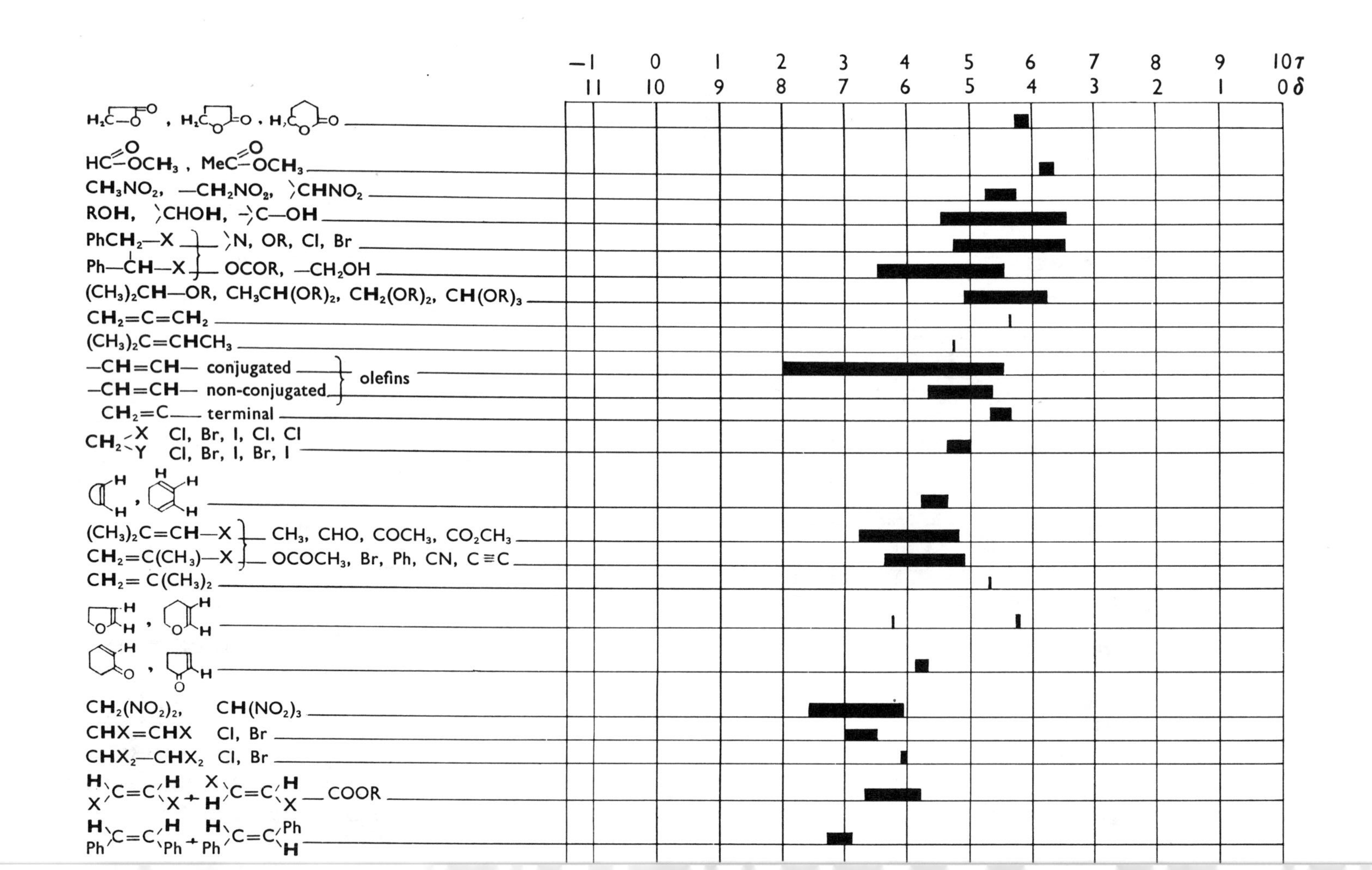

−1 0 1 2 3 4 5 6 7 8 9 10τ
11 10 9 8 7 6 5 4 3 2 1 0δ
HC(=O)—OCH₃ , MeC(=O)—OCH₃
CH₃NO₂, —CH₂NO₂, ⟩CHNO₂
ROH, ⟩CHOH, —⟩C—OH
PhCH₂—X
Ph—CH—X
⟩N, OR, Cl, Br
OCOR, —CH₂OH
(CH₃)₂CH—OR, CH₃CH(OR)₂, CH₂(OR)₂, CH(OR)₃
CH₂=C=CH₂
(CH₃)₂C=CHCH₃
—CH=CH— conjugated
—CH=CH— non-conjugated
olefins
CH₂=C— terminal
CH₂<X Y
Cl, Br, I, Cl, Cl
Cl, Br, I, Br, I
(CH₃)₂C=CH—X
CH₂=C(CH₃)—X
CH₃, CHO, COCH₃, CO₂CH₃
OCOCH₃, Br, Ph, CN, C≡C
CH₂= C(CH₃)₂
CH₂(NO₂)₂, CH(NO₂)₃
CHX=CHX Cl, Br
CHX₂—CHX₂ Cl, Br
COOR

Refs. p. 333.

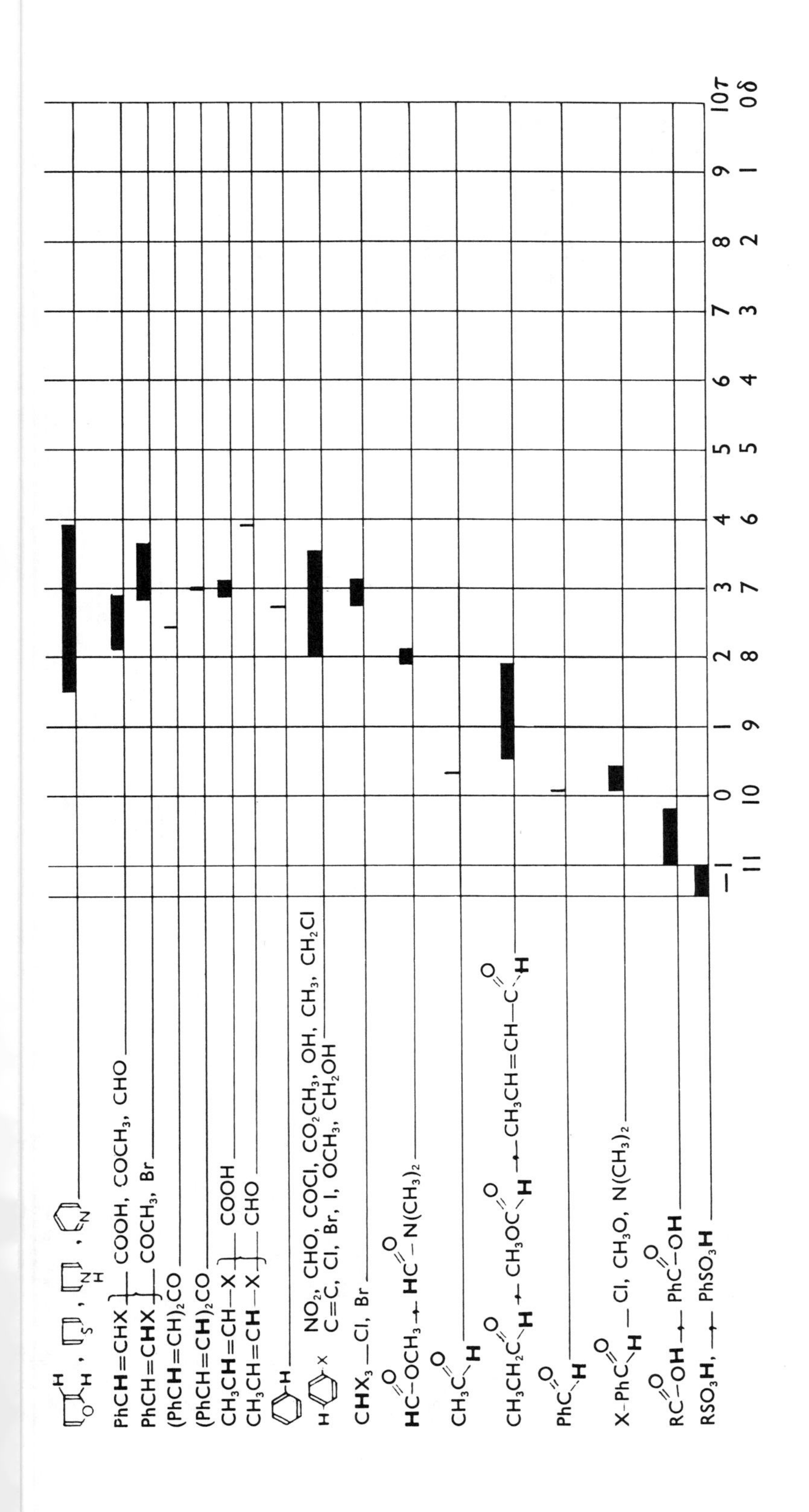

PhCH=CHX — COOH, COCH3, CHO
PhCH=CHX — COCH3, Br
(PhCH=CH)2CO
(PhCH=CH)2CO
CH3CH=CH—X — COOH
CH3CH=CH—X — CHO
NO2, CHO, COCl, CO2CH3, OH, CH3, CH2Cl
C=C, Cl, Br, I, OCH3, CH2OH
CHX3 — Cl, Br
HC(=O)—OCH3 → HC(=O)—N(CH3)2
CH3C(=O)H
CH3CH2C(=O)H → CH3OC(=O)H → CH3CH=CH—C(=O)H
PhC(=O)H
X-PhC(=O)H — Cl, CH3O, N(CH3)2
RC(=O)—OH → PhC(=O)—OH
RSO3H, → PhSO3H
10τ 9 8 7 6 5 4 3 2 1 0 −1
0δ 1 2 3 4 5 6 7 8 9 10 11

4. Chemical Shift Correlation Tables

The δ-values in the following tables were taken from numerous sources but primarily from references 2d and 4. In Section a., the values are often the average of several observations and may vary for any specific compound as much as ±3 or 4 in the last significant figure. (Even larger variation is possible in unusual cases.) The data in all of the tables are for the compound in dilute carbon tetrachloride or deuteriochloroform relative to internal TMS where such data were available. It is important to remember that solvent effects, especially in the case of aromatics, can cause significant variation in observed chemical shifts. A "reference independent" scale for solvent shifts has recently been proposed (13). As noted above, two excellent reviews of solvent effects have been published (11a, 11b).

a. Protons Bound to Acyclic sp^3 Carbon Atoms

i. Compounds with a Single Functional Group[a]

Group X	CH_3-X	C-CH_2-X	C-CH(C)-X	CH_3-C(-)(-)-X	C-CH_2-C(-)(-)-X	C-CH(C)-C(-)(-)-X
-H	0.233	0.86	1.33	0.86	1.33	1.56
-CH=CR_2	1.73	2.00	1.73	1.55	1.35	1.00
-C≡CR	1.75	2.15	2.7	1.15	1.50	1.80
-C≡N	1.98	2.35	2.8	1.30	1.6	2.00
-Ph	2.34	2.60	2.87	1.18	1.6	1.8
-CHO	2.17	2.4	2.4	1.13	1.65	
-COR	2.10	2.4	2.55	1.05	1.5	1.7
-COPh	2.5	2.9	3.4	1.18	1.6	2.0
-CO_2R	2.1	2.2	2.5	1.15	1.7	1.8
-$CONR_2$	2.05	2.23	2.4	1.1	1.6	1.8
-I	2.16	3.17	4.25	1.8	1.8	2.1
-Br	2.68	3.36	4.2	1.8	1.9	2.0
-Cl	3.05	3.44	4.1	1.5	1.8	2.0
-F	4.26	4.4	4.8	1.4	1.8	2.1
-OR	3.38	3.4	3.6	1.2		
-OPh	3.82	3.95	4.6	1.3	1.5	1.7
-OCOR	3.65	4.1	5.0	1.25	1.6	1.8
-OCOPh	3.82	4.2	5.1	1.5	1.7	1.9
-$OCOCF_3$	3.95	4.3		1.4	1.6	
-ONO		4.75		1.4		
-NR_2	2.3	2.6	2.9	1.05	1.45	1.7
-N^+R_3	~3.2	~3.1	~3.6	1.4	1.7	2.0
-NRPh	~2.7	~3.1	~3.6	1.1	1.5	1.8
-NHCOR	2.8	3.3	3.8	1.1	1.5	1.9
-NO_2	4.30	4.4	4.6	1.6	2.05	2.5
-N≡C	2.85		4.8	1.6		
-SR	2.09	2.5	3.0	1.25	1.60	1.9
-SSR	2.30	2.7		1.3	1.7	
-SOR	2.5	3.0	2.8	1.35	1.7	
-SO_2R	2.8	2.9	3.1	1.35	1.7	

[a]Chemical shift values refer to the H or H's written in the formulas for the column headings; dashes on the C's in the last three column headings refer to other hydrogens or saturated alkyl group(s).

Refs. p. 333.

ii. Compounds with Two Functional Groups, X-CH_2-Y. Values in parentheses were calculated by the empirical method of Shoolery as described below. All values refer to the methylene (CH_2) protons in XCH_2Y.

Y \ X	CH_3	C=C	C≡C	Ph	CF_3	CN	C(O)R	C(O)OR	C(O)NR_2	C(O)Ph	Cl	Br	I	NR_2	N_3	NHC(O)R	OH	OR	OPh	OC(O)R	SR
CH_3	1.34	1.97	2.14	2.63	(1.84)	2.31	2.47	2.33	2.23	2.99	3.57	3.40	3.20	2.50	(2.67)	3.26	3.70	3.36	4.04	4.12	2.53
C=C		2.73	3.39	3.30	(2.69)	3.15	3.13	(3.10)	(3.14)	(3.39)	4.08	3.93	3.87	3.30	(3.52)	(3.82)	4.13	3.97	(4.78)	4.68	3.08
C≡C			(3.11)	(3.52)	(2.81)	(3.37)	(3.37)	3.27	(3.26)	(3.51)	4.09	3.86	(3.49)	3.50	(3.64)	3.94	4.28	(4.03)	(4.90)	4.71	(3.31)
Ph				3.92	3.50	3.68	3.55	3.55	(3.67)	(3.92)	4.55	4.50	(3.90)	3.54	(4.05)	(4.35)	4.58	4.52	4.90	5.08	3.70
CF_3					(2.51)	(3.07)	(3.07)	(2.92)	(3.96)	(3.21)	(3.90)	(3.70)	3.56	(2.94)	(3.34)	(3.64)	3.93	(3.73)	(4.60)	(4.50)	(3.01)
CN						4.13	(3.63)	3.48	(3.52)	(3.77)	4.07	(4.26)	3.65	(3.50)	(3.90)	(4.20)	(4.49)	4.20	(5.16)	(5.06)	(3.57)
C(O)R							3.62	3.32	3.52	(3.77)	(4.46)	(4.26)	(3.75)	(3.50)	3.90	3.52	(4.49)	(4.29)	(5.16)	(5.06)	(3.57)
C(O)OR								4.22	(4.95)	(3.62)	4.24	3.70	3.72	4.22	(5.33)	3.87	(5.92)	4.13	4.60	(6.49)	(5.00)
C(O)NR_2									3.30	(3.66)	4.17	3.92	3.65	(3.39)	(3.79)	(4.09)	(4.38)	(4.18)	(5.05)	(4.95)	(3.46)
C(O)Ph										(3.91)	(4.60)	4.43	(3.89)	(3.64)	(4.04)	(4.34)	(4.63)	(4.43)	(5.30)	(5.20)	(3.71)
Cl											5.28	5.16	4.99	(4.33)	(4.73)	(5.03)	(5.32)	5.40	(5.99)	(5.89)	(4.40)
Br												4.94	(4.38)	(4.13)	(4.53)	(4.83)	(5.12)	(4.92)	(5.79)	(5.69)	(4.20)
I													3.89	(3.62)	(4.02)	(4.32)	(4.61)	(4.41)	(5.28)	(5.18)	(3.69)
NR_2														3.10	(3.77)	(4.07)	(4.36)	(4.16)	(5.03)	(4.93)	(3.44)
N_3															(4.17)	(4.47)	(4.76)	(4.56)	(5.43)	(5.33)	(3.84)
NHC(O)R																(4.77)	(5.06)	(4.86)	(5.73)	(5.63)	(4.14)
OH																	(5.35)	(5.15)	(6.02)	(5.92)	(4.43)
OR																		4.49	(5.82)	(5.72)	(4.23)
OPh																			(6.69)	(6.59)	(5.10)
OC(O)R																				(6.49)	(5.00)
SR																					(3.51)

Refs. p. 333.

b. Protons Bonded to Carbocyclic sp^3 Carbon Atoms

Compound	δ	Compound	δ	Compound	δ
Cyclopropane	0.22				
Cyclobutane	1.96		1.65		0.92
Cyclopentane	1.51				
Cyclohexane	1.44				
Cycloheptane	1.54	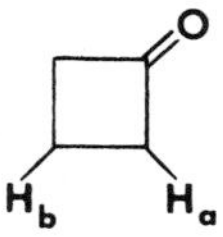	a 3.03 b 1.96	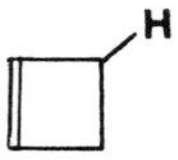	2.57
Cyclooctane	1.54				
Adamantane	1.78[a]	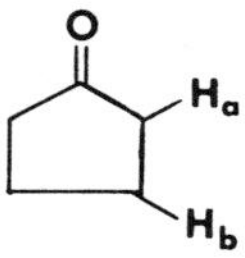		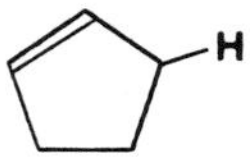	
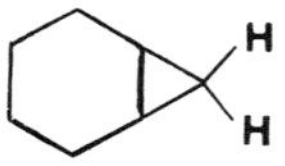	0.02		a 2.06 b 2.02		2.28
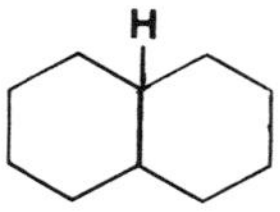	∿1.4	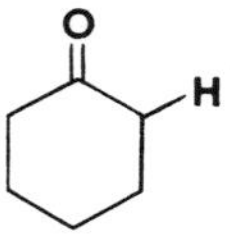	2.25	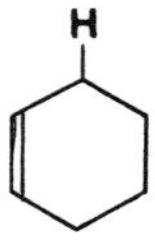	1.96
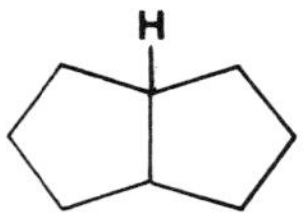	2.37	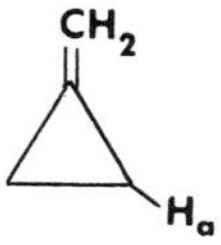	a 0.99	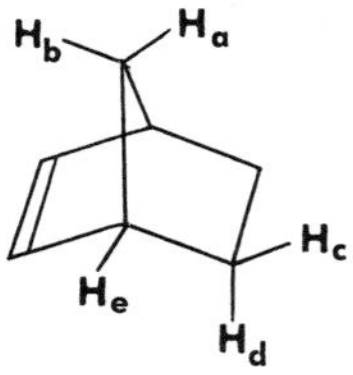	a 1.32 b 1.07 c 1.57 d 0.94 e 2.83
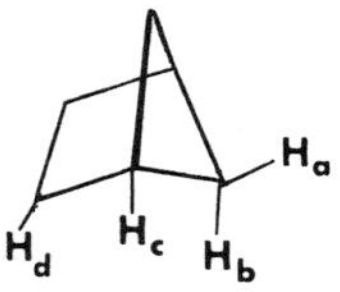	a 1.56 b 0.87 c 2.49 d 1.58	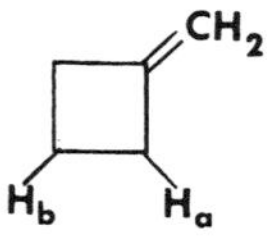	a 2.7 b 1.92	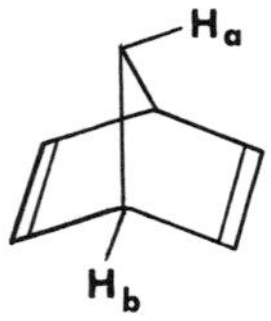	a 1.95 b 3.53
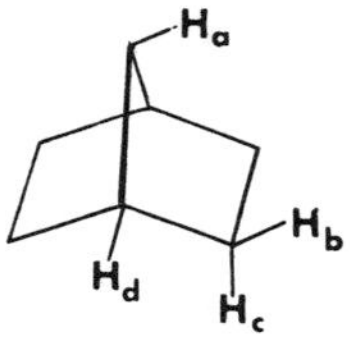	a 1.21 b 1.49 c 1.18 d 2.20	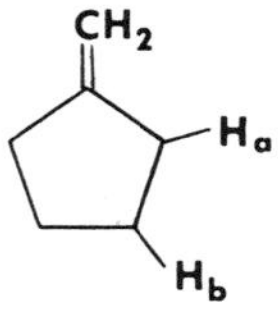	a 2.7 b 1.92	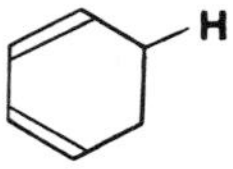	2.15
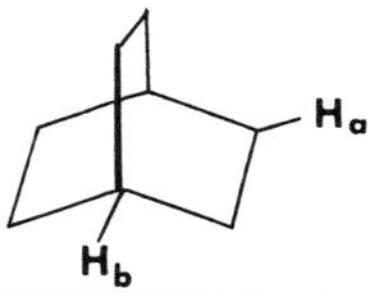	a 1.51 b∿2.1	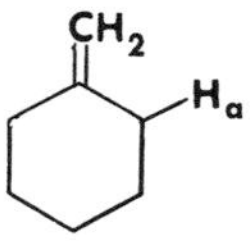	a 1.5	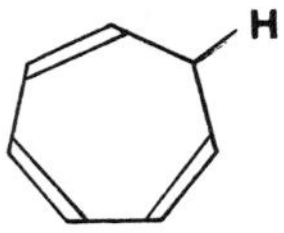	2.20

[a] All the protons absorb at 1.78 (accidental chemical shift equivalence).

Refs. p. 333.

c. <u>Protons Bonded to Heterocyclic sp^3 Carbon Atoms</u>

Compound	δ	Compound	δ	Compound	δ
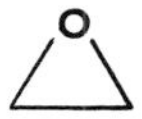	2.54	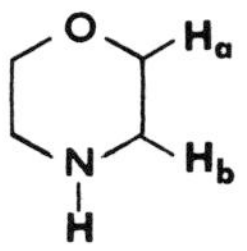	a 3.57 b 2.83	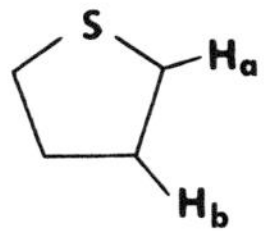	a 2.82 b 1.93
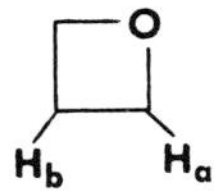	a 4.73 b 2.72	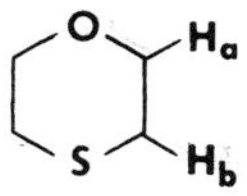	a 3.88 b 2.57	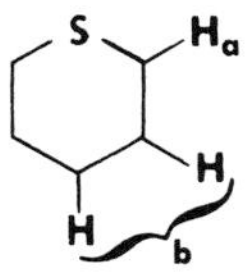	a 2.57 b 1.6
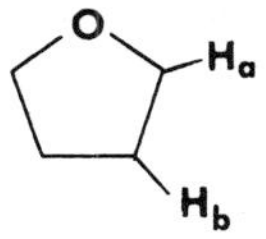	a 3.63 b 1.79	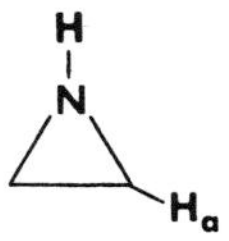	a 1.48	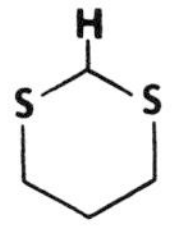	3.69
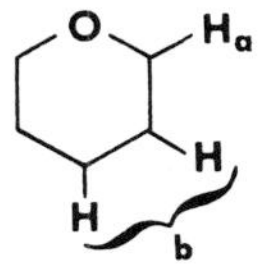	a 3.56 b 1.58	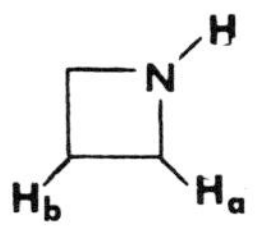	a 3.54 b 2.23		4.18
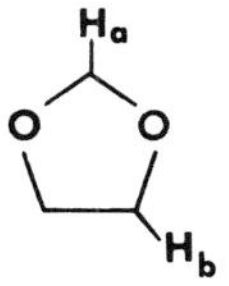	a 4.77 b 3.77	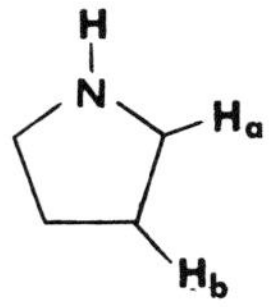	a 2.74 b 1.62	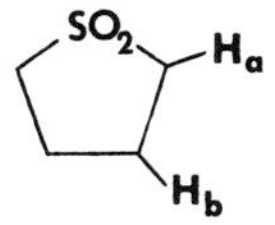	a 2.92 b 2.16
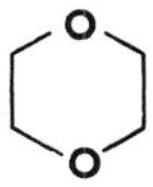	3.59	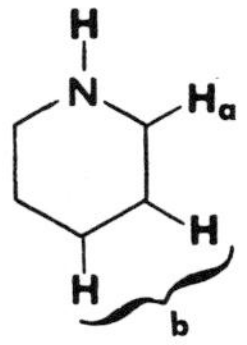	a 2.69 b 1.49	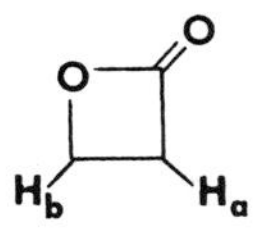	a 3.48 b 4.22
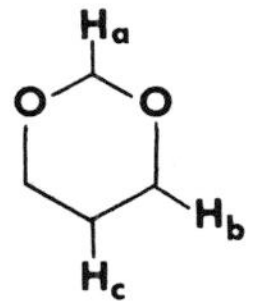	a 4.82 b 3.80 c 1.68	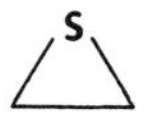	2.27	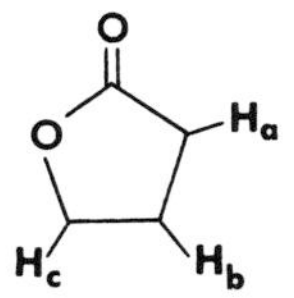	a 2.31 b 2.08 c 4.28
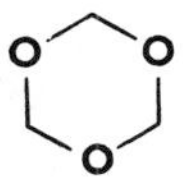	5.00	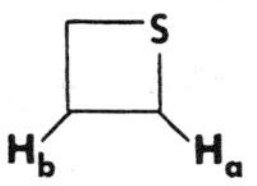	a 2.82 b 1.93	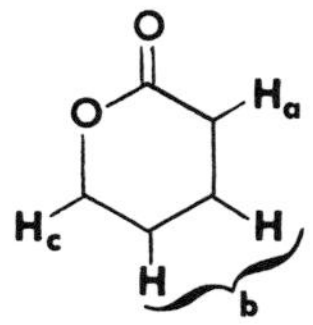	a 2.27 b 1.62 c 4.06

(Continued)

Refs. p. 333.

Compound	δ	Compound	δ	Compound	δ
H–N, O, H_a, H_b	a~2.3 b 3.4	H–N, O, H_a	a 3.17	H–N, O, H_a, N–H, O	a 1.44
CH_3^c–N, O, H_a, H_b	a~2.3 b 3.4 c 2.83				

d. Protons Bonded to sp^2 (Nonaromatic) Carbon Atoms

Compound	δ	Compound	δ	Compound	δ
$H_2C=CH_2$	5.33	X = CCl_3	a 6.41 b 5.30 c 5.78	$(H_b)(H_b)C=C=C(H_a)Br$	a 5.85 b 4.82
$(H_b)(H_c)C=C(H_a)R$	a 5.70 b 4.88 c 4.96	X = C≡N	a 5.53 b 6.05 c 5.91	$(H_b)(H_b)C=C=C(H_a)I$	a 5.62 b 4.46
$(H_b)(H_c)C=C(H_a)X$		X = CH_2OH	a 6.0 b 5.13 c 5.25	H	7.01
X = F	a 6.17 b 4.03 c 4.37	X = SCH_3	a 6.43 b 5.18 c 4.95	H	5.97
X = Cl	a 6.30 b 5.44 c 5.52	X = SO_2CH_3	a 6.70 b 6.13 c 5.95	H	5.60
X = Br	a 6.49 b 5.88 c 6.03	X = OCH_3	a 6.43 b 3.90 c 4.04	H	5.57
X = C_6H_5	a 6.69 b 5.21 c 5.71	$H_2C=C=CH_2$	4.55		
		$(H_b)(H_b)C=C=C(H_a)Cl$	a 5.76 b 5.05		

Refs. p. 333.

(Continued)

Compound	δ	Compound	δ	Compound	δ
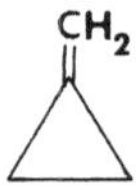 CH_2	5.38	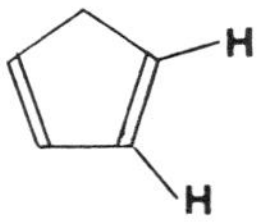 H, H	6.42	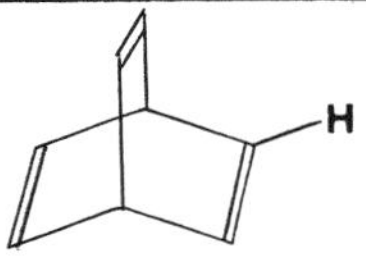 H	6.70
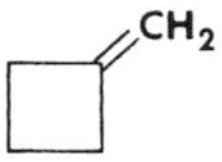 CH_2	4.70	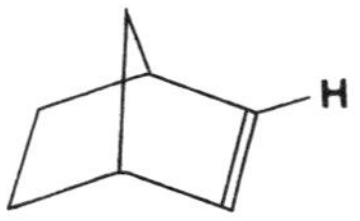 H	5.95	$CH_3-C(=O)-H$	9.72
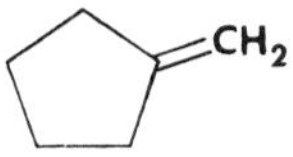 CH_2	4.82	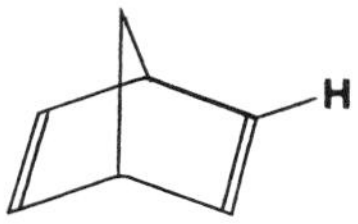 H	6.66	$Ph-C(=O)-H$	9.96
				$CH_2=CH-C(=O)-H$	9.48
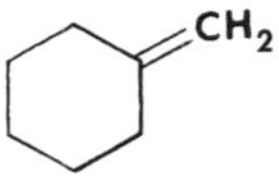 CH_2	4.55	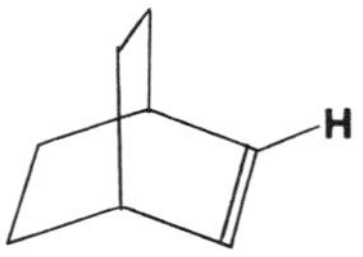 H	6.25	$(CH_3)_2N-C(=O)-H$	7.84
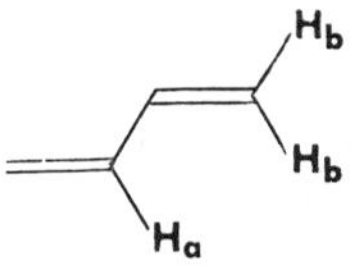 H_b, H_b, H_a	a 5.20 b 5.11	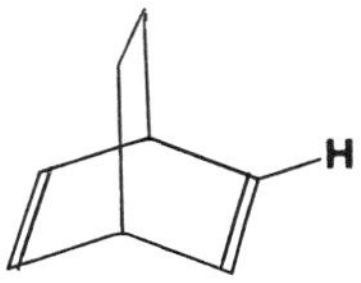 H	6.27	$CH_3O-C(=O)-H$	8.08

e. Protons Bonded to Aromatic Carbon Atoms

i. Monosubstituted Benzenes. These data were taken from references 14, 15, and 16 as well as the abovementioned general references.

Substituent	Ortho	Meta	Para
H	7.27	7.27	7.27
CH_3	7.07	7.07	7.07
CH_2CH_3	7.13	7.13	7.13
CH_2OH	7.28	7.28	7.28
CH_2Cl	7.32	7.32	7.32
$CHCl_2$	7.42	7.42	7.42
CCl_3	7.91	7.40	7.37
$CH=CH_2$	7.5	7.5	7.5
CHO	7.83	7.49	7.56
$COCH_3$	7.89	7.41	7.56
CO_2H	8.12	7.43	7.51
CO_2CH_3	7.98	7.38	7.48
COCl	8.11	7.49	7.63

Refs. p. 333.

(Continued)

Substituent	Ortho	Meta	Para
COBr	8.07	7.48	7.64
$CONH_2$	7.8	7.5	7.5
CN	7.63	7.45	7.55
F	6.99	7.24	7.08
Cl	7.30	7.25	7.18
Br	7.45	7.19	7.23
I	7.67	7.06	7.27
NH_2	6.52	7.02	6.62
$NHCH_3$	6.47	7.05	6.59
$N(CH_3)_2$	6.61	7.09	6.60
$NHCOCH_3$	7.7	7.1	7.0
NH_3+	7.7	7.5	7.5
NO	7.81	7.55	7.61
NO_2	8.22	7.53	7.65
OH	6.68	7.15	6.82
OCH_3	7.77	7.15	7.37
$OCOCH_3$	6.79	7.18	6.83
SCH_3	7.4	7.2	7.1
SO_2Cl	8.04	7.62	7.72
SO_3CH_3	7.87	7.53	7.60

ii. Para-Disubstituted Benzenes:

Where R_1 and R_2 differ, the shift given is for the proton ortho to R_1. These data are primarily from reference 14.

R_2 \ R_1	CH_3	CHO	$COCH_3$	COCl	CN	Cl	Br	I	NH_2	NO_2	OCH_3
CH_3	6.95	7.68	7.80	7.99	7.45	7.16	7.32	7.51	6.43	8.10	6.69
CHO	7.26	8.02	8.11	8.30	7.74	7.48	7.63	7.85	6.64	8.46	6.92
$COCH_3$	7.16	7.92	8.00	8.19	7.63	7.41	7.57	7.73	6.57	8.33	6.85
COCl	7.23	8.00	8.07	8.26	7.70	7.43	7.59	7.80	6.61	8.41	6.94
CN	7.23	7.93	8.01	8.20	7.77	7.37	7.53	7.74	6.57	8.35	6.86
Cl	7.03	7.78	7.84	8.03	7.47	7.24	7.36	7.56	6.46	8.17	6.75
Br	6.97	7.65	7.77	7.97	7.41	7.14	7.34	7.49	6.41	8.06	6.7
I	6.85	7.49	7.64	7.83	7.27	7.00	7.16	7.38	6.32	7.91	6.67
NH_2	6.84	7.48	7.64	7.83	7.27	7.00	7.15	7.34	6.37	7.90	6.61
NO_2	7.23	8.00	8.07	8.26	7.70	7.52	7.59	7.80	6.61	8.45	6.92
OCH_3	6.97	7.75	7.80	8.02	7.43	7.16	7.3	7.53	6.43	8.13	6.70

Refs. p. 333.

iii. Meta-Disubstituted Benzenes:

These data were taken from reference 17.

X	Y	2	4	5	6
Br	Br	7.66	7.40	7.07	7.40
Br	Cl	7.51	7.25	7.14	7.35
Br	OH	6.94	6.69	7.04	7.02
Br	CN	7.79	7.59	7.36	7.72
Br	$COCH_3$	8.00	7.81	7.30	7.62
Br	NO_2	8.35	8.17	7.45	7.81
Cl	NH_2	6.54	6.40	6.96	6.63
Cl	CH_2Cl	7.34	7.24	7.24	7.24
I	NH_2	6.92	6.48	6.71	6.99
I	$NHCH_3$	6.81	6.41	6.78	6.93
I	NO_2	8.53	8.18	7.29	8.00
OCH_3	OCH_3	6.31	6.37	7.04	6.37
OCH_3	OH	6.29	6.33	7.00	6.37
OH	t-Bu	6.77	6.86	7.06	6.56
NH_2	t-Bu	6.53	6.66	6.94	6.32
CO_2CH_3	CO_2CH_3	8.57	8.14	7.47	8.14

iv. Ortho-Disubstituted Benzenes:

These data were taken mostly from references 18 and 19.

X	Y	3	4	5	6	X	Y	3	4	5	6
CH_3	CH_3	7.01	7.01	7.01	7.01	Cl	Cl	7.37	7.11	7.11	7.37
CH_3	Cl	7.24	7.02	7.04	7.11	Cl	Br	7.54	7.03	7.16	7.37
CH_3	Br	7.44	6.95	7.09	7.13	Cl	NH_2	6.58	6.93	6.57	7.14
CH_3	NH_2	6.43	6.87	6.55	6.88	Cl	NO_2	7.81	7.41	7.50	7.52
CH_3	NO_2	7.87	7.29	7.44	7.29	Cl	OH	6.97	7.11	6.79	7.25
CH_3	OH	6.59	6.92	6.73	6.99	Cl	OCH_3	6.81	7.10	6.8	7.26
CH_3	OCH_3	6.66	7.01	6.72	6.99	Br	Br	7.54	7.08	7.08	7.54
CH_3	CN	7.50	7.23	7.42	7.27	Br	I	7.79	6.91	7.12	7.55
Ph	Ph	7.27	7.31	7.31	7.27	Br	NH_2	6.61	6.99	6.51	7.32
Ph	F	7.04	7.17	7.07	7.33	Br	t-Bu	7.35	7.12	6.92	7.49
Ph	Cl	7.38	7.16	7.19	7.24	I	I	7.81	6.96	6.96	7.81
Ph	Br	7.58	7.08	7.23	7.23	I	NH_2	6.58	7.01	6.36	7.53
Ph	I	7.87	6.92	7.27	7.21	NH_2	OCH_3	6.56	6.56	6.56	6.56
Ph	NH_2	6.53	6.98	6.67	6.98	NH_2	NO_2	8.07	6.67	7.30	6.79
Ph	NO_2	7.71	7.36	7.48	7.33	NH_2	CN	7.29	6.64	7.23	6.69
Ph	OH	6.83	7.11	6.84	7.10	NO_2	NO_2	8.16	8.01	8.01	8.16
Ph	CN	7.64	7.32	7.51	7.41	NO_2	OH	7.10	7.53	6.94	8.05
F	Cl	7.34	7.01	7.16	7.07	OH	OH	6.68	6.82	6.82	6.68
F	Br	7.48	6.94	7.24	7.04	t-Bu	t-Bu	6.96	7.46	7.46	6.96
F	I	7.69	6.83	7.24	6.99						

Refs. p. 333.

v. Miscellaneous Aromatic Carbocyclic Compounds. These data were taken from references 2d and 4.

Compound	δ	Compound	δ	Compound	δ
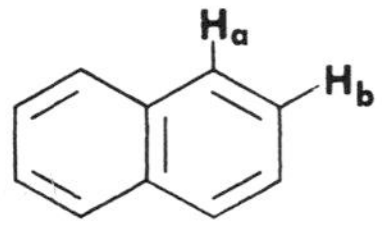	a 7.81 b 7.46	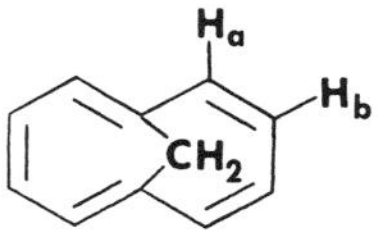	a 7.27 b 6.95	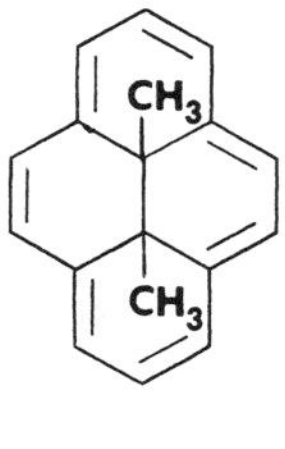	(CH_3's −4.25)
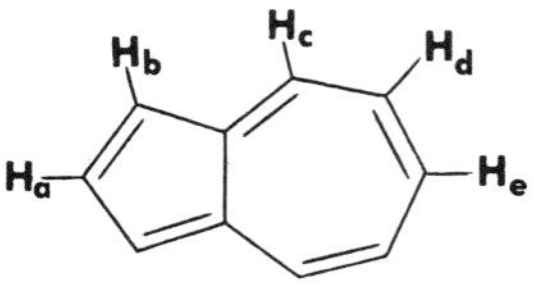	a 7.82 b 7.33 c 8.26 d 7.73 e 7.51	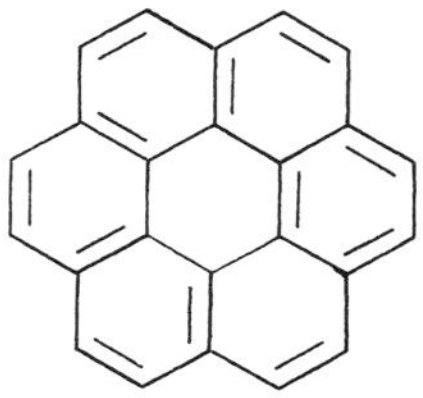	8.84	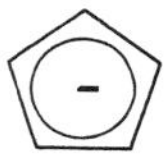	(THF) 5.57
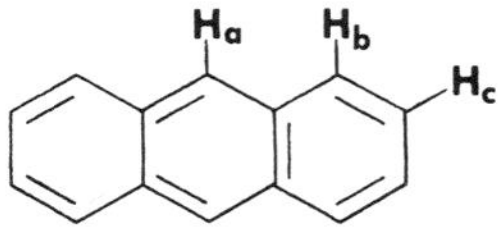	a 8.31 b 7.91 c 7.39			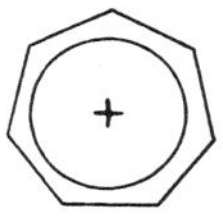	(SO_2) 9.28
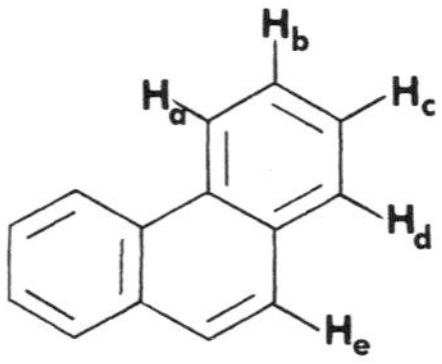	a 8.93 b 7.88 c 7.82 d 8.12 e 7.72	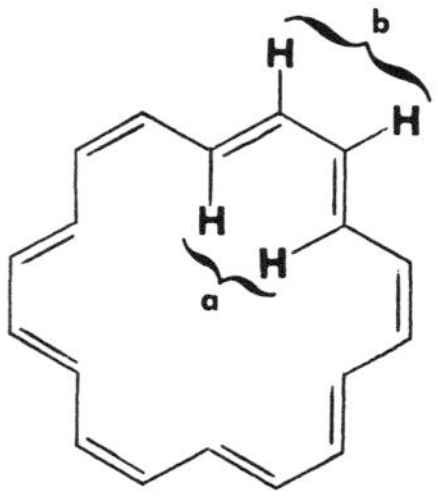	a −1.8 b 8.9	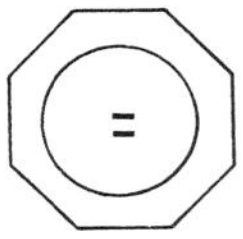	(THF) 5.69
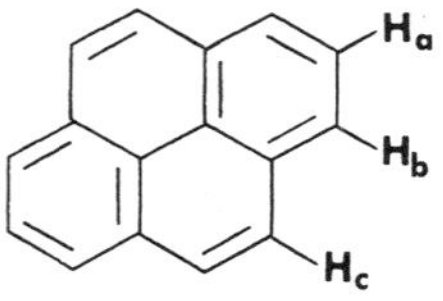	a 7.99 b 8.16 c 8.06	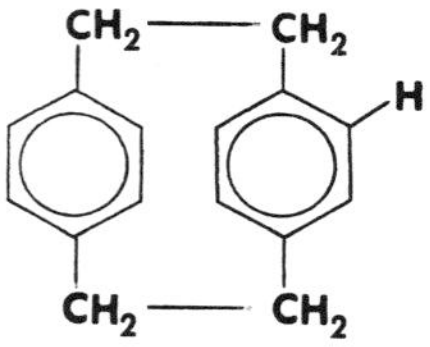	6.30	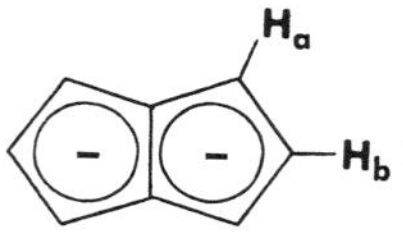 (THF)	a 4.98 b 5.73
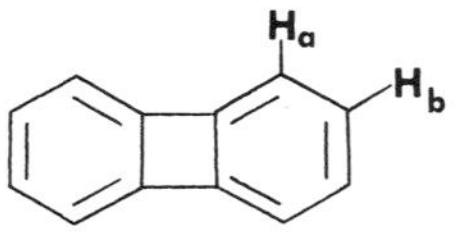	a 6.60 b 6.47	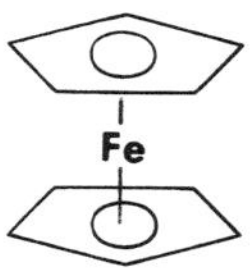	4.15	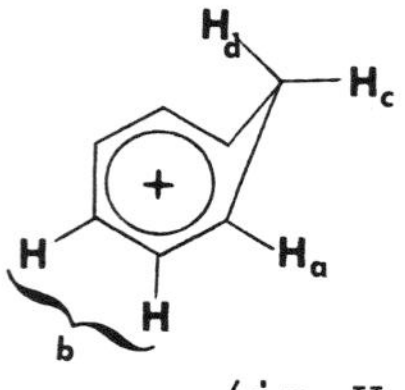 (in H_2SO_4)	a 6.6 b 8.6 c 5.2 d −0.6

Refs. p. 333.

vi. Heteroaromatic Compounds. These data were taken primarily from reference 2d.

Compound	δ	Compound	δ	Compound	δ
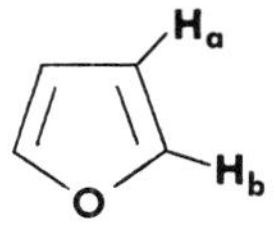	a 6.30 b 7.38	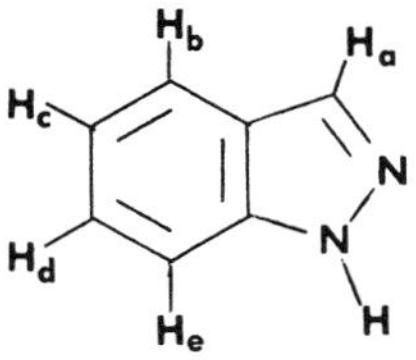	a 8.03 b 7.77 c 7.12 d 7.34 e 7.58	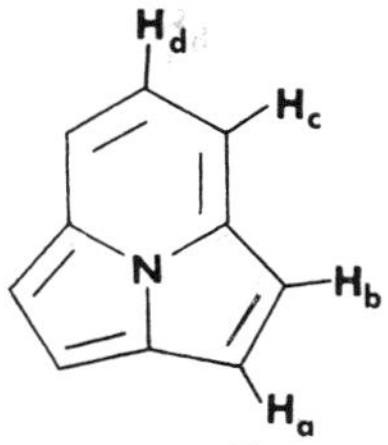	a 7.50 b 7.19 c 7.86 d 7.59
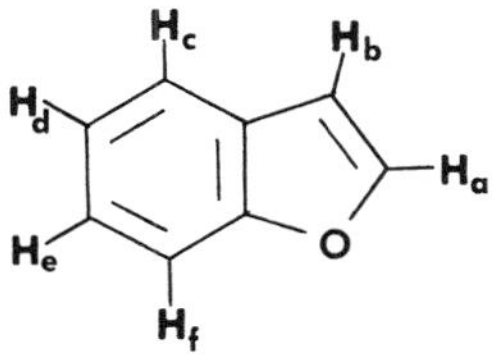	a 7.52 b 6.66 c 7.49 d 7.13 e 7.19 f 7.42	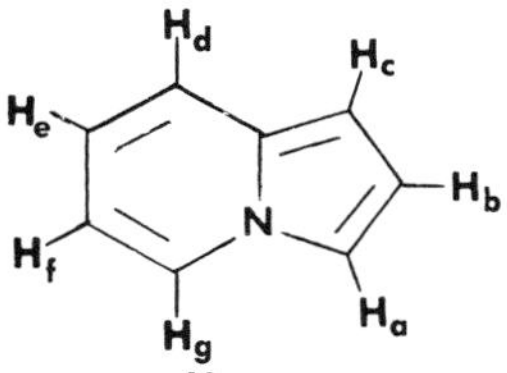	a 7.14 b 6.64 c 6.28 d 7.25 e 6.50 f 6.35 g 7.76	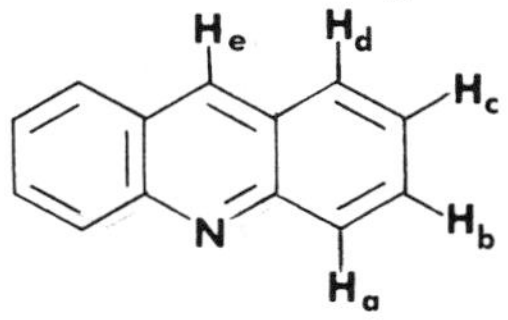	a 8.20 b 7.69 c 7.39 d 7.80 e 8.53
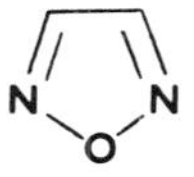	8.19	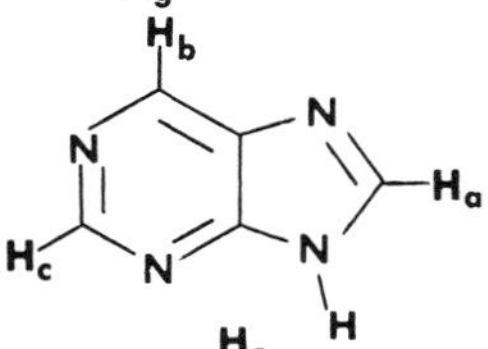	a 8.63 b 9.16 c 8.95		8.58
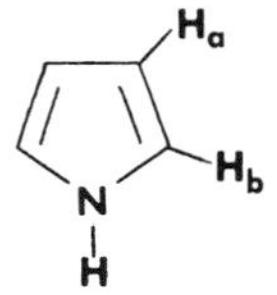	a 6.05 b 6.62	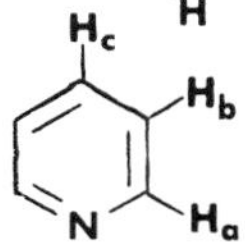	a 8.50 b 7.06 c 7.46	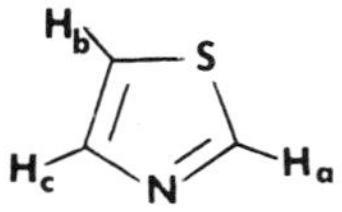	a 8.88 b 7.41 c 7.98
		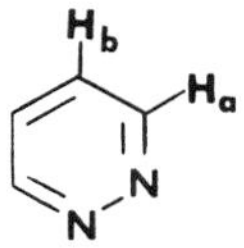	a 9.17 b 7.68	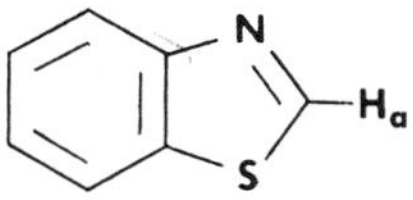	a 8.95
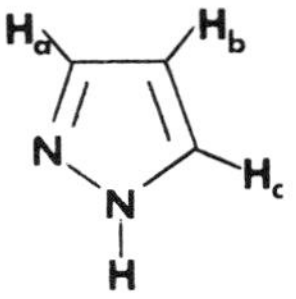	a,c 7.55 b 6.25	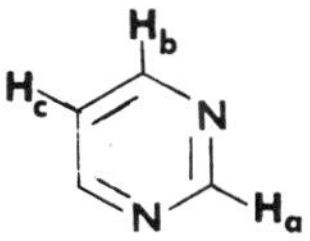	a 9.15 b 8.60 c 7.09	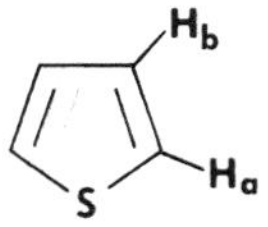	a 7.19 b 7.04
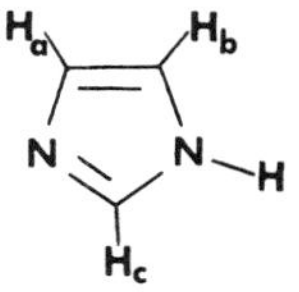	a,b 7.14 c 7.70	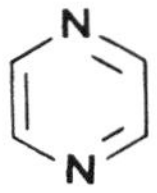	8.5	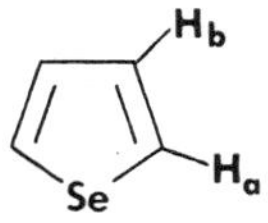	a 7.70 b 7.12
		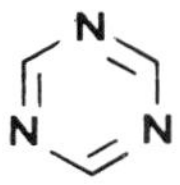	9.18		
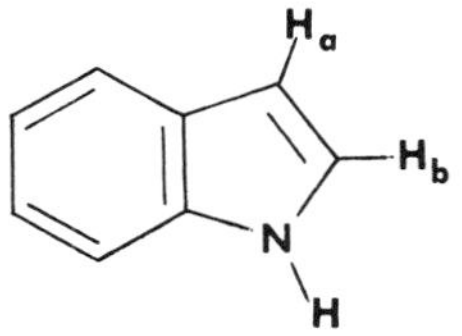	a 6.34 b 6.54	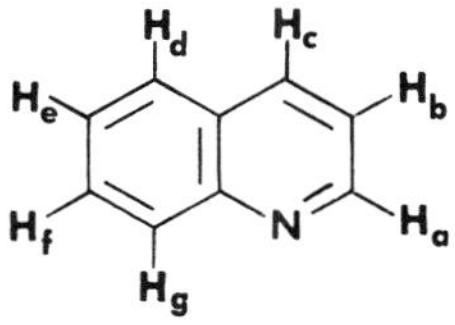	a 8.81 b 7.26 c 8.00 d 7.68 e 7.43 f 7.61 g 8.05	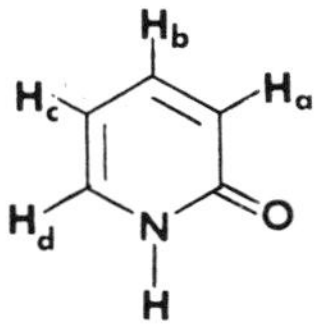	a 6.57 b 7.26 c 6.15 d 7.13

Refs. p. 333.

f. Protons Bonded to sp Carbon Atoms

Compound	δ	Compound	δ
$H-C{\equiv}C-\underline{H}$	1.80	$HOCH_2C{\equiv}C-\underline{H}$	2.33
$CH_3C{\equiv}C-\underline{H}$	1.80	(cyclohexane ring)-C(OH)-$C{\equiv}C\underline{H}$	2.48
$CH_3CH_2C{\equiv}C-\underline{H}$	1.76		
$Ph-C{\equiv}C-\underline{H}$	3.05		
$CH_2{=}CH-C{\equiv}C-\underline{H}$	2.92	$Cl-CH_2C{\equiv}C-\underline{H}$	2.40
$C_2H_5C{\equiv}C-C{\equiv}C-\underline{H}$	1.95	$Br-CH_2C{\equiv}C-\underline{H}$	2.33
$CH_3(C{\equiv}C)_2C{\equiv}C\underline{H}$	1.87	$I-CH_2C{\equiv}C-\underline{H}$	2.19
		$CH_3O-C{\equiv}C-\underline{H}$	1.33
$H-\overset{O}{\overset{\Vert}{C}}-C{\equiv}C-\underline{H}$	1.89	$CH_2{=}CH-O-C{\equiv}C-\underline{H}$	1.89

g. Protons Bonded to Oxygen, Nitrogen, and Sulfur. These values are highly dependent on both solvent and temperature, and in some cases concentration. Absorptions are usually broad. Presence of a trace of acid affects the chemical shift (by rapid, catalyzed proton exchange), which for alcohols and amines in water appears between the expected value and that for water (δ ∿ 4.7). Most of the protons included below are readily exchanged with D_2O. For extensive discussion and references, see page 215 of reference 2d.)

ROH (monomeric)[a]	0.5	RNH_2	1.1-1.8
(hydrogen bonded)	0.5-5.0	R_2NH	1.2-2.1
ArOH (monomeric)	4.5	$ArNH_2$	3.3-4.0
(hydrogen bonded)	4.5-9		
Enols (intramolecular hydrogen bonded)	15-19	ArNHR	3.1-3.8
Carboxylic acids	10-13	$R-\overset{O}{\overset{\Vert}{C}}-NH_2$, $AR\overset{O}{\overset{\Vert}{C}}NH_2$	5-6.5
Sulfonic acids	11-12	$R-\overset{O}{\overset{\Vert}{C}}-NHR$, $Ar\overset{O}{\overset{\Vert}{C}}NHR$	6-8.2
RSH	1-2		
ArSH	3-4	$R-\overset{O}{\overset{\Vert}{C}}-NHAr$, $Ar\overset{O}{\overset{\Vert}{C}}NHAr$	7.8-9.4
Oximes	7-11	R_3N^+H	7.1-7.7

[a]Information on the structure of alcohols is obtained from DMSO solution spectra [Chapman and King, J. Amer. Chem. Soc., 86, 1256 (1964)].

Refs. p. 333.

h. Charged Species. A number of long-lived carbonium ions and carbanions have been studied by nmr. These ionic species require solvents of high ionizing power such as the "super acids" (see p. 21), aqueous mineral acids, anhydrous hydrofluoric acid, sulphur dioxide, or tetrahydrofuran. Because of the insolubility of TMS in many of these solvents, spectra have often been reported with no standard mentioned or with reference to an external standard. The standard of choice in these solvents is the tetramethylammonium cation either as its tetrafluoroborate or hexafluoroantimonate salt (2d,31). Although its exact absorption position is dependent on solvent (31), the shift position of 3.10 ppm (downfield from TMS) has been accepted (2d,31). In the following selected examples it should be noted that the chemical shifts are quite dependent on solvent, temperature, counterion, concentration, and amount of ion pairing. The data were taken from reference 2d and references therein. (See also "Miscellaneous Aromatic Carbocyclic Compounds," p. 264).

Ion	Solvent	δ	Ion	Solvent	δ
$(C\underline{H}_3)_2C^+H$	SbF_5	5.06	$(C_6H_5)_2C^+CH_3$	H_2SO_4	CH_3 3.70 o 8.03 m 7.88 p 8.28
$(CH_3)_2C^+\underline{H}$	SbF_5	13.5			
$(CH_3)_3C^+$	SbF_5	4.35			
+ H_a H_b H_c	SO_2/SbF_5	a 4.50 b 5.42 c 2.67	$C_6H_5C^+(C\underline{H}_3)_2$	H_2SO_4	CH_3 3.57 o 8.80 m 7.97 p 8.45
a b	CH_3NO_2	a 0.9 b 10.1	H_1 H_1 H_2 H_3 H_4	SO_2	1 3.6 2 8.8 3 7.2 4 7.5
$\overset{+}{CH_2-CH-CH_2}$ a b	SO_2/SbF_5	a 8.97 b 9.64	CH_3CH_2Li	$C_6H_5CH_3$	CH_2 -0.83 CH_3 1.4
$(C_6H_5)_3C^+$	H_2SO_4	o 7.69 m 7.87 p 8.29	$(C_6H_5)_3CLi$	THF	o 7.31 m 6.52 p 5.96
$(C_6H_5)_2C^+H$	H_2SO_4	α 9.81 o 8.46 m 7.98 p 8.38	$(C_6H_5)_2\bar{C}$–(α)cyclopropyl K^+	THF	α ∼-0.27 o 7.06 m 6.66 p 3.78

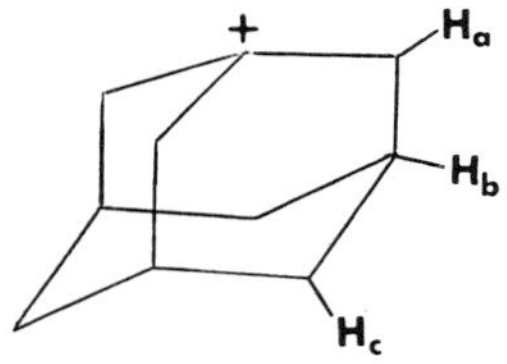

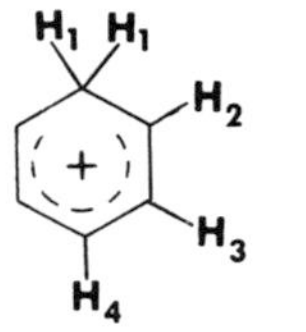

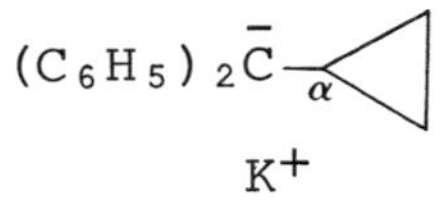

5. Empirical Correlations for Chemical Shifts

a. Shoolery's Constants. J. N. Shoolery has developed an empirical procedure for predicting the chemical shifts of substituted methylenes and

Refs. p. 333.

methanes. This procedure consists of adding "effective shielding constants" (σ_{eff}) for each substituent to the chemical shift of methane:

$$\delta = 0.23 + \Sigma\ \sigma_{eff}$$

For example, for $PhCH_2Br$ we calculate 0.23 + 1.85 + 2.33 = 4.41 as compared to an observed value of 4.50. The procedure has been used with trisubstituted methanes but much larger error (up to 1 ppm) is obtained. Following the table of constants is a table of randomly chosen compounds for which comparison of observed and calculated shifts is made.

Effective Shielding Constants (from references 2d and 20):

Group	σ_{eff}	Group	σ_{eff}	Group	σ_{eff}
$-CH_3$	0.47	-CO-Ph	1.84	-N=C=S	2.86
-C=C	1.32	-Ph	1.85	$-N_3$	1.97
-C≡CR	1.44	$-CF_2$	1.21	-NHCOR	2.27
-C≡CAr	1.65	$-CF_3$	1.14	-OH	2.56
-C≡C-C≡C-R	1.65	-Cl	2.53	-OR	2.36
-C≡N	1.70	-Br	2.33	-OCO-R	3.13
-COR	1.70	-I	1.82	-OPh	3.23
-CO-OR	1.55	$-NR_2$	1.57	-SR	1.64
$-CO-NR_2$	1.59	$-NO_2$	2.46	-SCN	2.30

Comparison of observed and predicted chemical shifts using Shoolery's constants:

Compound	Calculated	Observed	Difference	Compound	Calculated	Observed	Difference
$C\underline{H}_3CHO$	1.93	2.17	0.24	$ClCH_2CN$	4.20	4.09	0.11
$CH_3C\underline{H}_2CHO$	2.40	2.60	0.20	$CH_3C{\equiv}C{-}H$	1.67	1.80	0.13
CH_3Br	2.56	2.68	0.12	$HOCH_2C{\equiv}CH$	4.23	4.28	0.05
CH_3CH_2Br	3.03	3.36	0.30	CH_3OCH_2CN	4.29	4.20	0.09
$PhCH_2Br$	4.41	4.50	0.09	$NCCH_2CN$	3.63	4.13	0.50
$PhCH_3$	2.08	2.34	0.26	$NCCH_2CO_2CH_3$	3.48	3.48	--
$PhCH_2Ph$	3.93	3.92	0.01	$CH_3CO_2CH_3$	2.25	2.01	0.24
$PhCH_2Cl$	4.61	4.55	0.06	ICH_2CO_2H	3.60	3.72	0.12
$ClCH_2Cl$	5.29	5.28	0.01	ICH_3	2.05	2.16	0.11

b. Substituent Constants for Olefinic Protons. A procedure for calculating the chemical shifts of protons bonded to olefinic systems has been developed (21). In this case, each substituent has a different

Refs. p. 333.

constant depending on its orientation (cis, trans, geminal) to the hydrogen being calculated; the basic frequency is that of ethylene. Thus

$$\delta = 5.28 + \sigma_{gem} + \sigma_{cis} + \sigma_{trans}$$

For example, the chemical shift of the proton geminal to the bromo group in trans-1-phenyl-2-bromoethylene ($C_6H_5CH{=}CHBr$) is calculated as 5.28 + 0 + 1.04 + 0.37 = 6.69 as compared to 6.75 observed.

The following constants are taken from reference 21 with some recently published (22) new values in parentheses, which in some cases give better calculated values.

gem \ / trans
C=C
H / \ cis

Group	σ_{gem}	σ_{cis}	σ_{trans}
H	0	0	0
Alkyl	0.44	-0.26	-0.29
(CH_3)	(0.44)	(-0.32)	(-0.34)
Alkyl-ring	0.71	-0.33	-0.30
$-CH_2O-$, OCH_2I	0.67	-0.02	-0.07
$-CH_2S-$	0.53	-0.15	-0.15
$-CH_2Cl$, $-CH_2Br$	0.72	0.12	0.07
$-CH_2N$	0.66	-0.05	-0.23
-C≡C-	0.50	0.35	0.10
-C≡N	0.23 (0.30)	0.78 (0.75)	0.58 (0.53)
-C=C	0.98	-0.04	-0.21
-C=C (conj.)	1.26	0.08	-0.01
-C=O	1.10	1.13	0.81
-C=O (conj.)	1.06	1.01	0.95

Group	σ_{gem}	σ_{cis}	σ_{trans}
CO_2H	1.00	1.35	0.74
CO_2H (conj.)	0.69	0.97	0.39
CO_2R	0.84	1.15	0.56
CO_2R (R conj.)	0.68	1.02	0.33
-CHO	1.03	0.97	1.21
-CON<	1.37	0.93	0.35
-COCl	1.10	1.41	0.99
-OR (R aliph.)	1.18	-1.06	-1.28
-OR (R conj.)	1.14	-0.65	-1.05
-OCOR	2.09	-0.40	-0.67
-Ar (Ph)	1.35 (1.43)	0.37 (0.39)	-0.10 (0.06)
-Cl	1.00 (1.05)	0.19 (0.14)	0.03 (0.09)
-Br	1.04 (1.02)	0.40 (0.33)	0.55 (0.53)

(Continued)

Refs. p. 333.

gem, trans, cis, H — C=C

Group	σ_{gem}	σ_{cis}	σ_{trans}	Group	σ_{gem}	σ_{cis}	σ_{trans}
$-NR_2$ (R aliph.)	0.69	-1.19	-1.31	-SR	1.00	-0.24	-0.04
$-NR_2$ (R conj.)	2.30	-0.73	-0.81	$-SO_2-$	1.58	1.15	0.95

6. Coupling Constants

Spin-spin coupling is the interaction of the magnetic moments of two (or more) nuclei through the bonding electrons. This coupling is independent of the applied magnetic field but is dependent on the electronic structure and is thus a constant for a particular arrangement of atoms and electrons. The constant is designated as ^{n}J (in hertz), where n is the number of sigma bonds separating the interacting nuclei. The sign of the coupling constant (i.e., ±J) refers to whether the two interacting (i.e., coupled) nuclei are in a lower energy state when their magnetic moments are parallel (-J) or antiparallel (+J).

The sign makes an observable difference only in complex ("second order") spectra. (For a discussion of the analysis of these complex spectra see the section on special techniques below.)

An extensive review and collection of data on proton-proton coupling can be found in reference 23. It was from this reference as well as the two more specialized reviews mentioned below that most of the data were obtained. Signs are included only when they were given in these references. Where coupling is reported between magnetically equivalent protons (see p. 307), the constant is obtained generally through nonequivalent coupling with a ^{13}C atom in the molecule or through substitution of one of the protons with deuterium (J_{HD}/J_{HH} = 6.514). For example, the spectrum of dichloromethane consists of a single peak, but a coupling constant of -1.15 can be observed in $CHDCl_2$. This allows us to talk about an interaction (coupling) in the parent compound (J = -1.15 x 6.514 = -7.5).

For reviews of the theory of spin-spin coupling see reference 24. A series of articles on the calculation of spin-spin coupling using the self consistent field (SCF) method can be found in reference 25.

a. Geminal Coupling. Geminal coupling is defined as the interaction of two nuclei bound to the same atom. An extensive collection of geminal proton constants can be found in reference 26. For protons, the geminal coupling constants are generally in the range of +5 to -21 Hz.

Type	^2J	Type	^2J
C C C H H	-12 to -15	H H	-12.0 to -15.0
CH_4	-12.4		
$(CH_3)_4Si$	-14.2	H H O	-15.3 to -18.0
CH_3X	- 9.2 to -16.9		
X=I	- 9.2		
X=Br	-10.2		
X=Cl	-10.8	H H CH	-15.6
X=F	- 9.6		
X=OH	-10.8		
X=CN	-16.9		
X=NO	-13.2		
$X=COCH_3$	-14.9		
CH_2Br_2	- 5.5	H H O O	-16.6
$CH_2(CN)_2$	-20.3		
CH_2Cl_2	- 7.5	H H	- 5.4
H H	- 0.5 to -9.9		
		H H	- 3.1
O H H	+ 4.0 to +6.3		
		H H	-12.0 to -15.0
S H H	0 to -1.4		
		H H O	-19.0 to -19.5
H N H H Ph	+ 0.97		

Refs. p. 333.

(Continued)

Type	^2J	Type	^2J
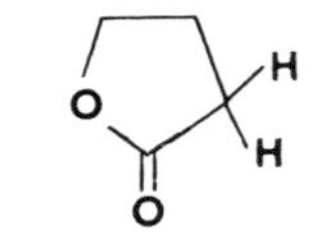	-17.0 to -18.9	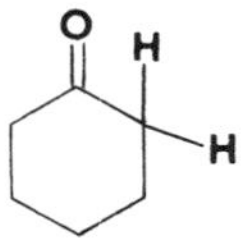	-12.0 to -16.0
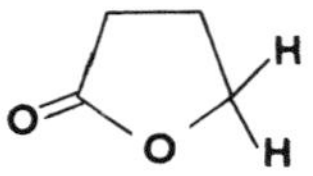	- 8.8 to -10.5	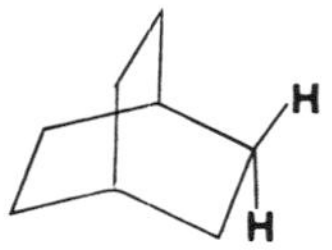	-13.6 to -13.8
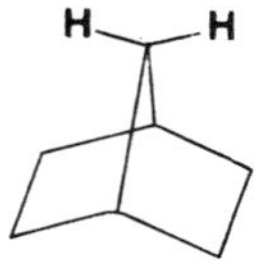	- 9.5 to -13.0	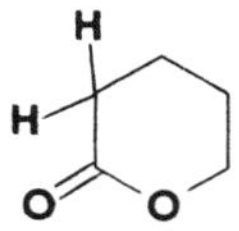	- 6.0 to -17.0
		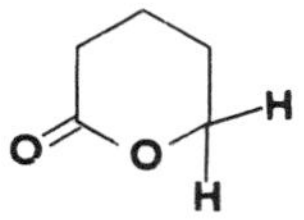	-11.0 to -13.4
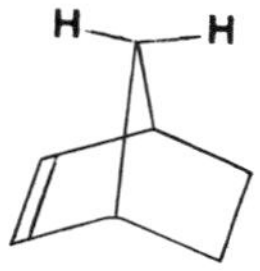	- 8.0 to -12.0	$\underline{H}_2C=O$	+41
		$\underline{H}_2C=C(H)X$	(See also p. 276)
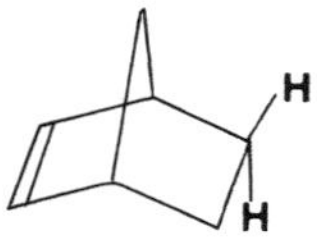	-10.4 to -13.7	X=Li	+ 7.1
		X=H	+ 2.3
		X=CH_3	+ 2.1
		X=Ph	+ 1.3
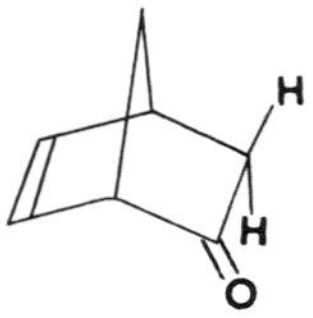		X=SCH_3	- 0.3
	-16.4 to -13.7	X=Cl	- 1.4
		X=Br	- 1.8
		X=OCH_3	- 2.1
		X=F	- 3.2
		$\underline{H}_2C=C=C<$	- 9.0
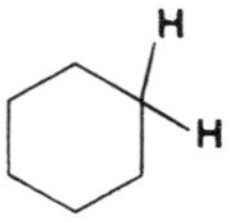	-11.6 to -15.0	$\underline{H}_2C=C=O$	-15.8
		$\underline{H}_2C=NR$	+ 8 to +16.5

b. Vicinal Coupling. Vicinal coupling is defined as the interaction of nuclei across three bonds such as H-C=C-H or H-C-C-H. The coupling in the saturated system has been found to be highly dependent on the angle between the carbon-hydrogen bonds, that is the dihedral angle, Φ:

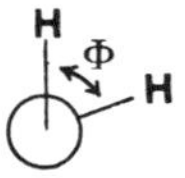

Refs. p. 333.

This dependence has been described in terms of valence bond theory by Karplus (27) and was found approximately to fit the equation

$$J_{vic} = A + B \cos \Phi + C \cos 2\Phi$$

where A = 4, B = -0.5, and C = 9.5 Hz. Bothner-By (23) has proposed a better set of constants, A = 7, B = -1 and C = +5 Hz, from empirical studies. It should be kept in mind that this treatment is only approximate and deviations may occur as a result of the following factors (23):

(1) Change in electronegativity (X) of substituents,

$$J_{vic} = J^{\circ}_{vic} \; (1 - 0.07 \; \Delta X)$$

where J°_{vic} is the vicinal coupling constant in ethane (+8.0) and ΔX is the difference in electronegativity for hydrogen and the substituent.

(2) Change in H-C-C bond angles, where an increase generally results in a decrease in coupling constant.

(3) Change in C-C bond lengths, where an increase results in a decrease in coupling constant.

(4) Variation in hybridization of the carbon atom, where H-C_{sp^3}-C_{sp^2}-H is generally smaller than for H-C_{sp^3}-C_{sp^3}-H.

The variation of vicinal and allylic (H-C=C-C-H) coupling with dihedral angle has been plotted (29) according to the Karplus equation as modified by Bothner-By (23) and Garbisch (28) and is reproduced here.

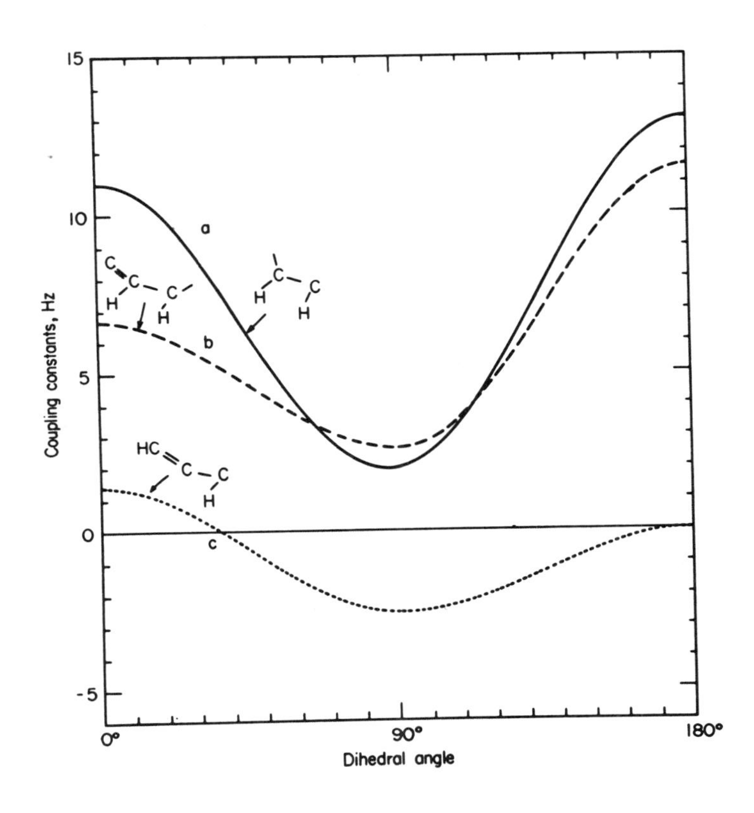

Refs. p. 333.

i. Across C-C Single Bonds. Where two or more conformations are rapidly interconverting, the nmr spectrum is an average of the spectra of each of the separate conformations, weighted as to the percent of each. Thus the observed coupling might be both temperature and solvent dependent.

Type		^3J	Type		^3J
CH_3CH_3		+8.0		a-a	6 to 14
				a-e	0 to 5
CH_3CH_2X				e-e	0 to 5
X=Li		8.96			
X=Ph		7.62	$CH\text{-}\overset{O}{\overset{\|}{C}}H$		1 to 3
X=CN		7.60			
$X=C(C_2H_5)_2$		7.53	$C=CH\text{-}\overset{O}{\overset{\|}{C}}H$		5 to 8
X=I		7.45			
$X=N(C_2H_5)_2$		7.4	CH-CH=C		4 to 10
$X=\overset{+}{N}(C_2H_5)_3$		7.4			
X=Br		7.33	$C=\overset{H}{C}\text{-}\overset{H}{C}=C$		9 to 11
$X=CH_3$		7.26			
X=Cl		7.23			6 to 7
X= C≡CH		6.97			
$X=OC_2H_5$		6.97			
	cis	8 to 10			9 to 10
	trans	4 to 6			
O	cis	2 to 3.5			3 to 4
	trans	3 to 5			
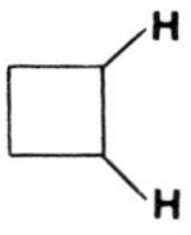	cis or trans	6 to 10			0 to 2
		0 to 7			2.5 to 5.0

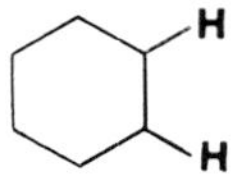

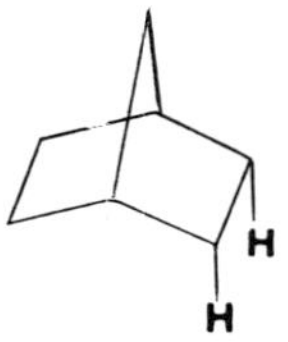

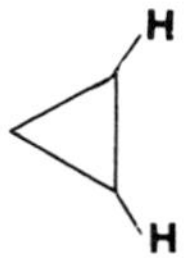

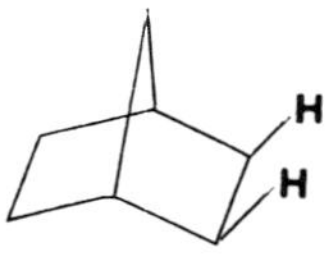

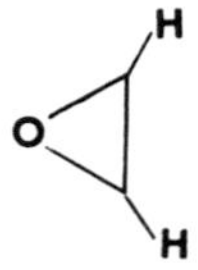

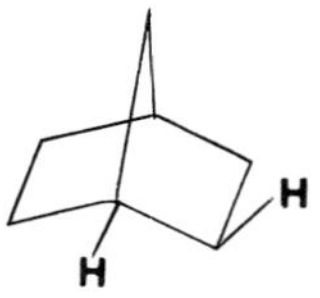

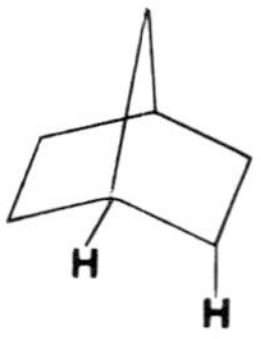

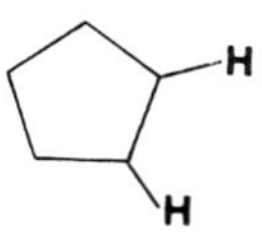

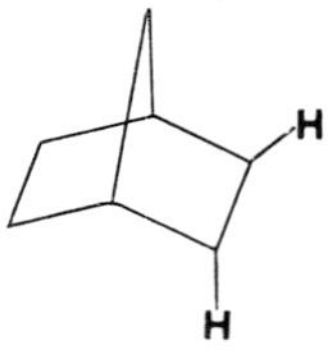

Refs. p. 333.

ii. Across C-C Double Bonds. For more detail on olefin couplings, see the next table.

Type		^3J
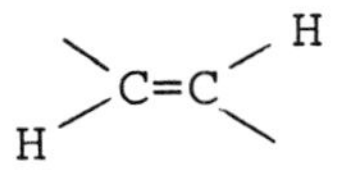 H C=C H		12 to 18 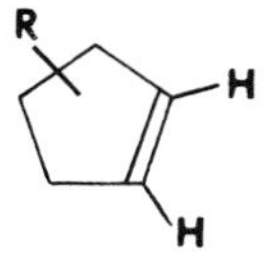
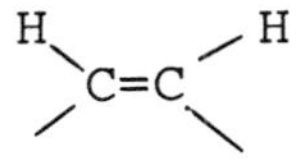 H C=C H		6 to 12
$H_2C{=}CH_2$	cis	11.5
	trans	19.0
FCH=CHF	cis	−2.0
	trans	+9.5
FCH=CHBr	cis	3.5
	trans	11.0
ClCH=CHCl	cis	5.2
	trans	12.2
$FCH{=}CHCH_3$	cis	+4.5
	trans	+11.1
BrCH=CHBr	cis	4.7
	trans	11.8
$C_2H_5O_2CCH{=}CHCO_2C_2H_5$	cis	11.9
	trans	15.5
$CH_3CH{=}CHCHO$	trans	15.45
$PhCH{=}CHCH_3$	trans	15.6
H H		0.5 to 1.5
H H		+2 to 4

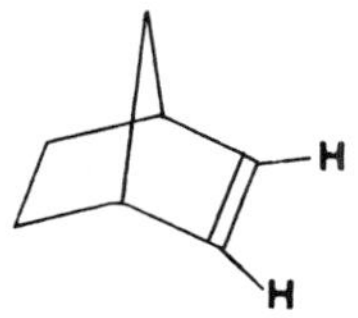

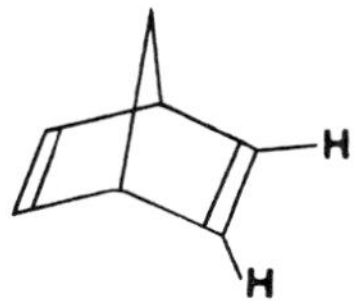

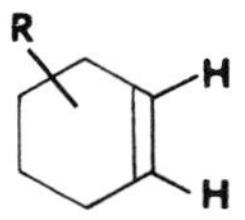

Type		^3J
R H H		5.1 to 7.0
Cyclopentene		+5.3
H H		5.8
H H		5.05
R H H		8.8 to 11
Cyclohexene		+8.8
Cyloheptene		10.8
cis-Cyclooctene		+10.3
Cyclooctatetraene		11.8
cis-Cyclononene		+10.7
	ortho	6.0 to 9.4
	meta	1.2 to 3.1
	para	0.2 to 1.5

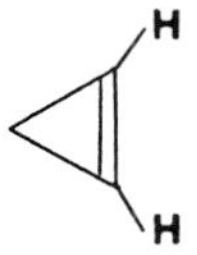

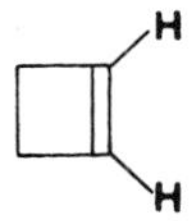

(Continued)

Refs. p. 333.

Type		^3J	Type		^3J
Pyridine (positions 2–6, N)	2-3	4 to 5.7	Thiophene (positions 2–5, S)	2-3	4.7 to 5.5
	3-4	6.8 to 9.1		3-4	3.3 to 4.0
	2-4	0 to 2.5		2-4	1.0 to 1.5
	3-5	0.5 to 1.8		2-5	2.8 to 3.5
	2-5	0 to 2.3			
	2-6	0 to 0.6			
Furan (positions 2–5, O)	2-3	1.7 to 2.0	Naphthalene (positions 1–4)	1-2	8.1 to 9.1
	3-4	3.3 to 3.8		2-3	6.0 to 7.0
	2-4	0.4 to 0.9		1-3	1.1 to 1.6
	2-5	1 to 2		1-4	0 to 1
Pyrrole (positions 2–5, N–H)	2-3	2 to 3			
	3-4	3 to 4			
	2-4	1 to 2			
	2-5	1.5 to 2.5			

c. Average Values of Coupling Constants in Vinyl Compounds, $CH_2=CHX$. The following table was taken from reference 2d, page 278, and is based on a paper by Schaeffer, Can. J. Chem., 40, 1 (1962).

X	E_x[a]	J_{gem}	J_{cis}	J_{trans}
F	3.95	-3.2	4.65	12.75
Cl	3.2	-1.4	7.3	14.6
Br	3.0	-1.8	7.1	15.2
OR	3.5	-1.9	6.7	14.2
OAr	3.5	-1.5	6.5	13.7
OCOR	3.5	-1.4	6.3	13.9
Phosphate	3.5	-2.3	5.8	13.2
NO_2	3.35	-2.0	7.6	15.0
NR_2	3.0	0	9.4	16.1
COOR	2.5	1.7	10.2	17.2
CN	2.5	1.3	11.3	18.2
COR	2.5	1.8	11.0	18.0
R	2.5	1.6	10.3	17.3
Ar	2.5	1.3	11.0	18.0
Py	2.5	1.1	10.8	17.5
Sulphone	3.0	-0.6	9.9	16.6
Sn	1.9	2.8	14.1	20.3
As	2.1	1.7	11.6	19.1
Sb	2.0	2.0	12.6	19.5

(Continued)

X	E_x[a]	J_{gem}	J_{cis}	J_{trans}
Pb	1.9	2.0	12.1	19.6
Hg	1.9	3.5	13.1	21.0
Al	1.5	6.3	15.3	21.4
Li	1.0	7.1	19.3	23.9

[a]Electronegativity of substituent X.

d. Long-Range Coupling. Long-range coupling is defined as the interaction of nuclei through four or more bonds and is usually in the range 0 to ±3 Hz. A recent review can be found in reference 30. There are three general types of structures in which long-range coupling might be expected. The first occurs through four sigma bonds when the five atoms are in the all-trans or "W" relationship. Several examples of this are included below. The second type is found in allylic and benzylic systems. Here coupling is often observed between the allylic hydrogens and the vinylic hydrogens:

H_1 / C-H_3

C=C ($J_{1,3}$ or $J_{2,3}$ generally 0 to -3Hz)

H_2

Long-range coupling also may be observed in extended π systems such as aromatics, acetylenes, allenes, and cumulenes.

i. Through Sigma Bonds

Type	J	Type	J
	1.0 to 1.4		1.0
	1.0 to 1.4		6.7 to 8.1
	1.7 to 2.6		8

Refs. p. 333. (Continued)

Type	J	Type	J
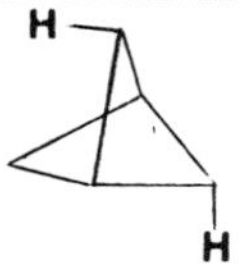 H, H	10	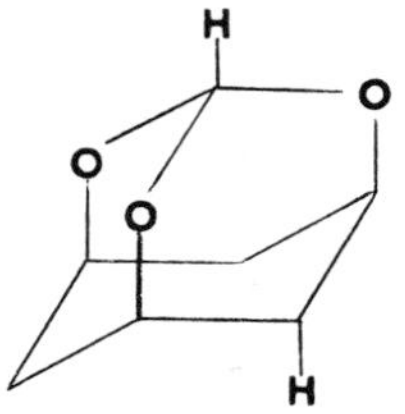 H, O, O, O, H	1.25
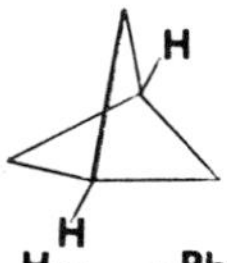 H, H	18	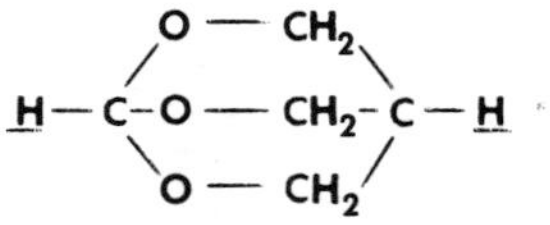 O — CH_2, H—C-O — CH_2-C—H, O — CH_2	1.7
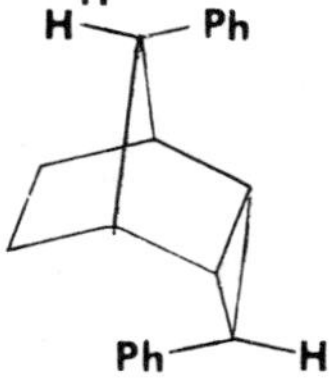 H, Ph, Ph, H	2.3		

ii. Allylic and Benzylic Coupling

Type	J	Type	J
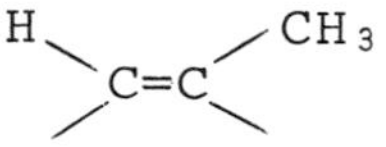 H, C=C, CH_3	-1.0 to -2.0	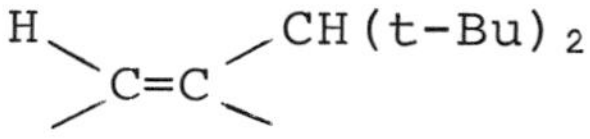 H, C=C, CH(t-Bu)$_2$	-0.63
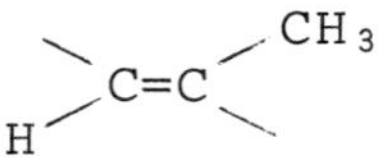 C=C, CH_3, H	-0.4 to -1.7	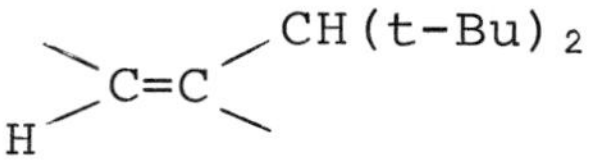 C=C, CH(t-Bu)$_2$, H	-0.1
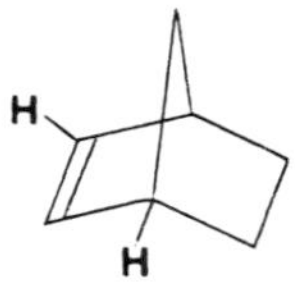 H, H	+0.5	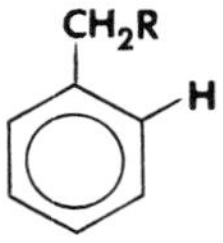 CH_2R, H	0.6 to 0.9
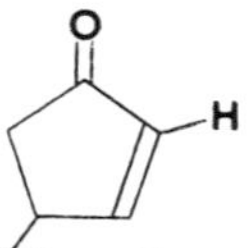 O, H, H	-2.1	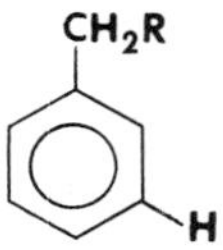 CH_2R, H	≃0.4
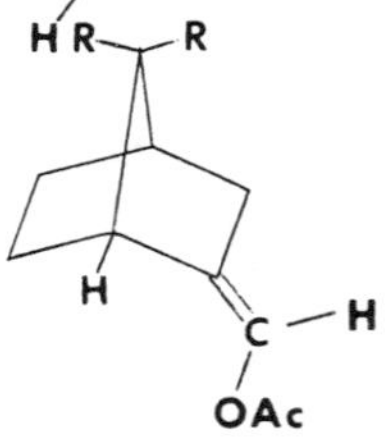 H R, R, H, C—H, OAc	0.7 to 0.8	 CH_2R, H	0.5 to 0.6

Refs. p. 333.

iii. Coupling Through Extended π-Systems

Type	J
$CH_2{=}C{=}CH_2$	-7.37
$CH_2{=}C{=}C{=}CH_2$	+7.01
$CH_2{=}C{=}C{=}C{=}CH_2$	-5.30
$C\underline{H}_2{=}C{=}CH{-}CH{=}C{=}C\underline{H}_2$	+1.73
H-C=C-C(=C)-H	-0.89
H-C≡C-C(=C)-H	-3.18
$\underline{H}$-C≡C-C$\underline{H}$=CH_2	-2.17
H-C=C-C=C-H	0.95
H-C≡C-C(=O)-C-H	0.59
H-C≡C-C(=C)-C-H	0.54
H-C≡C-C≡C-C-H	-1.07
H-C=C-C=C-C-H	-0.71
H-C=C-C=C-C=C-H	0.51
H-C≡C-C≡C-C≡C-H	0.77
H-C-C≡C-C≡C-C-H	0.73
H-$(C{\equiv}C)_3$-C-H	-0.44
H-$(C{=}C)_3$-C-H	-0.30
H-$(C{=}C)_4$-H	0.24
H-C-$(C{\equiv}C)_3$-C-H	0.31
H, H (cyclohexene ring)	1.04

Type		J
H, H (ring)		1.11
(benzene ring)	ortho	6.0 to 9.4
	meta	1.2 to 3.1
	para	0.2 to 1.5
(pyridine ring: 2, 3, 4, 5, 6, N)	2-3	4 to 5.7
	3-4	6.8 to 9.1
	2-4	0 to 2.5
	3-5	0.5 to 1.8
	2-5	0 to 2.3
	2-6	0 to 0.6
(furan ring: 2, 3, 4, 5, O)	2-3	1.7 to 2.0
	3-4	3.3 to 3.8
	2-4	0.4 to 0.9
	2-5	1 to 2
(pyrrole ring: 2, 3, 4, 5, N-H)	2-3	2 to 3
	3-4	3 to 4
	2-4	1 to 2
	2-5	1.5 to 2.5
(thiophene ring: 2, 3, 4, 5, S)	2-3	4.7 to 5.5
	3-4	3.3 to 4.0
	2-4	1.0 to 1.5
	2-5	2.8 to 3.5
(naphthalene: 1, 2, 3, 4)	1-2	8.1 to 9.1
	2-3	6.0 to 7.0
	1-3	1.1 to 1.6
	1-4	0 to 1

Refs. p. 333.

E. Carbon-13 Magnetic Resonance (CMR)

Carbon in its natural composition contains 1.108% of the magnetic isotope ^{13}C. Carbon-13 has a spin quantum number of 1/2 but has a much lower sensitivity than hydrogen (see "Magnetic Properties of the Elemental Isotopes" on page 313).

CMR has become a routine tool only through the recent development of more sensitive spectrometers (mostly through higher field strengths) combined with computer-averaged spectra, Fourier transform methods, and complete decoupling of all protons (resulting in enhancement of signal both through collapse of all multiplets to singlets and through the nuclear Overhauser effect). These techniques are defined in the "Special Techniques" section below.

Of course, it is also possible (and much easier) to study ^{13}C-enriched compounds. Basic compounds from which more complex substances can be synthesized are readily available with enrichment up to 90%. Basic suppliers are 5, 11, 12, 14, 16, 17, 18, 22, 23, and 24.

The rapid growth of cmr is evidenced by the numerous reviews and collections of data that are appearing in the literature (32).

1. Reference Compounds

Several different substances have been used as shift references for cmr; the most common of these is carbon disulfide. It now appears, however, that tetramethylsilane (TMS) will be the accepted standard. The following chemical shifts are given in reference to both CS_2 and TMS. In the rest of this section all shifts were converted to the TMS scale [δ (TMS)] using the following formula:

$$\delta_{(TMS)} = 192.8 - \delta_{(CS_2)}$$

where a positive value refers to a downfield shift (as in the case of proton resonance). The opposite is true when reference to CS_2 is made.

The following substances have been used as references; however, their chemical shifts are solvent-dependent and may vary as much as 2 to 3 ppm. For a discussion of solvent effect on ^{13}C chemical shift of TMS see reference 11c.

Compound	$\delta_{(TMS)}$	$\delta_{(CS_2)}$
Tetramethylsilane	0	192.8
Carbon disulfide	192.8	0
Dioxane	67	126
Benzene	128.7	64.1
Cyclohexane	27.7	165.1
Methyl iodide	-20.5	213.3
CH_3CO_2H	20	173
K_2CO_3 (aqueous)	172	21

2. General Regions of Chemical Shifts

The following general chart was constructed from a large number of observed shifts obtained from numerous sources. Most of the data had been collected by Dr. Marjorie Malmberg of the National Bureau of Standards who generously made them available to the authors. These values are all in ppm relative to TMS (bottom scale) and CS_2 (top scale). Attention is called to two references, one on cmr spectra of steroids (33) and one on cmr of amino acids (34).

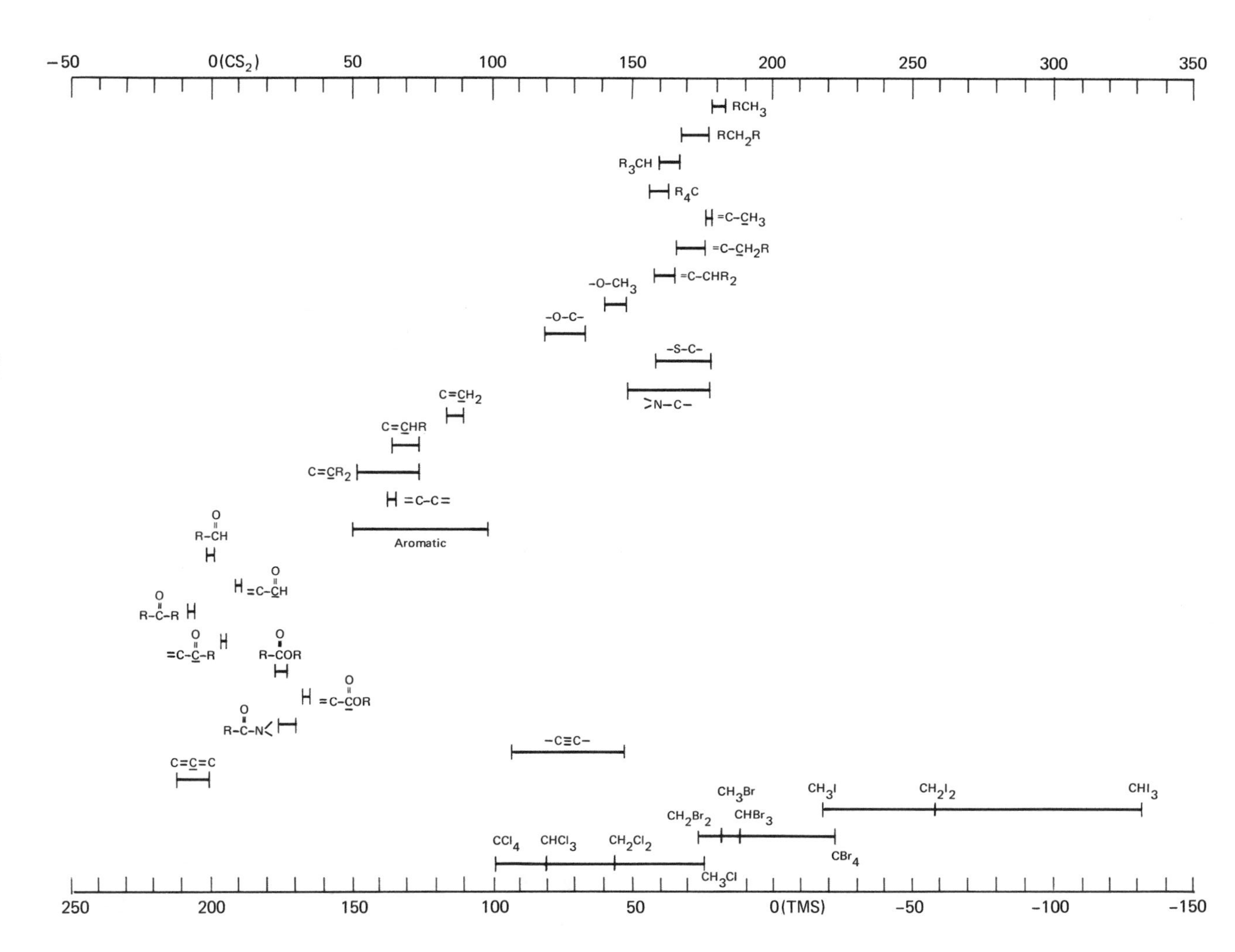

Refs. p. 333.

3. Chemical Shift Correlation Tables

The following tables are presented in order to illustrate some general trends and give absorption data for some common substances. Often the chemical shifts were reported in the literature with no mention of solvent (in many cases the spectra were obtained on neat liquids) or with no mention of (or no correction for) internal or external referencing. In those cases when TMS was not used as the standard, there is some error involved in transforming the reported shifts to shifts relative to TMS.

a. Sp³ Carbon

Compound	δ (TMS)	Compound	δ (TMS)
CH_4	2.1	$CHCl_3$	81
CH_3CH_3	5.9	CCl_4	98
$CH_3CH_2CH_3$	C1 15.6 C2 16.1	CH_3Br	14
$(CH_3)_3CH$	C1 24.3 C2 25.2	CH_3I	-21
$(CH_3)_4C$	C1 31.5 C2 27.9	CH_3OH	49.3
Heptane	C1 14.0 C2 23.0 C3 32.4 C4 29.6	CH_3OCH_3	59.4
Cyclopropane	-2.2	$\underline{C}H_3OPh$	54
Cyclohexane	27.7	$\underline{C}H_3OCOPh$	51.3
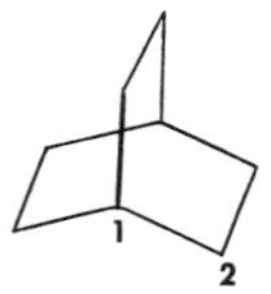 (bicyclic structure, positions 1, 2)	C1 24.7 C2 26.8	$\underline{C}H_3COPh$	24.9
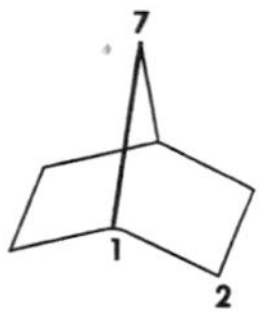 (bicyclic structure, positions 1, 2, 7)	C1 36.8 C2 30.1 C7 38.7	$\underline{C}H_3COCH_3$	29
CH_3Cl	22	$\underline{C}H_3Ph$	21.0
CH_2Cl_2	55	$\underline{C}H_3CN$	0.5
		$CH_3N(CH_3)_2$	47.5
		CH_3NO_2	57.3
		CH_3SCH_3	19.5
		CH_3SOCH_3	43.5
		1-Hexanol	C1 62.2 C2 33.1 C3 26.1 C4 32.3 C5 23.1 C6 14.5

Refs. p. 333.

(Continued)

Compound	δ (TMS)	Compound	δ (TMS)
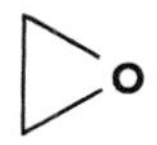 O	39.7	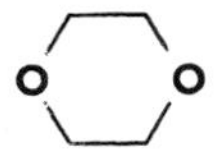 O O	67.8
 N—	28.7	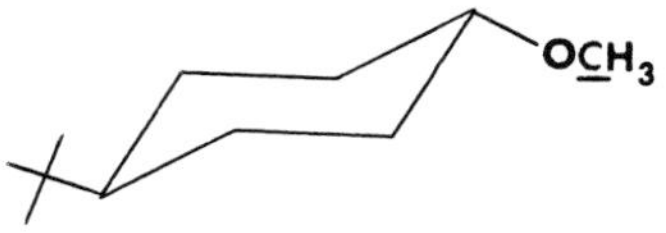 $O\underline{C}H_3$	80.1
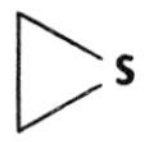 S	18.9	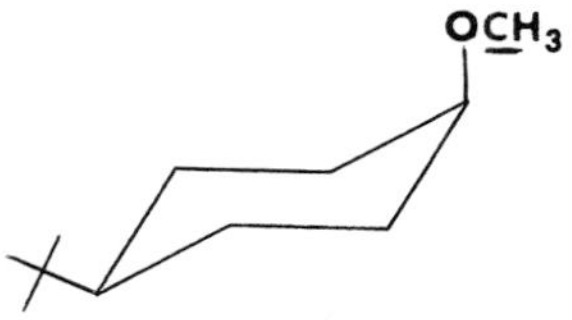 $O\underline{C}H_3$	75.2

b. Sp^2 Carbon

Compound	δ (TMS)	Compound	δ (TMS)
Alkenes			
$CH_3\underset{2}{C}H=\underset{1}{C}H_2$	C1 115.9 C2 136.2	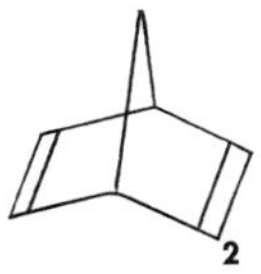 2	C2 144
EtCH=CHEt	cis 130.3 trans 130.4	$CH_2=\underset{2}{C}H-CH=\underset{1}{C}H_2$	C1 116.6 C2 137.2
CHCl=CHCl	cis 118.4 trans 120.2	Aromatics	
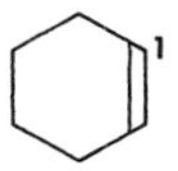 1	C1 127.2		128.7
cis-Cyclooctene trans-Cyclooctene	130 134	4, 3, 2, N	C2 149.7 C3 123.6 C4 135.5
2	C2 137		

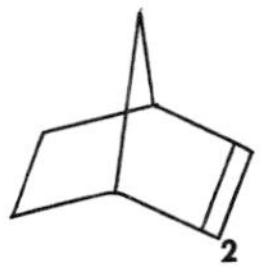

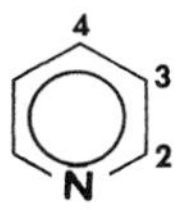

(Continued)

Refs. p. 333.

Compound	δ (TMS)
(pyrazine structure, N / N)	145.2
(1,3,5-triazine structure, N / N / N)	166.8
(furan structure, O, positions 2, 3)	C2 142.8 C3 109.8
(pyrrole structure, N, positions 2, 3)	C2 118.7 C3 108.4
(thiophene structure, S, positions 2, 3)	C2 125.6 C3 127.4
Carbonyls	
(cyclohexanone structure, O, position 1)	C1 208.9
(cyclohexenone structure, O, position 1)	C1 197.1
(cyclopentanone structure, O, position 1)	C1 217.2
(cyclobutanone structure, O, position 1)	C1 206.9

Compound	δ (TMS)
(norbornan-2-one structure, O, position 2)	C2 214.1
$CH_3\underline{\overset{O}{\overset{\parallel}{C}}}CH_3$	205.1
$CH_3CH_2\underline{\overset{O}{\overset{\parallel}{C}}}CH_3$	207.1
$CH_2{=}CH{-}CH_2\underline{C}OCH_3$	204.7
$CH_3CH{=}CH{-}\underline{C}OCH_3$	196.5
(cyclopropyl)$-\underline{\overset{O}{\overset{\parallel}{C}}}-CH_3$	205.9
$Ph\underline{C}OCH_3$	196.0
$CH_3\underline{C}O_2H$	179
$CH_3CH_2\underline{C}O_2H$	180.2
$Ph\underline{C}O_2H$	172.6
$CH_3CH_2\underline{C}O_2CH_3$	173.3
$CH_3\underline{C}O_2CH_3$	170.7
$Ph\underline{C}O_2CH_3$	165.8
$CH_3CH_2\underline{C}HO$	201.8
$Ph\underline{C}HO$	191.0
$CH_3\underline{C}OCl$	170.6
$CH_3\underline{C}OBr$	169.6
$CH_3\underline{C}ON(CH_3)_2$	169.4
$H_2NCH_2\underline{C}O_2H$	173.2
$H_2NCH(CH_3)\underline{C}O_2H$	176.5

(Continued)

Refs. p. 333.

Compound	δ (TMS)	Compound	δ (TMS)
Carbonium Ions		$(Ph)_3\underline{C}^+$ (SO_2/SbF_5)	210.9
$(CH_3)_3\underline{C}^+$ (SO_2ClF/SbF_5)	328.2	$(Ph)_2\underline{\overset{+}{C}}H$ (SO_2/SbF_5)	198.4
$(CH_3)_2\underline{\overset{+}{C}}H$ (SO_2ClF/SbF_5)	317.8		

c. Monosubstituted Benzenes: 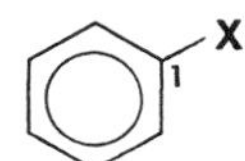

X	C-1	Ortho	Meta	Para
CH_3	137.8	129.0	129.0	125.9
CH_2OH	141.0	127.3	127.3	127.3
CHO	137.7	129.9	129.9	134.7
$COCH_3$	138.0	128.9	128.9	132.9
COCl	134.5	131.3	129.9	136.1
C_6H_5	138.7	129.1	129.1	129.1
F	163.8	114.4	129.6	124.3
Cl	135.1	128.9	129.7	126.7
Br	123.3	132.0	130.9	127.7
I	96.4	138.6	131.3	128.3
OH	155.6	116.4	130.4	121.4
OCH_3	158.9	114.0	129.6	120.6
OC_6H_5	157.9	119.3	130.3	123.6
$OCOCH_3$	151.7	122.3	130.0	126.4
O^-	168.3	120.5	130.6	115.1
NH_2	147.9	116.3	130.0	119.2
$N(CH_3)_2$	151.3	113.0	129.5	116.9
$N(C_2H_5)_2$	147.8	113.1	129.7	117.2
$NHCOCH_3$	139.8	118.8	128.9	123.1
NO_2	148.3	123.4	129.5	134.7

d. Sp Carbon

Compound	δ (TMS)	Compound	δ (TMS)
$C_2H_5\underset{2}{C}{\equiv}\underset{1}{C}-H$	C1 67.3 C2 85.0	$CH_3C{\equiv}\underline{C}CH_3$	73.9
$C_4H_9\underset{2}{C}{\equiv}\underset{1}{C}H$	C1 80.0 C2 85.8	$Ph\underset{2}{C}{\equiv}\underset{1}{C}H$	C1 77.6 C2 83.9

(Continued)

Refs. p. 333.

Compound	δ (TMS)	Compound	δ (TMS)
$C_4H_9\underset{2}{C}\equiv\underset{1}{C}Cl$	C1 56.7 C2 68.8	$CH_3CH=\underline{C}=CHCH_3$	206.2
		HCN	109.1
$C_4H_9\underset{2}{C}\equiv\underset{1}{C}Br$	C1 38.4 C2 79.8	$CH_3\underline{C}N$	117.8
$C_4H_9\underset{2}{C}\equiv\underset{1}{C}I$	C1 -3.3 C2 96.8	CN^-	168.8
		$Ph\underline{C}\equiv N$	118.7
$H_2C=\underline{C}=CHCH_3$	209.5		

4. Coupling Constants

Because of the low natural abundance of ^{13}C nuclei, the probability of finding two of them adjacent in the same molecule is very small and thus ^{13}C-^{13}C coupling is not usually observed. For an exception, see Ihrig and Marshall [J. Amer. Chem. Soc., 94, 1756 (1972)] who studied ^{13}C-7 labeled monosubstituted benzenes. On the other hand, coupling to hydrogen, J_{CH}, is often observed and has been extensively studied. For a theoretical treatment and a large collection of ^{13}C-H (as well as C-C, C-F, and C-N) coupling constants, see reference 35.

One bond C-H couplings ($^1J_{CH}$) have often been stated to be proportional to the hydridization of the carbon atom. This is approximately so, especially within a series of closely related compounds such as ethane, ethylene, and acetylene (see below). The following data are in Hz and are primarily from reference 35 as well as the abovementioned general references for cmr.

a. ^{13}C-H Coupling

System	$^1J_{CH}$	System	$^1J_{CH}$
CH_4	125.0	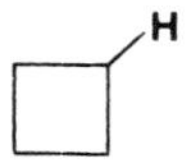	
CH_3CH_2-H	124.9	H	134
CH_2F-H	149.1		
CHF_2-H	184.5	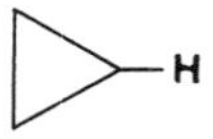	
CF_3-H	239.1	H	161
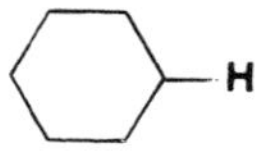 H	123	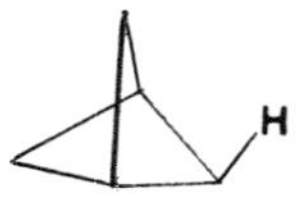 H	144

Refs. p. 333. (Continued)

System	$^{1}J_{CH}$	System	$^{1}J_{CH}$
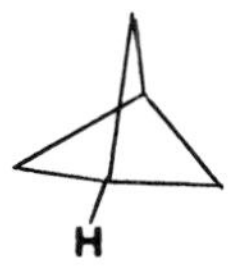 H	164	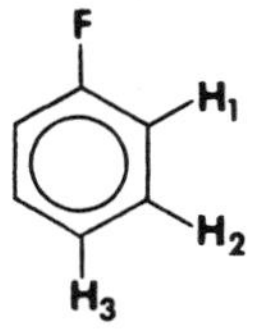 F, H_1, H_2, H_3	1 155 2 163 3 161
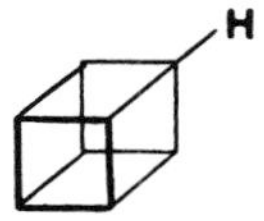 H	160	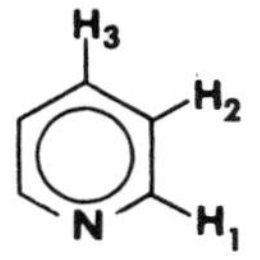 H_3, H_2, N, H_1	1 170.0 2 163.0 3 152.0
CH_2=CH-H	156.2		
CH_2=CF-H	200.2	HC≡C-H	249.0
H-CO-H	172.0	PhC≡C-H	251.0
CH_3CO-H	172.4	$HOCH_2$C≡C-H	248.0
$(CH_3)_2$NCO-H	191.2	N≡C-H	269.0
CH_3OC(=O)-H	226.2	HN^{+}≡C-H	320.0
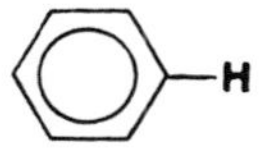 —H	158.5		

b. ^{13}C-X Coupling

System	$^{1}J_{C-C}$	System	$^{1}J_{C-F}$
CH_3CH_3	34.6	CH_3-F	157.5
CH_3CF_3	39.3	CFH_2-F	234.8
CH_3CO_2H	56.7	CF_2H-F	274.3
	57.0	CF_3-F	259.2
CH_3CN	56.5	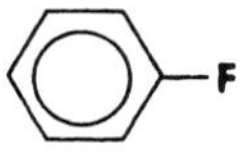 —F	244.0
CH_2=CH_2	67.6	CH_2=CF-F	271.0
HC≡CH	171.5	HC(=O)-F	369.0

(Continued)

Refs. p. 333.

System	$^{1}J_{C-^{15}N}$
CH_3NH_2	-4.5
$C_6H_5CH{=}N{-}CH_3$ (trans)	7.1
$CH_3C{\equiv}N$	-17.5
CN^-	5.9

F. Fluorine-19 Magnetic Resonance

Fluorine-19 (the only natural isotope of fluorine) has a spin quantum number of 1/2 and a high sensitivity (see "Magnetic Properties of the Elemental Isotopes" below) making it an easy element to study by nmr. ^{19}F absorbs over a much greater range than hydrogen, making it useful for detecting subtle structural variations. General discussions of ^{19}F nmr can be found in references 1b (p. 211), 1c (p. 317), 1d (p. 317), and 36. A comprehensive collection (covering the years 1951 to mid-1967) of ^{19}F chemical shifts has also been published (37a); another such collection (37b) extends through the end of 1967 and includes a review of the field. It was from these sources that the following data were taken.

1. Reference Compounds

There is no universally accepted standard for ^{19}F chemical shifts although most data are now reported referenced to trichlorofluoromethane. Chemical shifts in ppm referenced to CCl_3F are called Φ values if extrapolated to infinite dilution and Φ* if not (in which case Φ* and δ have the same meaning). Note that this results in positive values for upfield shifts with CCl_3F being at the low field end of the spectrum. The following are Φ* values except where noted otherwise.

Compound	δ (CCl_3F)
CCl_3F	0.00
CF_3CO_2H	76.53 (Φ)
C_6F_6	162.9
C_4F_8	134.92 (Φ)
CF_4	62.3
CCl_2FCCl_2F	67.78 (Φ)
$C_6H_5CF_3$	63.73 (Φ)
F_2	-422.9

Refs. p. 333.

2. General Regions of Chemical Shifts

A positive value refers to a higher field strength.

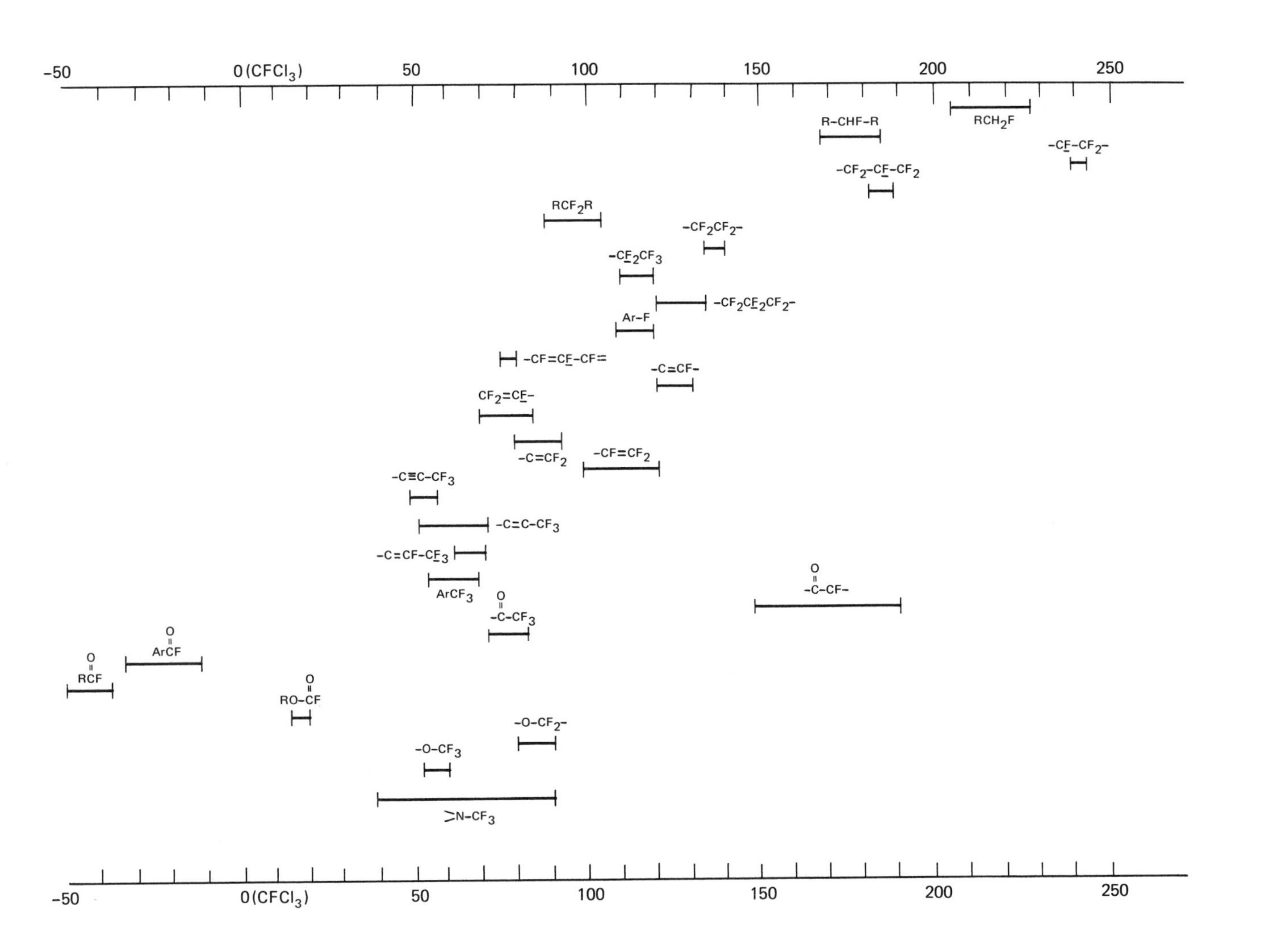

Refs. p. 333.

3. Chemical Shift Correlation Tables

The following table illustrates some general trends and gives absorption data for some common substances. These values are mostly Φ* values (see above) in CCl_3F (where the value is a result of extrapolation to infinite dilution, this is noted with a Φ). Positive values refer to higher field than for CCl_3F.

Compound	δ (CCl_3F)	Compound	δ (CCl_3F)
KF (aq.)	125.3	$CF_3(CF_2)_5CF_3$	F1 81.6 F2 126.7 F3 123.1 F4 122.2
$F_3B{\cdot}BH_3$	146.5		
$F_3B{\cdot}O(Et)_2$	153.0		
$F_3B{\cdot}O(Me)_2$	158.3	Cyclohexane-F_{12}	133.25 (Φ)
F_3SiPh	134.8	Cyclobutane-F_8	134.92
F_3SiEt	136.4	Cyclopropane-F_6	151
SiF_4	163.3	$CF_2{=}CF_2$	132.84 (Φ)
NF_3	-146.9	CHF=CHF	cis 165.0 trans 185.0
SOF_2	-74.5		
SF_6	-57.0	PhF	113.1
PF_5	72	$PhCF_3$	63.73 (Φ)
PF_3	34.0	C_6F_6	162.9
CH_3F	271.9	$CF_2{=}CF{-}\underset{2}{CF}{=}\underset{1}{CF_2}$	F1 105 F2 175
CH_2F_2	143.4	$C\underline{F}_3OF$	71.1
CHF_3	78.6	CF_3SH	31.9
CF_4	63.3	CF_3SPh	43.2
CH_3CF_3	65	CF_3OCF_3	58.3
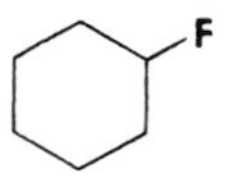 F	ax 186.0[a] eq 165.5[a] avg 175.0	CF_3OPh	58.4
		FCN	157
		FC≡CH	94.6
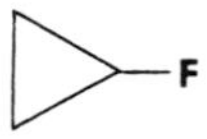 F	218.0		

[a]At -90°C.

4. Coupling Constants

Because of the large number of partially and fully fluorinated compounds, both ^{19}F-^{19}F and ^{19}F-^{1}H couplings are important.

Refs. p. 333.

a. $^{19}F-^{19}F$ Coupling. These constants are strongly dependent on geometry, although this dependence is not as well understood as in proton-proton couplings. J_{FF} does not necessarily decrease (in fact, it often increases) with larger number of bonds separating the fluorines. This is illustrated in several examples below. For recent evidence concerning long-range "through-space" H-F coupling, see E. Abushanab, J. Amer. Chem. Soc., 93, 6532 (1971), and I. Agranat et al., ibid., 94, 2889 (1972).

The expected signs of $^{19}F-^{19}F$ coupling have been summarized (1c) as

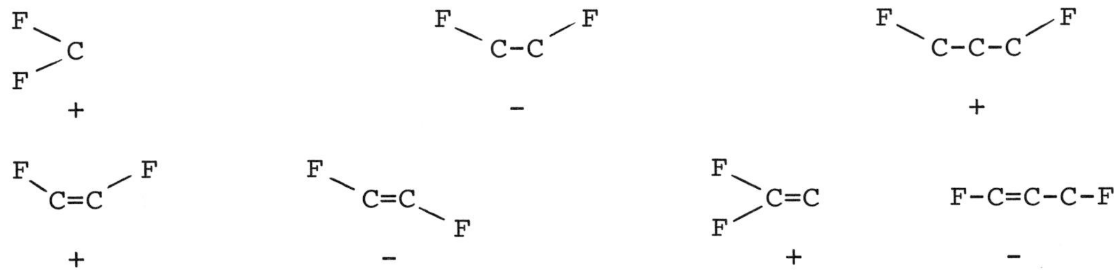

Structure	Sign
F(F)C (geminal)	+
F-C-C-F	-
F-C-C-C-F	+
F-C=C-F (cis)	+
F-C=C-F (trans)	-
$F_2C=C$	+
F-C=C-C-F	-

Mooney and Winson (36) have summarized J_{FF} values for fluorinated cyclohexanes as follows:

J_{FF} (gem)	= 290-305 Hz		$J_{F_1F_3}$ (diax)	∿	27 Hz
$J_{F_1F_2}$ (gauche)	∿ 14 Hz		$J_{F_1F_3}$ (dieq)	∿	9 Hz
$J_{F_1F_2}$ (trans)	= 0-3 Hz		$J_{F_1F_3}$ (ax-eq)	∿	1 Hz

Other J_{FF} values are illustrated below.

System	J_{FF}	
CF_3CF_2H (2, 1)	1-2	2.8
CF_3CFH_2 (2, 1)	1-2	15.5
Perfluoromethylcyclohexane ring: CF_3 (1), F (2), F_2 (3)	1-2	6.1
	1-3	14.2
$CF_3CF_2CF_2H$ (3, 2, 1)	1-2	≤ 1
	2-3	4
	1-3	≤ 1
$CF_3CF_2CF_3$ (3, 2, 1)	1-2	< 1
	1-3	7.3
$CF_3CF_2CF_2CO_2H$ (3, 2, 1)	1-2	≤ 1
	2-3	≤ 1
	1-3	9.0

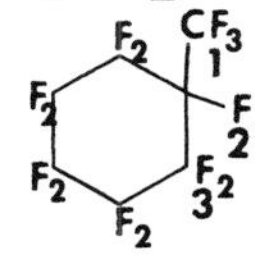

System	J_{FF}	
$FO-CF_2CF_2-C(=O)F$ (4, 3, 2, 1)	1-2	8.4
	2-3	3.2
	1-3	5.1
	3-4	7.0
	2-4	3.5
$C_6H_4F_2$ (F on benzene ring)	ortho	20.2-20.8
	meta	0- 7
	para	0-15
HFC=CF₂ (H, F / F, F)	gem	87
	cis	33
	trans	119
CF_3(4)(F 3)C=C(F 2)(F 1)	1-2	57
	1-3	39
	2-3	116
	1-4	8
	2-4	22
	3-4	13
	4-4	157

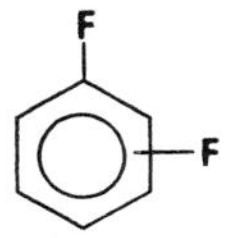

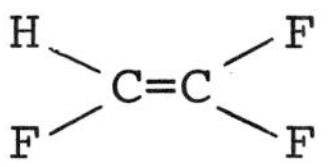

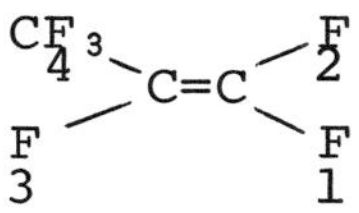

Refs. p. 333.

b. H-^{19}F Coupling. These couplings more closely resemble the proton-proton coupling in that they generally decrease with increasing number of bonds separating the nuclei. J_{HF} values are also strongly dependent on geometry. Mooney and Winson (36) have summarized ^{1}H-^{19}F couplings in fluorocyclohexanes, as follows:

J_{HF} (gem)	$\sim$ 45 Hz	$J_{H_1F_2}$ (trans)	$\sim$ 17 Hz
$J_{H_1F_2}$ (cis)	$\sim$ 6 Hz	$J_{H_1F_3}$	$\sim$ 4 Hz

The expected signs of ^{1}H-^{19}F couplings have been summarized (1c) as

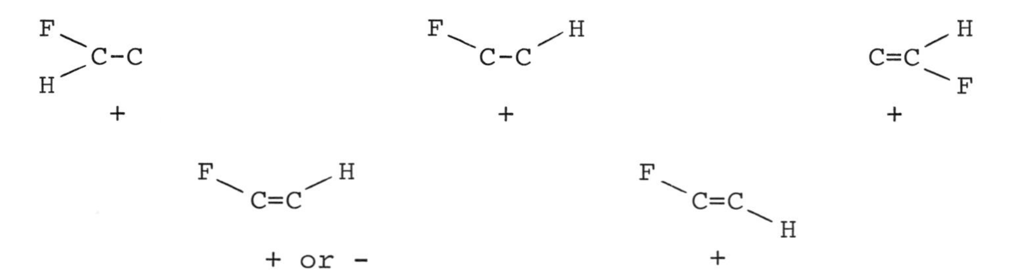

Additional J_{HF} values are illustrated below.

System	J_{HF}	System	J_{HF}
$\underset{2}{CH_3}\underset{1}{CH_2}F$	1-F 47.5 2-F 25.7	Fluorocyclohexane (F, H_1, H_2)	$1\text{-}F_{ax}$ 49 $2_{ax}\text{-}F_{ax}$ 43.5 $2_{eq}\text{-}F_{ax}$ 3 $1\text{-}F_{eq}$ 49 $2_{ax}\text{-}F_{eq}$ <3 $2_{eq}\text{-}F_{eq}$ <3
$\underset{2}{CH_3}\underset{1}{CHF_2}$	1-F 57.2 2-F 20.8		
CH_3CF_3	12.8		
$\underset{2}{CF_3}CH_2\underset{1}{F}$	H-F1 45.5 H-F2 8.0		
$\underset{2}{CF_3}\underset{1}{CF_2}H$	H-F1 52.6 H-F2 2.6	H_3, H_1 / C=C / H_2, F	1-F 84.7 2-F 20.1 3-F 52.4
CF_3CH_2Cl	8.5		
CF_3CH_2Br	9.0	H, F_2 / C=C / F_1, F_3	H-F1 72 H-F2 <3 H-F3 12
CF_3CH_2OH	8.9		

(Continued)

Refs. p. 333.

System	J_{HF}	System	J_{HF}
H(H)C=C(F)(F)	cis ∿1 trans 34	H–C_6H_4–F (ortho)	6.2-10.1
H(F)C=C(H)(F)	gem 72.7 trans 20.4	H–C_6H_4–F (meta)	6.2-8.3
H(F)C=C(F)(H)	gem 74.3 cis 4.4	H–C_6H_4–F (para)	2.1-2.3
CH_3 (3), H (2) C=C F, H (1)	1-F 89.9 2-F 41.8 3-F 2.6	F-C≡C-H	21
CH_3 (3), H (2) C=C H (1), F	1-F 84.8 2-F 19.9 3-F 3.3		

G. Phosphorus Magnetic Resonance

Phosphorus-31 (the only natural isotope of phosphorus) has a spin quantum number of 1/2 and a sensitivity of about 0.07 that of hydrogen, making it a relatively easy nucleus to study. This, combined with its importance in biochemistry, has made it the subject of numerous investigations. The only extensive review to appear is a recent book (43); however, a short general review (1c, Vol. 2, p. 1052) and a review of P-31 resonance in coordination compounds (44) have also been published. The magnetic resonance of phosphorus compounds using nuclei other than phosphorus has been published (45).

1. Reference Compounds

The most common standard for P-31 resonance has been 85% aqueous phosphoric acid. The high reactivity of this substance limits its use in almost all cases to an external standard. More recently P_4O_6 has come to be preferred (chemical shift = 112.5 ppm lower field than 85% H_3PO_4) and it is to this substance that the following values are referenced. Conversion from phosphoric acid standard was necessary in most cases.

Refs. p. 333.

2. General Regions of Chemical Shifts

A positive value refers to a higher field strength and all values are referenced to P_4O_6. This chart is reproduced from reference 1a.

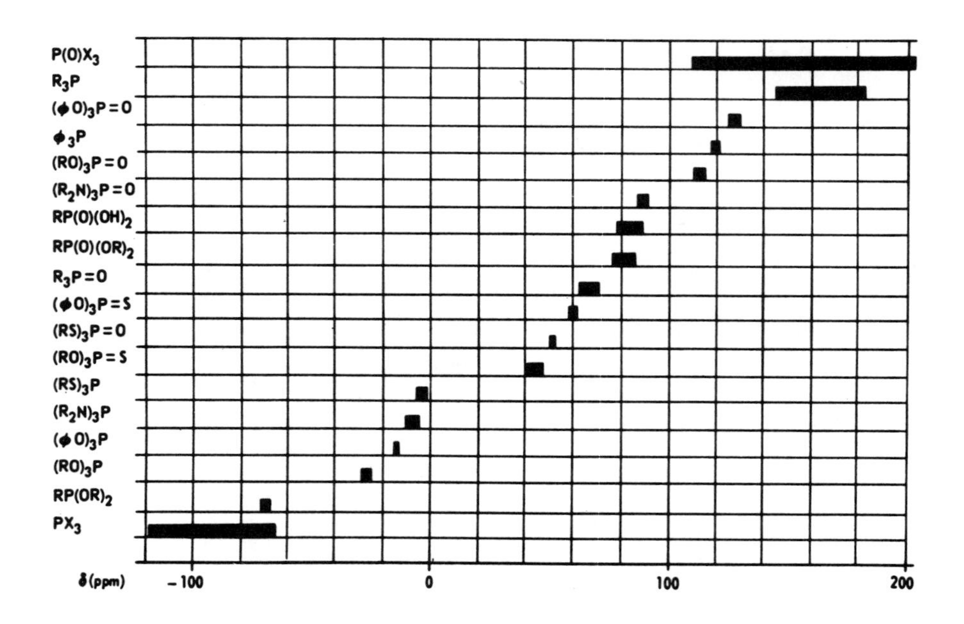

3. Chemical Shift Correlation Table

The following table illustrates some general trends and gives absorption data for some specific compounds. All values are referenced to P_4O_6 (correction from 85% H_3PO_4 was necessary in most cases), and a positive value refers to a higher field strength.

Compound	δ (ppm)	Compound	δ (ppm)
P_4	573	$(C_2H_5)_3P$	133
PH_3	352	$(Ph)_3P$	118
PF_3	16	$_{\alpha}P(-O-CH_2-)_3P_{\beta}$ (O-CH$_2$ / αP-O-CH$_2$-Pβ / O-CH$_2$)	α 23 β 180
PCl_3	-107		
PBr_3	-116		
PI_3	-65		
CH_3PH_2	276	$C_2H_5PCl_2$	-83
$C_2H_5PH_2$	241	$PhPCl_2$	-47
$(CH_3)_2PH$	212	CH_3PBr_2	-71
$(CH_3)_3P$	175		

(Continued)

Refs. p. 333.

Compound	δ (ppm)	Compound	δ (ppm)
$C_2H_5PBr_2$	-81	$(PhO)_3P$	-14
$PhPBr_2$	-40	POF_3	148
CH_3PI_2	-17	POF_2Cl	128
$(CH_3)_2PCl$	17	$POFCl_2$	113
$(C_2H_5)_2PCl$	-6	$POCl_3$	110
$(Ph)_2PCl$	33	$PO(CH_3)F_2$	86
$(CH_3)_2PBr$	23	$PO(Ph)F_2$	101
$(C_2H_5)_2PBr$	-3	$PO(CH_3)_2F$	47
$(Ph)_2PBr$	43	$PO(CH_3)_3$	76
$CH_3OP(Ph)_2$	-3	$PO(C_2H_5)_3$	64
$CH_3OP(CH_3)_2$	-88	$PO(Ph)_3$	88
$(C_2H_5O)_2P$-O-	-35	$PO(OCH_3)_2CH_3$	81
-O-P(OC_2H_5)-O-	-15	$HOP_{\alpha}(O)(H)O\left[P_{\beta}(O)(H)O\right]_m P_{\alpha}(O)(H)OH$	α 120 β 127
$(CH_3O)_3P$	-27		
$(C_2H_5O)_3P$	-24		

4. Coupling Constants

No general correlation of phosphorus coupling constants with structure or stereochemistry has yet been developed.

Compound	J (Hz)	Compound	J (Hz)
PH_3	PH 179	H_2PPH_2	PP 108.2 PH 186.5 PPH 11.9
CH_3PH_2	PH 207 PCH 4		
		CH_3PBr_2	PH 19.3
$(CH_3)_2PH$	PH 207		

(Continued)

Refs. p. 333.

Compound	J (Hz)	Compound	J (Hz)
$P(CH_3)_3$	PH 2.66	PF_2Cl	PF 1390
$O-CH_2$ $P-O-CH_2-P$ $O-CH_2$	PCH 8.9 POCH 2.6	$PFCl_2$	PF 1320
		POF_3	PF 1080
		$POFCl_2$	PF 1190
$P(C_2H_5)_3$	PCH 13.7 PCCH 0.5	$PO(CH_3)_2F$	PF 990
$P(CH=CH_2)_3$	PH alpha 11.7 PH trans 30.2 PH cis 13.6	$PO(Ph)F_2$	PF 1115
		$ClCH_2PF_2$	PF 1196
$CH_3OP(CH_3)_2$	PCH 8.3 POCH 10.9	$HPO_2PO_3^-$	PP 480
		$(C_2H_5O)_2P(O)H$	PH -686
PF_3	PF 1410		

H. Nitrogen Magnetic Resonance

Nitrogen is made up of two magnetic isotopes, ^{14}N(99.635%) and ^{15}N(0.365%). Severe difficulties are encountered in the nmr study of both. ^{14}N (spin quantum number = 1), although of very high natural abundance, has a very low sensitivity and a quadrupole moment which results in substantial line broadening (limiting the observation of coupling) and large errors in chemical shift measurements (over 50 ppm in some cases). On the other hand, coupling across ^{14}N between other nuclei can be observed; for instance, the N-methyl group in N-methylacetamide is clearly coupled both to the N-H and the other methyl.

The low natural abundance of ^{15}N (spin quantum number = 1/2) combined with its very low sensitivity almost precludes its detection in other than enriched compounds. On the other hand, the potential of nitrogen magnetic resonance in the study of biologically important molecules is such that the future of this technique looks promising. Many compounds are available with ^{15}N enrichment of up to 99% from the following suppliers; 5, 11, 12, 16, 17, 22, 23, and 24. Three recent reviews of nitrogen magnetic resonance have been published (38), and it is from these sources as well as the data collections in reference 39 that most of the following data were obtained.

1. Reference Compounds

No agreement has been reached on the standard compound that should be used for nitrogen magnetic resonance. The most common reference materials are listed here along with the recently suggested (40) tetramethyl and tetraethylammonium salts. The effect of solvent and anion on the chemical shift of these two substances is also discussed in reference 40. The shifts reported are all referenced to the nitrate resonance of acidified ammonium nitrate or nitric acid. Note that this results in positive values for upfield shifts.

Compound	$\delta (NO_3^-)$
NO_3^-	0.0
HNO_3	0
CH_3NO_2	0[a]
$C(NO_2)_4$	48
$(C_2H_5)_4N^+I^-$	312
$(CH_3)_4N^+I^-$	332
NH_3	376

[a]This value has also been reported as -3 (38b).

2. General Regions of Chemical Shifts

A positive value refers to a higher field strength. All values are referenced to nitrate ion and are good for ^{14}N and ^{15}N.

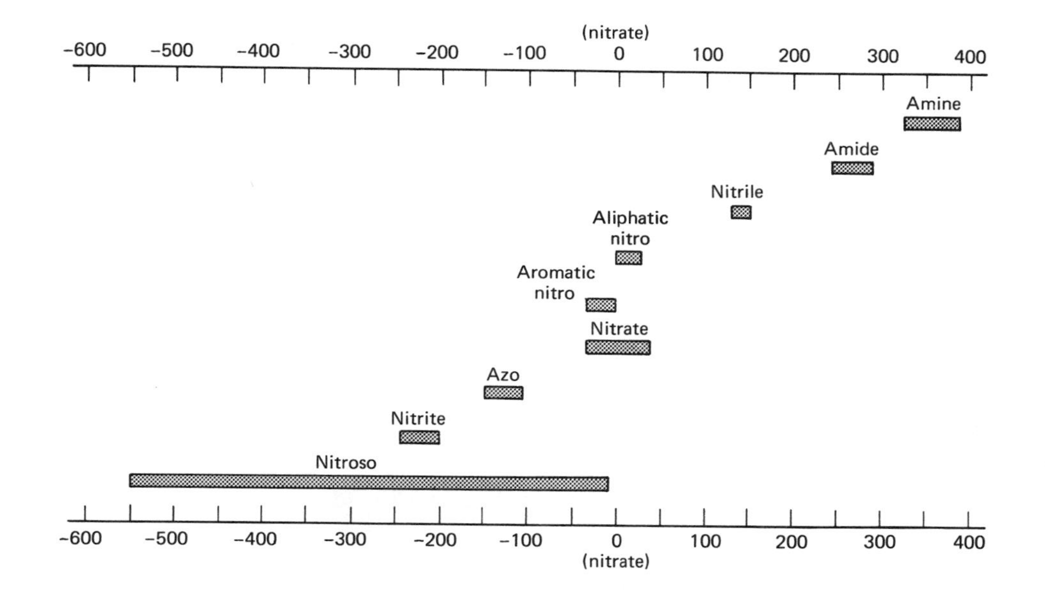

Refs. p. 333.

3. Chemical Shift Correlation Table

The following table illustrates some general trends and gives absorption data for some common substances. These values are all referenced to nitrate ion (correction from other standards was necessary in some cases), and a positive value refers to a higher field strength. The values will be the same for ^{14}N and ^{15}N within experimental error. It should be kept in mind that this experimental error is often several ppm.

Compound	δ	Compound	δ
N_2	14	▷N-H	387.2
NO_3^-	0	$(CH_3)_3N$	365
NO_2^-	-335	$(C_2H_5)_3N$	327
CN^-	112	H_2NNH_2	328.7
NH_3	383	CH_3NHNH_2 (N: 2 1)	1 321.9 2 299.4
CH_3NH_2	378	$HCONH_2$	267.8
$C_2H_5NH_2$	355	CH_3CONH_2	269.6
$C_3H_7NH_2$	355.9	$C_2H_5CONH_2$	272.8
$CH_2=CH-CH_2NH_2$	355.5	$HCONHCH_3$	264.1
▷—CH_2NH_2	352.8	$CH_3CONHCH_3$	268.7
$C_6H_5CH_2NH_2$	352.4	$C_2H_5CONHC_6H_5$	243.3
$(CH_3)_2CHNH_2$	332.2	CH_3CSNH_2	238.9
▷—NH_2	346.5	$CH_3CSNHC_6H_5$	210.9
⬠—NH_2	337.3	H_2NCONH_2	299.1
$(CH_3)_3CNH_2$	317.2	pyridine	74
$C_6H_5NH_2$	319	pyridinium ion	170
$(CH_3)_2NH$	371	pyrrole	230
$(C_2H_5)_2NH$	327.9	indole	251

(Continued)

Refs. p. 333.

Compound	δ	Compound	δ
pyrimidine	73	C_2H_5NC	203
pyrazine	41	CH_3NO_2	0
$^6NH_2$, 1N, 3N, 7N, 9N–R (purine)	1 144.7	$C_2H_5NO_2$	-12
	3 135.6	$(CH_3)_2CHNO_2$	-24
	6 282.8	$(CH_3)_3CNO_2$	-30
	7 129.5	$C(NO_2)_4$	48
	9 191.6	$C_6H_5NO_2$	8
Quinoline	72	C_2H_5ONO	37.5
Isoquinoline	64	C_6H_5NO	-546
Benzthiazole	60	$C_2H_5C(CH_3)=NOH$	80
CH_3CN	137	NH_4^+	353
C_2H_5CN	137	$CH_3NH_3^+$	342
$CH_2=CHCN$	124		
C_6H_5CN	124		
CH_3NC	218		

4. Coupling Constants

The quadrupole moment of ^{14}N precludes observations of spin-spin coupling to other nuclei in most cases. However, ^{15}N-X coupling can be converted to ^{14}N-X coupling by the equation $J(^{14}N\text{-}X) = -0.713 \times J(^{15}N\text{-}X)$, where -0.713 comes from γ_{14}/γ_{15}. The following data are all for ^{15}N, although the original observation may have been for ^{14}N (conversion by multiplication by -1.402). Nitrogen-hydrogen couplings are dependent on geometry (see reference 41 for a discussion of angular dependence of vicinal coupling in amino acids) and hybridization of the nitrogen (42).

System	$J_{^{15}N-H}$	System	$J_{^{15}N-H}$
NH_3	61.2	Pyridine (ring H labelled H_1, H_2, H_3)	N-H1 10.76 N-H2 1.53 N-H3 0.21
NH_4Cl	73.2	HCN	8.7
CH_3NH_2	N-H2 1.0	CH_3CN	2.8
$C_6H_5NH_2$	78.5	$H_2NCH_2CO_2H$ (H of CH_2 = 1)	N-H1
H_3–C(=O)–N(H_1)(H_2)	N-H1 88.0 N-H2 92.0 N-H3 19.0	pH 6.33	0.4
CH_3CONH_2	N-H1 1.3	pH 13.68	1.5
HO–N(H_1)(H_2)	N-H1 2.68 N-H2 13.8	$CH_3\overset{+}{N}H_3$ (2 1)	N-H1 75.6 N-H2 0.8
$H_2NC(=O)NH_2$	89	$(CH_3)_4N^+$	0.8
		$(CH_3CH_2)_4N^+$ (2 1)	N-H1 0.3 N-H2 2.7

I. Magnetic Resonance of Other Nuclei (59)

In principle, any of the more than 200 isotopes with magnetic moments (see p. 313) can be studied by nmr. In practice most of these have not been examined for various reasons (low abundance, low sensitivity, etc.). The following is a list of leading references to some of the more commonly studied elements. These references are not necessarily reviews and are meant only to provide a starting point in a search for more data. The best general reference for nmr of other nuclei is that by Emsley, Feeney, and Sutcliffe (1c) referred to below as EFS. In addition, there are compilations of nmr data for compounds of a specific element, not necessarily involving the magnetic resonance of that element. A case in point is The Analytical Chemistry of Sulfur and its Compounds. Part 3. NMR Data of Sulfur Compounds by N. F. Chamberlain and J. J. Reed, Wiley-Interscience, New York, 1971.

1. Boron

Boron consists of two magnetic nuclei, ^{10}B (18.8%, I = 3) and ^{11}B (81.2%, I = 3/2). The latter is studied more because of its higher abundance and greater sensitivity. There is some quadrupole broadening in ^{11}B spectra, but it is not serious. Reference compounds include $BF_3 \cdot O(C_2H_5)_2$ (most common; $\delta = 0.00$); $B(OCH_3)_3$ ($\delta = -18.3$); $BF_3 \cdot O(CH_3)_2$ ($\delta = -0.5$); BF_3 ($\delta = -11.5$); BCl_3 ($\delta = -47.0$); BBr_3 ($\delta = -40.0$ and $NaBH_4$ ($\delta = 42.6$). A positive delta is a shift to higher field. General discussions can be found in:

Refs. p. 333.

EFS (lc), p. 971.
NMR Studies of Boron Hydrides and Related Compounds, G. R. Eaton and W. N. Lipscomb, Benjamin, New York, 1969.
W. G. Henderson and E. F. Mooney, Ann. Rev. NMR Spectros., 2, 219 (1969).

2. Oxygen

Oxygen-17 has a natural abundance of only 0.037%, hence must be studied in enriched compounds. Enriched water and O_2 gas are available from suppliers 5, 11, and 17 (see list below). Water itself ($H_2{}^{17}O$) absorbs at the highest field value and thus makes an ideal reference. General discussions can be found in:

EFS (lc), p. 1042.
B. L. Silver and Z. Luz, Quart. Rev., 21, 458 (1967).

3. Silicon

The only naturally occurring magnetic isotope is ^{29}Si (4.70%, I = 1/2). Because of low abundance and low sensitivity only limited studies have been made. Polymethyldisiloxanes $[(CH_3)_2SiO]_x$ (δ = 0.00) and tetramethylsilane (δ = -22.0) have been used as reference compounds. Some leading references are:

EFS (lc), p. 1048.
G. Englehardt, H. Jancke, M. Mägi, T. Pehk, and E. Leppma, J. Organometal. Chem., 28, 293 (1971).
B. J. Aylett, Advan. Inorg. Chem. Radiochem., 11, 249 (1968).

4. Halogens

Chlorine-35 (75.53%, I = 3/2), chlorine-37 (24.47%, I = 3/2), bromine-79 (50.54%, I = 3/2), bromine-81 (49.46%, I = 3/2), and iodine-127 (100%, I = 5/2) (fluorine is discussed above) are all potentially interesting nuclei for nmr study. The main problem is the very large linewidths of their signals. In spite of this, a number of studies have been made and a general review can be found in:

C. Hall, Quart. Rev., 25, 87 (1971).

5. Lithium

Lithium consists of two magnetic isotopes, ^{6}Li (7.42%, I = 1) and ^{7}Li (92.58%, I = 3/2), but very few studies have been reported. Two recent references are:

R. H. Cox, E. G. Janzen, and W. B. Harrison, J. Mag. Res., 4, 274 (1971).
E. E. Genser, J. Chem. Phys., 54, 4612 (1971).

6. Sodium

Sodium consists of 100% ^{23}Na (I = 3/2) but it too has been studied very little. Two recent references are:

E. E. Genser (see lithium above).
D. H. Haynes, B. C. Pressman, and A. Kowalsky, Biochem., 10, 852 (1971).

Refs. p. 333. Suppliers p. 332.

7. Tin

Tin consists of three magnetic isotopes, ^{115}Sn (0.35%), ^{117}Sn (7.61%), and ^{119}Sn (8.58%) all with I = 1/2. A recent reference reports chemical shift data for 40 organotin compounds [P. G. Harrison, S. E. Ulrich, and J. J. Zuckerman, J. Amer. Chem. Soc., 93, 5398 (1971)].

J. Special Techniques and Applications

This section provides recent, leading references and descriptions of special techniques and applications that are not discussed above. Some of these techniques and applications are covered in the general references (1-3). In those cases, only more recent references or comprehensive reviews are mentioned.

1. Biological Applications

An enormous amount of work has been published on the use of proton magnetic resonance to study biologically important systems. The advent of instrumentation with the ability to study ^{13}C, ^{31}P, ^{14}N, ^{15}N, ^{23}Na, and other nuclei has opened the door to many new studies. Many of the techniques described below are important for these studies. For instance, when only small samples are available, Fourier transform methods combined with time-averaging and superconducting magnet systems become invaluable. Some references illustrating biological application are:

A. F. Casey, PMR Spectroscopy in Medicinal and Biological Chemistry, Academic Press, New York, 1972.
A. S. V. Burgen and J. C. Metcalf, J. Pharm. Pharmacol., 22, 153 (1970). Application of nmr to pharmacological problems.
J. J. M. Rowe, J. Hinton, and K. L. Rowe, Chem. Rev., 70, 1 (1970). Nmr studies on the biochemistry of biopolymers.
J. M. Baker, Solid State Biophys., 41 (1969). Double resonance (ENDOR and Overhauser effects; see below) in biological applications.
Magnetic Resonance in Biological Systems, A. Ehrenberg, B. G. Malmström, and T. Vänngard, Eds., Pergamon Press, New York, 1967. Proceedings of the Second International Conference held at the Werner-Gren Center in Stolkholm, June 1966.

2. Conformational Analysis and Variable Temperature Measurement

Nmr has been used to study the conformational equilibria of molecules by two general methods. If the interconversion of the conformations is slow relative to the nmr time scale, in theory (assuming only two conformations) two different spectra will be observed. Simply measuring the relative areas under appropriate absorption regions will give a direct measure of the relative concentrations of each conformer, and thus the free energy difference between them. This method is dependent on the ability to separate corresponding signals for the two conformers. For instance, the hydrogen alpha to the bromine in bromocyclohexane appears as two bands at -75° (all other portions of the two spectra are overlapping); from the areas under these bands a free energy difference of about 0.5 kcal/mole is derived by use of the usual equations:

$$K = \frac{X}{1-X} \quad \text{and} \quad \Delta G° = -RT \ln K$$

Refs. p. 333.

where X is the mole fraction of the more stable conformer (equatorial bromine). On the other hand, if the conformers are intercoverting rapidly, one must either cool the sample (as was done with bromocyclohexane) to the point where the equilibrium is slow enough to allow analysis by the method just mentioned or one must make use of the averaged spectrum. In the averaged spectrum, both chemical shifts and coupling constants are averaged based on the percent of each conformer. For example, bromocyclohexane at 25° exists approximately as 70% equatorial isomer; therefore, the weighted average chemical shift for the alpha hydrogen will be $\delta = (0.7\delta_e + 0.3\delta_a)/2$, where δ_e and δ_a are the shifts for pure equatorial and axial conformers, respectively. Additional discussion of this technique is on page 156. More coverage of nmr methods of conformational analysis, with many examples are found in the following:

W. A. Thomas, Ann. Rept. NMR Spectros., 3, 91 (1970). A review covering 1967 and 1968.

H. Booth, Progr. NMR Spectros., 5, 149 (1969). An extensive review (230 pages, 740 references) extending through the beginning of 1967.

For the variable temperature techniques required in conformational analysis (or in general), it is difficult to measure accurately the temperature inside the nmr sample probe. (Ideally, the temperature inside the nmr sample tube should be measured, but in practice this is extremely difficult and almost never attempted). The standard method for measuring temperature is to use samples of ethylene glycol (above room temperature) or methanol (below room temperature). The chemical shift difference between the hydroxyl protons and the hydrogens on the carbon in these compounds is highly temperature dependent. Equations for this dependence were developed by A. L. Van Geet (46) using Fisher reagent grade alcohols with an added trace (0.03% by volume) of concentrated aqueous hydrochloric acid (the amount of HCl is not critical). No special precautions in preparing the solutions are necessary (such as degassing or sealing).

For methanol the dependence is not quite linear and a quadratic equation is necessary for a fit with an RMS error of less than 1° over the range 200 to 330°K.

$$T = 406.0 - 0.551|\Delta\nu| - 63.4\left(\frac{\Delta\nu}{100}\right)^2$$

where $\Delta\nu$ is the difference in chemical shifts in hertz at 60 MHz and T is in °K. Thus, for a chemical shift difference of 110 Hz, the temperature would be 268.7°K. For more narrow ranges, straight line equations can be used:

220-270°K $\quad T = 498.4 - 2.083\ |\Delta\nu|$ (error ≃1.0°)

265-313°K $\quad T = 468.1 - 1.810\ |\Delta\nu|$ (error ≃0.5°)

For ethylene glycol, a straight line dependence was noted over the range of 295 to 415°K:

$$T = 466.4 - 1.705\ |\Delta\nu| \quad \text{(error} \simeq 0.4°\text{)}$$

3. Exchange Processes

Nmr spectroscopy can be used to study the kinetics of the interchange of two (or more) states or species as was illustrated in the conformational analysis section above. That illustration was for the two limiting cases of very slow exchange where spectra for the separate states could be observed, or for very fast exchange where an average spectrum is observed. At intermediate rates, line broadening is observed as the spectra collapse to a single, average spectrum. By analysis of the shape of the broadened lines, information can be obtained on the rates of interchange of the two (or more) states. Methods for this type of analysis are often quite involved and require computer assistance, especially in cases where each of the two interchanging states gives more than a single absorption peak.

Where only two peaks (singlets) of equal population are interchanging (such as the methyls in N,N-DMF), the rate of exchange at the point of coalescence (temperature at which the two bands become one) can be easily determined by the equations

$$\tau = [\pi\sqrt{2}(\nu_a - \nu_b)]^{-1} \quad \text{or} \quad k_c = \pi\sqrt{2}(\nu_a - \nu_b)/2$$

where ν_a and ν_b are the frequencies in hertz of states (species) a and b in the absence of exchange (much lower temperature), τ is the lifetime of the exchanging system at coalescence, and k_c is the exchange rate at coalescence ($k_c = 1/2\tau$).

Similarly, for an AB system (see p. 307) of equally populated interchanging nuclei (such as H-axial and H-equational on a cyclohexane carbon atom), the exchange rate at coalescence is obtained from

$$k_c = \frac{\pi\sqrt{2}}{2}[(\nu_a - \nu_b)^2 + 6\,J_{ab}]^{-1/2}$$

Since $\Delta\nu$ values are usually in the range 10 to 1000 Hz, processes with τ values of 10^{-2} to 10^{-4} sec can be studied. Further discussion can be found in the following references.

R. M. Lynden-Bell, Progr. NMR Spectros., 2, 163 (1967).
C. S. Johnson, Advan. Magn. Reson., 1, 33 (1965). Chemical rate processes by nmr.

4. Superconducting Magnets

The conventional iron core magnet has an upper limit field strength of about 23.5 kilogauss (kG) which allows proton resonance at 100 MHz. To achieve higher magnet field strengths, it is necessary to utilize superconducting solenoids. (They are called solenoids because there is no core, only the superconducting coils.) It is now possible to purchase (Varian Associates) an nmr spectrometer with a superconducting magnet operating at 70.5 kG (300 MHz for protons). Very high field strengths are desirable for two reasons: higher sensitivity and simplification of spectra by the enhancement of the chemical shift spread. The current state of nmr with superconducting magnets has been discussed in L. F. Johnson, Anal. Chem., 43, 28A (1971).

5. Time Averaging

One of the most serious problems with nmr spectroscopy is its inherently low sensitivity. There are several ways this has been overcome (e.g., see superconducting magnets above). One of the most common methods at present is by time averaging. This method involves repeated scanning and summing of the spectrum in a computer (Computer of Average Transients, CAT). The

Refs. p. 333.

signals always add a positive number and thus slowly build up while the noise randomly adds both positively and negatively and thus averages to a low value. The theoretical improvement in signal-to-noise ratio by the method is the square root of the number of scans. Thus an approximately 20-fold increase is obtained after 500 scans.

If normal spectrum scan rates are used, this can take considerable time. The time factor can be greatly reduced by the use of pulse techniques or Fourier transform methods (see below). These and other methods for improving the signal to noise ratio of nmr spectrometers are discussed in the two following reviews:

G. E. Hall, Ann. Rev. NMR Spectros., 1, 227 (1968).
R. R. Ernst, Advan. Magn. Reson., 2, 1 (1966).

6. Fourier Transform Spectroscopy

The common method for obtaining nmr spectra consists of sweeping the frequency (or field) and looking at only one point in the spectrum at a time. This normally requires about 5 to 10 minutes for a complete spectrum. When time averaging (see above) 100 to 500 scans of a spectrum, this takes considerable time. Here lies the principal advantage of Fourier transform spectroscopy. By exciting all of the nuclei at the same time and "listening" to the emitted frequencies as these nuclei fall back to their equilibrium energy distribution, an interference pattern containing all of the information of a normal spectrum can be obtained. The time required for this is on the order of 1 sec. Thus 500 scans can be averaged in the time it takes for one normal scan and the increase in sensitivity is about 20-fold. Unfortunately, one cannot interpret the obtained interference pattern as is; however, the mathematical operation of Fourier transformation will convert this pattern to a normal frequency sweep spectrum. This transformation requires a small computer and takes only a small additional time after the spectra have been accumulated. The increase in sensitivity obtained by this method allows observation of much smaller samples; perhaps more importantly, the method allows the observation of nuclei (in particular ^{13}C) which have low sensitivities and/or low natural abundance. An example of this application is given in reference b below.

The excitation of all of the nuclei is accomplished by means of a short (≈100-μsec) pulse of fairly high-power radiofrequency. Use of continuous random frequency noise for excitation has also been reported and is described in references e and f below.

The following is a general bibliography. General reviews are found in reference g and h, while a brief description is given in reference a.

a. R. R. Ernst, Advan. Magn. Reson., 2, 108 (1966).
b. M. H. Freedman, J. S. Cohen, and I. M. Chiaken, Biochem. Biophys. Res. Comm., 42, 1148 (1971). An application of Fourier transform methods to ^{13}C magnetic resonance of peptides.
c. R. L. Vold and S. O. Chan, J. Magn. Reson., 4, 208 (1971). A discussion of the theory of Fourier transform spin-echo spectroscopy.
d. R. R. Ernst, J. Magn. Reson., 4, 280 (1971). A description of a slightly modified Fourier transform technique that simplifies the method.
e. R. R. Ernst, J. Magn. Reson., 3, 10 (1970). A discussion of the random noise excitation method.
f. R. Kaiser, J. Magn. Reson., 3, 28 (1970). Similar to reference e.
g. T. C. Farrar and E. D. Becker, Pulsed and Fourier Transform NMR, Academic Press, New York, 1971.
h. D. G. Gillies and D. Shaw, Ann. Rept. NMR Spectros., 5, to be published.

Refs. p. 333.

7. Double Resonance

Simultaneous irradiation of a sample with two (or more) rf fields often yields information difficult to obtain otherwise. If one field is used for observation of resonance and the other(s) used to perturb the spin system, this method is called nuclear magnetic double resonance (nmdr) or multiple resonance. If the perturbing field is at the resonant frequency of one of a pair of coupled nuclei, and is of sufficient power, the observing field will show the second nucleus as though it were no longer coupled. This technique of spin decoupling often leads to much simpler spectra and aids interpretation. If the perturbing field is of fairly low power, the observing field may show the second nucleus with additional splitting and/or a change in relative intensity. This technique is called spin tickling and is of great value in the analysis of complex spectra (see below). The change in intensity is due to the nuclear Overhauser effect (see below).

If the two nuclei are of the same type (e.g., both protons), the method is called homonuclear double resonance. Heteronuclear double resonance is also possible and has an additional use referred to as the INDOR (internuclear double resonance) technique. This method makes possible the indirect observation of other nuclei (usually of lower sensitivity). The observing field is held fixed at the resonant frequency of one nucleus in order to monitor the intensity of absorption. The perturbing field is then swept through the frequency range of the coupled heteronucleus; when the resonance absorption of the heteronucleus is located, a change in intensity of the observed absorption occurs. Thus, indirectly one may observe the chemical shift (or the entire spectrum in some cases) of the heteronucleus. These techniques are discussed in more detail in the references (1) as well as below:

a. R. A. Hoffman and S. Forsen, Progr. NMR Spectros., 1, 15 (1966). A comprehensive review of the theory, techniques, and applications of double and multiple resonance.
b. V. J. Kowalewski, Progr. NMR Spectros., 5, 1 (1969). A review covering the INDOR technique.
c. R. A. Dwek, R. E. Richards, and D. Taylor, Ann. Rev. NMR Spectros., 2, 293 (1969). A review of electron-nucleus double resonance (ENDOR). (This technique is related to the nuclear Overhauser effect discussed below.)
d. W. McFarlane, Ann. Rev. NMR Spectros., 1, 135 (1968). A review of heteronuclear magnetic double resonance.
e. J. M. Baker, Solid State Biophys., 41 (1969).

8. Nuclear Overhauser Effect (NOE)

As mentioned in the double resonance section above, sometimes when a second rather weakly perturbing rf field is applied at the resonant frequency of one of a pair of coupled nuclei, the observed absorption peak for the second nucleus is either increased or decreased in intensity. This effect, arising from a change in population of the energy levels, is called the nuclear Overhauser effect. (Originally this effect was observed for interactions between electrons and nuclei about which more will be said below.) One of the most useful applications of NOE results from the fact that it can be observed for a pair of nuclei that are not spin coupled but only in close proximity. In fact, the effect is quantitatively proportional to the inverse of the sixth power of the distance between the nuclei (if they are not spin-coupled). The application to conformational analysis and stereochemistry is obvious and is discussed in more detail in references a and b below. References c and d are also recommended for general infor-

Refs. p. 333.

mation. A general procedure for obtaining relative internuclear distances from NOE measurements has recently been published along with the types of conclusions that can be drawn from the quantitative data; see R. E. Schirmer and J. H. Noggle, J. Amer. Chem. Soc., 94, 2947 (1972).

The nuclear Overhauser effect between coupled heteronuclei has been useful in the study of magnetic resonance of nuclei other than protons. For example in ^{13}C magnetic resonance, when all of the protons are spin decoupled from the ^{13}C nuclei, a nearly threefold enhancement of the ^{13}C signal has been observed. This enhancement has served not only to increase the sensitivity of the ^{13}C but also to aid in the spectral assignment.

a. R. E. Schirmes, J. H. Noggle, J. P. Davis, and P. A. Hart, J. Amer. Chem. Soc., 92, 3266 (1970). A discussion of the quantitative application of the NOE to the determination of molecular geometry.
b. W. A. Thomas, Ann. Rept. NMR Spectros., 3, 137 (1970). A short discussion of NOE in conformational analysis.
c. K. H. Hausser and D. Stehlik, Advan. Magn. Reson., 3, 79 (1968). General review of NOE.
d. J. H. Noggle and R. E. Schirmer, The Nuclear Overhauser Effect: Chemical Applications, Academic Press, New York, 1971.

9. Chemically Induced Dynamic Nuclear Polarization (CIDNP)

A technique based on an effect very similar to the nuclear Overhauser effect has recently become useful in the study of short-lived free radicals. When a radical is generated in a magnetic field, the magnetic moment of the electron tends to line up with the magnetic field, and the sample becomes magnetically polarized. This polarization can cause a change in the population of the energy levels of the nuclei to which the free electron is spin-coupled, resulting in a change in the intensity of the nmr absorption. This effect can be observed in a normal nmr scan for some time (usually on the order of seconds), even after the radical has reacted. It can be used to assign structures of free radicals under conditions where electron spin resonance could not be used. This technique is further discussed in:

H. Fisher and J. Bargon, Accts. Chem. Res., 2, 110 (1969).

10. Analysis of Spectra

a. Spin-System Nomenclature. In spectral analysis it is convenient to deal with nuclei in terms of spin systems. A spin system is a group of nuclei that are interacting through spin coupling. A given molecule may contain one or more spin systems. For example, ethyl chloride has one spin system of five nuclei while ethyl acetate has one spin system of five nuclei and another spin system of three nuclei. The latter three spins are equivalent and give rise to a single absorption. See A. Ault, J. Chem. Educ., 47, 812 (1970), for a review of spin systems classification.

The concept of "equivalence" of magnetic nuclei has special meaning. Nuclei are said to be chemically equivalent if they exhibit identical chemical shifts. This can arise in two ways--one merely by accident under the experimental conditions (the trivial case), the other (and more commonly) because of actual equivalence by virtue of molecular symmetry. An example of accidental equivalence is found in propyne, which shows a single line at 60 MHz (δ 1.80 in chloroform), but separates when recorded as a neat liquid. Furthermore, if nuclei within a set (A) are chemically equivalent, and in addition each nucleus within the set is coupled identically to each nucleus of some other spin group (B), such nuclei are said to be magnetically

Refs. p. 333.

equivalent (also referred to as isochronous nuclei). For example, in CH_3CH_2Cl each of the two groups shows chemical and magnetic equivalence.

Note that the observation of equivalent or nonequivalent nuclei in many molecules rests on kinetic/thermodynamic considerations associated with energy barriers (bond rotations, inversions, chemical exchange, etc.). For example, in cyclohexane the nonequivalent axial and equatorial protons appear as a singlet at room temperature.

There is a commonly accepted method of notation for spin systems. Chemically nonequivalent nuclei, whose chemical shift difference is comparable to the coupling between them ($\Delta\nu \lesssim 10J$, the usual second-order situation), are designated by letters close to each other in the alphabet such as A, B, C. Other coupled groups of spins in the same molecule, but in a greatly separated chemical shift region from the A, B, C group may be designated as X, Y, Z; another group in a moderately separated region are designated as K, L, M. Usually, the nuclei are represented in order of increasing shielding, in alphabetical progression (A, least shielded; B, more shielded; etc.). The following examples illustrate these rules:

CH_3CH_2Cl	A_2X_3	$ClCH{=}CH_2$	ABX (or ABM)
CH_3CH_2F	A_3M_2X	$RCH{=}CH_2$	ABC

It is possible for chemically equivalent nuclei to be magnetically nonequivalent if they have nonequivalent coupling constants to any other single nucleus. For example, in 1,1-difluoroethylene, the two protons are chemically equivalent (by virtue of symmetry) but are magnetically nonequivalent because of the difference in coupling (cis or trans) to one of the fluorines. The fact that the same difference exists in coupling to the other fluorine does not alter the magnetic nonequivalence. Chemically equivalent but magnetically nonequivalent nuclei are differentiated by use of a prime. 1,1-Difluoroethylene would then be an AA'XX' system; the ring protons in p-ethyltoluene would comprise an AA'BB' system.

Further discussion of equivalent and nonequivalent nuclei can be found in M. van Gorkom and G. E. Hall, Quart. Rev., 22, 14 (1968), and in K. Mislow and M. Raban, Topics in Stereochem., 1, 1 (1967).

b. First-Order Spectra. Within a spin system, if the chemical shift difference between the coupled nuclei is large compared to the coupling constant between them, ($\Delta\nu > 10J$), then the system is classified as first-order. In this case, the splitting can be determined by the following rules:

1. When a nucleus is coupled to n equivalent nuclei of spin I, it will give rise to a signal of $(2nI + 1)$ lines. For protons, this equals $(n + 1)$ lines. When the nucleus is coupled to more than one set of equivalent spin 1/2 nuclei (all of the chemical shift differences should be larger than 10 times the coupling constant between them), the nucleus will give rise to a signal of $(n + 1)\cdot(m + 1)$ peaks, where m and n are the numbers of nuclei in each set. In other words, in an AM_2X_3 proton system, the A nucleus will appear as a quartet of triplets (or triplet of quartets). This should be true regardless of the number of A nuclei.

2. The spacing of the lines within a multiplet is equal to the coupling constant for the interaction. This spacing is independent of field strength.

3. The relative intensities of peaks within a multiplet are determined by the coefficients in the binomial expansion and are symmetrically situated around the midpoint of the band. For example, a doublet would have

Refs. p. 333.

intensities of 1:1, a triplet 1:2:1, a quartet 1:3:3:1, a quintet 1:4:6:4:1, a sextet 1:5:10:10:5:1, a septet 1:6:15:20:15:6:1, and so on.

c. Second-Order Spectra. When nuclei are strongly coupled (i.e., $\Delta\nu < 10J$), second-order characteristics are usually evident. Second-order spectra do not show regular splitting patterns and often contain many more lines than would be predicted by the simple rules given above. (An AA'BB' spectrum may have up to 24 lines.) Only the simplest of these second-order spectra can be analyzed without the aid of a computer (see below).

For the strongly coupled two-spin systems (AB), analysis is straightforward. This system gives rise to a four-line spectrum (AB quartet), the outer two lines of which (1 and 4) are of lower intensity than the inner two (2 and 3). Thus it might appear to be a quartet; in some cases (when $\Delta\nu = \sqrt{J_{AB}}$), an AB "quartet" is indistinguishable from a 1:3:3:1 quartet. The separation between the two lines in either pair is equal to the coupling constant, that is, $J_{AB} = (3 - 4) = (1 - 2)$. The difference in chemical shifts ($\Delta\nu$ or simple ν_{AB}) can be calculated by:

$$\nu_{AB} = \sqrt{(1-4)(2-3)}$$

An A_2B system can also be analyzed by hand using a straightforward, but more difficult method (1).

d. Very Complex Spectra. There are several methods for analysis of more complex cases. Experimentally, spin decoupling is very helpful and can often simplify a spectrum to the point where first-order analysis is possible. It is also sometimes possible to compare the spectrum visually with tables of calculated spectra for different values of coupling constants and chemical shifts. Three references that are useful for this purpose are:

P. L. Corio, Chem. Rev., 60, 363 (1960).
K. B. Wiberg and B. J. Nist, Interpretation of NMR Spectra, Benjamin, New York, 1962.
Reference 1b, Appendix D.

A third method of analysis of complex spectra involves subspectral analysis. It may be possible to divide a spectrum into noninteracting (but usually overlapping) subspectra associated with different symmetry properties. These subspectra can then be analyzed more easily than the entire spectrum. This method is discussed in detail in:

P. Diehl, R. K. Harris, and R. G. Jones, Progr. NMR Spectros., 3, 1 (1967).

The spectrum of a three spin system can be calculated, hence analyzed exactly. This calculation can be done by hand but is quite lengthy and tedious. Discussion of this method can be found in:

S. Castellano and J. S. Waugh, J. Chem. Phys., 34, 295 (1961), and 35, 1900 (1961).

For the analysis of more complex spectra, computer assistance is necessary. There are two generally accepted computer techniques for these analyses. The first of these, developed by J. D. Swalen and C. A. Reilly (47), consists of two programs, NMREN (available in two variations, NMREN1 and NMREN2) and NMRIT (a more recent version is NMRIT-IV). NMREN calculates energy levels from an observed spectrum using an assignment of the observed transitions to their respective energy levels. It is this assignment that is the most difficult step in the method. NMRIT has two capabilities.

Refs. p. 333.

First it can calculate line position and intensities from a set of assumed chemical shifts and coupling constants; second it can adjust these chemical shifts and coupling constants until agreement is reached (if possible) with the energy levels calculated from the observed spectrum by NMREN. These programs can calculate systems of from 3 to 7 nuclei of spin 1/2. Further discussion and copies of these programs can be obtained from sources mentioned below.

The second general method for computer analysis of nmr spectra was developed by A. A. Bothner-By and S. M. Castellano (48). This computer program, LAOCN4 (earlier versions were LAOCN3 and LAOCOONI and II) also has two capabilities; first, as in NMRIT, it can calculate line positions and intensities from a set of assumed chemical shifts and coupling constants; second, if the calculated spectrum resembles the observed spectrum closely enough for line assignments to be made, it can, by iterative calculation, yield a new set of chemical shifts and coupling constants for the observed spectrum. Again, the most difficult step is the assignment of calculated lines to observed ones. This program can calculate systems of two to seven nuclei with spin 1/2. There is a recent modification of this method called UEAITR (49) which allows calculation of larger systems in which symmetry factoring is possible (where there is at least one set of equivalent nuclei). It should be noted that neither of the methods necessarily gives unique solutions. To be sure of an analysis, solution at two different field strengths, results of spin tickling, or other additional data should be obtained.

These programs, with the exception of UEAITR, are all reproduced and discussed in:

Computer Programs for Chemistry, Vol. I, D. F. Detar, Ed, W. A. Benjamin, New York, 1968.

Copies of these programs on magnetic tape are also available from W. A. Benjamin, Inc., 2 Park Avenue, New York, N.Y. 10016.

All of the programs, including UEAITR, are available on cards or magnetic tape from the Quantum Chemistry Program Exchange, Chemistry Department, Room 204, Indiana University, Bloomington, Ind. 47401. (For more information see Computer Programs, p. 480). A program that facilitates calculation of AA'BB' spectra with the LAOCOON III Program is also available; see G. Ng and C. R. Dillard, J. Chem. Educ., 48, 607 (1971).

Reference 3f and the following describe analyses of very complex spectra:

E. O. Bishop, Ann. Rev. NMR Spectros., 1, 91 (1968).
J. D. Swalen, Progr. NMR Spectros., 1, 205 (1966).
E. Garbisch, Jr., J. Chem. Educ., 45, 311, 402, 480 (1968).

11. Paramagnetic Shift Reagents

a. Background. The effect of paramagnetism on nuclear resonances has been studied extensively and is reviewed in the following two references:

E. de Boer and H. van Willigen, Progr. NMR Spectros., 2, 111 (1967).
D. R. Eaton and W. D. Phillips, Advan. Magn. Reson., 1, 103 (1965).

In both references the larger resonance shifts that occur in the presence of paramagnetic species are noted. While potentially very useful for simplification of complex spectra or for structural assignments (as will be seen below) these shifts are almost always accompanied by line broadening of such a degree as to limit their use greatly. Recently, however,

Refs. p. 333.

Hinckley (50) reported the ability of tris(2,2,6,6-tetramethyl-3,5-heptanedionato)europium(III), Eu(thd)$_3$ (also known as trisdipivalomethanatoeuropium (III), Eu(DPM)$_3$) as a dipyridine complex, to cause large downfield shifts in the resonance values of protons in the vicinity of the cholesterol OH group (to which the reagent becomes coordinated) without appreciable line broadening. The structure of this reagent is:

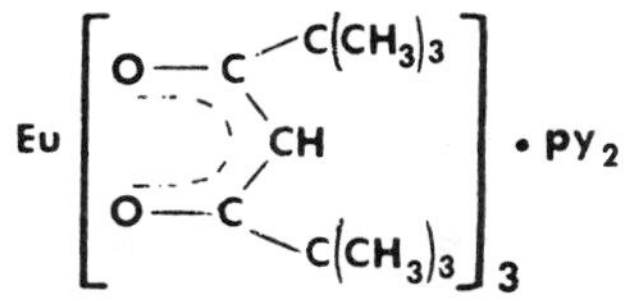

$$\mathrm{Eu}\left[\begin{array}{l}\mathrm{O{-}C{-}C(CH_3)_3} \\ \quad\quad \mathrm{CH} \\ \mathrm{O{-}C{-}C(CH_3)_3}\end{array}\right]_3 \cdot \mathrm{py_2}$$

Sanders and Williams (51) later reported that the complexing ability of this reagent was improved considerably by elimination of the pyridine. They illustrated the utility of this reagent by adding it to a solution of n-hexanol in CCl_4 leading to downfield shifts large enough to result in a completly first-order spectrum for the hexanol. (Without the reagent only the methylene adjacent to the hydroxyl was resolved from the methylene envelope.)

Hart et al. (52) reported that the praseodymuim (III) complex, Pr(thd)$_3$ (or Pr(DPM)$_3$) also gave large shifts with little line broadening but *upfield* instead of downfield and about three times as large. They reported a first-order spectrum for n-pentanol with all resonances coming *above* TMS and the hydroxyl methylene appearing at highest field.

Rondeau and Sievers (53) have recently reported improved versions of both "shift reagents," using a partially fluorinated ligand: 1,1,1,2,2,3,3-heptafluoro-7,7-dimethyl-4,6-octanedione. These reagents, termed Eu(fod)$_3$ and Pr(fod)$_3$, have the structure:

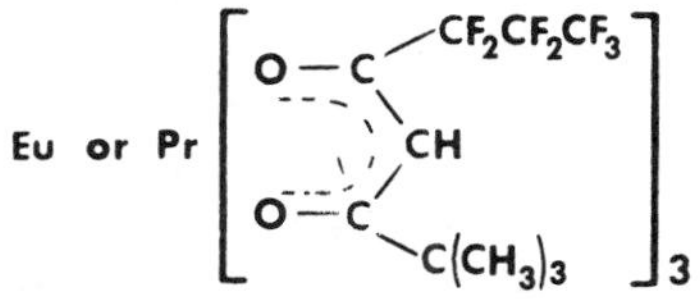

$$\mathrm{Eu\ or\ Pr}\left[\begin{array}{l}\mathrm{O{-}C{-}CF_2CF_2CF_3} \\ \quad\quad \mathrm{CH} \\ \mathrm{O{-}C{-}C(CH_3)_3}\end{array}\right]_3$$

They have the advantages of higher solubility and improved complexing ability (due to their being stronger Lewis acids). These are also available in deuterated form from supplier 16. A water-soluble shift reagent is under development by supplier 16 and is expected to be available in 1972. For a recent survey of lanthanide shift reagents, see W. de W. Horrocks, Jr. and J. P. Sipe, III, J. Amer. Chem. Soc., 93, 6800 (1971).

b. *Use*. The shifts produced by these reagents are temperature and concentration dependent; they also show a dependency on the strength of complex formed with the compound being studied. This is due to the existence of a rapid equilibrium between complexed and uncomplexed reagents. The concentration dependence is approximately linear up to a 1/1 mole ratio of shift reagent to compound and thus can be used to extrapolate absorption assignments back to the spectrum with no shift reagent added. This is illustrated in reference 54. Coupling constants do not appear to be affected by the shift reagents; however, conformational changes as a result of complexation could result and indirectly cause coupling constant changes.

The shifts induced by these reagents are dependent both on the distance between complexation site and the nucleus being observed and on the angle between the complex and that nucleus. In fact, an upfield shift can be observed with Eu(thd)$_3$ at certain angles (55).

The shift reagents can be very useful in assignment of stereochemistry as illustrated in the above mentioned references (50-55). An optically

Refs. p. 333. Suppliers p. 332.

active shift reagent, tris[3-(tert-butylhydroxymethylene)d-camphorato] europium(III), has been used (56) for determination of enantiomeric purity. Recent refinements in the use of chiral shift reagents to differentiate enantiotopic protons are reported in reference 58.

The Lewis basicity of the complexing functional group determines the complex stability and can influence the complexation equilibrium. This results in larger shifts for the more basic groups, approximately in the order NH_2>OH>NHCO>C=O>-O->CO_2R>CN. Other groups with which complexation has been observed are sulfoxides, carboxylic acids, aldehydes, thiocarbamate esters, and oximes.

These reagents can be purchased from suppliers 3,9,10,16,18,20,21,23, 26, and 27. A fairly thorough bibliography of papers dealing with shift reagents can be found in W. Horrocks, Jr. et al., J. Amer. Chem. Soc., 93, 5258 (1971).

Paramagnetic shift reagents have also been used in ^{13}C nmr spectroscopy [see E. Wenkert et al., J. Amer. Chem. Soc., 93, 6271 (1971)]. Recently, iron (II) phthalocyanines have been reported useful as nmr shift reagents for amines [J. E. Maskasky et al., J. Amer. Chem. Soc., 94, 2132 (1972)].

12. Liquid Crystal Solvents

Nmr spectra of substances in the gas or liquid phase are much simpler than in the solid phase due to averaging of magnetic anisotropies as a result of rapid tumbling. For instance, the chemical shift of a proton on benzene is highly dependent on the orientation of benzene to the magnetic field. This dependence is not important in the liquid or gas phase, however, because of the averaging just mentioned, with the result that all the protons absorb at the same frequency. A different kind of coupling also becomes important in systems where the molecules are not free to tumble rapidly enough for averaging. (Thus we see line broadening in very viscous liquids.) This coupling arises from a direct interaction through space of the magnetic dipoles of the nuclei. It is dependent on the angle that an internuclear line makes with the magnetic field, and averages to zero in liquids or gases. These dipole couplings, D_{ij}, decrease as $1/r^3_{ij}$ (where r_{ij} is the distance between i and j nuclei) and, thus, can be used to determine accurately these distances. For example, the dipole coupling in benzene has been determined (57) as D_{ortho} = -639.45 Hz, D_{meta} = -123.06 Hz, and D_{para} = -79.93 Hz. From these data, it can be determined that benzene must be a regular hexagon. The usefulness of determining these couplings is obvious.

The anisotropic effects are so complex in solids that we see only a very broad envelope of absorption. However, the use of liquid crystals as solvents overcomes this difficulty by allowing rapid tumbling about only two of the three axes, resulting in some averaging but still allowing observation of dipole coupling, as well as chemical shift anisotropies. It was by this method that the parameters for benzene were obtained. It should also be noted that this restricted tumbling results in nonequivalence of the chemical shifts of the protons in benzene and normal coupling constants can be obtained (57):

$$J_{ortho} = +6.0 \text{ Hz}, \quad J_{meta} = +2.0 \text{ Hz}, \quad \text{and} \quad J_{para} = 1.0 \text{ Hz}$$

The following references give more detailed discussion on the theory and the experimental aspects of liquid crystal nmr spectroscopy. Further

Refs. p. 333. Suppliers p. 332.

discussion of liquid crystals in general, including commercial suppliers, can be found on page 46.

A Saupe, in Magnetic Resonance, C. K. Coogan, N. S. Ham, S. N. Stuart, J. R. Pilbrow, and G. V. H. Wilson, Eds., Plenum Press, New York, 1970, p. 339.
P. Diehl and C. L. Khetrapal, in NMR. Basic Principles and Progress, Vol. 1, P. Diehl, E. Fluck, and R. Kosfeld, Eds., Springer-Verlag, New York, 1969, p. 1.
A. Saupe, Angew. Chem., Intern. Ed. Engl., 7, 107 (1968).
G. R. Lockhurst, Quart. Rev., 22, 179 (1968).
A. O. Buckingham and K. A. McLauchlan, Progr. NMR Spectros., 2, 63 (1967).
S. Meiboom and L. C. Snyder, Accts. Chem. Res., 4, 81 (1971).

K. Magnetic Properties of the Elemental Isotopes

The basic resonance frequencies at a field strength of 10 kG were taken, where possible, from the table compiled by Varian Associates: NMR Table, 5th ed., 1965. The frequency values at other field strengths and the field strength values at the various frequencies were all calculated from the basic frequencies at 10 kG. The magnetic moments (μ, reported in units of the nuclear magneton, $eh/4\pi Mc$) and the quadrupole moments (Q, in units of barns, 10^{-24} cm^2) were taken from G. H. Fuller and V. W. Cohen, Nuclear Spins and Moments, in Nuclear Data Tables, Section A, Vol. 5, p. 433, March 1969.

The sensitivities relative to hydrogen at constant field strength (H) and at constant frequency (ν) were calculated, respectively, from the equations

$$S_H = 7.651 \times 10^{-3} \mu^3 \frac{I + 1}{I^2}$$

$$S_\nu = 0.23871 \frac{I + 1}{\mu}$$

using the reported values for the magnetic moments, μ, and spin, I.

An asterisk (*) after the isotope symbol indicates radioactivity; m indicates a metastable excited state.

Refs. p. 333.

Isotope	Natural Abund-ance	I	NMR Frequency (MHz) at Field Strength (kG)				
			10.000	14.092	21.139	23.487	51.672
$_{0}n^{1}$*	-	1/2	29.167	41.104	61.655	68.506	150.713
$_{1}H^{1}$	99.985	1/2	42.5759	60.0000	90.0000	100.0000	220.0000
$_{1}H^{2}$	0.015	1	6.53566	9.21037	13.81555	15.35061	33.77134
$_{1}H^{3}$*	-	1/2	45.4129	63.9980	95.9971	106.6634	234.6595
$_{2}He^{3}$	0.00013	1/2	32.433	45.706	68.559	76.177	167.589
$_{3}Li^{6}$	7.42	1	6.2653	8.8294	14.244	14.7156	32.3743
$_{3}Li^{7}$	92.58	3/2	16.546	23.317	34.976	38.862	85.497
$_{4}Be^{9}$	100	3/2	5.9834	8.4321	12.6481	14.0535	30.9177
$_{5}B^{10}$	19.7	3	4.5754	6.4479	9.6718	10.7465	23.6422
$_{5}B^{11}$	80.3	3/2	13.660	19.250	28.875	32.084	70.585
$_{6}C^{13}$	1.108	1/2	10.7054	15.0866	22.6298	25.1443	55.3174
$_{7}N^{13}$	-	1/2	4.91	6.92	10.38	11.53	25.37
$_{7}N^{14}$	99.635	1	3.0756	4.3343	6.5014	7.2238	15.8924
$_{7}N^{15}$	0.365	1/2	4.3142	6.0798	9.1197	10.1330	22.2925
$_{8}O^{17}$	0.037	5/2	5.772	8.134	12.201	13.557	29.825
$_{9}F^{19}$	100	1/2	40.0541	42.3537	63.5305	94.0769	206.9692
$_{10}Ne^{21}$	0.257	3/2	3.3611	4.7366	7.1049	7.8944	17.3676
$_{11}Na^{22}$*	-	3	4.436	6.251	9.377	10.419	22.922
$_{11}Na^{23}$	100	3/2	11.262	15.871	23.806	26.452	58.193
$_{12}Mg^{25}$	10.13	5/2	2.6054	3.6717	5.5075	6.1194	13.4627
$_{13}Al^{27}$	100	5/2	11.094	15.634	23.451	26.057	57.325
$_{14}Si^{29}$	4.70	1/2	8.4578	11.9191	17.8787	19.8652	43.7035
$_{15}P^{31}$	100	1/2	17.235	24.288	36.433	40.481	89.057

Refs. p. 333.

Field Value (kG) at Frequency			Relative Sensitivity				
4 MHz	10MHz	16MHz	Constant H	Constant ν	μ	Q	Isotope
1.371	3.429	5.486	0.321	0.685	-1.9131	-	$_{0}n^{1}$*
0.940	2.349	3.758	1.00	1.00	2.79278	-	$_{1}H^{1}$
6.120	15.30	24.48	9.65×10^{-3}	0.409	0.85742	0.0028	$_{1}H^{2}$
0.881	2.202	3.523	1.21	1.07	2.9789	-	$_{1}H^{3}$*
1.233	3.083	4.933	0.442	0.762	-2.1276	-	$_{2}He^{3}$
6.384	15.96	25.54	8.50×10^{-3}	0.392	0.82202	0.0008	$_{3}Li^{6}$
2.42	6.04	9.67	0.294	1.94	3.2564	-0.04	$_{3}Li^{7}$
6.685	16.713	26.74	0.0139	0.703	-1.1776	0.05	$_{4}Be^{9}$
8.742	21.86	34.97	0.0199	1.72	1.8007	0.08	$_{5}B^{10}$
2.928	7.321	11.71	0.165	1.60	2.6885	0.04	$_{5}B^{11}$
3.736	9.341	14.946	0.0159	0.252	0.7024	-	$_{6}C^{13}$
8.15	20.4	32.6	1.5×10^{-3}	0.115	±0.3221	-	$_{7}N^{13}$
13.01	32.51	52.02	1.01×10^{-3}	0.193	0.4036	0.01	$_{7}N^{14}$
9.272	23.18	37.09	1.04×10^{-3}	0.101	-0.2831	-	$_{7}N^{15}$
6.93	17.3	27.7	0.0291	1.58	-1.8937	-0.026	$_{8}O^{17}$
0.999	2.497	3.994	0.834	0.941	2.6288	-	$_{9}F^{19}$
11.90	29.75	47.60	2.50×10^{-3}	0.395	-0.6618	0.09	$_{10}Ne^{21}$
9.02	22.5	36.1	0.0181	1.67	1.746		$_{11}Na^{22}$*
3.552	8.879	14.21	0.0927	1.32	2.2175	0.14	$_{11}Na^{23}$
15.35	38.38	61.41	2.68×10^{-3}	0.714	-0.8551	0.22	$_{12}Mg^{25}$
3.606	9.014	14.42	0.207	3.04	3.6414	0.15	$_{13}Al^{27}$
4.729	11.82	18.92	7.84×10^{-3}	0.199	-0.55477	-	$_{14}Si^{29}$
2.321	5.802	9.284	0.0665	0.405	1.1317	-	$_{15}P^{31}$

(Continued)

Refs. p. 333.

Isotope	Natural Abund-ance	I	NMR Frequency (MHz) at Field Strength (kG)				
			10.000	14.092	21.139	23.487	51.672
${}_{16}S^{33}$	0.76	3/2	3.2654	4.6018	6.9026	7.6696	16.8731
${}_{16}S^{35}$*	-	3/2	5.08	7.16	10.74	11.932	26.250
${}_{17}Cl^{35}$	75.53	3/2	4.1717	5.8790	8.8184	9.7983	21.5562
${}_{17}Cl^{36}$*	-	2	4.8931	6.8956	10.3434	11.4927	25.2838
${}_{17}Cl^{37}$	24.47	3/2	3.472	4.893	7.339	8.155	17.941
${}_{19}K^{39}$	93.22	3/2	1.9868	2.7999	4.1999	4.6665	10.2663
${}_{19}K^{40}$*	0.118	4	2.470	3.481	5.221	5.801	12.763
${}_{19}K^{41}$	6.77	3/2	1.0905	1.5368	2.3052	2.5613	5.6349
${}_{19}K^{42}$*	-	2	4.345	6.123	9.185	10.205	22.452
${}_{20}Ca^{43}$	0.145	7/2	2.8646	4.0369	6.0554	6.7282	14.8021
${}_{21}Sc^{45}$	100	7/2	10.343	14.576	21.863	24.293	53.445
${}_{21}Sc^{46}$*	-	4	5.77	8.13	12.20	13.55	29.82
${}_{22}Ti^{47}$	7.28	5/2	2.4000	3.3822	5.0733	7.9439	17.4766
${}_{22}Ti^{49}$	5.51	7/2	2.4005	3.3829	5.0743	5.6382	12.4040
${}_{23}V^{49}$*	-	7/2	9.71	13.68	20.53	22.81	50.17
${}_{23}V^{50}$	0.25	6	4.2450	5.9823	8.9734	9.9704	21.9349
${}_{23}V^{51}$	99.75	7/2	11.19	15.77	23.65	26.28	57.82
${}_{24}Cr^{53}$	9.55	3/2	2.4065	3.3914	5.0870	5.6523	12.4350
${}_{25}Mn^{52}$*	-	6	3.907	5.506	8.259	9.177	20.188
${}_{25}Mn^{53}$*	-	7/2	11.0	15.5	23.3	25.8	56.8
${}_{25}Mn^{55}$	100	5/2	10.501	14.799	22.198	24.664	54.261
${}_{26}Fe^{57}$	2.19	1/2	1.3758	1.9388	2.9083	3.2314	7.1091
${}_{27}Co^{56}$*	-	4	7.34	10.34	15.52	17.24	37.93

Refs. p. 333.

Field Value (kG) at Frequency			Relative Sensitivity				
4 MHz	10 MHz	16 MHz	Constant H	Constant ν	μ	Q	Isotope
12.25	30.62	49.0	2.26×10^{-3}	0.384	0.6533	-0.055	$_{16}S^{33}$
7.87	19.7	31.5	8.50×10^{-3}	0.597	1.00	0.04	$_{16}S^{35}$*
9.588	23.97	38.35	4.72×10^{-3}	0.490	0.82183	-0.079	$_{17}Cl^{35}$
8.175	20.44	32.70	0.0122	0.920	1.285	-0.017	$_{17}Cl^{36}$*
11.5	28.8	46.1	2.72×10^{-3}	0.408	0.68411	-0.062	$_{17}Cl^{37}$
20.13	50.33	80.53	5.10×10^{-4}	0.234	0.3914	0.055	$_{19}K^{39}$
16.2	40.5	64.8	5.23×10^{-3}	1.55	-1.298	-0.07	$_{19}K^{40}$*
36.68	91 70	146.7	8.44×10^{-5}	0.128	0.2149	-0.067	$_{19}K^{41}$
9.21	23.0	36.8	8.52×10^{-3}	0.817	-1.141		$_{19}K^{42}$*
13.96	34.91	55.85	6.42×10^{-3}	1.41	-1.317		$_{20}Ca^{43}$
3.867	9.668	15.47	0.302	5.11	4.7564	-0.22	$_{21}Sc^{45}$
6.93	17.3	27.7	0.0665	3.6	3.03	0.12	$_{21}Sc^{46}$*
16.67	41.67	66.67	2.09×10^{-3}	0.659	-0.7883	0.29	$_{22}Ti^{47}$
16.67	41.66	66.65	3.78×10^{-3}	1.18	-1.1039	0.24	$_{22}Ti^{49}$
4.12	10.3	16.5	0.256	4.8	±4.5		$_{23}V^{49}$*
9.423	23.56	37.69	0.0558	5.59	3.3470	±0.06	$_{23}V^{50}$
3.57	8.94	14.3	0.384	5.53	5.149	-0.05	$_{23}V^{51}$
16.62	41.55	66.49	9.08×10^{-4}	0.283	-0.4744	±0.03	$_{24}Cr^{53}$
10.2	25.6	40.9	0.0422	5.1	±3.05		$_{25}Mn^{52}$*
3.6	9.1	14.5	0.353	5.38	±5.01		$_{25}Mn^{53}$*
3.809	9.523	15.24	0.175	2.88	3.444	0.4	$_{25}Mn^{55}$
29.07	72.69	116.3	3.37×10^{-5}	3.2×10^{-2}	0.0902	-	$_{26}Fe^{57}$
5.45	13.6	21.8	0.134	4.6	±3.83		$_{27}Co^{56}$*

(Continued)

Refs. p. 333.

Isotope	Natural Abund-ance	I	NMR Frequency (MHz) at Field Strength (kG)				
			10.000	14.092	21.139	23.487	51.672
$_{27}Co^{57}$*	-	7/2	10.1	14.2	21.4	23.7	52.2
$_{27}Co^{58}$*	-	2	15.4	21.7	32.6	36.2	79.6
$_{27}Co^{59}$	100	7/2	10.054	14.169	21.253	23.614	51.951
$_{27}Co^{60}$*	-	5	5.793	8.164	12.246	13.606	29.934
$_{28}Ni^{61}$	1.19	3/2	3.8047	5.3618	8.0426	8.9363	19.6598
$_{29}Cu^{63}$	69.09	3/2	11.285	15.903	23.855	26.506	58.312
$_{29}Cu^{64}$*	-	1	3.1	4.4	6.6	7.3	16.0
$_{29}Cu^{65}$	30.91	3/2	12.089	17.036	25.555	28.394	62.467
$_{30}Zn^{65}$*	-	5/2	2.345	3.305	4.957	5.508	12.117
$_{30}Zn^{67}$	4.11	5/2	2.663	3.753	5.629	6.255	13.760
$_{31}Ga^{68}$*	-	1	0.0892	0.126	0.189	0.210	0.461
$_{31}Ga^{69}$	60.4	3/2	10.22	14.40	21.60	24.00	52.81
$_{31}Ga^{71}$	39.6	3/2	12.984	18.298	27.447	30.496	67.091
$_{31}Ga^{72}$*	-	3	0.33591	0.4734	0.7101	0.7890	1.7357
$_{32}Ge^{71}$*	-	1/2	8.4	11.8	17.8	19.7	43.4
$_{32}Ge^{73}$	7.76	9/2	1.4852	2.0930	3.1395	3.4884	7.6744
$_{33}As^{75}$	100	3/2	7.2919	10.2761	15.4141	17.1268	37.6790
$_{33}As^{76}$*	-	2	3.45	4.86	7.29	8.10	17.83
$_{34}Se^{77}$	7.58	1/2	8.118	11.440	17.160	19.067	41.948
$_{34}Se^{79}$*	-	7/2	2.22	3.13	4.69	5.21	11.47
$_{35}Br^{76}$*	-	1	4.18	5.89	8.84	9.82	21.60
$_{35}Br^{79}$	50.54	3/2	10.667	15.032	22.549	25.054	55.119
$_{35}Br^{81}$	49.46	3/2	11.498	16.204	24.305	27.006	59.413

Refs. p. 333.

Field Value (kG) at Frequency			Relative Sensitivity				
4MHz	10 MHz	16 MHz	Constant H	Constant ν	μ	Q	Isotope
4.0	9.9	15.8	0.277	5.0	±4.62		$_{27}Co^{57}$*
2.6	6.5	10.4	0.376	2.9	4.03		$_{27}Co^{58}$*
3.979	9.946	15.91	0.277	5.0	4.62	0.4	$_{27}Co^{59}$
6.90	17.3	27.6	0.099	5.4	3.78		$_{27}Co^{60}$*
10.51	26.28	42.05	3.57×10^{-3}	0.447	-0.7487	±0.16	$_{28}Ni^{61}$
3.545	8.861	14.18	0.0934	1.33	2.223	-0.180	$_{29}Cu^{63}$
13.0	32.0	52.0	1.54×10^{-4}	0.10	-0.216		$_{29}Cu^{64}$*
3.309	8.272	13.24	0.115	1.42	2.382	-0.195	$_{29}Cu^{65}$
17.1	42.6	68.2	1.95×10^{-3}	0.64	0.769	-0.026	$_{30}Zn^{65}$*
15.0	37.6	60.1	2.87×10^{-3}	0.731	0.8754	0.17	$_{30}Zn^{67}$
450	1,100	1,800	2.45×10^{-8}	5.6×10^{-3}	±0.0117	±0.031	$_{31}Ga^{68}$*
3.91	9.78	15.7	0.0697	1.203	2.016	0.19	$_{31}Ga^{69}$
3.081	7.702	12.32	0.143	1.529	2.562	0.12	$_{31}Ga^{71}$
119.1	297.7	476.3	7.86×10^{-6}	0.126	-0.1322	0.59	$_{31}Ga^{72}$*
5.0	12.0	19.0	7.47×10^{-3}	0.20	0.546	-	$_{32}Ge^{71}$*
26.93	67.33	107.7	1.41×10^{-3}	1.15	-0.8792	-0.28	$_{32}Ge^{73}$
5.486	13.71	21.94	0.0253	0.859	1.439	0.29	$_{33}As^{75}$
11.6	29.0	46.0	4.25×10^{-3}	0.648	-0.905		$_{33}As^{76}$*
4.93	12.3	19.7	6.99×10^{-3}	0.191	0.534	-	$_{34}Se^{77}$
18	45	72	2.98×10^{-3}	1.1	-1.02	0.8	$_{34}Se^{79}$*
9.6	24	38	2.52×10^{-3}	0.26	±0.548	±0.25	$_{35}Br^{76}$*
3.750	9.375	15.00	0.0794	1.26	2.106	0.31	$_{35}Br^{79}$
3.479	8.697	13.92	0.0994	1.35	2.270	0.26	$_{35}Br^{81}$

(Continued)

Isotope	Natural Abund- ance	I	NMR Frequency (MHz) at Field Strength (kG)				
			10.000	14.092	21.139	23.487	51.672
$_{35}Br^{82}$*	-	5	2.479	3.494	5.240	5.823	12.810
$_{36}Kr^{83}$	11.55	9/2	1.638	2.308	3.463	3.847	8.464
$_{36}Kr^{85}$*	-	9/2	1.6956	2.390	3.584	3.983	8.762
$_{37}Rb^{83}$*	-	5/2	4.33	6.10	9.15	10.17	22.37
$_{37}Rb^{84}$*	-	2	5.03	7.09	10.63	11.81	25.99
$_{37}Rb^{85}$	72.15	5/2	4.1108	5.7931	8.6897	9.6552	21.2415
$_{37}Rb^{86}$*	-	2	6.44	9.08	13.61	15.13	33.28
$_{37}Rb^{87}$	27.85	3/2	13.931	19.632	29.448	32.720	71.985
$_{38}Sr^{87}$	7.02	9/2	1.8452	2.6003	3.9005	4.3339	9.5346
$_{39}Y^{89}$	100	1/2	2.0859	2.9396	4.4093	4.8993	10.7784
$_{39}Y^{90}$*	-	2	6.17	8.70	13.04	14.49	31.88
$_{39}Y^{91}$*	-	1/2	2.49	3.51	5.26	5.85	12.87
$_{40}Zr^{91}$	11.23	5/2	3.97249	5.5982	8.3973	9.3304	20.5268
$_{41}Nb^{93}$	100	9/2	10.407	14.666	21.999	24.443	53.776
$_{42}Mo^{95}$	15.72	5/2	2.774	3.909	5.864	6.515	14.334
$_{42}Mo^{97}$	9.46	5/2	2.832	3.991	5.986	6.652	14.634
$_{43}Tc^{99}$*	-	9/2	9.5830	13.5048	20.2572	22.5080	49.5177
$_{44}Ru^{99}$	12.72	3/2	1.44	2.03	3.04	3.38	7.44
$_{44}Ru^{101}$	17.07	5/2	2.1	3.0	4.0	4.9	10.9
$_{45}Rh^{103}$	100	1/2	1.3401	1.885	2.8328	3.1476	6.9246
$_{46}Pd^{105}$	22.23	5/2	1.95	2.75	4.12	4.58	10.08
$_{47}Ag^{105}$*	-	1/2	1.54	2.17	3.26	3.62	7.96
$_{47}Ag^{107}$	51.82	1/2	1.7229	2.4280	3.6420	4.0467	8.9026

Refs. p. 333.

Field Value (kG) at Frequency			Relative Sensitivity				
4 MHz	10 MHz	16 MHz	Constant H	Constant ν	μ	Q	Isotope
16.1	40.3	64.5	7.89×10^{-3}	2.33	±1.626	±0.70	$_{35}Br^{82*}$
24.4	61.1	97.7	1.90×10^{-3}	1.27	-0.970	0.26	$_{36}Kr^{83}$
23.59	58.98	94.36	2.11×10^{-3}	1.32	±1.005	0.43	$_{36}Kr^{85*}$
9.2	23	37	0.01	1.2	1.4		$_{37}Rb^{83*}$
8.0	20	32	0.013	0.95	-1.32		$_{37}Rb^{84*}$
9.731	24.33	38.92	0.0105	1.13	1.3524	0.26	$_{37}Rb^{85}$
6.2	16	25	0.0277	1.21	-1.691		$_{37}Rb^{86*}$
2.871	7.178	11.49	0.177	1.64	2.7500	0.12	$_{37}Rb^{87}$
21.68	54.19	86.71	2.71×10^{-3}	1.44	-1.093	0.3	$_{38}Sr^{87}$
19.18	47.94	76.71	1.19×10^{-4}	4.90×10^{-2}	-0.1373	-	$_{39}Y^{89}$
6.5	16	26	0.025	1.17	-1.63	-0.15	$_{39}Y^{90*}$
16	40	64	2.0×10^{-4}	0.059	±0.164	-	$_{39}Y^{91*}$
10.07	25.17	40.28	9.48×10^{-3}	1.09	-1.303		$_{40}Zr^{91}$
3.844	9.609	15.37	0.487	8.10	6.167	-0.22	$_{41}Nb^{93}$
14.4	36.0	57.7	3.26×10^{-3}	0.763	-0.9133	±0.12	$_{42}Mo^{95}$
14.1	35.3	56.5	3.47×10^{-3}	0.779	-0.9325	±1.1	$_{42}Mo^{97}$
4.174	10.435	16.70	0.38	7.46	5.68	0.3	$_{43}Tc^{99*}$
28	69	111	2.0×10^{-4}	0.4	-0.63		$_{44}Ru^{99}$
19	48	76	1.4×10^{-3}	0.6	-0.69		$_{44}Ru^{101}$
29.85	74.62	119.4	3.16×10^{-5}	3.16×10^{-2}	-0.0883	-	$_{45}Rh^{103}$
20.5	51	82	1.1×10^{-3}	0.54	-0.642	+0.8	$_{46}Pd^{105}$
26	65	104	4.7×10^{-5}	0.036	±0.101	-	$_{47}Ag^{105*}$
23.22	58.04	92.87	6.71×10^{-5}	4.06×10^{-2}	-0.1135	-	$_{47}Ag^{107}$

(Continued)

Refs. p. 333.

Isotope	Natural Abund-ance	I	NMR Frequency (NHz) at Field Strength (kG)				
			10.000	14.092	21.139	23.487	51.672
$_{47}Ag^{109}$	48.18	1/2	1.9807	2.7913	4.1869	4.6522	10.2348
$_{47}Ag^{110m}$*	-	6	4.557	6.423	9.633	10.703	23.547
$_{47}Ag^{111}$*	-	1/2	2.21	3.11	4.67	5.19	11.42
$_{48}Cd^{109}$*	-	5/2	2.529	3.564	5.346	5.940	13.068
$_{48}Cd^{111}$	12.75	1/2	9.028	12.723	19.084	21.205	46.650
$_{48}Cd^{113}$	12.26	1/2	9.445	13.310	19.966	22.184	48.805
$_{48}Cd^{113m}$*	-	11/2	1.51	2.13	3.19	3.55	7.80
$_{48}Cd^{115}$*	-	1/2	9.862	13.898	20.847	23.163	50.959
$_{48}Cd^{115m}$*	-	11/2	1.447	2.039	3.059	3.399	7.477
$_{49}In^{113}$*	4.28	9/2	9.3099	13.1200	19.6800	21.8666	48.1065
$_{49}In^{113m}$*	-	1/2	3.209	4.522	6.783	7.537	16.582
$_{49}In^{114m}$*	-	5	7.2	10.1	15.2	16.9	37.2
$_{49}In^{115m}$*	95.72	9/2	9.3301	13.1484	19.7226	21.9140	48.2109
$_{50}Sn^{115}$	0.35	1/2	13.992	19.718	29.577	32.864	72.300
$_{50}Sn^{117}$	7.61	1/2	15.168	21.375	32.063	35.626	78.377
$_{50}Sn^{119}$	8.58	1/2	15.869	22.363	33.545	37.272	81.999
$_{51}Sb^{121}$	57.25	5/2	10.189	14.359	21.538	23.931	52.649
$_{51}Sb^{122}$*	-	2	7.24	10.20	15.30	17.00	37.41
$_{51}Sb^{123}$	42.75	7/2	5.5176	7.7757	11.6635	12.9594	28.5108
$_{52}Te^{123}$	0.87	1/2	11.16	15.73	23.59	26.21	57.67
$_{52}Te^{125}$	6.99	1/2	13.45	18.95	28.43	31.59	69.50
$_{53}I^{125}$*	-	5/2	9.0	12.7	19.0	21	47
$_{53}I^{127}$	100	5/2	8.5183	12.0044	18.0066	20.0073	44.0161

Refs. p. 333.

Field Value (kG) at Frequency			Relative Sensitivity				
4 MHz	10 MHz	16 MHz	Constant H	Constant ν	μ	Q	Isotope
20.19	50.49	80.78	1.02×10^{-4}	4.67×10^{-2}	-0.1305	-	$_{47}Ag^{109}$
8.78	21.9	35.1	0.0696	6.02	3.604		$_{47}Ag^{110m}$*
18	45	72	1.40×10^{-4}	0.052	-0.145	-	$_{47}Ag^{111}$*
15.8	39.5	63.3	2.42×10^{-3}	0.691	-0.8270	0.8	$_{48}Cd^{109}$*
4.43	11.1	17.7	9.64×10^{-3}	0.213	-0.5943	-	$_{48}Cd^{111}$
4.24	10.6	16.9	0.0110	0.223	-0.6217	-	$_{48}Cd^{113}$
26	66	106	2.11×10^{-3}	1.69	-1.087	-0.8	$_{48}Cd^{113m}$*
4.06	10.1	16.2	0.0125	0.232	-0.6477	-	$_{48}Cd^{115}$*
27.6	69.1	111	1.85×10^{-3}	1.61	-1.040	-0.6	$_{48}Cd^{115m}$*
4.297	10.74	17.19	0.350	7.25	5.523	0.82	$_{49}In^{113}$*
12.5	31.2	49.9	4.3×10^{-4}	0.075	-0.210	-	$_{49}In^{113m}$*
5.6	14	22	0.2	6.7	4.7		$_{49}In^{114m}$*
4.287	10.72	17.15	0.352	7.27	5.534	0.83	$_{49}In^{115m}$*
2.859	7.147	11.44	0.036	0.329	-0.918	-	$_{50}Sn^{115}$
2.637	6.593	10.55	0.046	0.358	-1.000	-	$_{50}Sn^{117}$
2.521	6.302	10.08	0.0525	0.375	-1.046	-	$_{50}Sn^{119}$
3.926	9.815	15.70	0.162	2.81	3.359		$_{51}Sb^{121}$
5.5	14	22	0.0394	1.4	-1.90	0.69	$_{51}Sb^{122}$*
7.250	18.12	29.00	0.0464	2.74	2.547	-0.37	$_{51}Sb^{123}$
3.58	8.96	14.3	0.018	0.264	-0.7359	-	$_{52}Te^{123}$
2.97	7.43	11.9	0.0320	0.318	-0.8871	-	$_{52}Te^{125}$
4	11	18	0.116	2.5	3.0	-0.89	$_{53}I^{125}$*
4.696	11.74	18.78	0.0948	2.35	2.808	-0.79	$_{53}I^{127}$

(Continued)

Refs. p. 333.

Isotope	Natural Abund-ance	I	NMR Frequency (MHz) at Field Strength (kG)				
			10.000	14.092	21.139	23.487	51.672
$_{53}I^{129}$*	-	7/2	5.6694	7.9896	11.9844	13.3160	29.2952
$_{53}I^{131}$*	-	7/2	5.963	8.403	12.605	14.006	30.812
$_{54}Xe^{129}$	26.44	1/2	11.777	16.597	24.895	27.661	60.855
$_{54}Xe^{131}$	21.18	3/2	3.4911	4.9198	7.3797	8.1997	18.0394
$_{55}Cs^{131}$*	-	5/2	10.72	15.11	22.66	25.18	55.39
$_{55}Cs^{132}$ *	-	2	8.46	11.92	17.88	19.87	43.71
$_{55}Cs^{133}$	100	7/2	5.58469	7.8702	11.8053	13.1170	28.8574
$_{55}Cs^{134}$*	-	4	5.666	7.985	11.977	13.308	29.278
$_{55}Cs^{135}$*	-	7/2	5.9096	8.3281	12.4921	13.8802	30.5363
$_{55}Cs^{137}$*	-	7/2	6.1459	8.6611	12.9916	14.4352	31.7574
$_{56}Ba^{135}$	6.59	3/2	4.2296	5.9606	8.9408	9.9343	21.8554
$_{56}Ba^{137}$	11.32	3/2	4.7315	6.6679	10.0018	11.1131	24.4488
$_{57}La^{138}$	0.089	5	5.6171	7.9159	11.8738	13.1931	29.0249
$_{57}La^{139}$	99.911	7/2	6.0144	8.4758	12.7137	14.1263	31.0779
$_{58}Ce^{141}$*	-	7/2	2.1	3.0	4.4	4.9	10.9
$_{59}Pr^{141}$	100	5/2	12.5	17.6	26.4	29.4	64.6
$_{59}Pr^{142}$*	-	2	1.1	1.6	2.3	2.6	5.7
$_{60}Nd^{143}$	12.17	7/2	2.315	3.262	4.894	5.437	11.962
$_{60}Nd^{145}$	8.30	7/2	1.42	2.00	3.00	3.34	7.34
$_{60}Nd^{147}$*	-	5/2	1.77	2.49	3.74	4.16	9.15
$_{61}Pm^{147}$*	-	7/2	5.62	7.92	11.88	13.20	29.04
$_{61}Pm^{148}$*	-	1	16	23	34	38	83
$_{62}Sm^{147}$	14.97	7/2	1.76	2.48	3.72	4.13	9.09

Refs. p. 333.

Field Value (kG) at Frequency			Relative Sensitivity				
4 MHz	10 MHz	16 MHz	Constant H	Constant ν	μ	Q	Isotope
7.055	17.64	28.22	0.050	2.81	2.617	-0.55	$_{53}I^{129}$*
6.71	16.8	26.8	0.0578	2.9	2.74	-0.40	$_{53}I^{131}$*
3.397	8.491	13.59	0.0215	0.278	-0.7768	-	$_{54}Xe^{129}$
11.46	28.64	45.83	2.80×10^{-3}	0.412	0.6908	-0.12	$_{54}Xe^{131}$
3.73	9.33	14.9	0.19	3.0	3.54	-0.57	$_{55}Cs^{131}$*
4.7	12	19	0.0628	1.6	2.22	0.46	$_{55}Cs^{132}$*
7.162	17.906	28.650	0.0482	2.77	2.578	-3×10^{-3}	$_{55}Cs^{133}$
7.06	17.6	28.2	0.0639	3.57	2.990	0.36	$_{55}Cs^{134}$*
6.769	16.92	27.07	0.0571	2.93	2.729	0.044	$_{55}Cs^{135}$*
6.508	16.27	26.03	0.0642	3.05	2.838	0.045	$_{55}Cs^{137}$*
9.457	23.64	37.83	4.90×10^{-3}	0.498	0.8365	0.18	$_{56}Ba^{135}$
8.454	21.13	33.82	6.96×10^{-3}	0.558	0.9357	0.28	$_{56}Ba^{137}$
7.121	17.80	28.48	0.0935	5.31	3.707	±0.8	$_{57}La^{138}$
6.651	16.63	26.60	0.060	2.98	2.778	0.22	$_{57}La^{139}$
19	48	76	2.05×10^{-3}	1.0	±0.9		$_{58}Ce^{141}$*
3.2	8.0	13	0.3	3.6	4.3	-0.07	$_{59}Pr^{141}$
36	91	150	9×10^{-5}	0.18	±0.25	±0.03	$_{59}Pr^{142}$*
17.3	43.2	69.1	3.5×10^{-3}	1.2	-1.08	-0.48	$_{60}Nd^{143}$
28	70	110	8×10^{-4}	0.71	-0.66	-0.25	$_{60}Nd^{145}$
23	56	90	8×10^{-4}	0.49	±0.59		$_{60}Nd^{147}$*
7.1	18	28	0.06	2.9	2.7	0.7	$_{61}Pm^{147}$*
2.5	6.3	10	0.1	1.0	2.0	0.2	$_{61}Pm^{148}$*
23	57	91	1.5×10^{-3}	0.87	-0.813	-0.20	$_{62}Sm^{147}$

(Continued)

Refs. p. 333.

Isotope	Natural Abundance	I	NMR Frequency (MHz) at Field Strength (kG)				
			10.000	14.092	21.139	23.487	51.672
$_{62}Sm^{149}$	13.83	7/2	1.40	1.97	2.96	3.29	7.23
$_{63}Eu^{151}$	47.82	5/2	10.559	14.880	22.320	24.800	54.561
$_{63}Eu^{152}$*	-	3	4.848	6.846	10.269	11.410	25.103
$_{63}Eu^{153}$	52.18	5/2	4.6627	6.5609	9.8564	10.9515	24.0933
$_{63}Eu^{154}$*	-	3	5.084	7.165	10.747	16.829	37.023
$_{64}Gd^{155}$	14.73	3/2	1.6	2.3	3.4	5.4	11.9
$_{64}Gd^{157}$	15.68	3/2	2.0	2.8	4.2	4.7	10.3
$_{65}Tb^{159}$	100	3/2	9.66	13.61	20.42	22.69	49.92
$_{65}Tb^{160}$*	-	3	4.1	5.8	8.7	9.6	21.2
$_{66}Dy^{161}$	18.88	5/2	1.4	2.0	3.0	3.3	7.2
$_{66}Dy^{163}$	24.97	5/2	2.0	2.8	4.2	4.7	10.3
$_{67}Ho^{165}$	100	7/2	8.73	12.30	18.45	20.50	45.11
$_{68}Er^{167}$	22.94	7/2	1.23	1.73	2.60	2.89	6.36
$_{68}Er^{169}$*	-	1/2	7.8	11.0	16.5	18.3	40.3
$_{69}Tm^{169}$	100	1/2	3.52	4.96	7.44	8.27	18.19
$_{69}Tm^{170}$*	-	1	2.0	2.8	4.2	4.7	10.3
$_{69}Tm^{171}$*	-	1/2	3.46	4.88	7.31	8.13	17.88
$_{70}Yb^{171}$	14.31	1/2	7.4990	10.5680	15.8519	17.6133	38.7492
$_{70}Yb^{173}$	16.13	5/2	2.0659	2.9114	4.3670	4.8523	10.6750
$_{71}Lu^{175}$	97.41	7/2	4.86	6.85	10.27	11.41	25.11
$_{71}Lu^{176}$*	2.59	7	3.4	4.8	7.2	8.0	18
$_{71}Lu^{177}$*	-	7/2	4.84	6.82	10.23	11.37	25.01
$_{72}Hf^{177}$	18.50	7/2	1.3	1.8	2.7	3.1	6.7

Refs. p. 333.

Field Value (kG) at Frequency			Relative Sensitivity				
4 MHz	10 MHz	16 MHz	Constant H	Constant ν	μ	Q	Isotope
29	71	114	8.4×10^{-4}	0.72	-0.670	0.058	$_{62}Sm^{149}$
3.788	9.471	15.15	0.178	2.89	3.464	1.1	$_{63}Eu^{151}$
8.23	20.6	32.9	0.0242	1.84	1.924	±3.0	$_{63}Eu^{152}$*
8.579	21.45	34.31	0.0153	1.28	1.530	2.8	$_{63}Eu^{153}$
7.87	19.7	31.5	0.0272	1.91	±2.000		$_{63}Eu^{154}$*
25	63	100	1.39×10^{-4}	0.15	-0.254	1.3	$_{64}Gd^{155}$
20	50	80	3.3×10^{-4}	0.20	-0.339	2	$_{64}Gd^{157}$
4.1	10	17	0.067	1.2	±1.99	1.3	$_{65}Tb^{159}$
9.8	24	39	0.016	1.6	1.68	3.0	$_{65}Tb^{160}$*
29	71	110	4.0×10^{-4}	0.4	-0.46	2.3	$_{66}Dy^{161}$
20	50	80	1.1×10^{-3}	0.5	0.64	2.5	$_{66}Dy^{163}$
4.6	11	18	0.20	4.4	4.12	3.0	$_{67}Ho^{165}$
33	81	130	5.04×10^{-4}	0.61	-0.564	2.8	$_{68}Er^{167}$
5.1	13	21	6.2×10^{-3}	0.18	0.513	-	$_{68}Er^{169}$*
11	28	45	5.7×10^{-4}	0.083	-0.232	-	$_{69}Tm^{169}$
20	50	80	2.3×10^{-4}	0.12	±0.246	±0.59	$_{69}Tm^{170}$*
12	29	46	5.5×10^{-4}	0.082	±0.229	-	$_{69}Tm^{171}$*
5.334	13.34	21.34	5.46×10^{-3}	0.176	0.4919	-	$_{70}Yb^{171}$
19.36	48.41	17.45	1.33×10^{-3}	0.566	-0.6776	3.0	$_{70}Yb^{173}$
8.2	21	33	0.031	2.4	2.23	5.6	$_{71}Lu^{175}$
12	29	47	0.040	6.1	3.18	8.0	$_{71}Lu^{176}$*
8.3	21	33	0.032	2.4	2.24	5.4	$_{71}Lu^{177}$*
31	77	120	6×10^{-4}	0.7	0.61	3	$_{72}Hf^{177}$

(Continued)

Isotope	Natural Abund-ance	I	NMR Frequency (MHz) at Field Strength (kG) 10.000	14.092	21.139	23.487	51.672
$_{72}Hf^{179}$	13.75	9/2	0.80	1.13	1.69	1.88	4.13
$_{73}Ta^{181}$	99.988	7/2	5.096	7.182	10.772	11.969	26.332
$_{74}W^{183}$	14.40	1/2	1.7716	2.4966	3.7449	4.1610	9.1543
$_{75}Re^{185}$	37.07	5/2	9.5855	13.5083	20.2625	22.5139	49.5306
$_{75}Re^{186*}$	-	1	13.17	18.56	27.84	30.93	68.05
$_{75}Re^{187*}$	62.93	5/2	9.6837	13.6467	20.4701	22.7446	50.0380
$_{75}Re^{188*}$	-	1	13.55	19.10	28.64	13.83	70.02
$_{76}Os^{187}$	1.64	1/2	0.98059	1.3819	2.0728	2.3032	5.0669
$_{76}Os^{189}$	16.1	3/2	3.3034	4.6553	6.9830	7.7588	17.0695
$_{77}Ir^{191}$	37.3	3/2	0.7318	1.0313	1.5469	1.7188	3.7814
$_{77}Ir^{193}$	62.7	3/2	0.7968	1.1229	1.6843	1.8715	4.1173
$_{78}Pt^{195}$	33.8	1/2	9.153	12.899	19.348	21.498	47.296
$_{79}Au^{195*}$	-	3/2	0.742	1.046	1.568	1.743	3.834
$_{79}Au^{196*}$	-	2	2.3	3.2	4.8	5.4	11.9
$_{79}Au^{197}$	100	3/2	0.729188	1.0276	1.5414	1.7127	3.7679
$_{79}Au^{198*}$	-	2	2.227	3.138	4.708	5.231	11.507
$_{79}Au^{199*}$	-	3/2	1.358	1.914	2.871	3.190	7.017
$_{80}Hg^{197}$	-	1/2	7.9	11.1	16.7	18.6	40.8
$_{80}Hg^{199}$	16.84	1/2	7.59012	10.6964	16.0445	17.8273	39.2200
$_{80}Hg^{201}$	13.22	3/2	2.8099	3.9598	5.9398	6.5997	14.5194
$_{80}Hg^{203*}$	-	5/2	2.5	3.5	5.3	5.9	12.9
$_{81}Tl^{201*}$	-	1/2	24.1	34.0	51.0	56.6	125
$_{81}Tl^{203}$	29.50	1/2	24.332	34.290	51.435	57.150	125.73

Refs. p. 333.

Field Value (kG) at Frequency			Relative Sensitivity				
4 MHz	10 MHz	16 MHz	Constant H	Constant ν	μ	Q	Isotope
50	130	200	2×10^{-4}	0.6	-0.47	3	$_{72}Hf^{179}$
7.85	19.6	31.4	0.037	2.54	2.36	4.2	$_{73}Ta^{181}$
22.58	56.45	90.31	7.3×10^{-5}	0.042	0.117	-	$_{74}W^{183}$
4.173	10.43	16.69	0.137	2.65	3.172	2.7	$_{75}Re^{185}$
3.04	7.59	12.1	0.079	0.83	±1.73		$_{75}Re^{186}$*
4.131	10.33	16.52	0.141	2.68	3.204	2.6	$_{75}Re^{187}$*
2.95	7.38	11.8	0.086	0.85	1.78		$_{75}Re^{188}$*
40.8	102	163	1.22×10^{-5}	0.023	0.0643	-	$_{76}Os^{187}$
12.11	30.27	48.43	2.40×10^{-3}	0.392	0.6566	0.8	$_{76}Os^{189}$
54.7	137	219	2.6×10^{-5}	0.087	0.145	1.3	$_{77}Ir^{191}$
50.2	126	201	3.4×10^{-5}	0.094	0.158	1.2	$_{77}Ir^{193}$
4.37	10.9	17.5	0.0102	0.217	0.6060	-	$_{78}Pt^{195}$
53.9	135	216	2.7×10^{-5}	0.088	±0.147		$_{79}Au^{195}$*
17	43	70	0.001	0.4	+0.58 or -0.62		$_{79}Au^{196}$*
54.9	137	219	2.58×10^{-5}	0.086	0.14486	0.58	$_{79}Au^{197}$
18.0	44.9	71.8	1.2×10^{-3}	0.42	0.590		$_{79}Au^{198}$*
29.5	73.6	118	1.7×10^{-4}	0.16	0.270		$_{79}Au^{199}$*
5.1	13	20	6.6×10^{-3}	0.19	0.524	-	$_{80}Hg^{197}$
5.27	13.18	21.08	5.83×10^{-3}	0.180	0.5027	-	$_{80}Hg^{199}$
14.24	35.59	56.94	1.47×10^{-3}	0.332	-0.5567	0.45	$_{80}Hg^{201}$
16	40	64	2.5×10^{-3}	0.7	0.84		$_{80}Hg^{203}$*
1.7	4.1	6.6	0.19	0.57	1.60	-	$_{81}Tl^{201}$*
1.64	4.11	6.58	0.192	0.577	1.6115	-	$_{81}Tl^{203}$

(Continued)

Isotope	Natural Abund-ance	I	NMR Frequency (MHz) at Field Strength (kG)				
			10.000	14.092	21.139	23.487	51.672
$_{81}Tl^{204}$*	-	2	0.34	0.48	0.72	0.80	1.8
$_{81}Tl^{205}$	70.50	1/2	24.570	34.625	51.938	57.709	126.959
$_{82}Pb^{207}$	22.6	1/2	8.9077	12.5532	18.8297	20.9219	46.0282
$_{83}Bi^{205}$*	-	9/2	9.3	13.1	19.7	21.8	48.1
$_{83}Bi^{209}$*	100	9/2	6.84178	9.6418	14.4626	16.0696	35.3531
$_{83}Bi^{210}$*	-	1	0.337	0.475	0.712	0.792	1.741
$_{84}Po^{207}$*	-	5/2	0.82	1.16	1.73	1.93	4.24
$_{85}At$							
$_{86}Rn$							
$_{87}Fr$							
$_{88}Ra$							
$_{89}Ac^{227}$*	-	3/2	5.6	7.9	11.8	13.2	28.9
$_{90}Th^{229}$*	-	5/2	1.2	1.7	2.5	2.8	6.2
$_{91}Pa^{231}$*	-	3/2	9.96	14.04	21.05	23.39	51.47
$_{91}Pa^{233}$*	-	3/2	17	24	36	40	88
$_{92}U^{233}$*	-	5/2	1.6	2.3	3.4	3.8	8.3
$_{92}U^{235}$*	0.72	7/2	0.76	1.07	1.61	1.79	3.93
$_{93}Np^{237}$*	-	5/2	18	25	38	42	93
$_{94}Pu^{239}$*	-	1/2	3.05	4.30	6.45	7.16	15.76
$_{94}Pu^{241}$*	-	5/2	2.09	2.95	4.42	4.91	10.80
$_{95}Am^{241}$*	-	5/2	4.82	6.79	10.19	11.32	24.91
$_{95}Am^{242}$*	-	1	2.90	4.09	6.13	6.81	14.99
$_{95}Am^{243}$*	-	5/2	4.79	6.75	10.12	11.25	24.75

Refs. p. 333.

Field Value (kG) at Frequency			Relative Sensitivity				
4 MHz	10 MHz	16 MHz	Constant H	Constant ν	μ	Q	Isotope
120	290	470	4.1×10^{-6}	0.06	±0.089		$_{81}Tl^{204}$*
1.63	4.07	6.51	0.198	0.583	1.6274	-	$_{81}Tl^{205}$
4.491	11.23	17.96	9.40×10^{-3}	0.211	0.5895	-	$_{82}Pb^{207}$
4.3	11	17	0.35	7	5.5		$_{83}Bi^{205}$*
5.846	14.62	23.39	0.141	5.35	4.080	-0.33	$_{83}Bi^{209}$*
120	300	470	1.32×10^{-6}	0.021	0.0442	±0.13	$_{83}Bi^{210}$*
49	120	200	8×10^{-5}	0.2	0.27	0.28	$_{84}Po^{207}$*
							$_{85}At$
							$_{86}Rn$
							$_{87}Fr$
							$_{88}Ra$
7.1	18	29	0.01	0.06	1.17	1.7	$_{89}Ac^{227}$*
33	83	130	2.3×10^{-4}	0.3	0.38	4.6	$_{90}Th^{229}$*
4.0	10	16	0.066	1.2	±1.98		$_{91}Pa^{231}$*
2	6	9	0.33	2.0	3.4	-3.0	$_{91}Pa^{233}$*
25	63	100	6.7×10^{-4}	0.5	0.54	3.5	$_{92}U^{233}$*
53	130	210	1.2×10^{-4}	0.4	-0.35	4.1	$_{92}U^{235}$*
2.2	5.6	8.9	0.15	3	3.3		$_{93}Np^{237}$*
13	33	52	3.7×10^{-4}	0.072	0.200	-	$_{94}Pu^{239}$*
19	48	77	1.7×10^{-3}	0.6	-0.73		$_{94}Pu^{241}$*
8.3	21	33	0.017	1.3	1.59	4.9	$_{95}Am^{241}$*
14	34	55	8.5×10^{-4}	0.18	±0.382	±2.8	$_{95}Am^{242}$*
8.4	21	33	0.01	1.2	1.4	4.9	$_{95}Am^{243}$*

Refs. p. 333.

L. NMR Suppliers

The nmr related materials stocked by each of the following companies are indicated after each address. Several references to this list are also made in the preceding text.

1. Aldrich Chemical Co., Inc., 940 West St. Paul Ave., Milwaukee, Wis. 53233 (deuterated solvents, liquid crystals).
2. Anderson Development Co., 1415 East Michigan Street, Adrian, Mich. 49221 (pmr standards).
3. Apache Chemical Inc., Department AL, P. O. Box 17, Rockford, Ill. 61105 (shift reagents).
4. J. T. Baker Chemical Co., Phillipsburg, N. J. 08865 (pmr standards, deuterated solvents).
5. Bio-Rad Laboratories, 32nd and Griffen Avenues, Richmond, Calif. 94804 (deuterated solvents, ^{13}C, ^{15}N, and ^{17}O-labeled compounds).
6. Brinkman Instruments Inc., Cantiague Road, Westbury, N. Y. 11590 (pmr standards).
7. Chemical Procurement Labs, Inc., 18-17 130th Street, College Point, N. Y. 11356 (pmr standards).
8. Columbia Organic Chemicals Co., Inc., 912 Drake Street, Columbia, S. C. 29205 (pmr standards, deuterated solvents).
9. Diaprep, Inc., P. O. Box 1844, Atlanta, Ga. 30301 (deuterated solvents, shift reagents, tetramethylsilane).
10. E. M. Laboratories, Inc., 500 Executive Boulevard, Elmsford, N. Y. 10523 (pmr standards, shift reagents).
11. Isomet Corp., 433 Commercial Avenue, Palisades, N. J. 07650 (deuterated solvents, ^{13}C, ^{15}N, and ^{17}O-labeled compounds).
12. Isotopes, A Teledyne Company, 50 Van Buren Avenue, Westwood, N. J. 07675 (deuterated solvents, ^{13}C, and ^{15}N-labeled compounds).
13. Kontes Glass Co., Vineland, N. J. 07650 (sample tubes, microsample tubes).
14. Mallinckrodt Chemical Works, Science Products Division, P. O. Box 5439, St. Louis, Mo. 63160 (deuterated solvents, ^{13}C and ^{15}N-labeled compounds).
15. Marshallton Research Laboratories, Inc., Box 305, West Chester, Pa. 19380 (pmr standards).
16. Merck Chemical Division, Merck and Co., Inc. 111 Central Avenue, Teterboro, N. J. 07065 (deuterated solvents, ^{13}C and ^{15}N-labeled compounds, pmr standards, shift reagents).
17. Miles Laboratories, Inc., Research Products Division, Elkhart, Ind. 46514 (deuterated solvents, ^{13}C, ^{15}N, and ^{17}O-labeled compounds).
18. Norell Chemical Company, Inc., P. O. Box D1200, Landing, N. J. 07850 (deuterated solvents, pmr standards, shift reagents, ^{13}C-labeled compounds, sample tubes, recorder chart paper).
19. Nuclear Magnetic Resonance Specialities, Inc., 1410 Greensburg Road, New Kensington, Pa. 15068 (deuterated solvents, standards for ^{1}H, ^{13}C, ^{11}B, ^{19}F, and ^{31}P, liquid crystals, sample tubes, microsample tubes, recorder chart paper, other accessories).
20. PCR, Inc., P. O. Box 1466, Gainsville, Fla. 32601 (deuterated solvents, standards for ^{1}H, ^{11}B, ^{13}C, and ^{19}F, shift reagents).
21. Pierce Chemical Co., P. O. Box 117, Rockford, Ill. 61105 (shift reagents).
22. Prochem Ltd., P. O. Box 555, Lincoln Park, N. J. 07035 (deuterated solvents, pmr standards, ^{13}C and ^{15}N-labeled compounds).
23. Stohler Isotope Chemicals, Inc., 49 Jones Road, Waltham, Mass. 02154 (deuterated solvents, ^{13}C and ^{15}N-labeled compounds, sample tubes, recorder chart paper, other accessories).

Refs. p. 333.

24. Thompson Packard Inc., 48 Notch Road, Little Falls, N. J. 07424 (deuterated solvents, standards for ^{1}H, ^{11}B, ^{13}C, ^{19}F, and ^{31}P, shift reagents, ^{13}C and ^{15}N-labeled compounds, sample tubes, recorder chart paper, other accessories).
25. Wilmad Glass Co., Inc., U. S. Route 40 and Oak Road, Buena, N. J. 08310 (sample tubes, microsample tubes, recorder chart paper, other accessories).
26. Ventron Alpha Inorganics, P. O. Box 159, Beverly, Mass. 01915 (shift reagents).
27. Willow Brook Laboratories, Inc., P. O. Box 526, Waukesha, Wis. 53186 (shift reagents).

M. References

1. General books: (a) E. D. Becker, High Resolution NMR, Academic Press, New York, 1969; (b) F. A. Bovey, NMR Spectroscopy, Academic Press, New York 1969; (c) J. W. Emsley, J. Feeney, and L. H. Sutcliffe, High Resolution NMR, Vols. 1 and 2, Pergamon, New York, 1966; (d) J. A. Pople, W. G. Schneider, and H. J. Bernstein, High Resolution NMR, McGraw-Hill, New York, 1959; (e) J. D. Roberts, Nuclear Magnetic Resonance, McGraw-Hill, New York, 1959.
2. Books with emphasis on application: (a) N. S. Bhacca and D. H. Williams, Applications of NMR Spectroscopy in Organic Chemistry, Holden-Day, San Francisco, 1964; (b) R. H. Bible, Introduction to NMR Spectroscopy, Plenum Press, New York, 1965; (c) R. H. Bible, Guide to the NMR Empirical Method, Plenum Press, New York, 1967; (d) L. M. Jackman and S. Sternhell, Application of Nuclear Magnetic Resonance in Organic Chemistry, 2nd ed., Oxford, New York, 1969; (e) D. W. Mathieson, Ed., NMR for Organic Chemistry, Academic Press, New York, 1967.
3. Books with emphasis on theory: (a) A. Carrington and A. D. McLachlen, Introduction to Magnetic Resonance, Harper and Row, New York, 1966; (b) P. L. Corio, Structure of High Resolution NMR Spectra, Academic Press, New York, 1967; (c) J. D. Memory, Quantum Theory of Magnetic Resonance Parameters, McGraw-Hill, New York 1968; (d) C. P. Slichter, Principles of Magnetic Resonance, Harper and Row, New York, 1963; (e) R. T. Schumacher, Introduction to Magnetic Resonance, W. A. Benjamin, New York, 1970; (f) R. J. Abraham, The Analysis of High Resolution NMR Spectra, Elsevier, New York, 1971.
4. Compilations of data (in addition to some of the books listed above): (a) H. A. Szymanski and R. E. Yelin, NMR Band Handbook, IFI/Plenum, New York, 1968; (b) F. A. Bovey, NMR Data Tables for Organic Compounds, Wiley-Interscience, New York, 1967; (c) W. Brügel, NMR Spectra and Chemical Structure, Academic Press, New York, 1967; (d) Sadtler Standard NMR Spectra, Sadtler Research Laboratories, Philadelphia, Pa., 1967; (e) M. G. Howell, A. S. Kende, and J. S. Webb, Formula Index to NMR Literature Data, Vols. 1 and 2, Plenum Press, New York, 1966; (f) H. M. Hershenson, NMR and ESR Spectra Index, Academic Press, New York, 1965; (g) A Catalogue of the NMR Spectra of Hydrogen in Hydrocarbons and their Derivatives, Humble Oil and Refining Co., Baytown, Texas; (h) Selected NMR Spectral Data, Chemical Thermodynamics Center, Texas A. and M., College Station, Texas; (i) Varian High Resolution NMR Spectra Catalog, Vols. 1 and 2, Varian Associates, Palo Alto, Calif., 1963.
5. Serial publications: Advances in Magnetic Resonance (Advan. Magn. Reson.), Vols. 1-4, J. S. Waugh, Ed., Academic Press, New York, 1965-

1970; Annual Review of NMR Spectroscopy (Ann. Rev. NMR Spectros., starting with volume 3 this series is called Annual Reports on NMR Spectroscopy), Vols. 1-4, E. F. Mooney, Ed., Academic Press, New York, 1968-1971; Journal of Magnetic Resonance (J. Magn. Reson.) Vols. 1-, W. S. Brey, Ed., Academic Press, New York, 1969-; NMR, Basic Principles and Progress (NMR), Vols. 1-, P. Diehl, E. Fluck, and R. Kosfeld, Eds., Springer-Verlag, New York, 1969-; Organic Magnetic Resonance (Org. Magn. Reson.) Vols. 1-, E. F. Mooney, Ed., Heyden and Son Ltd., London, 1969-; Progress in NMR Spectroscopy (Progr. NMR Spectros.) Vols. 1-7, J. W. Emsley, J. Feeney, and L. H. Sutcliffe, Eds., Pergamon Press, New York, 1966-1971.

6. D. G. Howery, J. Chem. Educ., 48, A327 and A389 (1971).
7. F. H. A. Rummens, Org. Magn. Reson., 2, 209 (1970).
8. L. Pohl and M. Eckle, Angew. Chem., Intern. Ed. Engl., 8, (a) 380 and (b) 381 (1969).
9. A Indriksons and R. West, J. Amer. Chem. Soc., 92, 6704 (1970).
10. D. N. Hentz and S. Vary, Chem. Ind., 1783 (1967).
11. (a) J. Ronayne and D. H. Williams, Ann. Rev. NMR Spectros., 2, 83 (1969); (b) P. Laszlo, Progr. NMR Spectros., 3, 231 (1967); (c) M. Bacon, G. E. Maciel, W. K. Musker, and R. Scholl, J. Amer. Chem. Soc., 93, 2537 (1971).
12. E. Mohacsi, Analyst, 91, 57 (1966).
13. J. K. Becconsall, G. D. Daves, and W. R. Anderson, J. Amer. Chem. Soc., 92, 431 (1970).
14. P. W. Hickmott and O. Meth-Cohn, in An Introduction to Spectroscopic Methods for the Identification of Organic Compounds, Vol. 1, F. S. Scheinmann, Ed., Pergamon Press, New York, 1970.
15. S. Castellano, C. Sun, and R. Kostelnik, Tetrahedron Letters, 5205 (1967).
16. K. Hayamizu and O. Yamamoto, J. Mol. Spectros., 28, 89 (1968).
17. N. VanMeurs, Rec. Trav. Chim. Pays-Bas, 87, 145 (1968).
18. S. Castellano and R. Kostelnik, Tetrahedron Letters, 5211 (1967).
19. W. B. Smith, A. M. Ihrig, and J. L. Roark, J. Phys. Chem., 74, 812 (1970).
20. R. M. Silverstein and G. C. Bassler, Spectrometric Identification of Organic Compounds, 2nd ed., Wiley, New York, 1967.
21. C. Pascual, J. Meier, and W. Simon, Helv. Chim. Acta., 49, 164 (1966).
22. S. W. Tobey, J. Org. Chem., 34, 1281 (1969).
23. A. B. Bothner-By, Advan. Magn. Reson., 1, 195 (1965).
24. (a) J. N. Murrell, Progr. NMR Spectros., 6, 1 (1970); (b) M. Barfield and D. M. Grant, Advan. Magn. Reson., 1, 149 (1965).
25. C. E. Maciel, J. W. McIver, Jr., N. S. Ostlund, and J. A. Pople, J. Amer. Chem. Soc., 92, 4151, 4497 and 4506 (1970).
26. R. C. Cookson, T. A. Crabb, J. J. Frankel, and J. Hudec, Tetrahedron, Supplement # 7, 355 (1966).
27. M. Karplus, J. Chem. Phys., 30, 11 (1959); J. Amer. Chem. Soc., 85, 2870 (1963).
28. E. W. Garbisch, Jr., J. Amer. Chem. Soc., 86, 5561 (1964).
29. R. J. Highet, personal communication to E. D. Becker in reference 1a.
30. M. Barfield and B. Chakrabarti, Chem. Rev., 69, 757 (1969).
31. N. C. Deno et al., J. Amer. Chem. Soc., 85, 2991 (1963).
32. Reviews of ^{13}C magnetic resonance: (a) J. Stothers, C-13 NMR Spectroscopy, Academic Press, New York, 1972; (b) P. S. Pregosin and E. W. Randall, in Determination of Organic Structures by Physical Methods, Vol. 3, F. C. Nachod and J. S. Zuckerman, Eds., Academic Press, New York, 1971; (c) D. M. Grant, Chem. Rev., in press; (d) E. F. Mooney and P. H. Wilson, Ann. Rev. NMR Spectros., 2, 153 (1969); (e) J. B. Stothers, Quart. Rev., 19, 144 (1965); (f) L. F. Johnson and

W. C. Jankowski, Carbon-13 NMR Spectra, Wiley-Interscience, New York, 1972 (a collection of 500 actual spectra with indexes). (g) G. C. Levy and G. L. Nelson, Carbon-13 Nuclear Magnetic Resonance for Organic Chemists, Wiley-Interscience, New York, 1972.

33. H. J. Reich, M. Jautelat, M. T. Messe, F. J. Weigert, and J. D. Roberts, J. Amer. Chem. Soc., 91, 7445 (1969).
34. H. Horsley, H. Sternlicht, and J. S. Cohen, ibid., 92, 680 (1970).
35. G. E. Maciel, J. W. McIver, Jr., N. S. Ostlund, and J. A. Pople, ibid., 92, 1 and 11 (1970).
36. (a) E. F. Mooney, An Introduction to ^{19}F NMR Spectroscopy, Heyden and Son Ltd., London, and Sadtler Research Laboratories, Inc., Philadelphia, 1970; (b) E. F. Mooney and P. H. Winson, Ann. Rev. NMR Spectros., 1, 243 (1968).
37. (a) C. H. Dungan and J. R. VanWazer, Compilation of F^{19} Chemical Shifts, Wiley-Interscience, New York, 1970; (b) J. W. Emsley and L. Phillips, Progr. NMR Spectros., 7, 1 (1971).
38. (a) R. L. Lichter, in Determination of Organic Structures of Physical Methods, Vol. 3, F. C. Nachod and J. S. Zuckerman, Eds., Academic Press, New York, 1971; (b) E. W. Randall and D. G. Gillies, Progr. NMR Spectros., 6, 119 (1971); (c) E. F. Mooney and P. H. Winson, Ann. Rev. NMR Spectros., 2, 125 (1969); (d) G. W. A. Milne et al., J. Amer. Chem. Soc., 93, 6536 (1971).
39. D. Herbison-Evans and R. E. Richards, Mol. Phys., 8, 19 (1964).
40. E. D. Becker, J. Magn. Reson., 4, 142 (1971).
41. R. L. Lichter and J. D. Roberts, J. Org. Chem., 35, 2806 (1970).
42. T. Axenrod, M. J. Wieder, G. Berti, and P. L. Barilli, J. Amer. Chem. Soc., 92, 6066 (1970).
43. M. M Crutchfield, C. H. Dungan J. H. Letcher, V. Mark and J. R. Van Wazer, P^{31} Nuclear Magnetic Resonance, Topics in Phosphorous Chemistry, Vol. 5, Wiley-Interscience, New York, 1967.
44. J. F. Nixon and A. Pidcock, Ann. Rev. NMR Spectros, 2, 346 (1969).
45. G. Manel, Progr. NMR Spectros. 1, 251 (1966).
46. A. L. VanGeet, Abstracts of the 10th Experimental NMR Conferences, March, 1969, Mellon Institute, Pittsburgh, Pa.
47. J. D. Swalen and C. A. Reilly, J. Chem. Phys., 37, 21 (1962).
48. S. Castellano and A. A. Bothner-By, ibid., 41, 3863 (1964).
49. R. B. Johannesen, J. A. Ferretti and R. K. Harris, J. Magn. Reson., 3, 84 (1970).
50. C. C. Hinckley, J. Amer. Chem. Soc., 91, 5160 (1969).
51. J. K. M. Sanders and D. H. Williams, Chem. Comm., 422 (1970).
52. J. Briggs, G. H. Frost, F. A. Hart, G. P. Moss, and M. L. Stanforth, Chem. Comm., 749 (1970).
53. R. E. Rondeau and R. E. Sievers, J. Amer. Chem. Soc., 93, 1522 (1971).
54. P. V. Demarco, T. K. Elzey, R. B. Lewis, and E. Wenkert, J. Amer. Chem. Soc., 92, 5735 (1970).
55. T. H. Siddall, Chem. Comm., 452 (1971).
56. G. Whitesides and D. W. Lewis, J. Amer. Chem. Soc., 92, 6979 (1970).
57. L. C. Snyder and E. W. Anderson, ibid., 86, 5024 (1964).
58. R. R. Fraser, M. A. Petit, and M. Miskow, ibid., 94, 3253 (1972).
59. Several applications of metal and donor atom nmr to the study of coordination compounds of Hg, Co, Cr, Ir, W, As, Ru, Rh, Pt, and B are reviewed in J. W. Dawson and L. M. Venanzi, Trans. N.Y. Acad. Scis., 32, 304 (1970).

IX. ELECTRON PARAMAGNETIC RESONANCE SPECTROSCOPY

A. Introduction

Electron paramagnetic resonance or epr (also known as electron spin resonance or esr) is a means of studying the environment of unpaired electrons in paramagnetic substances. The most common application has been the study of organic free radicals, although paramagnetic metal ion complexes and photoexcited or stable triplet states have also been investigated. A current bibliography of the subject is found at the end of this section.

The basic method for epr is similar to that for nmr. The free electron has a spin quantum number of 1/2; thus, in a magnetic field, it exists in two states of different energy. This energy difference is given by

$$E = h\nu = g\beta_e H_o$$

where β_e is the Bohr magneton, H_o is the applied magnetic field and g is the spectroscopic splitting factor--the ratio of the magnetic moment to the angular momentum with a value of 2.002319 for an unbound (free) electron. This g factor is only slightly different for most free radicals. It can be seen that the frequency at which resonance occurs is determined by the applied magnetic field (usually 3000 to 3400 G, resulting in a resonance frequency of 9000 to 9500 MHz) and by the g factor (the Bohr magneton is a constant = 9.2732×10^{-21} erg/G). Furthermore, the total area under an absorption is proportional to the number of unpaired electrons. In practice, absolute calculation is difficult, but direct comparison is made conveniently with a standard containing a known number of radical molecules; area ratios then provide concentrations for the unknown. A common standard is diphenylpicrylhydrazyl [DPPH-$(C_6H_5)_2N\text{-}\dot{N}\text{-}C_6H_2(NO_2)_3$] which exists 100% as radicals (1.53×10^{21} unpaired electrons per gram).

The g factor for an electron is analogous to the shielding factor in nmr. That is, unpaired electrons will have different g values depending on the chemical environment. For liquids, a single g value is observed due to averaging by rapid tumbling, while in a solid different g values may be observed along different axes. As in nmr, the use of liquid crystals as solvents has yielded valuable information. This technique is discussed in:

G. R. Lockhurst, RIC Rev., 3, 63 (1970).
J. R. Morton, Chem. Rev., 64, 453 (1964).

Organic radicals in solution all possess g values close to that of an unbound electron. The following representative g values were taken from the references given below:

Radical	g Value
Methyl	2.00255
Ethyl	2.00260
Isopropyl	2.003
t-Butyl	2.003
Vinyl	2.00220

(Continued)

Radical	g Value
Allyl	2.00254
1,4-Benzosemiquinone	2.00468
Anthracene cation	2.00249
Triphenylphosphine cation	2.00554
Naphthalene anion	2.00263
Anthracene anion	2.00266
Benzene anion	2.00276

B. Spin-Spin Coupling

It is possible to observe spin-spin coupling between the electron and any nuclei (with a magnetic moment) with which the electron is either wholly or partially associated. This coupling, just as in pmr gives rise to a splitting of the resonance line (hyperfine splitting) according to the same rules for first-order coupling in nmr. The number of peaks is $2nI + 1$, where n is the number of equivalent nuclei of spin I, with the relative intensities determined by the coefficients in the binomial expansion; the spacing is equal to the coupling constant, a_i. For details, see the rules covering first-order coupling in nmr (p. 308).

These coupling constants are directly proportional to the probability of finding the electron at the position of the given nucleus; in other words, the value is a measure of the spin density at that nucleus. The coupling constant for the hydrogen atom is 508 G, and thus experimental values of coupling to hydrogen atoms, a_H can be interpreted in terms of the electron spin density, ρ_H, at that hydrogen by the equation

$$\rho_H = \frac{a_H}{508}$$

Likewise, in conjugated π-systems, a_H values may be interpreted in terms of the electron spin density at the carbon atom by the approximate relationship

$$a_H = Q\rho$$

where Q is a constant with a value between 22.5 and 30 G, usually taken as 24.

The following are typical hyperfine splitting constants and refer to hydrogen coupling unless otherwise indicated.

Radical	Coupling (G)	Ref.
H•	508	a
Li•	81	a
Na•	632	a

(Continued)

Radical		Coupling (G)				Ref.
K•		165				a
Cs•		3280				a
$^{14}NO\cdot$		(^{14}N) 14.3				a
$\cdot C^{14}N$		(^{14}N) 174				a
$CH_3\cdot$		(^{1}H) 22.8	(^{13}C) 41			a,b
$CH_3CH_2\cdot$		α-22.3	β-26.8			b
$CH_3CH_2CH_2\cdot$		α-22.1	β-30.3	γ-0.3		b
$CH_3\dot{C}HCH_3$		α-21.2	β-24.7			c
$(CH_3)_3C\cdot$			22.7			c
$PhCH_2^{\bullet}$		α-16.3	o-5.2	m-1.8	p-6.2	c
$CH_2{=}CH{-}CH_2CH_2\cdot$		α-22.2	β-28.5	γ-0.6	δ-0.35[e]	b
Allyl radical (Hα, Hβ, Hγ)		α-14.81	β-13.90	γ-4.06		c
$CH_2{=}CH\cdot$		α-16	cisβ-68[f]	transβ-34[f]		a
$H{-}C{\equiv}C\cdot$			16.1			a
Cyclopropyl radical		α-6.5	β-23.4			a
Cyclobutyl radical		α-21.3	β-36.8	γ-1.1		c
Cyclopentadienyl		6.0				c
Cyclohexadienyl		CH_2-47.7	2H-8.99	3H-2.65	4H-13.0	a
Cycloheptatrienyl		3.9				
Benzene	+	2.89				d
	−	(^{1}H)-3.75	(^{13}C)-2.8			a
Naphthalene	+	α-4.90	β-1.83			d
	−	α-4.95	β-1.87			d

(Continued)

Radical		Coupling (G)				Ref.
N, N	+·	(^{1}H)3.26	(^{14}N)7.6			d
	−·	(^{1}H)2.6	(^{14}N)7.1			
	−·	3.23				
—CH_3	−·	(CH_3)−0.79	o−5.12	m−4.45	p−0.59	d
—NO_2	−·	(^{14}N)−9.70	o−3.36	m−1.07	p−4.03	d

[a]E. M. Kosower, Physical Organic Chemistry, Wiley, New York, 1968, p. 418. [b]J. K. Kochi and P. J. Krusic, J. Amer. Chem. Soc., 91, 3940 (1969). [c]P. J. Krusic and J. K. Kochi, ibid., 90, 7155, 7157 (1968). [d]D. J. Pasto and C. R. Johnson, Organic Structure Determination, Prentice-Hall, Englewood Cliffs, N. J., 1969, p. 221 [e]Coupling to only one δ-H. [f]Cis or trans to the hydrogen.

C. Spin Labeling

One of the most useful applications of epr is in "spin labeling" of complex biomolecules such as enzymes. This method consists of attaching a stable radical species to the molecule and observing the effect on the epr spectrum. Changes in the physical or chemical environment of the spin label can reflect changes in conformation or structure of the biomolecule. Spin labels are available from Synvar Associates, Palo Alto, Calif., as a kit of four derivatives of 2,6-tetramethylpiperidino-1-oxyl radical.

D. Bibliography

R. S. Alger, Electron Paramagnetic Resonance, Techniques and Applications, Wiley-Interscience, New York, 1968.

C. P. Poole, Electron Spin Resonance, A Comprehensive Treatise on Experimental Techniques, Wiley-Interscience, New York, 1967.

H. M. Assenheim, Introduction to Electron Spin Resonance, Plenum Press, New York, 1967.

P. B. Ayscough, Electron Spin Resonance in Chemistry, Methuen and Co., London, 1967.

A. Carrington and A. D. McLachlen, Introduction to Magnetic Resonance, Harper and Row, New York, 1967.

M. Bersohn and J. C. Baird, An Introduction to Electron Paramagnetic Resonance, Benjamin, New York, 1966.

Paramagnetic Resonance, Proceedings of the First International Conference held in Jerusalem, 1962, Vols. I and II, W. Low, Ed., Academic Press, New York, 1963.
L. Kevan, in Methods in Free Radical Chemistry, Vol. I, E. S. Huyser, Ed., Marcel Dekker, New York, 1969.
D. H. Geske, Progr. Phys. Org. Chem., 4, 123 (1967).
F. Schneider, K. Möbius, and M. Plato, The Use of Electron Paramagnetic Resonance in Organic Chemistry, Angew. Chem. Intern. Ed. Engl., 4, 856 (1965).
C. Thompson, Electron-Spin Resonance Studies of the Triplet State, Quart. Rev., 22, 45 (1968).
E. T. Kaiser and L. Kevan, Ed., Radical Ions, Interscience, New York,1968.
K. W. Bowers, Electron Spin Resonance of Radical Ions, Advances in Magnetic Resonance, 1, 317 (1965).
A Carrington, Electron-Spin Resonance Spectra of Aromatic Radicals and Radical Ions, Quart. Rev., 17, 67 (1963).
D. J. E. Ingram, Biological and Biochemical Applications of Electron Spin Resonance, Plenum Press, New York, 1969.
A Ehrenberg, B. G. Malmstron, and T. Vänngård, Eds., Magnetic Resonance in Biological Systems; Proceedings, Pergamon Press, New York 1967.
T. F. Yen, Ed., Symposium on ESR of Metal Chelates, Cleveland, 1968, Plenum Press, New York, 1969.
P. W. Atkins and M. C. R. Symons, The Structure of Inorganic Radicals, An Application of Electron Spin Resonance to the Study of Molecular Structure, Elsevier, New York, 1967.
A. Hudson and G. R. Lockhurst, The Electron Resonance Line Shape of Radicals in Solutions, Chem. Rev., 69, 191 (1969).

X. NUCLEAR QUADRUPOLE RESONANCE SPECTROSCOPY

Nuclei with spin quantum number of 1 or greater (^{2}H, ^{14}N, ^{35}Cl, ^{37}Cl, ^{59}Co, ^{79}Br, ^{81}Br, e.g.) possess, in addition to a magnetic moment, an electric quadrupole moment (see p. 314 for a list of quadrupole moments) that can be thought of as two equal and opposing (antiparallel) electric dipoles. This is a result of the nonspherical distribution of nuclear charge. If the nucleus is in a uniform electric field, both moments experience equal but opposite torques, and there is no effect. This is the case for chloride ion or the sufur in $^{33}SF_6$; neither of these gives rise to an NQR signal even though both nuclei possess a quadrupole moment. If however, the nucleus is in a nonuniform electric field such as in HCl or H_2S, different force will be felt at each of the antiparallel dipoles and the nucleus will have (2I + 1) allowed orientations. The energy difference between these orientations can be observed by absorption of radiofrequency (resonance) in the 2 to 350 MHz range. These energy differences are determined by the charge (electron) distribution within the molecule, hence the resonance frequency for a nucleus can be said to be analogous to the chemical shift in nmr spectroscopy. This resonance frequency is dependent on both structure and conformation.

The resonance frequency is usually expressed in terms of the quadrupole coupling constant e^2qQ/h where -e is the charge on an electron, eq is the electric field gradient, Q is the nuclear quadrupole moment, and h is Planck's constant. This coupling constant is an expression of the energy of interaction between the quadrupole moment and the field gradient and is thus unique for a nucleus in a particular environment. It should be noted, however, that more than one frequency might be observed for this nucleus because of unequal energy spacing between the allowed orientations. These frequencies will be proportional to the quadrupole coupling constant. For example, ^{35}Cl has a spin of 3/2 and thus (2I +1) or 4 energy levels. In a molecule with axial symmetry, these energy levels will occur as two sets of two degenerate levels, and thus only one frequency will be observed: $\nu = e^2qQ/2h$. On the other hand, ^{127}I (spin 5/2) in HI has two resonance frequencies: $\nu_1 = 3e^2qQ/20h$ and $\nu_2 = 3e^2qQ/10h$. In cases of nonaxial symmetry, the degeneracy is removed, and three frequencies may be observed for a particular ^{35}Cl nucleus.

NQR spectra are observed only in the solid phase, and thus liquid samples must be frozen. Since resonant frequencies are temperature dependent, a standard temperature for measurement is helpful, and most spectra are reported at liquid nitrogen temperature, 77°K.

In a crystalline sample, different positions of the nucleus in the crystal lattice will result in different resonant frequencies, and thus information on the crystal structure may be obtained.

The following table of typical resonant frequencies for several nuclei is reproduced with permission from Schultz, Appl. Spectros., 25, 293 (1971) (see bibliography below).

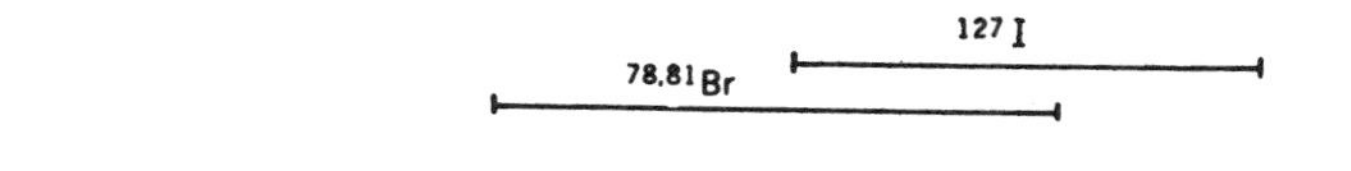

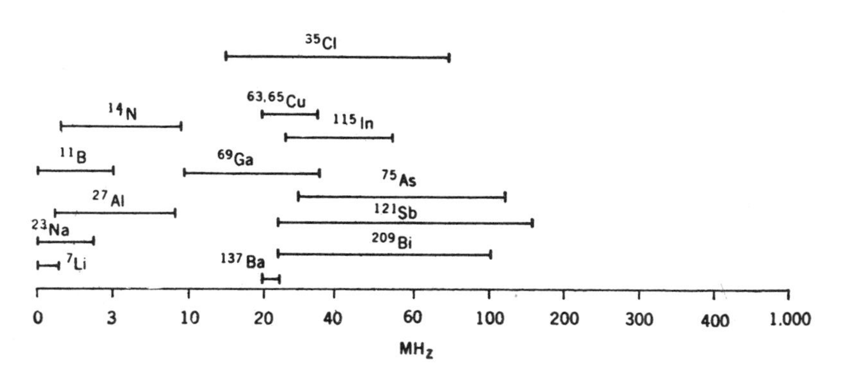

Further discussion of theory and applications of NQR spectrometry can be found in the references listed below. A simple introduction to NQR along with a description of a spectrometer, a bibliography (through 1966), and a table of NQR values appear in a pamphlet (Model NQR-1A Nuclear Quadrupole Resonance Spectrometer) distributed by Wilk Scientific Corporation, South Norwalk, Conn. 06856.

H. D. Schultz, Applications of Nuclear Quadrupole Resonance Spectrometry to Analytical Chemistry, Appl. Spectros., 25, 293 (1971).

J. A. S. Smith, J. Chem. Educ., 48, 39, A77, A147, and A243 (1971).

E. A. C. Lucken, Nuclear Quadrupole Coupling Constants, Academic Press, New York, 1969.

T. P. Dos and E. L. Hahn, Nuclear Quadrupole Resonance Spectroscopy, Academic Press, New York, 1958.

I. P. Biryukov, M. G. Voronkov, and I. A. Safin, Tables of Nuclear Quadrupole Resonance Frequencies, International Scholarly Book Services, Inc. (Portland, Oregon), 1969. (A translation from the 1968 Russian original, by the Israel Program for Scientific Translation.)

XI. BIBLIOGRAPHY OF SPECTRAL DATA COMPILATIONS

The following references provide detailed spectral data or indexes to such data for specific compounds and structural types. For addresses of publishers, see p. 506.

A. Infrared Spectra

1. Catalog of Infrared Spectral Data, American Petroleum Institute Research Project 44 (APIRP 44) and Manufacturing Chemists Association Research Project, Chemical and Petroleum Research Laboratories, Carnegie Institute of Technology, Pittsburgh, Pa., to June 30, 1960; Chemical Thermodynamics Properties Center, Texas A. and M., College Station, Texas, form July 1, 1960. An extensive collection of complete spectra (mainly hydrocarbons and simple derivatives).
2. Catalog of Infrared Spectrograms, Sadtler Research Laboratories, Inc., 3322 Spring Garden Street, Philadelphia, Pa. 19104. Various compendia of prism and grating spectra for the infrared, near-infrared, and far-infrared, including a special collection of spectra for commercial compounds. In all, over 120,000 spectra as of October 1969; additions being made regularly. Sadtler also offers an ir information system (IRIS) based on a computer data-retrieval method using more than 120,000 spectra in the coded data base. Eastman Kodak (Rochester, N. Y. 14650) has a similar retrieval service; ask for Publication JJ-165).
3. The Coblentz Society Spectra; high-quality spectra for several thousand pure compounds. Available from Sadtler.
4. An index of Published Infrared Spectra, M. B. Thomas, Ed., Vols. 1 and 2, British Information Service, 45 Rockefeller Plaza, New York, 1960 (complete to 1957); for more recent indices, contact H. M. Stationary Office, London, England (Ministry of Aviation Technical Information and Library Services, Ed.).
5. Documentation of Molecular Spectroscopy, (DMS) Butterworth's Scientific Publications, London and Verlag-Chemie GMBH, Weinheim/Bergstrasse, West Germany. Spectra on coded punch cards (needle sorting); also a separate card set containing abstracts of articles relating to ir spectrometry. Coding based on basic skeletal features, functional groups and number of carbons. A similar collection is IRDC Cards, Infrared Data Committee of Japan (Nankodo Co., Haruki-Cho, Tokyo).
6. Infrared Band Handbook, H. A. Szymanski, Plenum Press, New York, Vol. 1, 1963 and later supplements; 2nd ed., 2 volumes, 1970.
7. Infrared Absorption Spectra of Steroids. An Atlas, K. Dobriner, E. Katzenellenbogen, and R. Jones, Interscience, New York, Vol. 1, 1953; Vol. 2, 1958.
8. Infrared Spectra and Characteristic Frequencies of Inorganic Ions, F. Miller and C. Wilkins, Anal. Chem., 24, 1253 (1952).
9. Infrared Absorption Spectra: Indexes, H. M. Hershenson, Academic Press, New York. Collections of indexes to the literature for spectra of specific compounds. Covers years 1945-1957 (published 1959) and 1958-1962 (published 1964).
10. Infrared Spectra of Cellulose and Its Derivatives, R. G. Zhbankov, Consultants Bureau, New York., 1966 (translated from Russian).
11. Selected Raman Spectral Data, APIRP 44, Vols. I and II covering 1948-1964. Chemical Thermodynamics Properties Center, Texas A. and M., College Station, Texas.
12. IR Analysis of Polymers, Resins and Additives. An Atlas, D. O. Hummel and F. Scholl, Eds., Wiley-Interscience, New York, 1969.

13. Infrared Spectra of Selected Chemical Compounds, R. Mecke and F. Langenbucher, Heyden and Son, Ltd., London, 1967. About 2000 selected spectra in 8 volumes of pure organic and inorganic compounds.
14. IRSCOT-SYSTEM (Infrared Structural Correlation Tables and Data Cards) R. G. J. Miller and H. A. Willis, Heyden and Son, Ltd., London, 1967 and later issues. A reference system of tables directing researcher to one or more specific cards containing structural information.
15. The Aldrich Library of Infrared Spectra, C. J. Pouchert, Aldrich Chemical Co., 940 West St. Paul Avenue, Milwaukee, Wis. 53233. More than 8000 grating (mainly) spectra of pure compounds in one book.
16. IRIS, A computer search program of Sadtler Research Laboratories and University Computing Company (Palo Alto, Calif.). Compares on individual basis (via time sharing) ir spectra of unknowns to those of >100,000 reference compounds (see reference 2 above).
17. Infrared Spectra and X-Ray Spectra of Hetero-Organic Compounds, G. F. Bolshakov et al., Khemia Publishing House, Moscow, 1967.

B. Electronic Spectra

1. Organic Electronic Spectral Data, a multivolume collection of uv and vis data covering literature from 1946-1953 (Vol. I, M. Kamlet, Ed.), 1953-1955 (Vol. II, H. Ungnade, Ed.), 1955-1958 (Vol. III, O. Wheeler and L. Kaplan, Eds.), 1958-1959 (Vol. IV, J. Phillips and F. Nachod, Eds.), 1960-1961 (Vol. V, L. Phillips and Jones), 1962-1963 (Vol. VI, J. P. Phillips et al., Eds.), 1964-1965 (Vol. VII, J. P. Phillips et al.), Interscience, New York.
2. Catalog of Ultraviolet Absorption Spectrograms, API 44 and MCA Research Projects, Carnegie Institute of Technology, Pittsburgh, Pa.
3. Selected Ultraviolet Spectral Data, APIRP 44, Vols. I-III covering 1945-1964 (Vols. I and II, see reference 2 above; Vol. III, Texas A. and M.).
4. Handbook of Ultraviolet and Visible Absorption Spectra of Organic Compounds, K. Hirayama, Plenum Press, New York, 1967.
5. Ultraviolet and Visible Absorption Spectra: Indexes, H. Hershenson, Academic Press, New York. Covers literature for 1930-1954 (published 1956), 1955-1959 (published 1961), and 1960-1963 (published 1966).
6. Ultraviolet Spectra of Aromatic Compounds, R. Friedel and M. Orchin, Wiley, New York, 1958.
7. Sadtler Standard UV Spectra, Sadtler Research Laboratories, 1969 (26,000 spectra in 62 volumes thus far).
8. Interpretation of the UV Spectra of Natural Products, A. I. Scott, Pergamon Press, New York, 1964.
9. U.V. Atlas of Organic Compounds, Butterworth, London, and Plenum Press, New York (published in collaboration with Photoelectric Spectrometry Group, London and Institut für Spektrochemie und Angewandte Spektroskopie, Dortmund, Germany). Volumes 1-4 (1966-1969). A continuing collection.
10. Spectra of Nucleic Acid Compounds, T. Venkstern and A. Baev, Plenum Press, New York, 1968.
11. Absorption Spectra in the Ultraviolet and Visible Region, L. Láng, Ed., Academic Press, New York and London, and Publishing House of Hungarian Academy of Sciences, Budapest. A series starting in 1959 (Vol. 1).
12. Ultraviolet Spectra of Elastomers and Rubber Chemicals, by V. S. Fikhtengolts et al., Plenum Press, New York, 1966 (translation from the Russian).
13. Atlas of Protein Spectra in the Ultraviolet and Visible Regions, D. M. Kirschenbaum, Ed., IFI/Plenum, New York, 1972.

14. Absorption Spectra of Minor Bases: Their Nucleosides, Nucleotides, and Selected Oligoribonucleotides, T. Vladimirovna et al., Plenum Press, New York, 1965 (translation from the Russian).

C. Nuclear Magnetic Resonance Spectra

1. Nuclear Magnetic Resonance and Electron Spin Resonance Spectra: Index 1958-1963, H. Hershenson, Academic Press, New York, 1965.
2. High Resolution NMR Spectra Catalog, N. S. Bhacca et al., Varian Associates, Palo Alto, Calif. (Vol. I, 1962; Vol. 2, 1963).
3. Formula Index to NMR Literature Data, M. Howell, A. Kende, and J. Webb, Vols. I and II, Plenum Press, New York, 1965.
4. A Catalogue of the NMR Spectra of Hydrogen in Hydrocarbons and Their Derivatives, Humble Oil and Refining Co., Baytown, Texas.
5. Selected NMR Spectral Data, APIRP 44, Chemical Thermodynamics Properties Center, Texas A. and M., College Station Texas (Vols. I-III covering 1959-1969).
6. NMR Data Tables for Organic Compounds, Vol. I, F. Bovey, Interscience, New York, 1967.
7. NMR Spectra and Chemical Structures, W. Brügel, Academic Press, New York, 1967.
8. NMR Band Handbook, H. Szymanski and R. Yelin, Plenum Press, New York, 1968.
9. Sadtler NMR Spectra, Sadtler Research Laboratories, Inc., 3322 Spring Garden Street, Philadelphia, Pa. 19104 (12 volumes of 8000 spectra thus far).
10. Nuclear Magnetic Resonance Abstracts, continuously updated punched card sets of abstracts dealing with all aspects of nmr. Preston Technical Abstracts Co., 2101 Dempster Street, Evanston, Ill. 60201 (system begun in 1964).
11. Compilation of Reported F^{19} NMR Chemical Shifts. 1951 to mid 1967, C. H. Dungan and J. R. Van Wazer, Wiley-Interscience, New York, 1970.
12. NMR Studies of Boron Hydrides and Related Compounds, G. R. Eaton and W. N. Lipscomb, Benjamin, New York, 1969. Contains extensive data and discussion of ^{11}B nmr.
13. Carbon-13 NMR Spectra, L. F. Johnson and W. C. Jankowski, Wiley-Interscience, New York 1972. A collection of 500 spectra and appropriate indexes.

D. Electron Spin Resonance Spectra

1. Atlas of Electron Spin Resonance Spectra, B. Bielski and J. Gebicki, Academic Press, New York, 1967 (first of a continuing series).
2. See reference C1 above.
3. Atlas of Electron Spin Resonance Spectra, Y. Lebedev et al., Plenum Press, New York, Vol. 1, 1963; Vol. 2, 1964. Theoretically calculated multicomponent symmetrical spectra.

E. Mass Spectra

1. Selected Mass Spectral Data, APIRP 44, Vols. I-VI, covering 1947-1969 and another set, Vols. I-VII, covering 1943-1967.
2. Index of Mass Spectral Data, L. E. Kuentzel, Ed., Wyandotte-ASTM Punched Card Project, ASTM, Philadelphia, 1963 (ASTM Special Publication No. 356); also AMD II-Index of Mass Spectral Data, ASTM, 1969.

3. Compilation of Mass Spectral Data, A. Cornu and R. Massot, Heyden and Sons, London, 1966 and supplements.

4. Mass Spectrometry Bulletin, Mass Spectrometry Data Center, AWRE, Aldermaston, Berkshire, England. A monthly bulletin covering all aspects of mass spectrometry with a continuously updated master list of spectra on file. These spectra--on digital tape--may be purchased individually or subscribed to regularly.
5. Atlas of Mass Spectral Data, E. Stenhagen, S. Abrahamsson, and F. McLafferty, Interscience, New York, 1969 (3 volumes). Spectral data (low resolution) for >6000 compounds (printout of computer-based storage and retrieval system at University of Göteborg).

F. X-Ray Data (See Also Reference A17 Above)

1. A Reference List of Organic Structures Whose Absolute Configurations Have Been Determined by X-Ray Methods, F. Allen and D. Rogers, Parts 1-3, J. Chem. Soc. (D), 838 (1966); 308 (1968); and 452 (1969).
2. Crystal Structures, R. W. Wyckoff, Wiley-Interscience, New York. First edition of 4 volumes (1948, 1951, 1953, 1957) and supplements (1951, 1959, 1960); second edition of 6 volumes, 1963-1967.
3. Index to Crystallographic Data Compilations, ASTM. An index to the "Powder Diffraction Data File (PD-1)," ASTM (for organic and inorganic substances). For details write Joint Committee on Powder Diffraction Standards, 1845 Walnut Street, Philadelphia, Pa. 19103.
4. Crystal Data, J. D. H. Donnay, General Editor, Polycrystal Book Service, Box 11567, Pittsburgh, Pa. 15238. Monograph 5 (1963) of the American Crystallographic Association; covers organic, inorganic, and protein substances.
5. Molecular Structures and Dimensions. Volume 1, Bibliography 1935-1969. Solid State Classes 1-59, and Volume 2, Bibliography 1935-1969. Solid State Classes 60-86, O. Kennard and D. G. Watson, Eds. One of the most complete compilations of crystal structure data available. Published for the Crystallographic Data Centre (University Chemical Laboratory, Cambridge, England) and the International Union of Crystallography (1 rue Victor-Cousin, Paris 5, France) by N. V. A. Oosthoek's Uitgevers Mij., Utrecht, Netherlands). Available through Polycrystal Book Service (see reference 4 above).
6. Structure Reports, a continuing series [most recent is Vol. 26 (1969) covering data reported in 1960-1961 literature] providing a critical assessment of accuracy of X-ray determinations, along with listings of atomic coordinates and many diagrams. Published for the International Union of Crystallography by Oosthoek's (see reference 5 above).
7. International Tables for X-Ray Crystallography, a multivolume series, compiling essential data for crystallographers. Kynock Press, Birmingham, England.

G. Mössbauer Effect Spectra

1. Mössbauer Effect Data Index, Vol. 1, covering literature from 1958-1965, A. H. Muir, Jr., K. J. Ando, and H. M. Coogan, Eds., Interscience, New York, 1966, and Vol. 2, covering literature of 1969, J. G. Stevens and V. E. Stevens, Eds., IFI/Plenum, New York, 1970. A summary of nuclear and Mössbauer properties of each isotope and a thorough bibliography.

2. Mössbauer Spectrometry, J. R. Devoe and J. J. Spijkerman, Anal. Chem., 38, 382R-393R (1966); 40, 472R-489R (1968); 42, 366R-388R (1970). A continuing biennial review series.
3. The Mössbauer Effect, Bibliographical Series No. 16, International Atomic Energy Agency, Vienna, 1965. A bibliography of 1958-1964 publications, with abstracts of each.
4. Index of Publications in Mössbauer Spectroscopy of Biological Materials, L. May, Department of Chemistry, Catholic University of America, Washington, D. C. 20017. Issued every six months as a peek-a-boo card system.

H. Miscellaneous

1. Atlas of Steroid Spectra, W. Neudert and H. Repke, Eds., Springer-Verlag, New York, 1965 (ir, uv, nmr, and other physical data for steroids).
2. Thermodynamic Research Center Data Projects, Texas A. and M. University, College Station, Texas. These collections of loose-leaf sheet spectra are the extensions of the various APIRP 44 and MCA Research Projects. Included are:
   (1) Selected IR Spectral Data, Vols. I and II, 1959-1969.
   (2) Selected UV Data, Vol. I, 1959-1965.
   (3) Selected NMR Spectral Data, Vols. I-III, 1960-1969.
   (4) Selected Raman Spectral Data, Vol. I, 1956-1966.
   (5) Selected Mass Spectral Data, Vols. I and II, 1959-1969.
3. Documentation of Molecular Spectroscopy, (DMS) literature services. Issued by Butterworth Publishers, London. Continuously published literature lists, author indexes, general indexes; abstracts printed on edge-notched cards (needle-sorting); cover literature from 1963 on. Included are:
   (1) NMR, NQR, EPR Current Literature Lists.
   (2) IR, Raman, Microwave Current Literature Service.
4. National Standard Reference Data Series-National Bureau of Standards (NSRDS-NBS), U.S. Government Printing Office, Washington, D.C. 20402. An elaborate series of reference works on all aspects of physical and chemical data, covering thermal, spectroscopic, kinetic, and other properties. For a description and catalog, send $0.15 to the address above for "NSRDS-NBS 1." This series has recently become a separate journal published jointly by the ACS and APS (see page 494 for a discussion).
5. ASTM indexes of ir, mass spectra, and others (some on magnetic tape). Contains compilations of data from other sources. American Society for Testing and Materials, 1916 Race Street, Philadelphia, Pa. 19103. A time-shared computer search service for ir spectra in the ASTM collection is offered by Singer Technical Services, Inc., New York.
6. Spectral Data of Natural Products, K. Yamaguchi, Vol. I (1970), Vol. II (in preparation), American Elsevier, New York.
7. Spectral Data and Physical Constants of Alkaloids, J. Holubek and O. Strouf, Heyden and Son, Ltd., 1967. Data on cards; include uv, ir, and physical constants.
8. A Handbook of Alkaloids and Alkaloid-Containing Plants, R. F. Raffauf, Wiley-Interscience, New York 1970.
9. Physical Properties of the Steroid Hormones, L. L. Engel, Ed., Macmillan Co., New York, 1963. Uv and fluorescence data along with physical constants.

10. Tables of Frequencies of Nuclear Quadrupole Resonance, I. P. Biryukov, et al., Khemia Publishing House, Moscow, 1968. Available in an English translation from International Scholarly Book Service, Inc. (Portland, Oregon), 1969.
11. Spectroscopic Properties of Inorganic and Organometallic Compounds, Chemical Society, London. A series that reviews the literature published in a given year. For example, Vol. 4, 1971 covers 1970.
12. Atlas of Spectral Data and Physical Constants for Organic Compounds, CRC Press, Cleveland, 1972. A large collection (about 1500 pages) of critically reviewed and evaluated data for about 10,000 compounds. Includes CAS number, density, n_D, specific rotation, Wiswesser line notation, mp, bp, solubility, ir, uv, nmr, and mass spectral data.

PHOTOCHEMISTRY

I. ELECTRONIC ENERGY STATE DIAGRAM (JABLONSKI DIAGRAM)

Details on all aspects of photochemistry are covered thoroughly in the suggested references 1 through 10. Of particular interest to experimentalists is the new series on creation and detection of the excited state (7e). The Jablonski diagram (2,4,8) on the next page summarizes the various energy levels and transitions commonly available to ground and excited states of molecules. Horizontal lines represent vibrational levels within the electronic state (S = singlet, T = triplet). Solid arrows indicate absorption or emission; wavy lines indicate radiationless (thermal) processes. Abbreviations used are defined as follows (for absorption laws and terminology governing the uv-vis region, see page 211).

Symbol	Photophysical Process	$k\ (sec^{-1})$[a]
A	Photon absorption[e]	b
F	Fluorescence: singlet→singlet radiative emission[e]	$10^6 - 10^9$ [c]
P	Phosphorescence: triplet→singlet radiative emission[e]	$10^{-2} - 10^3$ [c]
ISC	Intersystem crossing: nonradiative spin inversion (change in multiplicity, usually from excited S→T or T→S)[d]	$<10^{-1} - 10^2$
IC	Internal conversion: radiationless (thermal) relaxation relaxation of a molecule from one level to a lower level within the same multiplicity (e.g., $S_2 \rightarrow S_1$)[d]	10^{11}
R	Vibrational-rotational relaxation from one vibrational level of a state, usually to the zero-point level of the same state [e.g., $(S_1)_{v_4} \rightarrow (S_1)_{v_0}$]	$10^{12} - 10^{14}$

[a]Approximate first order rate constant for the process; typical ranges for molecules in condensed phases.

[b]Virtually instantaneous; a unit of absorption often used is an "einstein" of light, which refers to an Avogadro's number of protons.

[c]Values of the radiative lifetimes (τ) of the excited state from which emission occurs may be calculated as 1/k; the lifetime for the state corresponding to a particular absorption band can be obtained from

$$1/\tau = 2.9 \times 10^{-9}\ n^2\ \bar{\nu}^2_{max} \int_0^{\infty} \varepsilon\ d\bar{\nu}$$

where n = refractive index of medium, $\bar{\nu}_{max} = cm^{-1}$ at band maximum, and ε = molar absorptivity (integration under the absorption curve).

[d]ISC and IC are "isoenergetic," that is, they occur between different electronic, but identical energy levels (see diagram below).

[e]A so-called 0-0 (zero-zero) absorption or emission band refers to the process occurring between the zeroth (ground state) vibrational levels of the respective electronic states.

Refs. p. 367.

Excited States and Photophysical Transitions Between States for Typical Polyatomic Singlet Molecules*

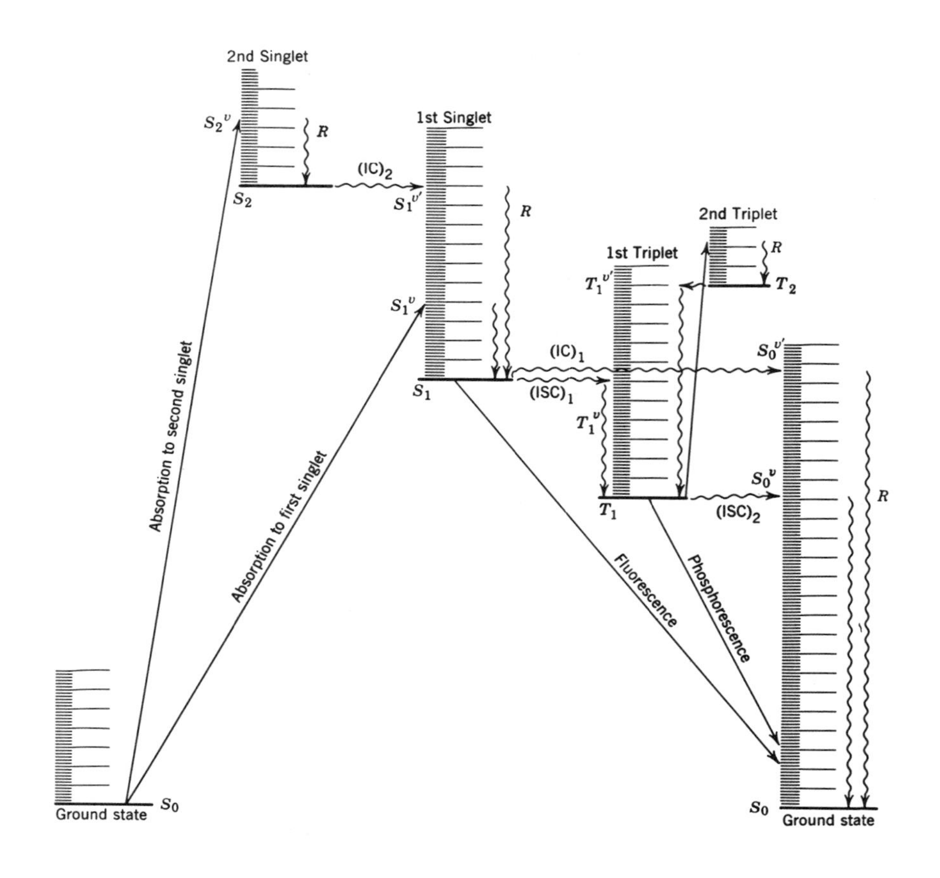

*Reprinted with slight alteration by permission of the copyright owners [(2), p. 244].

II. EXCITED STATE ENERGY TRANSFER: SENSITIZERS AND QUENCHERS

A molecule in an excited state may be capable of transferring its energy as a donor to a ground state acceptor molecule ($D^* + A \longrightarrow D + A^*$); such energy transfer is efficient only if the energy level of the excited state of the donor is higher than that of the acceptor. Singlet-singlet transfer takes place by a long-range "resonance" mechanism, whereas triplet-triplet transfer is usually a diffusion-controlled (see p. 137) bimolecular process requiring overlap of donor and acceptor electronic wave functions (interaction between A and D at close range with no dipole-dipole coupling involved). Triplet-triplet transfer refers to a donor-triplet giving rise to acceptor-triplet, and so on. Triplet-singlet transfer is also possible [see Galley and Stryer, Biochemistry, 8, 1831 (1969)].

The following tables provide triplet and singlet excited state energies for a variety of substances. The values are defined as follows:

E_T = triplet excitation energy in kcal/mole; refers mainly to the 0-0 maximum of phosphorescence at 77°K in a hydrocarbon glass (usually methylcyclohexane-isopentane, 5/1 by volume).

Refs. p. 367.

τ = phosphorescence lifetime in seconds; for recent details on triplet lifetimes of cyclic ketones, see P. J. Wagner and R. W. Spoerke, J. Amer. Chem. Soc., 91, 4437 (1969).

E_S = lowest excited singlet state energy, kcal/mole.

Φ = quantum yield for intersystem crossing, S→T (in benzene as solvent).

A "Photosensitizer and Quencher Kit" containing 24 organic compounds with E_T from 39 to 74 kcal/mole is available from J. T. Baker, Department HYT, Phillipsburg, N. J. 08865. Recently a method (the Ullman color test) has been developed for measuring triplet energies of molecules with τ shorter than $\sim 10^{-7}$ sec and E_T < 64 kcal/mole; also described is a routine measurement of Φ_{ISC} using cis-trans olefin isomerization (cis-piperylene, usually). See reference 8, pages 341-343, for references and more details.

The data in the tables below have been adapted from collections in reference 2, pages 283-313; reference 4, Chapter 5; and reference 8, Chapters 2 and 3.

A. Ketones*

Compound	E_T	$\tau \times 10^3$	E_S	Φ
Acetone	∿78	∿1	∿84	
Propiophenone	74.6	3.8	80.1	1.00
Xanthone	74.2(70.9)[a]	20[a]	77.2[a]	
Phenyl cyclopropyl ketone[a]	74			
Di-isopropyl ketone	74			
Acetophenone	73.6	2.3	78.6	0.99
1,3,5-Triacetylbenzene[a]	73.3			
Isobutryophenone	73.1			
4-Methylacetophenone	73[g]			
1,3-Diphenyl-2-propanone[b]	72.2			
4-Chloracetophenone[a]	72			
4-Methoxyacetophenone[a]	72			
Anthrone[a]	71.9	1.5	77.3	
4-Bromoacetophenone[a]	71			
3,5-Dimethylacetophenone[a]	71			
2-Acetylpyridine[a]	71			
3-Acetylpyridine	71			
Triphenylmethyl phenyl ketone	70.8			
Sodium Benzophenone-3-sulfonate[a]	70.2			
4-Acetylpyridine[a]	70			
3-Cyanoacetophenone	70			
4,4-Dimethoxybenzophenone[a]	70			
4,4-Dimethylbenzophenone[a]	69			
4,4-Diphenylcyclohexadienone	69			
2-Benzoylbenzophenone	68.7			
Benzophenone	68.5	4.7[a]	74.4[a]	1.00

(Continued)

Refs. p. 367.

Compound	E_T	$\tau \times 10^3$	E_S	Φ
4,4'-Dichlorobenzophenone	68			
4-Trifluoromethylbenzophenone	68			
4-Hydroxybenzophenone	68			
Pyruvic acid (CH_3COCO_2H)	68		80	
4-Acetylacetophenone	67.6			
4-Aminobenzophenone	67			
9-Benzoylfluorene	66.8			
4-Cyanobenzophenone	66.4			
Thioxanthone	65.5		∿76	
4-Aminoacetophenone	65			
Ethyl pyruvate	65			
Phenylglyoxylic acid ($C_6H_5COCO_2H$)	63		76	
Phenylglyoxal	62.5			
Anthraquinone	62.4			0.90
α-Naphthoflavone	62.2			
Flavone	62.0			
Ethyl phenylglyoxylate	61.9			
Michler's ketone [4,4'-bis-(dimethylamino)benzophenone]	61.0		∿77	1.00
4-Acetylbiphenyl[a]	61			
β-Naphthyl phenyl ketone	59.6			
β-Acetonaphthone	59.3		∿79	0.84
α-Naphthyl phenyl ketone	57.5			
Ethyl naphthylglyoxylate	57			
α-Acetonaphthone	56.4			
5,12-Naphthacenequinone	55.8			
Biacetyl	54.9	1	65	
Acetylpropionyl ($CH_3COCOC_2H_5$)	54.7			
Benzil	53.7	2	65	0.92
Fluorenone	53.3			0.93
Camphorquinone	50	3	57	
Phenanthrenequinone	49		55	
3-Acetylpyrene	46[g]			
Thiobenzophenone	40			

*Footnotes appear on page 355.

B. Aldehydes

Compound	E_T	$\tau \times 10^3$	E_S
Benzaldehyde[a]*	71.3	1.5	76.5
2-Hydroxybenzaldehyde[a]	71		
4-Chlorobenzaldehyde[a]	70.8		
3-Iodobenzaldehyde[a]	70.8	0.65	75.1

(Continued)

Refs. p. 367.

Compound	E_T	$\tau \times 10^3$	E_S
2-Chlorobenzaldehyde[a]	69.6		
Acrolein	∿69		
2-Naphthaldehyde	59.5		
1-Naphthaldehyde	56.3		
Glyoxal	56		63
9-Anthraldehyde[a]	40		

*Footnotes appear on page 355.

C. Acids and Derivatives

Compound[a]*	E_T	E_S
Acetic Acid		135
Methyl acetate		135
Acetamide		140
Phenol	82	
Benzamide	79	90
Benzoic acid	78	140
Benzonitrile	77	
Sodium triphenylene-2-sulfonate	65	
2-Naphthoic acid	60	
Disodium naphthalene-2,6-disulfonate	60.0	
1-Naphthoic acid	58	

*Footnotes appear on page 355.

D. Aromatic Hydrocarbons, Heterocycles and Derivatives

Compound	E_T	$\tau \times 10^3$	E_S	Φ
Benzene	85	7,000		0.24
Benzene-d_6		26,000		
Pyridine	82		100	
Aniline[a]*	77			
Carbazole[a]	70.3	7,600	84.4	0.36
Diphenylene oxide	70.1			
Hexachlorobenzene	70			
Dibenzothiophene	69.7			

(Continued)

Compound	E_T	$\tau \times 10^3$	E_S	Φ
Thiophene[a]	69			
Fluorene	67.6			0.31
Triphenylene	66.6	4,900	83.5	0.95
Triphenylene-d_{12}		23,000		
Biphenyl-d_{10}[a]	66.0	11,300	96.3	
Biphenyl[a]	65.7	3,100	95.6	
Phenanthrene[a]	62.0	3,300	83.0	0.76
Phenanthrene-d_{10}[a]		16,400		
Quinoline[a]	62.0	1,400	91.3	0.32
Naphthalene-d_8	61.2	22,000	90.9	0.38
Naphthalene[a]	60.8	2,300	90.8	0.39
1-Methylnaphthalene[a]	60.0	2,100	90.1	0.48
Nitrobenzene[a]	60			
Acenaphthene	59.3		89.1	
1-Chloronaphthalene[a]	59.2	290	89.7	
1-Bromonaphthalene[a]	59.0	18	89.5	
1-Iodonaphthalene[a]	58.6	2.0	88.9	
Chrysene	56.6		79.4	0.67
Coronene[a]	55			
1,2,5,6-Dibenzanthracene[a]	52.2			0.89
1,2,3,4-Dibenzanthracene[a]	50.8			
Pyrene	48.7	200	77.0	
Pyrene-d_{10}		3,200		
Pentaphene[a]	48			
1,2-Benzanthracene	47			
11,12-Trimethylenetetraphene	46			
1,12-Benzperylene	46			
Acridine	45.3			
Phenazine	44			
3,4-Benzpyrene	42			
Anthracene	42.0	60	75.5	0.75
Anthracene-d_{10}		100		
Azulene[a]	31-39			
Naphthacene	29			

*Footnotes appear on page 355.

E. Conjugated Aromatics and Olefins

Compound	E_T	Compound	E_T
Phenylacetylene	72	2,4-Hexadien-1-ol	59.5
Diphenylacetylene	62.5	1,3-Butadiene	59.3
Styrene	61.5	trans-1,3-Pentadiene	58.8
Isoprene	60.1	Chloroprene	58.6

(Continued)

Refs. p. 367.

Compound	E_T	Compound	E_T
Cyclopentadiene	58.3	1,4-Diphenyl-1,3-butadiene	52
1-Chlorobutadiene	57.4	trans-4-Nitrostilbene	50
cis-Stilbene	57	trans-Stilbene	49
cis-1,3-Pentadiene	56.9	trans,trans-1,3,5-Hexatriene	47.5
1-Methoxybutadiene	56.6	1,3,5,7-Octatetraene (all trans)	39.0
1,3-Cyclohexadiene	53.5		

F. Unsaturated Nitrogen Compounds

Compound	E_T	E_S	Compound	E_T	E_S
Phenyl azide	78	99	N=N (bicyclic)	63	84
Aniline	77		Nitrobenzene	60	
Benzonitrile	77		O-N=N-O	45	67
$C_6H_5CH{=}NC_6H_5$ *		82	CH_3ONO		73
Diphenylamine[c]	72		CH_2N_2		67
Triphenylamine[d]	70.1				

*Footnotes appear at the end of the table which follows.

G. Miscellaneous

Compound[a]	E_T	Compound[a]	E_T
Diphenyl sulfide	74	Eosin	43
Diphenyl selenium	72	Crystal violet	39
Acridene yellow	58	Oxygen (O_2)	23[e]
Fluoroscein	51		

[a]Measurements in ether-alcohol or other polar glass (solvent); values do not differ significantly (a maximum of ±3 kcal/mole) from those in hydrocarbon glass or solvent.

[b]Dibenzyl ketone. There is some question about the value for E_T according to recent work [P. S. Engle, J. Amer. Chem. Soc., 92, 6074 (1970); W. K. Robbins and R. H. Eastman, ibid., 92, 6076, 6077 (1970)].

[c]$\Phi_{ISC} = 0.38$

[d]$E_S = 80.2$, $\tau = 0.7$ sec, $\Phi = 0.88$.

[e]Lowest triplet-singlet transition (ground state is T).

Refs. p. 367.

III. PHOTOCHEMISTRY LIGHT SOURCES AND EQUIPMENT

The commercially available equipment described below is mainly for laboratory-scale use; some suppliers manufacture industrial-scale equipment (e.g., suppliers 1 and 8).

A. Definitions (2)

Immersion lamp: light source in sealed glass envelope (usually quartz) is surrounded by reaction medium.
Point source: usually very small, high-pressure, very intense lamp (not immersion type).
Radiated power: radiation output of lamp, usually in watts (0.239 cal/sec = 1 W).
Low-pressure Hg arc: operates at vapor pressure of Hg, $\sim 10^{-3}$ mm, at room temperature. Emits mainly at 2536.5 and 1849 Å. Useful for initiating Hg-photosensitized reactions. External cooling not necessarily required.
Medium-pressure Hg arc: under 1 to several atmospheres inert gas pressure; emits from about 2200 to 14,000 Å, with most radiation from 3100 to 10,000 Å. Most intense "lines"--3650, 4358, 5461, 5780.
High pressure Hg arc: under 100 to several hundred atmospheres; emission almost continuous from 2200 to 14,000 Å.

B. Properties and Suppliers of Lamps[a]

Lamp[b]	Supplier[c,d]	Energy (W) Input	Output[e]	Comments
1. Low-Pressure				
NK6/20	1	7	∿1	(Hanau lamp)
NFUV	2	Variable (variac); maximum input 50.		
11 SC-1	3,4	Output = 42 x 10^{-4}/cm^2 at 1 in. "Pencil lamps" or "pen-ray" 0.5-5 in. long.		
11 SC-1L	3,4	Emits at 3600 Å. ("black light" lamp, ∿5 in. long).		
PCQ-024	4	Coil form (for immersion or to surround reaction tube); 9-W output.		
PCQ-023	4	10-in. square, flat grid lamp; 20-W output.		
G15T8	5,6	16.5	3.0	Heated cathode.
G30T8	5,6	34.0	7.5	Heated cathode.
ST46A22	6	24.0	3.4	Cold cathode.
ST430A32	6	36.0	7.0	Cold cathode.
WL-782L-30	7	20.5	5.2	Cold cathode.
2852Q	8	13.5	3.3	Cold cathode.

(Continued)

Refs. p. 367.

Suppliers p. 366.

Lamp[b]	Supplier[c,d]	Energy (W) Input	Output[e]	Comments
2. Medium-Pressure[f]				
SH(616A)	8	100	6.3	Hanovia sells quartz immersion wells (water cooling) to house all these lamps. Pyrex, corex, and vycor sleeves (tubes) are also available for use as filters. Useful lifetime of these and most other medium-pressure lamps is ≥1000 hr.
S(654A)	8	200	10.5	
L(679A)	8	450	83.7	
A(673A)	8	550	94.9	
LL(189A)	8	1200	347.	
BMS(47A)	8	3500	874.	
MMS(77A)	8	4500	1047	
HST(40B)	8	7500	1601	
UA-2	5	285	30.6	
UA-3	5	419	38.8	
UA-11	5	1395	240.	
UA-15	5	3160	810.	
3. High-Pressure (Immersion)				
AH6	5	1176	195	
Q81	1	100	∿10	
Capillary lamps	10	500-12,000	(intense radiation from 1-12 in. long tubes).	
4. High-Pressure (Point Sources)[g]				
537B9	8	1000	87	Xe-Hg
C-45-61-2	9	150	Xe arc (Osram XBO 150 W/1)	
C-46-61-2	9	200	Hg arc (Osram HBO 200 W/2)	
C-47-35-2	8,9	200	Xe-Hg (Hanovia 901-B1)	
C-45-35-51	8,9	1000	Xe (Hanovia 976-C1)	
C-46-78-5	9	1000	Hg (Ushio 1005D)	
C-47-35-51	8,9	1000	Xe-Hg (Hanovia 977-B1)	

(Available with these items from supplier 9 are power supplies, housings, optics and other accessories)[g]

PEK lamps 10

X-series (Xe) available as 35, 75, 150, 300, 1600, and 2500 W input; strong near-uv continuum, several near-ir peaks, ir continuum, visible region like sunlight.

PEK-series (Hg) available as 100, 200, and 500 W; vis, uv, near-ir radiation of about the same intensity.

5. Miscellaneous

a. Supplier 12 carries a nearly complete line of low to medium-pressure Osram lamps (supplier 13) as follows (see also supplier 16); wattage in parentheses: Cd(18), Cs(12), He(65), Hg(50), Hg-Cd(30), K(18), Na(22), Ne(45), Rb(14), Tl(13), Zn(20), HgS(50). For complete description, including spectrograms, write to supplier 12. The same supplier also carries quartz-iodine, zirconium, and tungsten-halogen lamps.

(Continued)

Lamp[b]	Supplier[c,d]	Energy (W) Input	Output[e]	Comments
b. Vacuum (far) uv sources: for radiation in the region 2000-1000 Å and beyond, resonance lines of H_2, N_2, He, Ne, Ar, Kr, and Xe are used (fired by microwave generator (electrodeless) discharge -- supplier 14). For reviews, see reference 2, page 705 and reference 7a, Vol. 3, page 157 (1964).				
c. Various self-contained Hg, Xe, and Na vapor lamps. Supplier 16.				

[a]Data obtained principally from reference 2, Chapter 7; reference 8, Chapter 4; and individual company catalogs and data sheets. The list is by no means complete but is representative of available equipment. Not all types and models of a particular manufacturer are given.

[b]All are Hg-vapor immersion lamps unless indicated otherwise. Designation given is the manufacturer's.

[c]Suppliers are listed at the end of this Chapter; in some cases both manufacturers and distributors are cited.

[d]If lamp operates at other than line voltage (110, 220, etc), voltage transformers are necessary. If they are not available from the lamp supplier (they usually are), any large electrical equipment company can meet the requirements (e.g., Jefferson Electric Company, Bellwood, Ill.). The same applies to voltage regulators (stabilizers).

[e]Radiated output in uv region (∿200-380 nm); in general, for low-pressure lamps, a linear relationship exists between spectral output and power input for all lines.

[f]Many references are made in the literature to Hanovia and other "high-pressure" lamps; by the adopted definition (2) given above they are strictly "medium-pressure" lamps.

[g]Supplier 11 provides a variety of power supplies and other accessories (housing, igniters, etc.) for compact Hg and Xe arcs.

C. Laser Systems

Although hundreds of spectral lines are available from the many lasing systems investigated to date, very few are produced with sufficient intensity for photochemical and related applications. This section summarizes the most useful lasers. It would be nearly impossible to list the suppliers of lasers and laser equipment; the reader is referred to the excellent directories already available (12). The data below were adapted from several sources (13,14,15), and the choice is somewhat arbitrary. A promising approach to wavelength tunability in the ir and vis regions is the liquid (organic dye) laser (16). However, this and the area of chemical lasers in general (17) are not yet fully developed. Recently the Chromatrix Company (1145 Terra Bella Avenue, Mountain View, Calif. 94040) has developed dye lasers tunable from 0.265 to 0.340 μm, and from 0.530 to 0.680 μm. The company also sells a lithium niobate/neodymium-YAG system, tunable from 0.54 to 3.7 μm (see Wallace and Harris, Laser Focus, November 1970). For a related development using a ruby laser down to 3139 Å, see E. S. Yeung and C. B. Moore, J. Amer. Chem. Soc., 93, 2059 (1971).

Medium	λ (μm)[a]	Pulsed (P) or CW	Pulse Width (nsec)	Photons per pulse[b]	Photons per second[b]
		Semiconductor Lasers			
GaAs	0.905 ± 0.050	P	2×10^{2}	10^{14}	10^{19}
	0.845 ± 0.050	P	2×10^{3}	10^{13}	10^{17}
In Sb	5.2	P	10^{2}	10^{11}	10^{14}
In As	3.15	P	2	10^{9}	10^{12}
		Crystalline and Glass Host Lasers			
Ruby (Cr^{+3}/Al_2O_3)	0.6943	P	5×10^{2}	10^{21}	10^{21}
Nd^{+3}/glass	1.06	P	10^{3}	10^{21}	10^{21}
Nd^{+3}/YAG	1.06	CW			10^{21}
		Gas Lasers			
N_2	0.3373	P	10	10^{15}	10^{17}
He-Ne	0.63282	CW			10^{17}
	1.1523	CW			10^{17}
	3.3913	CW			10^{17}
Ar	0.4880	CW			10^{19}
Kr	0.4762	CW			10^{17}
	0.5208	CW			10^{17}
	0.5682	CW			10^{17}
	0.6471	CW			10^{19}
CO_2	10.6	CW			10^{22}

[a] 1 μm = 10^{4} Å = 10^{-3} nm.

[b] Approximate intensities for the most powerful lasers available, calculated from data in references 13, 14, and 15. Photons/sec for pulsed lasers is, of course, an "average" flux; the number of actual pulses/sec is easily calculated from the data given. Typical pulse widths are given in nanoseconds (nsec). For CW (continuous wave), photons/sec is also an average.

D. Photochemical Reactors (see reference 2 for general information)

1. "Rayonet." Self-contained, air-cooled chamber whose inner wall is lined with lamps (held vertically); three different lamps are available (2537, 3000, and 3500 Å). Models RPR-100 and RPR-208 can be used with a "Merry-Go-Round" attachment (rotates reaction vessels). Quartz flasks and tubes also available. Characteristics of the Rayonet systems are summarized in the following table. Supplier 15.

Refs. p. 367. Suppliers p. 366.

Model	Inner Dimensions	Radiated Power at 2537 Å (W)
RPR-100	15 in. deep x 10 in. diameter	
RPR-204	24 in. deep x 5 in. diameter	60
RPR-208	24 in. deep x 12 in. diameter	120

2. Reaction flasks for Hanovia and Nester/Faust lamps. Suppliers 2 and 8.

3. Student (homemade) reactor. R. Silberman, J. Chem. Educ., 47, 122 (1970).

E. Light Filters

Filter systems are used to isolate specific radiation regions (narrow or broad) from a source; such filters are generally called bandpass filters. The medium used to absorb the unwanted radiation is usually a plate of treated glass, chemically impregnated plastic sheet, a chemical solution, or a suspension (called a Christiansen filter). For general discussion, see reference 2, page 728, and reference 11; light absorption properties of common materials and solvents may be found on pages 167-182.

1. Glass

Complete lines of glass filters covering uv through near ir are available (with spectral properties given) from suppliers 9, 12, and 17. Such filters are especially useful for medium-pressure Hg arcs.

2. Chemical

For some applications, such as protection of uv-sensitive materials, specially substituted benzophenones are available that efficiently absorb most radiation below 350 nm without serious decomposition. Supplier 18.

The following table of solution filters summarizes the more complete discussions given by Calvert and Pitts (2). These filter systems are best for isolating lines or regions from medium-pressure Hg lamps. The solutions are normally kept in quartz or pyrex cells (pyrex absorbs all light below 280 nm). Most of the filter systems consist of two or more component solutions, each one in its own cell. Path lengths are variable (usually 1 to 10 cm); suggested path lengths (2) are given for each component of the filter. In some cases, improved transmittance is achieved by preirradiation of the filter system. The %T column gives approximate transmittance values taken from data presented by Calvert and Pitts (2). Some filters having higher %T than the $NiSO_4$ solutions given below are described in W. W. Wladimiroff, Photochem. Photobiol., 5, 243 (1966).

Line or Region Transmitted (nm)	%T	Filter Components
184.9 (from low-pressure lamp)		A 1-mm path of deoxygenated cyclohexane solution of 9,10-dimethylanthracene (3.5×10^{-4} M = 72 mg/l).
253.7	15	$NiSO_4 \cdot 6H_2O$ (27.6 g/100 cc H_2O) - 5 cm; $CoSO_4 \cdot 7H_2O$ (8.4 g/100 cc H_2O) - 5 cm; I_2(0.108 g) + KI (0.155 g) in 1000 cc H_2O - 1 cm; Cl_2 (gas) at 1 atm, 25° - 5 cm.
265.2-265.5	25	Same as 253.7 except that I_2/KI component is changed to 0.170 g KI/100 cc H_2O - 1 cm.
312.6-313.2	18	$NiSO_4$ (p.178 M, aq) - 5 cm; K_2CrO_4 (5.0×10^{-4} M, aq = 97.1 mg/l) - 5 cm; potassium biphthalate (0.0245 M, aq = 5.00 g/l) - 1 cm; Corning filter 7-54 (9683) - 0.3 cm.
334.1	18	$NiSO_4 \cdot 6H_2O$ (10.0 g/100 cc H_2O) - 5 cm; Corning filter 7-51 (5970) - 0.5 cm; naphthalene (1.28 g) in isooctane (100 cc) - 1 cm.
365.0-366.3	25	$CuSO_4 \cdot 5H_2O$ (5.0 g/100 cc H_2O - 10 cm; Corning filter 7-37 (5860) - 0.5 cm; 2,7-dimethyl-3, 6-diazacyclohepta-1,6-diene perchlorate (0.01 g/100 cc H_2O) - 1 cm (perchlorate available from K and K Labs., 121 Express Street, Plainview, N.Y. 11803).
404.5-407.8	48	$CuSO_4 \cdot 5H_2O$ (0.44 g/100 cc of 2.7M NH_4OH) - 10 cm; I_2 (0.75 g/100 cc CCl_4) - 1 cm; quinine hydrochloride (2.0 g/100 cc H_2O) - 1 cm.
435.8	60	$CuSO_4 \cdot 5H_2O$ (0.44 g/100 cc of 2.7M NH_4OH) - 10 cm; $NaNO_2$ (7.5 g/100 cc H_2O) - 10 cm.
435.8, 700.0	80, 98	9,10-Dibromoanthracene (20 mg/100 cc toluene) - 1 cm pyrex cell; crystal violet (2.5 mg/100 cc 95% ethanol) - 5 cm.
546.1	55	$CuCl_2 \cdot 2H_2O$ (20 g) + $CaCl_2$ (27 g) in 100 cc H_2O (acidified slightly with HCl) - 1 cm; neodymium nitrate (60 g/100 cc H_2O) - 5 cm.
577.0-579.0	31	$CuCl_2 \cdot 2H_2O$ (10 g) + $CaCl_2$ (30 g) in 100 cc H_2O (acidified slightly with HCl) - 1 cm; $K_2Cr_2O_7$ (3 g/100 cc H_2O) - 10 cm.

Refs. p. 367.

IV. CHEMICAL ACTINOMETRY: QUANTUM YIELD

The actual light intensity from a source (photons/sec at a given λ) is usually measured by physical means (thermopile-galvanometer systems and phototubes) (reference 2, page 769 ff). For most cases, a secondary standard--some chemical reaction--is more convenient, efficient, and reproducible. When specifying quantum yield, the following definitions (2) should be adhered to, using a typical general reaction:

$$A_o + h\nu \rightarrow A^* \begin{cases} \nearrow [B + C + \ldots] \rightarrow X + Y + \ldots \\ \searrow A_o,\ A^*\text{-triplet} \ldots \end{cases}$$

Primary quantum yield: For the i-th process it is the fraction of molecules absorbing which decompose by that process. The process may be chemical (rearrangement, bond breaking, etc.) or physical (fluorescence, ISC, etc.); that is, the yield of chemical species (ground state, radical, etc.) produced directly from the initially formed excited state [said species is thus not necessarily the end photoproduct (X, Y, ... in the case above), but may be an intermediate].

$$\Phi_B = \frac{d[B]/dt}{I_a} = \frac{\text{number of species B formed/unit concentration/unit time}}{\text{number of quanta absorbed by } A_o\text{/unit concentration/unit time}}$$

Product quantum yield:

$$\Phi_X = \frac{d[X]/dt}{I_a} = \frac{\text{molecules of X formed/unit concentration/unit time}}{\text{number of quanta absorbed by } A_o\text{/unit concentration/unit time}}$$

One can also refer to quantum yields for fluorescence, ISC, sensitization (molecules of product from energy acceptors/quanta absorbed by energy donors) and so on. A special apparatus has been described for measuring quantum yields and rates of photochemical reactions (18). The following actinometers are the most widely used.

A. Ferrioxalate (2,8)[a]

$$[Fe^{III}(C_2O_4)_3]^{-3} \xrightarrow{h\nu} [Fe^{II}(C_2O_4)_2]^{-2} + 2\ CO_2$$

λ (nm)	$[K_3Fe(C_2O_4)_3]$, M[b]	$\Phi_{Fe^{+2}}$[c]
577,579	0.15	0.013
546	0.15	0.15
509	0.15	0.86
480	0.15	0.94
468	0.15	0.93
436	0.15	1.01

(Continued)

Refs. p. 367.

λ (nm)	$[K_3Fe(C_2O_4)_3]$,M[b]	$\Phi_{Fe^{+2}}$[c]
416	0.006	1.12
405	0.006	1.14
392	0.006	1.13
366	0.006	1.21(1.26)
358	0.006	1.25
334	0.006	1.23
313	0.006	1.24
302	0.006	1.24
297	0.006	1.24
253.7	0.006	1.25

[a]The best solution actinometer.

[b]Prepared as follows: mix 3 volumes of 1.5M (aq) $K_2C_2O_4$ and 1 volume of 1.5M (aq) $FeCl_3$; recrystallize the precipitated pure green $K_3Fe(C_2O_4)_3 \cdot 3H_2O$ three times from warm water and dry in a warm air current (45°). Store in the dark. To prepare a 0.006M solution, add 2.947 g in 800 ml H_2O to 100 ml 1.0N H_2SO_4 and dilute to 1 liter. For a 0.15M solution use 73.68 g, with everything else the same. Work in a darkroom. A 0.006M solution is convenient for $\lambda \leq 430$ mm; in a 1 cm path it absorbs >99% of light up to 390 nm and 50% at 430 nm. At λ >430 nm, a 0.15M solution is necessary.

[c]The $[Fe^{2+}]$ formed is determined as directed by Calvert and Pitts (2, p. 785): "Hatchard and Parker recommended that a standard calibration graph for analysis of the Fe^{2+} complex be prepared for use with the particular spectrophotometer or colorimeter available in the laboratory. To do this a standard solution (a), about 0.4 x 10^{-6}mole of Fe^{2+}/ml and 0.1 N in H_2SO_4, is freshly prepared from a standardized 0.1 M $FeSO_4$ solution which is 0.1 M in H_2SO_4 by dilution with 0.1 N H_2SO_4. Also necessary for the calibration and for subsequent use in the actinometry are a solution (b) containing 0.1% (by weight) 1,10-phenanthroline in water, and a buffer solution (c) prepared from 600 ml of 1 N NaO_2CCH_3 and 360 ml 1 N H_2SO_4 diluted to 1 liter. Now to a series of eleven 20 (or 25) ml volumetric flasks add with analytical pipets 0, 0.5, 1.0, 1.5.....4.5, 5.0 ml of standard solution (a). Add sufficient 0.1 N H_2SO_4 to each flask to make the total volume of solution in each about 10 ml. Add approximately 2 ml of (b) and about 5 ml of (c), dilute to volume, mix well, and allow to stand for at least 60 min; the solution may stand for several hours (in the dark) before measurement if desired. Then determine the transmission of each solution at 5100 Å in a 1 cm cell, using the blank iron-free solution which you have prepared, in the reference beam. A linear plot of log I_o/I versus molar concentration of Fe^{2+} complex is found usually; instrumental variations may alter slightly the apparent molar extinction coefficient (slope of plot), but it would be surprising if a value greatly different from $\varepsilon = 1.11 \times 10^4$ liters/mole-cm were obtained with a modern well-adjusted spectrophotometer."

If needed the intensity, I_i, incident just within the photolysis cell front window is calculated as follows for a given λ.

$$I_i = \frac{n_{Fe^{2+}}}{(1 - 10^{\varepsilon \ell [A]})\ t\ \Phi_{Fe^{2+}}} \quad \text{quanta/sec} \quad (n = \text{moles formed})$$

where t = irradiation time, Φ is taken from the table above, the term in parentheses = fraction of light absorbed by length, ℓ, of the actinometer solution (A = Fe^{3+}) (this term should be measured for the actual solution, and comes directly from Beer's law: absorbance = $\varepsilon \ell [A] = -\log(I/I_o)$; fraction transmitted, $I/I_o = 10^{-\varepsilon \ell [A]}$). However, quantum yield for a product X from the reaction of interest is easily calculated as

$$\Phi_x = \frac{n_x t_x \Phi_{Fe^{2+}}}{n_{Fe^{2+}} t_{Fe^{2+}}} \quad (t = \text{time of irradiation, } n = \text{moles formed})$$

B. Uranyl Oxalate (2)

$$UO_2C_2O_4 \xrightarrow{h\nu} U^{4+} + CO + CO_2 + HCO_2H + H_2O$$

λ (nm)	Φ of UO_2^{+2} Loss[a]	Φ^b_{CO}
435.8	0.58	0.31
405.0	0.56	
366.0	0.49	0.26
265.0	0.58	
253.7		0.33
245.0	0.61	
208.0	0.48	

[a]For 0.01M UO_2SO_4 and 0.05M $H_2C_2O_4$ at 25°; UO_2^{+2} loss determined titrimetrically with $KMnO_4$. See reference 11, page 295, for details.
[b]CO measured by vpc (19).

C. Reinecke's Salt (8, 20)

$$Cr(NH_3)_2(NCS)_4^- + H_2O \rightarrow Cr(NH_3)_2(NCS)_3(H_2O) + NCS^-$$

Useful mainly in the vis region (316 to >600 nm). Concentrations given in the table that follows are those that absorb all the light at λ <600 nm; corrections are needed above 600 nm.

Refs. p. 367.

λ (nm)	ε	$[KCr(NH_3)_2(NCS)_4]$, M	Φ_{NCS^-} (23°)
360	11,000	0.0011	0.291
350	>100	0.003	0.388
392	93.5	0.005	0.316
416	67.5	0.008	0.310
452	31.2	0.010	0.311
504	97.5	0.005	0.299
520	106.5	0.004	0.286
545	90.5	0.005	0.282
585	43.8	0.010	0.270
600	29.0	0.025	0.276
676	0.75	0.045	0.271
713	0.35	0.046	0.284
735	0.27	0.045	0.302
750	0.15	0.048	0.273

D. Benzophenone-Benzhydrol (8, 21)

$$(C_6H_5)_2CO + (C_6H_5)_2CHOH \xrightarrow{h\nu} (C_6H_5)_2\underset{}{\overset{HO}{C}}-\overset{OH}{C}(C_6H_5)_2$$

This actinometer is independent of any previously determined quantum yield. Several solutions of constant benzophenone and varying benzhydrol concentrations are irradiated simultaneously. The reaction itself follows the rate law:

$$\frac{1}{\Phi_B} = 1 + \frac{k_d}{k_r[BH_2]}$$

where Φ_B = quantum yield for benzophenone loss and $[BH_2]$ = concentration of benzhydrol. If one sample is chosen arbitrarily as the actinometer (Φ_{act}), and the relative amount of benzophenone loss for each of the other samples is measured, we have

$$\frac{\Phi_{act}}{\Phi_B} = \Phi_{act} + \frac{k_d\Phi_{act}}{k_r[BH_2]}$$

From a plot of Φ_{act}/Φ_B versus $1/[BH_2]$, the intercept is Φ_{act}.

E. Acetone Vapor (2)

$$CH_3COCH_3 \xrightarrow{h\nu} CO + \text{other}$$

Refs. p. 367.

At T >125° and p $\leq$ 50 mm, Φ_{CO} = 1.0 (over the range 250 to 320 nm, independent of λ, light intensity, p, and T).

F. Hexafluoroacetone Vapor (22)

$$CF_3COCF_3 \xrightarrow{h\nu} CO + \text{other}$$

At λ = 147.0 nm and p $\leq$ 50 mm, Φ_{CO} = 1.0 (room temp.). Not useful for the uv region; CO is the only noncondensable. Compound is available from supplier 19.

G. Nitrous Oxide (2)

$$N_2O \xrightarrow{h\nu} N_2 + \text{other}$$

Useful in range 147.0 to 184.9 nm. At 184.9 nm, ε_{N_2O} = 36. At moderate p near room temp., Φ_{N_2} = 1.44.

V. SUPPLIERS

1. Quartzlampen Gesellschaft, MBH, Hanau, Germany. Also, Brinkmann Instruments, Cantiague Road, Westbury, N. Y. 11590.
2. Nester/Faust Manufacturing Corp., 2401 Ogletown Road, Newark, Del. 19711.
3. Spectronics Corp., 29 New York Avenue, Westbury, N.Y. 11590
4. Ultra-Violet Products, Inc., 5114 Walnut Grove Avenue, San Gabriel, Calif. 91178.
5. General Electric Corp., Lamp Division, Cleveland, Ohio.
6. Sylvania Electric Products, Inc., Salem, Mass.
7. Westinghouse Electric Corp., Lamp Division, Bloomfield, N.J.
8. Hanovia Lamp Division, 100 Chestnut Street, Newark, N.J. 07105; also Ace Glass, Inc., Box 688, Vineland, N.J. 08360.
9. Oriel Optics Corporation, 1 Market Street, Stamford, Conn. 06902.
10. PEK, 825 East Evelyn Avenue, Sunnyvale, Calif. 94086.
11. Quartz Radiation Corp., 1275 Bloomfield Avenue, Fairfield, N.J. 07006.
12. Edmund Scientific Co., 150 Edscorp Building, Barrington, N.J. 08007.
13. Osram Gesellschaft MBH, Berlin and Müchen, Germany.
14. Raytheon Co., Sorenson Operation, Production Equipment Department, Richards Avenue, Norwalk, Conn. 06856.
15. The Southern New England Ultraviolet Company, 954 Newfield Street, Middletown, Conn. 06457.
16. George W. Gates and Co., Inc., Hempstead Turnpike and Lucille Avenue, Franklin Square, N.Y. 11010.
17. Corning Glass Works, Color Filter Sales Section, Corning, N.Y. 14830 (ask for Catalog CF-3); Baird-Atomic, Inc., 33 University Road, Cambridge, Mass. 02138; Bausch and Lomb, Inc., 60968 Bausch Street, Rochester, N.Y. 14602; Eastman Kodak Co., 343 State Street, Rochester, N.Y. 14650 ("Wratten" filters); Farrand Optical Co., 535 South Fifth Avenue, Mt. Vernon, N.Y. 10550; Ilford Ltd. (England); Jena Optical Works (Germany). Interference filters are available from some of these

Refs. p. 367.

suppliers as well as Photovolt Corp., 1115 Broadway, New York, N.Y. 10010; Barr and Stroud, Inc., Glasgow, Scotland.

18. Antara Chemicals, 435 Hudson Street, New York, N.Y. ("Uvinul" uv absorbers).
19. Peninsular Chem. Research, P. O. Box 14318, Gainesville, Fla. 32601.

VI. REFERENCES

1. For an early and very complete historical review, see C. Ellis, A. A. Wells, and F. Heyroth, The Chemical Action of Ultraviolet Rays, Reinhold, New York, 1941.
2. The most comprehensive book on the theory, practice, and recent literature of photochemistry is J. G. Calvert and J. N. Pitts, Photochemistry, Wiley, New York, 1966.
3. Easy-to-read review articles on the general types of excited state reactions are those by G. S. Hammond and N. J. Turro, Science, 142, 1541 (1964); and P. A. Leermakers and G. V. Vesley, J. Chem. Educ., 41, 535 (1964).
4. An introductory treatment for organic chemists is N. J. Turro, Molecular Photochemistry, Benjamin, New York, 1965.
5. Two books deal extensively with the mechanistic aspects of organic photochemistry.
   (a) R. O. Kan, Organic Photochemistry, McGraw-Hill, New York, 1966.
   (b) D. C. Nekkers, Mechanistic Organic Photochemistry, Reinhold, New York, 1967.
6. The best books for synthetic organic photochemistry are
   (a) A. Schönberg, Preparative Organic Photochemistry, Springer-Verlag, New York, 1967.
   (b) A. Schönberg, Präparative Organische Photochemie, Springer-Verlag, Berlin, 1958.
7. There are several continuing series dealing with contemporary research topics in photochemistry.
   (a) Advances in Photochemistry, Interscience, New York (Vol. 1, 1963).
   (b) Organic Photochemistry, Marcel Dekker, New York (Vol. 1, 1967).
   (c) The journal, Photochemistry and Photobiology, Pergamon Press (Vol. 1, 1962).
   (d) Annual Survey of Photochemistry, Wiley, New York (Vol. 1, 1969).
   (e) Creation and Detection of the Excited State, Marcel Dekker, New York, (Vol. 1, 1971, Ed. A. A. Lamola).
8. A recent, excellent monograph is A. A. Lamola and N. J. Turro, Eds., Energy Transfer and Organic Photochemistry, Vol. XIV of Technique of Organic Chemistry, P. A. Leermakers and A. Weissberger, Eds., Interscience, New York, 1969.
9. S. P. McGlynn, T. Azumi, and M. Kinoshita, Molecular Spectroscopy of the Triplet State, Prentice-Hall, Englewood Cliffs, N.J., 1969.
10. L. Heidt, R. S. Livingston, E. Rabinowitch, and F. Daniels, Photochemistry in the Liquid and Solid States, Wiley, New York, 1960. Based on a NAS-NRC sponsored symposium (1957) on photochemical energy storage.
11. C. R. Masson et al., in Vol. II, Technique of Organic Chemistry, 2nd ed., A. Weissberger, Ed., Interscience, New York, 1956, pp. 281-284.
12. Laser Marketers' and Buyers' Guide, published by the magazine Laser Focus, Advanced Technological Publications, 246 Walnut Street, Newtonville, Mass. 02160; Guide to Scientific Instruments, published by Science; Laboratory Guide to Instruments, Equipment and Chemicals, published by Analytical Chemistry (A.C.S.).

13. A. Bloom, Gas Lasers, Wiley, New York, 1968.
14. S. L. Marshall, Laser Technology and Applications, McGraw-Hill, New York, 1968; see also the new Handbook of Lasers, R. J. Pressley, Ed., CRC Co., Cleveland, Ohio, 1971.
15. A. F. Haught, Ann. Rev. Phys. Chem., 19, 343 (1968); B. Bova, Res./ Dev., December 1967, pp. 30-34.
16. M. Kagan, G. Farmer, and B. Huth, Laser Focus, Septemer 1968, pp. 26-33; P. Sorokin, J. Lankard, V. Moruzzi, and E. Hammond, J. Chem. Phys., 48, 4726 (1968).
17. G. C. Pimentel et al., J. Chem. Phys., 49, 5190 (1968), and other papers in the series. See also journals of IEEE, and Appl. Phys. Letters for current developments. Recent reviews include K. L. Kompa, Chemical Lasers, Angew. Chem. Intern. Ed. Engl., 9, 773 (1970), and F. P. Schäfer, Organic Dyes in Laser Technology, ibid., 9, 9 (1970).
18. K. Weiss et al., Photochem. Photobiol. 6, 321 (1967).
19. K. Porter and D. Volman, J. Amer. Chem. Soc., 84, 2011 (1962); D. Volman and J. Seed, ibid., 86, 5095 (1964).
20. E. Wegener and A. Adamson, ibid., 88, 394 (1966).
21. W. Moore et al., ibid., 83, 2789 (1961); 84, 1368 (1962).
22. J. Magenheimer and R. B. Timmons, J. Chem. Phys., 52, 2790 (1970).

CHROMATOGRAPHY

I. FUNDAMENTAL TYPES OF CHROMATOGRAPHY AND BASIC DEFINITIONS (1,2)

A. Adsorption

A finely divided solid, stationary phase functioning as an adsorbing surface, with a liquid (column adsorption and thin layer chromatography, tlc) or a gas (gas-solid chromatography) serving as the mobile phase.

B. Partition

Distribution (partitioning) of a solute between two or more liquid phases (liquid-liquid extraction) or between a stationary liquid phase and a gas phase (gas-liquid chromatography, glc). One of the liquid phases may be stationary as a film or layer (paper and thin-layer partition chromatography) or dispersed throughout an inert, solid support (column partition chromatography). "Normal" partition means the more polar solvent is held on the support; in "reversed" phase, reverse is true.

C. Ion Exchange

A stationary, insoluble phase composed of a polymeric ion exchange resin (acidic or basic), with an ionic solution (aq acid, base, or salt) as the mobile phase.

D. Electron Exchange

A stationary, insoluble phase in the form of a poly(oxidizing) or poly (reducing) agent (e.g., redoxite resins having hydroquinone or methylene blue units) capable of selective oxidation or reduction on components of the mobile phase.

E. Electrophoresis

Components of a mixture of ions on a carrier (solid support) migrate at different rates and separate into zones upon application of an electric field (ac or dc) across the carrier (e.g., filter paper or a packed column, saturated with a conductive buffer). Low voltage (5 to 20 V/cm) used for proteins, high voltage (50 to 200 V/cm) for amino acids and peptides.

F. Gel Filtration and Gel Permeation

Separation based on molecular size (a molecular sieving process) using gels made from substances of known pore size or porosity (such as Sephadex, a cross-linked poly(dextran); Bio-Gel, a poly(acrylamide); or Sepharose, a non-ionic galactan from agar).

G. R_f-value

Distance traveled by a spot (to its center) divided by total distance traveled by the solvent front. Values are dependent on temperature, type of stationary support, solvent system, amount of solute, and so on.

Refs. p. 405.

H. Mesh Sizes and Particle Diameter

Particle grain size is characterized by meshes (number of meshes/inch in the finest sieve that will pass the grains); effective particle diameter is the width of a mesh opening (micrometers, μm). Column adsorbents (alumina, silica, etc.) are generally 80 to 200 mesh (U.S.), while those for tlc are >250 and are too fine for column work. Uniformity in particle size (narrow mesh range) leads to better packing and more efficient chromatography.

Opening (μm)	U.S. Sieve[a] Mesh No.	British[b] Mesh No.	Japanese[c] Mesh No.	French and German Standards[d]	
				Mesh No.	Opening (μm)
4000	5				
2000	10	8	9.2	34	2000
841	20	18	20	30	800
595	30	25	28	28	500
420	40	36	36	27	400
297	50	52	48	26	315
250	60	60	55	25	250
210	70	72	65	24	200
177	80	85	80	23	160
149	100	100	100	22	125
125	120	120	120	21	100
105	140	150	145	20	80
88	170	170	170	19	63
74	200	200	200	18	50
63	230	240	250	17	40
53	270	300	280		
44	325	350	325		
37	400				

[a]U.S. Standard Sieve Series, ASTM Specification E-11-61. Identical to Canadian Standard Sieve Series, 8-GP-16.
[b]British Standards Institution, London BS-410-62.
[c]Japanese Standard Specification JIS-Z-8801.
[d]French Standard AFNOR X-11-501; German Standard DIN-4188.

II. ADSORPTION CHROMATOGRAPHY

A. Adsorbents for Column, Thin-Layer, and Gas-Solid Chromatography

Adsorbent	pH Range[a]	Application[b]	Comments
Silica[c]	Weakly A[d]	col., tlc	Maximum activity if heated at 150-160° (few hours) before use (activated material used to separate hydrocarbons). Most often used as "deactivated" (10-20% H_2O), its usual state as purchased. An all-purpose

(Continued)

Refs. p. 405.

Adsorbent	pH Range[a]	Application[b]	Comments
			adsorbent for most types of functional groups and ionic or nonionic compounds, including alkaloids, sugar esters or DNPH's, glycosides, dyes, alkali metal cations, lipids, glycerides, steroids, terpenoids, and plasticizers. Methanol and ethanol as eluants decrease silica activity somewhat.
Silica[c]	Weakly A[d]	gsc	Useful particle size ~30-120 mesh; activate by heat (150°) under vacuum, if possible . Useful for low-boiling hydrocarbons, O_2, N_2, oxides of N and C.
Alumina[e]	B (9-10)	col., tlc	The usual (ordinary) form of alumina, containing 2-3% H_2O. Suitable for basic and neutral compounds that are stable to alkali, as well as for alcohols, hydrocarbons, steroids, alkaloids, and natural pigments. Can cause polymerization, condensation, and dehydration reactions; do not use acetone or ethyl acetate as eluants, the latter being subject to saponification.
Alumina[e]	N	col., tlc	Useful for aldehydes, ketones, quinones, esters, lactones, glycosides; it is considerably less active than the basic form.
Alumina[e]	A (4-5)	col., tlc	Weakest and least often used type. Useful for acid pigments (natural and synthetic) and strong acids (which chemisorb to neutral and basic alumina).
Alumina[e]	N	gsc	Useful for low-boiling hydrocarbons, oxides of N and C.
Magnesium silicate[f]	A	col., tlc	Somewhat like acidic alumina in strength, but many substances chemisorb. Good for steroids, esters, lactones, glycerides, alkaloids, some carbohydrates. For a review, see reference 3.

(Continued)

Refs. p. 405.

Adsorbent	pH Range[a]	Application[b]	Comments
Magnesia[g]	B	col.	Resembles alumina, but superior for olefins and aromatics. Best adsorbent for separating compounds differing only in the number of C-C double bonds; excellent for separating saturated from unsaturated compounds. For review, see reference 4.
Carbon	N	col	Activated carbon, mainly charcoal (250-350 mesh) of two types--polar (oxidized) and nonpolar (graphitized). Relative sample adsorption for both is determined mainly by sample molecular size (mw). Commercially available charcoal lies somewhere in between the two forms; useful for carbohydrates, peptides, amino acids, alkane homologs, polymers. Definite adsorption preference for aromatic (not olefinic) unsaturation. Carbon is usually mixed with diatomaceous earth (1:1) for column use to avoid very slow flow rate. For review, see reference 5.
Carbon	N	gsc	Useful nonpolar adsorbent for separation by mw. Good for CO, Kr, CH_4, oxides of N.
Diatomaceous earth[h]	N	col.	Usually used as inert diluent in admixture with other adsorbents for faster flow. Also good for very polar substances like chlorophylls, porphyrins, carbohydrates, and strongly hydrophilic compounds. Recommended for very labile molecules or those normally handled by paper chromatography.
Molecular sieves	N	col., gsc	For information, see page 447.
Porous beads	N	col., tlc, gsc	Recently developed variety of substances used for liquid chromatography (adsorption, partition, gel) as well as gas chromatography. Come as porous polystyrene ("Poragel"--separation of polymers, mw ~1000-50,000);[i] porous silica layer on solid glass beads

(Continued)

Refs. p. 405.

Adsorbent	pH Range[a]	Application[b]	Comments
			("Corasil"--particles 37-50 μm),[i] handled like any adsorbent or used as a support for a liquid phase in ordinary column chromatography (activated at 110° overnight); porous Vycor (silica) for tlc plates (Corning 7235);[i,j] liquid phase chemically bonded to porous silica for ordinary column use (e.g., "Durapak").[i] (Also, see Permeation Gels, p. 390).
Special adsorbents	Varies	col., tlc	Custom adsorbents for specific functional groups (e.g., $AgNO_3$-impregnated magnesium silicate for olefins; poly(caprolactam) for phenols on tlc; hydroxyapatite (form of calcium phosphate) for preparative biochemistry. For certain applications, sucrose, cellulose, and others may be used. For details see reference 6.

[a]A = acidic, B = basic, N = neutral.
[b]Col. = column, tlc = thin layer, gsc = gas-solid chromatography.
[c]Also known as silica-gel or silicic acid in various forms.
[d]If pure, will not chemisorb basic compounds or act as acid catalyst ordinarily.
[e]See the next section for activity grades of alumina; the less water present, the more active (and reactive). A, B, and N forms are available commercially, but can be made from each other if desired (see reference 1).
[f]Commercial names: Florisil, Magnesol, Magnesium Trisilicate.
[g]Magnesium oxide.
[h]Commercial names: Celite, Kieselguhr, Hyflow (Super Cel).
[i]Waters Associates, Framingham, Mass.
[j]Corning Lab. Products, Corning, N.Y.

B. Activity Grades of Alumina

Acidic, basic or neutral alumina can be prepared in various activity grades (I to V) according to the Brockmann scale, by addition of H_2O to Grade I (prepared by heating alumina at 400 to 450° until no more H_2O is lost).

Water added (wt. %):	0	3	6	10	15
Activity grade:	I	II	III	IV	V
R_f (p-aminoazobenzene):	0.0	0.13	0.25	0.45	0.55

Refs. p. 405.

Suppliers p. 403.

A convenient way to determine the approximate activity of a given adsorbent batch is to measure the R_f for p-aminoazobenzene using benzene as eluant; small capillary tubes filled with adsorbent can be used.

C. Eluotropic Series: Relative Solvent Strengths (1a; 1b; 6 Ch. 8)

There is no universal series of solvent strengths; relative elution properties depend not only on the adsorbent, but in many cases on the compound types being separated. In addition, there appears to be no uniform relationship between solvent property (dielectric constant, dipole moment, e.g.) and eluting power; nor is there a relationship between relative solubility of a compound and relative adsorbability. However, the following empirical series have been found useful for most applications [in order of increasing "polarity" (eluting power)--read down]. The entries are those for which actual comparative measurements have been made.

For Alumina

Fluoroalkanes
Pentane
Isooctane
Petroleum ether (light)
Hexane
Cyclohexane
Cyclopentane
Carbon tetrachloride
Carbon disulfide
Xylene
Di-i-propyl ether
Toluene
1-Chloropropane
Chlorobenzene
Benzene
Ethyl bromide
Diethyl ether
Diethyl sulfide
Chloroform
Methylene chloride
Tetrahydrofuran
1,2-Dichloroethane
Ethyl methyl ketone
1-Nitropropane
(Acetone)
1,4-Dioxane
Ethyl acetate
Methyl acetate
1-Pentanol
Dimethyl sulfoxide
Aniline
Diethylamine
Nitromethane
Acetonitrile
Pyridine
Butyl cellosolve
2-Propanol
1-Propanol
Ethanol
Methanol
Ethylene glycol
Acetic acid

For Silica

Cyclohexane
Heptane
Pentane
Carbon tetrachloride
Carbon disulfide
Chlorobenzene
Ethylbenzene
Toluene
Benzene
2-Chloropropane
Chloroform
Nitrobenzene
Di-i-propyl ether
Diethyl ether
Ethyl acetate
2-Butanol
Ethanol
Water
Acetone
Acetic acid
Methanol
Pyruvic acid

For Magnesium Silicate (Florisil)

Pentane
Carbon tetrachloride
Benzene
Chloroform
Methylene chloride
Diethyl ether

(Continued)

Refs. p. 405.

For Magnesia

Petroleum ether (light)	Cyclohexane	Acetone
Hexane	Carbon tetrachloride	Benzene
Heptane	Diethyl ether	Pyridine
Decane	Carbon disulfide	
Isoöctane	Triethylamine	

For Charcoal (Nonpolar)

Water	Acetone	Ethyl acetate
Methanol	1-Propanol	Hexane
Ethanol	Diethyl ether	Benzene

In binary solvent mixtures, addition of small amounts of one solvent (0 to 40% of the mixture) to another less polar solvent rapidly increases solvent strength, after which a leveling off occurs. The further away in the series the pair is, the more drastic the change (and the less miscible).

D. Column Chromatography Hints

1. Weight of adsorbent to be used per gram of adsorbed material: 25 to 50 g (100-1000 g if material is poorly adsorbed or when working with very small amounts (milligrams, e.g.) of adsorbed material)).

2. Amount of adsorbent for a given height, h, of actual packing in a cylindrical column, diameter, d (d and h in cm). (Note: height of wet-packed column will be ∿10% lower than that of dry-packed.)

	Bulk density (g/cc)	Weight (g)	Volume (cc)
Alumina	∿1	$0.8d^2h$	$0.8d^2h$
Silica	∿0.3	$0.25d^2h$	$0.8d^2h$

3. Tlc as a model for column conditions. A tlc system (adsorbent type and solvent) may be used as a model for the larger scale column process; for effective column separation, it is important that most of the solutes to be separated have tlc R_f-values not greater than ∿0.3. Also, adsorbents should come from the same manufacturer (particle size differences are relatively unimportant in the scale-up).

4. Dry-column chromatography. Using the same adsorbent system as for tlc, deposit mixture to be separated on top of dry column; allow solvent to move through the dry column by capillary action (ascending or descending development); when solvent front reaches column bottom, extrude the adsorbent and separate fractions by cutting. (Alternatively, thin-wall nylon tubes can be used and cut through without extruding). The technique apparently gives separations comparable to those obtained by tlc [see B. Loev et al., Chem. Ind., 15 (1965); 2026 (1967)].

Refs. p. 405.

E. Thin Layer Chromatography Hints (7)

1. Preparation of microscope-slide thin layers. Such tlc plates are efficiently and easily made using the slurries prepared as follows from $CHCl_3$-MeOH mixtures. Pure $CHCl_3$ alone is often satisfactory.

Adsorbent[a]	Amount (g)[b]	$CHCl_3$ (ml)	MeOH (ml)
Alumina	60	70	30
Silica gel G[c]	35	67	33
Cellulose[d]	50	50	50
Magnesium silicate[e]	55	70	30

[a]Use any commercial adsorbent which may or may not have binder [G stands for $CaSO_4$ (plaster of Paris) or phosphor incorporated].
[b]Amounts including binder (usually about 20-30% of powder mixture).
[c]A variation uses 50 g in $CHCl_3$-MeOH-H_2SO_4 (70-30-1 ml); heating the plate after development visualizes spots by charring.
[d]Forms satisfactory layers without binder.
[e]Prepare by mixing adsorbent with 1 ml HOAc and minimal $CHCl_3$. Dilute to proportions shown.

Mix slurry thoroughly by shaking. Dip two microscopic slides (back-to-back) into slurry with a brief paddle motion; remove slowly and let drain. Separate slides and allow to dry; store in dry atmosphere. Heating before use may be desirable to be sure of activation. The developed and visualized chromatogram may be stored by transferring the layer to a piece of transparent (cellophane) tape by pressing the tape carefully and firmly over the layer, which will adhere. Another piece of tape on the underside seals the chromatogram for storage. Note: precoated glass, plastic, and "paper" tlc plates are commercially available. See suppliers, page 402.

2. Visualization of tlc spots (1a,1d,8). All solution reagents described in the table below should be applied as a spray; observations apply to organic compounds unless mentioned otherwise.

Reagent	Compound Type	Preparation and Observation
a. I_2	General use	I_2 vapor-chamber, or 1% methanol solution. Compounds appear as brown spots; limited sensitivity. Reversible reaction.
b. H_2SO_4 (50-98%)	General use	After sprayed plate is heated at 100-150° for several minutes, compounds appear as black spots.

(Continued)

Refs. p. 405.

Reagents	Compound Type	Preparation and Observation
c. $H_2SO_4/Cr_2O_7^{=}$ (Na^+ or K^+)	General use	Cleaning solution or 5-10% dichromate in 40-50% H_2SO_4. Same effect as for b, but stronger (5-10% HNO_3 in concentrated H_2SO_4 used similarly).
d. Uv light (250-400 nm)	Fluorescent compounds	Exposure of plate to light from typical "black" light lamp develops fluorescent spots against neutral background.
e. Uv light (250-400 nm)	Compounds that quench fluorescence or phosphorescence	Used with commercially available (or homemade) phosphor-containing adsorbents; dark spots appear on fluorescent background.
f. 2',7'-Dichloro (or dibromo) fluorescein (0.2% in 95% EtOH)	Lipids, lipophiles	Uv light (254 nm) shows yellow fluorescent spots on dark backgrounds.
g. Fluorescein (0.04% aq Na salt)	Conjugated systems	Uv light shows yellow spots on pink background.
h. Indicator dyes	Carboxylic acids	0.1-0.5% EtOH solutions made slightly alkaline with NH_4OH or NaOH (e.g., bromcresol green or purple; bromphenol blue; bromthymol blue); yellow spots on green, purple or blue background.
i. $SbCl_3$, 50% in HOAc; 25% in CCl_4; or saturated $CHCl_3$ solution	Steroids, vitamins, lipids, carotenoids	Various colors.
j. $FeCl_3$ (1% aq)	Phenols, enols	Various colors.
k. Ninhydrin	Amino acids, amino sugars, amino-phosphatides	0.3% in n-butanol containing 3% HOAc. After heat (125°, 10 min.), various blue colors appear.
l. 2,4-Dinitro-phenylhydrazine	Aldehydes and ketones	0.5% in 2N HCl. Red to yellow spots.

(Continued)

Refs. p. 405.

Reagent	Compound Type	Preparation and Observation
m. Anisaldehyde	Carbohydrates	0.5 ml in 0.5 ml concentrated H_2SO_4 + 9 ml 95% EtOH + few drops HOAc. Heat sprayed plate (100-110°, 20-30 min.). Various blue colors.
n. Dragendorff's reagent	Alkaloids; organic bases	Solution (a): 1.7 g $BiONO_3$ (bismuth subnitrate) in 100 ml H_2O-HOAc (80-20). Solution (b): 40 g KI in 100 ml H_2O. Use 5 ml (a) and 5 ml (b) in 20 ml HOAc + 70 ml H_2O. Orange spots.
o. 8-Hydroxyquinoline	Inorganic cations	Expose developed plate to NH_3; spray with reagent (0.5% in 60% EtOH). Uv light shows various colors.
p. $AgNO_3$ - fluorescein	Halide ions	Solution (a): 1% $AgNO_3$, made basic with NH_4OH. Solution (b): 0.1% fluorescein in EtOH. Spray (a), then (b) separately.

III. PAPER CHROMATOGRAPHY (9)

A. Mixotropic Solvent Series (1a,1b)[a]

Water[b]	1,4-Dioxane	Cyclohexanol
Lactic acid	Propionic acid	i-Amyl alcohol
Formamide	Tetrahydrofuran	1-Pentanol
Morpholine	t-Butanol[b]	Benzyl alcohol
Formic acid	i-Butyric acid	Ethyl acetate
Acetonitrile	2-Butanol	1-Hexanol
Methanol	Ethyl methylketone	s-Collidine
Acetic acid	Cyclohexanone	Pentanoic acid
Ethanol	Phenol	Ethyl formate
2-Propanol	t-Amyl alcohol	i-Valeric acid
Acetone	1-Butanol	Furan
1-Propanol	m-Cresol	Diethyl ether

(Continued)

Refs. p. 405.

1-Octanol	Chloroform	Carbon tetrachloride
Diethoxymethane	Di-i-amyl ether	Carbon disulfide
Caproic acid	1,2-Dichloroethane	Decalin
Butyl acetate	Bromobenzene	Cyclopentane
Di-i-propoxymethane	1,1,2-Trichloroethane	Cyclohexane
Nitromethane	1,2-Dibromoethane	Hexane
1-Bromobutane	Bromoethane	Heptane
Di-i-propyl ether	Benzene	Kerosene
Butyl butyrate	1-Chloropropane	Petroleum ether
1-Bromopropane	Trichloroethylene	Petroleum
Dibutyl ether	Toluene	Paraffin oil
Methylene chloride	Xylenes	

[a]Similar to the eluotropic series for adsorption; this series applies generally to partition separations on paper, column, or tlc. Reading down, the series goes from most to least hydrophilic.

[b]Up to t-butanol all solvents are completely water miscible.

B. Important Chromatographic Papers (10)[a]

Paper	Speed[b]	Properties[c,d]	Paper	Speed[b]	Properties[c,d]
Whatman			Schleicher-Schüll		
1	M	Std	2040a	F	
2	S	Std	2040b	M	Smooth
3	M	Prep	2043a	S	Std
3MM	M	Prep	2043b	M	Std
4	F		2071	S	Prep
31ET	VF	Prep			
54	F		Eaton Dikeman		
540	F		048	VF	
			248		Std
			613		Std
			320		Prep

[a]See page 403 for suppliers. There are other brands, especially outside the United States (e.g., Ederol; Macherey-Nagel; Filtrak-Niederschlag). Ion-exchange papers are also available; see page 385.

[b]Based on descending movement of 1-butanol-acetic acid-water (4-1-5) over 35 cm length. F (fast): 1-10 hr; M (medium): 10-15 hr; S (slow): 16 hr; V = very.

[c]Std = standard (all purpose) paper; Prep used for preparative purposes.

[d]Many of the types shown also come acid washed; with especially smooth texture; modified for use as ion-exchange papers; lipophile free; etc.

C. Recommended General Solvent Systems

The following are among the most successful systems for general use and should be tried in the absence of details for a specific separation.

Solvents[a]	Ratio[b]
HOAc-H_2O	15-85
2-Propanol-NH_3(conc)-H_2O	9-1-2
1-Butanol-HOAc-H_2O[c]	4-1-5
Phenol-H_2O	d
1-Butanol-1.5N NH_3	1-1
Ethyl acetate-H_2O	d
Butyl acetate-H_2O	d
Formamide/chloroform[f]	e
Formamide/C_6H_6-chloroform	d,g
Formamide/benzene[f]	e
H_2O-HOAc/carbon tet.	5-3/10
Formamide/C_6H_6-cyclohexane[f]	e
DMF/cyclohexane[h]	e
Kerosene/methanol-1-butanol-H_2O	/80-5-15[i]
Kerosene/1-propanol-H_2O	/88-12[i]
Paraffin oil/DMF-methanol-H_2O	/10-10-1[j]

[a]From most polar (hydrophilic) to least polar (hydrophobic). Each system written as: stationary phase/mobile phase; no / means single-phase system.

[b]By volume.

[c]Most universal system; stable up to ∿1 wk.

[d]First solvent (phase) saturated with second.

[e]Second solvent saturated with first.

[f]For all formamide systems, impregnate paper with 40% ethanolic formamide. To characterize acidic substances, 5% ammonium formate or 0.5% H_3PO_4 is added to stationary phase.

[g]Mobile phase composition varies from 1-9 to 9-1 depending on substances analyzed.

[h]Paper impregnated with 50% ethanolic DMF.

[i]Reversed phases; paper impregnated with kerosene. Ratio refers only to mobile phase.

[j]Same as previous footnote, but paper impregnated with 10% paraffin oil in benzene.

IV. COLUMN AND THIN LAYER PARTITION CHROMATOGRAPHY

A. Common Supports and Separations[a,b]

Used to Separate	Stationary Phase	Mobile Phase[c]
	Silica Gel	
Lipids	H_2O	Varies
N-Acetylaminoacids	H_2O	$CHCl_3$-1-butanol (100-1)

Refs. p. 405. (Continued)

Used to Separate	Stationary Phase	Mobile Phase[c]
N-Acetylpeptides	H_2O	$EtOAc-H_2O$
Acids (saturated and unsaturated)	0.05N H_2SO_4	$CHCl_3$, then $CHCl_3$-1-butanol mixtures up to 35% butanol
Aromatic acids	H_2O	$CHCl_3$, then $CHCl_3$-1-butanol mixtures up to 35% butanol
Monoalcohols	H_2O	CCl_4, then mixtures with $CHCl_3$, up to $CHCl_3$-HOAc(9-1)
Phenols	H_2O	Cyclohexane
17-Oxosteroids	H_2O	CH_2Cl_2-pet. ether
Alkanes, cycloalkanes	Aniline	2-Propanol-benzene
Aldehyde semicarbazones	H_2O	1-Butanol-$CHCl_3$
	Diatomaceous Earth	
C_1-C_4 alcohols	H_2O	$CHCl_3$ or CCl_4
C_2-C_{10} acids	30N H_2SO_4	Benzene or benzene-pet. ether
Di- and triols	H_2O	EtOAc or benzene-1-butanol
Ribonuclease	H_2O	$(NH_4)_2SO_4$-H_2O-cellosolve
Corticosteroids	H_2O	EtOH-CH_2Cl_2
Alkaloids	H_2O	Hexane-methanol-water-1,2-dichloroethane
Nitroanilines	50%(wt.) of 8N H_2SO_4	$CHCl_3$
Dinitrophenylamino acids	H_2O (buffer)	H_2O-EtOAc
Penicillins	Citrate buffer (pH5.7)	Et_2O-$(i\text{-}Pr)_2O$ (1-1)
Pentose nucleosides and nucleic acids	H_2O	1-Butanol
Insecticides	DMF or CH_3CN	Hexane
	Cellulose Powder	
Amino acids	H_2O	EtOH-H_2O; 1-butanol; 1-butanol-HOAc-H_2O (3-1-1); or phenol-H_2O (3-1)
Flavones	H_2O	Varies (pet. ether to ethyl acetate to 1-butanol to CH_3OH-H_2O to H_2O)
Monosaccharides	1-Butanol-ethanol-water (4-1-5)	
Monosaccharides	H_2O	H_2O saturated phenol
Methylated sugars	H_2O	Pet. ether-1-butanol (3-2)
Phenols	H_2O	1-Butanol-MeOH-$CHCl_3$
Metal cations	H_2O	Acetone-HCl
	Starch	
Amino acids	H_2O	1-Propanol or 1-butanol-HCl
Purines and pyrimidines	H_2O	1-Propanol-HCl (0.5N)

[a]Typical systems; see reference 1. For most cases in column chromatography, mobile phase mixtures are altered by gradient elution (stepwise change in mixture). [b]See also Porous Beads, page 373. [c]Saturated with stationary phase before use.

Refs. p. 405.

B. Column Partition Chromatography Hints

1. Use columns whose diameter-to-length ration is ≥ 20.
2. The finer the particle size (in a range covering somewhere between ∿100-400 mesh), the shorter the column required.
3. Weight of solid support/g solute: 500 to 1000 (greater than for typical adsorption columns).
4. Use the mixotropic solvent series for eluant guide (p. 379).

V. ION-EXCHANGE CHROMATOGRAPHY

A. General Resin Applications[a]

Type	Exchanging Group	Effective in pH Range	Approximate Exchange Capacity (meq/g dry resin)
		Cation Exchange	
Strong acid	$-SO_3H$	1-14	4
Weak acid	$-CO_2H$	5-14	9-10
Chelating	$-N(CH_2CO_2H)_2$	10-14	
		Anion Exchange[b]	
Strong base	$-CH_2\overset{+}{N}R_3$	1-15	4
Weak base	$-CH_2NR_1R_2$	1-9	4

[a]Adapted from reference 1c, page 33.
[b]$R = CH_3$ usually; R_1 and R_2 usually CH_3 and/or H.

B. Ion Affinity Sequence (Lyotropic Series) (1a,1b,1c)

For dilute, aqueous solutions at room temperature, cations and anions may be arranged in an approximate selectivity sequence for general purpose ion exchangers (in order of most to least strongly held).

1. Cations ($M^{4+} > M^{3+} > M^{2+} > M^{1+}$)

M^{1+}: Ag > Tl > Cs > Rb > NH_4 > K > Na > Li > R_4N

M^{2+}: Pb > Ba > Sr > Ca > Mg > Be

M^{2+}: Zn > Cu > Ni > Co > Fe > Ba > Sr > Ca > Mg

Refs. p. 405.

M^{3+}: La > Ce > Pr > Nd > Pm > Sm > Eu > Gd > Tb > Dy > Y > Ho > Er > Tm > Yb > Lu > Sc > Al

Mixed: Fe^{3+} > Al^{3+} > Ca^{2+} > Mg^{2+} > K^{+} > Na^{+} > H^{+} > Li^{+}

(Position of proton near Li for strong acid exchangers, but close to Ba for carboxylic or phenolic type.)

2. Monoanions

ClO_4 > SCN > I > NO_3 > Br > CN > HSO_4,HSO_3 > NO_2 > Cl > HCO_3 > CH_3CO_2 > F

(Position of hydroxyl: between F and CH_3CO_2 for strong-base exchangers, but more firmly held on weak-base type.)

C. Commercially Available Ion-Exchange Resins

The variety available is too large for comprehensive description and collection here. The following tables summarize the most common types; manufacturers and suppliers of these and numerous others are listed on page 403. Purified ("analytical grade") forms of "commercial" resin types ("Dowex" especially) are available from Bio-Rad Laboratories. Many of the resins come in a variety of particle sizes (mesh) in the form of beads and/or grains (gel).

1. Commercial Names and Manufacturers

If address is not given, consult the alphabetical listing on page 403.

Name (Symbol)	Manufacturer
	STANDARD RESINS
Acuolite	Louis Kelly and Co. (Argentina)
Amberlite, Amberlyst	Rohm and Haas (USA)
Anex; Katex	Institute of Synthetic Resins and Lacquers (Pardubice, USSR)
Bio-Rex; AG-series	Bio-Rad Laboratories (USA)
Diaion	Mitsubishi Kasei Kogyo (Japan)
Dowex	Dow Chemical Co. (USA)
Duolite	Diamond Shamrock Chemical Co. (USA)
Montecatini (Resina)	Montecatini Chemical (Italy)
Nalcite	National Aluminate Corp. (USA)

(Continued)

Refs. p. 405.

Suppliers p. 403.

Name (Symbol)	Manufacturer
Permutit, De-acidite	Permutit Co. (USA); Ionac Chemical Co. (USA)
Staionit	Závody Československo-sovětského přátelstvi (Zaluzi near Most, USSR)
Wofatit	Veb. Farbenfabrik Wolfen (East Germany)
Zerolit	Zerolit Limited (England)
Zeo-Karb	The Permutit Co. (England)
	CHELATING RESINS
Chelex-100	Bio-Rad Laboratories (USA)
Dowex A-1	Dow Chemical Co. (USA)

ION EXCHANGE CELLULOSE

The nomenclature used to identify the type is in terms of the cellulose derivative (exchange function): AE--aminoethyl; CM--carboxymethyl; CT--citrate; DEAE--diethylaminoethyl; ECTEOLA--mixed amines; GE--guanidoethyl; P--phosphonic acid; PAB--p-aminobenzyl; PEI--polyethyleneimine; SE--sulfoethyl ($-OC_2H_4-SO_3^-$); SM--sulfomethyl; TEAE--triethylaminoethyl ($-OC_2H_4-\overset{+}{N}Et_3$). They are classified as strong-base exchange groups (DEAE, GE, TEAE); intermediate-base (AE, ECTEOLA, PEI); weak-base (PAB); strong-acid (SE, SM); intermediate-acid (CT, P); weak-acid (CM). Most of these types are available as powders and as ion-exchange papers from the following [see also p. 389 (Gels)]:

Powder	Serva-Entwicklungslaboratorium (Germany)
Paper, powder	Schleicher and Schüll (Switzerland)
Paper, powder	Whatman (H. Reeve Angel-England, USA)
Paper	Rohm and Haas (USA) (Amberlite papers)
Powder	Bio-Rad (USA) ("Cellex" series); in size range appropriate for tlc

ION EXCHANGE GELS

(See also Gel Filtration, p. 389)

Bio-Gel P-2, CM-2	Bio-Rad Laboratories (USA)
Sephadex derivatives (CM, DEAE, SE)	Pharmacia Fine Chemicals (Sweden)

ION RETARDATION

AG-11A8	Bio-Rad Laboratories (USA)

2. Common Resins

The following table is reproduced with the permission of Bio-Rad Laboratories, Richmond, Calif. Resins with the same active groups made by different manufacturers may in many cases be used interchangeably, although they are not necessarily identical in composition or physical properties. Some resins have been discontinued by a particular manufacturer and are listed here for reference purposes only.

Type and Exchange Group	Bio-Rad Analytical Grade Ion Exchange Resins	Page No.	Dow Chem. Company "Dowex"	Diamond-Shamrock "Duolite"	Rohm & Haas Co. "Amberlite"	Permutit Company (England)	Permutit Company (U.S.A.)	Nalco Chemical Co. "Nalcite"
CATION EXCHANGE RESINS								
Strong Acidic, Phenolic type								
$R\text{-}CH_2SO_3^- H^+$	Bio Rex 40	5		C-3[ab]		Zeocarb 215		
Strong Acidic, Polystyrene type								
$\phi\text{-}SO_3^- H^+$	AG 50W-X1	5	50-X1[a]					
	AG 50W-X2	5	50-X2[a]					
	AG 50W-X4	5	50-X4[a]		IR-112	Zeocarb 225 (X4)		
	AG 50W-X5	5	50-X5[a]	C-25D				
	AG 50W-X8	5	50-X8[ab]	C-20	IR-120	Zeocarb 225	Permutit Q	HCR
	AG 50W-X10	5	50-X10[a]	C-20X10	IR-122		Q-100	HGR
	AG 50W-X12	5	50-X12[ab]	C-20X12	IR-124		Q-110	HDR
	AG 50W-X16	5	60-16[a]				Q-130	
Intermediate Acid, polystyrene type								
$\phi\text{-}PO_3^- (Na^+)_2$	Bio Rex 63	5		ES-63				X-219
Weakly Acidic, acrylic type								
$R\text{-}COO^- Na^+$	Bio Rex 70	5		CC-3	IRC-50 IRC-84	Zeocarb 226	Q-210	
Weakly Acidic chelating resin, polystyrene type								
$\phi\text{-}CH_2N(CH_2COO^-H^+)_2$	Chelex 100	5	A-1[a]					
ANION EXCHANGE RESINS								
Strongly basic, polystyrene type								
$\phi\text{-}CH_2N^+(CH_3)_3Cl^-$	AG 1-X1	3	1-X1[a]			DeAcidite FF (lightly crosslinked)	S-100	
	AG 1-X2	3	1-X2[a]					
	AG 1-X4	3	1-X4[ab]	A-101D	IRA-401			
	AG 1-X8	3	1-X8[ab]		IRA-400	DeAcidite FF		SBR
	AG 1-X10	3	1-X10[a]					SBR-P
	AG 21K	3	21K[a]					
$\phi\text{-}CH_2N^+(CH_3)_2(C_2H_4OH)\ Cl^-$	AG 2-X4	3	2-4[a]	A-102D			S-200	
	AG 2-X8	3	2-X8[a]		IRA-410			
	AG 2-X10	3						SAR
(pyridinium) NH^+	Bio-Rex 9	3					S-180	
Intermediate base, epoxypolyamine								
$R\text{-}N^+(CH_3)_2Cl^-$ and $R\text{-}N^+(CH_3)_2(C_2H_4OH)\ Cl^-$	Bio Rex 5	3		A-30[ab] A-30B		F	S-310 S-380	
Weakly basic, polystyrene or phenolic polyamine								
	AG 3-X4A	3	3-X4[a]	A-2	IR-45	G	S-300	WBR
$R\text{-}N^+H(R)_2Cl^-$		NL		A-6	IR-4B			
$R\text{-}N^+H(R)_2Cl^-$		NL		A-7[b]	IRA-68			
				A-4F			S-350	
MIXED BED RESINS								
$\phi\text{-}SO_3^-H^+$ & $\phi\text{-}CH_2N^+(CH_3)_3OH^-$	AG 501-X8	5		GPM-331 G	MB-1	Bio Demineralit	M-100	
$\phi\text{-}SO_3H^+$ & $\phi\text{-}CH_2N^+(CH_3)_3OH^-$ indicator dye	AG 501-X8 (D)	5				Indicator Bio-Demineralit		
$\phi\text{-}SO_3^-H^+$ & $\phi\text{-}CH_2N^+(CH_3)_2(C_2H_4OH)OH^-$	Supplied in Reactor Grade only	9		GPM-331A	MB-3		M-103	

(a) denotes the commercial resin selected by Bio–Rad for further sizing and purification to an Analytical Grade Ion Exchange Resin.

(b) denotes a resin also supplied by Bio–Rad in bulk commercial grade at manufacturers' list price.

Refs. p. 405.

Suppliers p. 403.

Refs. p. 405.

VI. GEL FILTRATION AND GEL PERMEATION CHROMATOGRAPHY (11-13)

By definition (11a), gel filtration involves aqueous solvents and hydrophilic gels while gel permeation involves organic solvents and hydrophobic gels. The former is used for biochemical and natural product applications, the latter for synthetic macromolecules. Chromatography or "filtration" with molecular sieves fall into these categories (12). The gel type is characterized by the size of the molecules (molecular weight range) that can be separated efficiently.

A recently developed application of gel technology is polyacrylamide gel electrophoresis (PAGE), a high-resolution method for fractionation and characterization of molecules on the basis of size, conformation, and net charge (as for proteins, nucleic acids, and so on); see the review by Chrambach and Rodbard in Science, 172, 440 (1971).

A. Types and Properties of Filtration Gels[a]

Type	Water Regain[b] (g/g dry gel)	Dry Particle Size (μm)	Range or Upper Limit (mw) for Fractionation of: Globular Proteins and Peptides	Dextran Fractions (Polysaccharides)
SEPHADEX[c,i]				
G-10	1.0	40-120	0-700	0-700
G-15	1.5	40-120	0-1,500	0-1,500
G-25 (coarse)	2.5	100-300	1,000-5,000	100-5,000
G-25 (medium)	2.5	50-150		
G-25 (fine)	2.5	20-80		
G-25 (superfine)[f]		10-40		100-5,000
G-50 (coarse)	5.0	100-300	500-30,000	500-10,000
G-50 (medium)	5.0	50-150		
G-50 (fine)	5.0	20-80		
G-50 (superfine)[f]		10-40		500-10,000
G-75	7.5	40-120	3,000-70,000	1,000-50,000
G-75 (superfine)[f]		10-40		1,000-50,000
G-100	10	40-120	4,000-150,000	1,000-100,000
G-100 (superfine)[f]		10-40		1,000-100,000
G-150	15	40-120	5,000-400,000	1,000-150,000
G-150 (superfine)[f]		10-40		
G-200	20	40-120	5,000-800,000	1,000-200,000

(Continued)

Suppliers p. 403.

Type	Water Regain[b] (g/g dry gel)	Dry Particle Size (μm)	Range or Upper Limit (mw) for Fractionation of Globular Proteins and Peptides	Dextran Fractions (Polysaccharides)
G-200 (superfine)[f]		10-40		1,000-200,000
LH-20[k]	2.1	25-100		
		SEPHAROSE[d,j]		
6B(6)		40-210	4×10^6	10^6
4B(4)		40-190	20×10^6	5×10^6
2B(2)		60-250	40×10^6	20×10^6
		BIO-GEL P[h]		
2	1.5	e	200-1,800	
4	2.4		800-4,000	
6	3.7		1,000-6,000	
10	4.5	f	1,500-20,000	
30	5.7	f	2,500-40,000	
60	7.2	f	3,000-60,000	
100	7.5	f	5,000-100,000	
150	9.2	f	15,000-150,000	
200	14.7	f	30,000-200,000	
300	18.0	f	60,000-400,000	
		BIO-GEL A[d]		
0.5m (10)		g	$<10^4-5 \times 10^5$	
1.5m (8)			$<10^4-1.5 \times 10^6$	
5m (6)			$10^4-5 \times 10^6$	
15m (4)			$4 \times 10^4-15 \times 10^6$	
50m (2)			$10^5-50 \times 10^6$	
150m (1)			$10^6- >150 \times 10^6$	
		SAG (AGO-GEL)[d,l]		
2		70-140	$50 \times 10^4-1.5 \times 10^8$	
4			$20 \times 10^4-15 \times 10^6$	
6			$5 \times 10^4-2 \times 10^6$	
8			$2.5 \times 10^4-70 \times 10^4$	
10			$10^4-25 \times 10^4$	

Footnotes appear on page 389.

[a]For addresses of suppliers, see page 403.
[b]Amount of water taken up (±10%) by dry gel material.
[c]A cross-linked dextran (polysaccharide) from Pharmacia.
[d]Gel medium containing agarose; % agarose in parentheses after type.
[e]All Bio-Gel P types come in 50-100, 100-200, 200-400, and 400 mesh.
[f]Suitable for use in thin layer gel filtration.
[g]All Bio-Gel A types come in 50-100, 100-200, and 200-400 mesh.
[h]A polyacrylamide from Bio-Rad Labs.
[i]A calibration kit for protein mw determination is available from Pharmacia.
[j]Ficoll, a copolymer of sucrose and epichlorohydrin, is useful for separation and isolation of cells and subcellular particles, primarily in centrifugation in dense media (from Pharmacia).
[k]An alkylated G-25 of uniform swelling properties in different solvents; 25-100 μm particle size.
[l]Product of Seravac Laboratories (PTY) Ltd. Available as "Ago-Gel" in the United States from Schwarz-Mann Bioresearch.

B. Types and Properties of Ion-Exchange Gels

(See also p. 385, Ion Exchange.)

Type	Functional Group	Capacity (meq/g)	Range (Upper Limit, mw)
		SEPHADEX[a]	
Anion			
DEAE A-25	$-NEt_2$	3.5	10,000
DEAE A-50	$-NEt_2$	3.5	>10,000
QAE A-25		3.0	
QAE A-50		3.0	
Cation			
CM C-25	$-CO_2H$	4.5	10,000
CM C-50	$-CO_2H$	4.5	>10,000
SE C-25	$-SO_3H$	2.3	10,000
SE C-50	$-SO_3H$	2.3	>10,000
		BIO-GEL	
CM-2[b]	$-CO_2H$	6.0	10,000
Carboxy P-2[c]	$-CO_2H$		

[a]All have 40-120 μm particle size. Ionic form: anion as Cl^-, cation as Na^+.
[b]A poly(acrylic acid); 100-200 mesh.
[c]A carboxylated Bio-Gel P-2, 100-200 mesh.

C. Types and Properties of Permeation Gels

Type	Operating mw Range	Exclusion Limits		
		Polystyrenes	Dextrans	
	BIO-GLAS[a,b]			
200[c]	3,000-30,000	30,000		
500	10,000-100,000	100,000		
1000	50,000-500,000	500,000		
1500	400,000-2,000,000	2,000,000		
2500	800,000-9,000,000	9,000,000		
	BIO-BEADS[a,d]			Swollen Form: ml Benzene/g Bio-Beads
S-X1	600-14,000	14,000		9.8
S-X2	100-2,700	2,700		6.2
S-X3	Up to 2,000	2,000		5.1
S-X4	Up to 1,400	1,400		4.2
S-X8	Up to 1,000	1,000		3.9
SM-1	600-14,000	14,000		3.1
SM-2	600-14,000	14,000		2.9
	MERCK-O-GEL-OR[e]			
750		750		
1,500		1,500		
5,000		5,000		
20,000		20,000		
100,000		10^5		
1,000,000		10^6		
	MERCK-O-GEL-Si[f,g]			
150[c]		50,000		
500		400,000		
1,000		10^6		
	CORNING CONTROLLED PORE GLASS (CPG)[g,h]			
75[c]	300-28,000	10,200	28,000	
125	650-48,000	33,000	48,000	
175	1,050-68,000	68,000	68,000	
240	1,150-95,000	160,000	200,000	
370	5,000-150,000	250,000	400,000	
700	15,000-300,000	900,000	1,000,000	
1250	40,000-550,000	3,000,000	2,000,000	
2000	120,000-1,200,000	12,000,000	8,000,000	

(Continued)

Refs. p. 405.

Suppliers p. 403.

Type	Operating mw Range	Exclusion Limits	
		Polystyrenes	Dextrans
	PORASIL[i]		
60		60,000	
250		250,000	
400		400,000	
1000		10^6	
1500		1.5×10^6	
2000		2.0×10^6	
	STYRAGEL[j]		
60	50-1,000	1,000	
100	100-3,000	3,000	
500	100-10,000	10,000	
10^3	500-50,000	50,000	
10^4	1,000-700,000	700,000	
10^5	$50,000-2 \times 10^6$	2×10^6	
10^6	10^5-10^7	5×10^7	
Aquapak A-440		10^5	

[a]From Bio-Rad Labs.
[b]Highly porous silicate, in mesh sizes 50-100, 80-100, 100-120, 100-200, 200-325, > 325.
[c]Number refers to pore diameter (Å).
[d]Porous cross-linked poly(styrenes). S-X in 200-400 mesh, S-M in 20-50.
[e]A poly(vinyl acetate) from Merck AG.
[f]Porous silica from Merck AG.
[g]Available from Waters Associates.
[h]Made by Corning Lab Products.
[i]Porous silica from Waters.
[j]Cross-linked poly(styrenes) from Waters in 200-400 mesh. All have HETP = 0.3 mm; see also, Poragel, page 373, under Porous Beads.

VII. AUTOMATED LIQUID CHROMATOGRAPHY

Aside from automatic fraction collection devices (1b, Chapter 7), there are high-speed liquid chromatographs and custom columns for analytical and preparative work in adsorption, partition, and gel chromatography; the equipment also includes refractive index and uv detectors for high-resolution, high-speed fractionation. Many companies make prepacked columns and accessories, particularly for gel chromatography. For extensive lists of suppliers of liquid chromatographic equipment see (11b,12). For a recent review on liquid chromatography detectors, see H. Veening, J. Chem. Educ., 47, A549 (1970). A concise discussion on "Contemporary Liquid Chromatography" is available in Vol. 11, No. 2, 1971 of Research Notes (Analabs, Inc., 80 Republic Drive, New Haven, Conn. 06473). A recent book summarizing recent developments is also available: J. Kirkland, "Modern Practice of Liquid Chromatography," Wiley-Interscience, New York, 1971.

Refs. p. 405. Suppliers p. 403.

VIII. ELECTROPHORESIS

In addition to treatises (21,22) and good introductory chapters of the subject (1a,1b), there is available a comprehensive bibliography with abstracts, covering recent developments (23).

In chromatography, solute movement and separation are effected by means of a moving solvent. In electrophoresis, the solvent (a buffer solution) is stationary, while the solute (charged particles) migrate under the influence of an applied electric field. References 21 and 22 provide recommended experimental conditions for many types of compounds, including the pertinent buffers.

Free Boundary Electrophoresis--also called "moving boundary" or "Tiselius" electrophoresis. The method has been used mainly for quantitative analysis of protein mixtures, and unlike zone electrophoresis, provides the intrinsic electrophoretic mobility of migrating species.

Zone electrophoresis--the use of a stabilizing support medium on or through which the ions migrate. The medium may be in the form of strips (paper), columns, disks, thin layers, and so on. The most commonly used is filter paper (such as Whatman No. 1 and No. 3MM), but others include cellulose acetate, agar gel, starch gel, and polyacrylamide gel (24,25). The electrophoresis may be low voltage (applied voltage <1000 V) or high voltage (≈ 1000 to 10,000 V). In continuous electrophoresis (a low-voltage preparative procedure), sample is applied continuously to the medium (especially Whatman 3MM or Schleicher and Schuell 2230 papers). High-voltage electrophoresis is ordinarily carried out on paper; it is not good for large molecules, but is very good for amino acids and other small molecules.

Immunoelectrophoresis--after electrophoretic separation in agar, starch gel, or cellulose acetate, a protein mixture is exposed to a solution of antiserum. This antiserum contains specific antibodies against individual components of the separated protein mixture. Where antigens and antibodies meet, they precipitate one another, forming curved lines of precipitation. Each of the lines corresponds to a different protein in most favorable cases, thus identifying the protein serologically as well as physicochemically (1a).

Mobility--the electric mobility, u, is the rate of migration of an ion of net charge, Θ, and radius, r, measured under unit field strength in a solution of viscosity, η: $u = \Theta/6\pi r\eta$ (cm^2/V-sec).

Potential gradient--electric field conditions, expressed as the voltage per cm separating the electrodes (V/d).

IX. VAPOR PHASE CHROMATOGRAPHY (2,14)

This section concentrates on gas-liquid (partition) chromatography; see page 371 (Adsorbents) for information on gas-solid chromatography. For recent details on many aspects of glc and gsc, consult the Advances in Chromatography series, J. C. Giddings and R. A. Keller, Eds., Marcel Dekker, New York [among others (2)].

A. Useful Equations and Definitions

1. Retention Volume, V (26)

A function of amount of sample and other factors; the most significant values are those extrapolated to zero sample size. While peaks are commonly identified by their retention times, t_R, retention volume is a more accurate measure of a component under the column conditions. In the following formulas for retention volume, the terms are defined as follows:

t_R = retention time (from point of injection to peak maximum)

t_o = retention time of nonsorbed substance (air, noble gas)

F = volumetric flow rate of carrier gas at the temperature and outlet pressure (P_o) of the column

V_M = gas holdup (dead volume) = V_R of a nonsorbed sample (air)

j = pressure gradient (compressibility) factor for a homogeneously packed column of uniform diameter = $[3(P_i/P_o)^2 - 1]/[2(P_i/P_o)^3 - 1]$, where P_i = carrier gas pressure at column inlet

T = absolute temperature of column

W_L = weight of stationary phase (solvent). Subscript s refers to some internal standard, usually an n-alkane, while x refers to a given component of a sample

Uncorrected: $V_R = t_R F$ Adjusted: $V_R' = V_R - V_M$

Corrected: $V_R^\circ = jV_R$ Net: $V_N = jV_R'$

Specific: $V_g = 273\ V_N/W_L T$ (equivalent to V_N at 0° per g liquid phase)

Relative: $\alpha = V_{N_x}/V_{N_s} = V_{g_x}/V_{g_s} = (t_{R_x} - t_o)/(t_{R_s} - t_o)$

2. Retention Index (Kováts Index)

An empirical system for characterization of organic compounds by gc on nonpolar stationary phases. Retention index (RI) is related to compound boiling point; it is independent of column conditions and apparatus type; its temperature dependence is small and linear; it can provide information about the chemical nature of a compound (15).

$$RI = \frac{100 \log \alpha_{x,N}}{\log \alpha_{N+1,N}} + 100\ N$$

where $\alpha_{x,N}$ is the specific retention (V_g) of x relative to an n-alkane of N-carbon atoms, and $\alpha_{N+1,N}$ is that of an n-alkane with N + 1 carbon atoms relative to one of N-carbon atoms. (For application of this index and use of "Rohrschneider" constants, see footnote a, p. 401).

3. Reaction Gas Chromatography, Carbon Skeleton Chromatography, Pyrolysis Gas Chromatography

Controlled thermal and/or catalytic decomposition of substances at the entrance to (or within) a column; gc record of fragments can provide information about starting materials. See reference 16. These techniques are helpful in determining structure with very small amounts of compounds (microgram level) by performing pyrolysis "chemistry" in the gc column.

4. Golay or Capillary Columns (Open-Tubular Columns)

Made by coating the inside walls of capillary tubes (stainless steel, glass, etc.), 0.01 to 0.05 in. i.d., up to several hundred feet in length

Refs. p. 405.

with ∿10 to 50 mg stationary phase per 100 ft. Requires the use of very small sample size (<1 μg) but achieves excellent separation of even closely related isomers; columns are capable of >100,000 theoretical plates. For reviews, see D. Desty, Advan. Chromatog., 1, 199 (1965), and I. Halász and E. Heine, ibid., 4, 207 (1967) (packed capillary columns).

5. Silylation and Other Derivatization Procedures

Silylation is a technique for converting hydroxy and amino compounds (sugars, steroids, phenols, amino acids, peptides) into volatile silane derivatives (e.g., trimethylsilyl) that lend themselves to vpc (as well as mass spectral) determinations. For recent reviews and silylation reagents, write for Handbook of Silylation, Pierce Chemical Co., Box 117, Rockford, Ill. 61105 (free) or the more detailed book by A. E. Pierce, Silylation of Organic Compounds, Pierce Chem. Co., 1970. A similar procedure using n-butylboronic acid ($n\text{-BuB(OH)}_2$) has been described for α-hydroxy fatty acids and 1-glyceryl ethers [see C. Brooks and I. Maclean, J. Chrom. Sci., 9, 18 (1971)]. Other common derivatives of nonvolatile OH and NH compounds are the acetate and trifluoroacetate.

B. Carrier Gas Properties and Flow Rates

Impurities in carrier gases have negligible effect on retention properties, but markedly affect detector stability and response. The following refer to highly purified gases; the values are taken or calculated from reference 14b, International Critical Tables, and sources cited on page 36.

Gas	Mw	C_p[a]	Viscosity[b] (cP x 10^2)	Thermal Conductivity[c] 0°	100°
Air	28.98	7.0	1.71	5.8	7.5
Hydrogen	2.016	6.8	0.84	41.6	53.4
Helium	4.003	5.1	1.87	34.8	41.6
Nitrogen	28.02	6.9	1.66	5.8	7.5
Oxygen	32.00	7.0	2.04 (23°)	5.9	7.6
Argon	39.95	5.0	2.10	4.0	5.2
Carbon dioxide	44.01	8.8	1.37	3.5	5.3
Carbon monoxide	28.01	6.9	1.66	5.6	7.2
Methane	16.04	8.5	1.02	7.2	10.9
Ethane	30.07	11.6	0.85	4.3	7.3
Propane	44.10			3.6	6.3

[a]Heat capacity in cal/deg-mole at 15°, 1 atm.
[b]At 0°, 1 atm in centipoise (cP).
[c]In 10^{-5} cal/sec-cm^2/(°C/cm).

Refs. p. 405.

Influence of Carrier Gas Type and Flow Rate

Carrier Property	Effect
Molecular weight	Sharper separation with gases of higher molecular weight (less diffusion of compounds being separated). However, with thermal conductivity detector, gases of lower molecular weight give better sensitivity because of higher thermal conductivity.
Flow rate	Column efficiency best if flow rate is kept constant; optimum flow rate is that giving rise to a minimum HETP (maximum number of plates). Useful rates are 75-100 ml/min for 1/4 in. o.d. and 25-50 ml/min for 1/8 in. o.d. columns.

C. Properties of Detectors

Different detectors do not produce the same response (peak area, e.g.) for the same compound, nor do equimolar amounts of different compounds produce equal response with the same detector. Each detector and each compound mixture require calibration (particularly for a TC detector) to determine correction factors (response factors) for quantitative analysis. Many response factors for FI and TC detectors are published [W. Dietz, J. Gas Chromatog., 5, 68 (1967)]. (See also reference 14a.) For a review of ionization methods for detection, see A. Karmen, Advan. Chromatog., 2, 293 (1966). The following table was constructed from information in references 1c, 1d, 14a, and 14b.

Type	Suggested Carrier Gases	Temperature Limit[a]	Lower Detection Limit (μg)[b]	Linear Range[c]	Use[d]
Thermal conductivity[e]	H_2, He(N_2)	450	2-5	10^4	General
Flame ionization[f]	He, N_2	400	10^{-5}-10^{-6}	10^6-10^7	Organic[f]
Electron capture[g]	N_2, Ar[h]	225	10^{-7}	500	Variable[g]
Helium[i]	He	225	10^{-6}	10^4	i
Cross-section ionization[j]	H_2, He[k]	225	20	5×10^5	General

(Continued)

Refs. p. 405.

Type	Suggested Carrier Gases	Temperature Limit[a]	Lower Detection Limit (μg)[b]	Linear Range[c]	Use[d]
Argon[l]	Ar	225			1
Phosphorus[m]	He, N_2	300	10^{-5}	10^4	Organo-P compounds
Gas density balance[n]	N_2, CO_2, Ar	150	Variable	10^3	General

[a]Upper operating T (°C).

[b]Approximate minimum amount detectable (μg); may depend on compound type.

[c]A measure of the concentration range over which detector response is linear. Such behavior is determined from a plot of response versus concentration; the slope of such a plot would be 1.0 for an ideal detector. In practice, values are <1, and calibration plots are suggested for best results over large concentration ranges. Linear range = ratio of the largest to the smallest concentration within which linearity is observed.

[d]Types of substances that give good detector response.

[e]Thermal conductivity (also called a katharometer). For maximum sensitivity, maintain high filament current, low block temperature, and constant carrier flow (H_2 best, but He used most often). Filaments are unstable in presence of air, halogens, hydrogen halides, alkyl halides. TC is best detector for water determination.

[f]Flame ionization detector. Effluent gas mixed with H_2 and burned with O_2 (air) to produce ionizing flame. Detection limit (μg) should be read as μg/sec (FI response is proportional to mass flow rate); to convert to mass, multiply by peak width (sec) at baseline. FI is excellent for quantitative studies. FI will not respond well, or at all, to noble gases, H_2O, O_2, N_2, NO, N_2O, NO_2, NH_3, COS, CS_2, CO, CO_2, SO_2, $HSiCl_3$, SiF_4, HCO_2H.

[g]Electron capture. Uses tritium source. Very sensitive to water (carrier gas must be dry); relatively insensitive to weakly electron-capturing compounds (hydrocarbons, ethers, carbonyl compounds, alcohols) but excellent for halocarbons.

[h]Argon containing 10% methane.

[i]All compounds having a lower ionization potential than He(19.8 eV) give a positive signal. Good for trace analysis of fixed gases. He used must be very pure and dry.

[j]Cross-section ionization detector. One of least sensitive, but most general (responds to all substances in any concentration in any carrier gas) and highly linear.

[k]He containing 3% methane.

[l]Argon ionization detector. Very similar to CSI[j], but much more sensitive. Substances with ionization potential >11.7 eV not detected (e.g., H_2, N_2, O_2, CO_2, $CO(CN)_2$, H_2O, fluorocarbons; very little response from CH_4, C_2H_6, CH_3CN, C_2H_5CN).

[m]Phosphorus. A normal flame detector modified by addition of an alkali salt pellet on the quartz burner jet. Highly specific to P-compounds; used extensively for pesticide residue analysis.

[n]Gas density balance. Sensitivity depends on density difference between sample and carrier (N_2 preferred, except for CO, C_2H_4, C_2H_2

Refs. p. 405.

determinations). Do not use H_2 or He as carriers. GDB is exceptional in that no calibration is needed for quantitative analyses; response is directly proportional to concentration and molecular weight [molecular weights can be measured with GDB (17)].

D. Comparison of Chromatogram Integration Methods

For discussions of various methods, see references 14a, 14b, and 18. There is some disagreement in the literature concerning the relative merits of methods, especially the manual ones (which depend on user skill).

Method[a,i]	Time (min)[b]	Accuracy[c]	Precision[d]
Peak Height[e]	1-2	--	--
Disk[g]	2-4	2	2
Triangulation[h]	4-8	3	4
Planimeter	4-8[f]	5	5
Electronic[j]	0.5-1	1	1
Cut-and-Weigh[k]	5-10	4	3

[a]In order of most to least often used according to one survey (14a); the first three are used by >65% of users polled.

[b]Approximate time required per typical peak for complete workup of a trace (individual peak areas and % composition). From references 14a, 18.

[c]Relative ranking; 1 is most accurate. Refers to how well the area methods perform compared to the true (real) values.

[d]Relative ranking; 1 is most precise. Refers to reproducibility (relative deviation) of several area measurements on a single peak.

[e]Recommended only if bandwidth is narrow and peaks are fairly small.

[f]At least two area determinations recommended per peak.

[g]Electromechanical device made by Disc Instruments, Inc., 2701 South Halladay Street, Santa Anna, Calif. 92705 (available with automatic print-out attachment).

[h]Most accurate variation [for Gaussian (symmetrical) peaks]: peak height x width at half height. See Condal-Bosch (18) for more details.

[i]For planimetry, triangulation, and cut-and-weigh methods, increasing the chart speed increases accuracy and precision of area measurements.

[j]May be analog or digital. Very expensive, but best and most efficient (e.g., Varian Aerograph , Walnut Creek, Calif.; Infotronics Corp., 7800 Westglen Drive, Houston, Texas 77042; Chemicon Co., 839 Beacon Street, Boston, Mass. 02215; Vidar Corp., 77 Ortega Avenue, Mt. View, Calif. 94040).

[k]Make xerox (or other photocopy) for cutting (preserves original trace and provides thicker, more uniform paper for weighing).

E. Solid Supports

Particle sizes fall within the range of 10-140 mesh (U.S.) (see p. 371 for mesh table); for maximum efficiency (better packing, less clogging), narrow ranges are highly recommended (e.g., 60/80). Most columns are prepared with 60/80, 80/100, or 100/120 mesh supports. Note that separation efficiency increases as grain size decreases (thus packing density

Refs. p. 405.

increases); at the same time, however, pressure drop across the column (resistance to carrier flow) increases, thereby increasing retention times. Most supports are made from diatomaceous earth (diatomite, diatomaceous silica, Kieselguhr)--a form of microamorphous hydrous silica containing metal oxide impurities (~10%)--and firebrick--similar to diatomaceous earth (high in silica and in some cases alumina, with metal oxide impurities). The latter typically has a larger surface area and is preferred for very long columns, but is also more prone to catalytic and adsorption effects (causes peak tailing). Tailing and other effects are reduced by acid washing and by treatment with silanizing vapors to tie up active sites (dimethyldichlorosilane (DMCS) used in most cases). The white supports (Chromosorb-W, Celite, etc.) are the most common; red (pink) are diatomaceous earth mixed with binder and calcined at high temperature (Chromosorb-P, C-22 Firebrick, Sterchamol, etc.). The latter are recommended for separation of weakly polar compounds. Chromosorb A is a diatomite specifically for use in preparative scale glc and holds up to 25% liquid phase. Listed below are trade names for common supports, most of which are available in many grain sizes, as well as in nontreated or treated forms (acid and/or base washed; DMCS treatment). Where possible, the manufacturer is given; most of the types listed are available from most vpc suppliers and from some chemical supply companies. For a recent review on supports, see D. Ottenstein, Advan. Chromatog., 3 (1967).

Support	Manufacturer[a]	Support	Manufacturer[a]
	Diatomaceous Earth and Firebrick		
Aeropak	Varian Aerograph	GC 32 (Celatom)	Eagle-Picher
Anakrom[b]	Analabs	GC Super Support	Coast Engineering Lab.
C-22 Firebrick	Johns-Manville	Hyflo	Johns-Manville
Celite 545	Johns-Manville	S-80	R. Grutzmacher
Chromosorb	Johns-Manville	Sil-o-cel Brick	Griffin and George
Diatom	Debton Co.	Sterchamol	Sterchamol-Werke
Diatoport S	Hewlett-Packard	Supelcoport	Supelco
Embacel	May and Baker	Varaport (see Aeropak)	
Gas Chrom	Applied Science Labs		
	Halocarbons[c]		
Anaport[d]	Analabs	Haloport F[e]	Hewlett-Packard
Chromosorb T[e]	Johns-Manville	Kel-F[d]	3M Company
Fluoropak 80[f]	Fluorocarbon Co.	Tee Six[e]	Analabs
	Porous Polymers[g]		
Chromosorb 100 Series	Johns-Manville	Porapak	Waters Assoc.

(Continued)

Refs. p. 405.

Suppliers p. 403.

Support	Manufacturer[a]	Support	Manufacturer[a]
Glass and Silica[h]			
Anaport	Analabs	Corning Porous Glass[i]	Corning Glass Works[i]
Cera Beads	Analabs	Porasil[j]	Waters Assoc.
Corning GLC-Beads	Corning Glass Works	Glassport	Hewlett-Packard

[a]For addresses, see page 403.
[b]Also "Anaprep" for preparative separations.
[c]Especially good for aqueous solutions, corrosive and highly polar compounds.
[d]Kel-F, a chlorofluorocarbon polymer; types 300 LD and 6051 used.
[e]Teflon 6 (Du Pont).
[f]A fluorocarbon polymer.
[g]Some form of poly(vinylbenzene), especially good for short chain, highly polar compounds (acids, amines, alcohols, etc.).
[h]Glass beads good for separation of high molecular weight compounds at lower temperatures.
[i]Made from 7930 Vycor glass.
[j]Certain liquid phases can be chemically bound to Porasil; sold as Durapak (Waters Assoc.).

F. Stationary Phases and Column Preparation

There are more than 400 commercially available liquid stationary phases (also called solvent phases) of recognized utility. For one of the most comprehensive lists and convenient bibliographies, write for Analabs Guide to Stationary Phases for Gas Chromatography, Analabs, Inc., 80 Republic Drive, North Haven, Conn. 06473 (a booklet published annually). Most vpc supply companies carry fairly complete collections; some chemical suppliers (Aldrich and Matheson Scientific, e.g.) also stock a large variety.

For most applications, columns are prepared using 1 to 30% by weight of liquid phase (15 to 20% most common; high proportions (≥20% are usually used in preparative columns). Generally, the best operation is experienced with a low percent liquid phase. Choice of a good phase for a particular application, especially a complex mixture, may be subject to trial and error; generally, the principle "like dissolves like" is a useful guide. Extensive bibliographies are available to assist in column choice (19). The following table lists, according to chemical structure, the most common and versatile liquid phases.

Stationary Phase	Use[a]	Solvent[b]	Maximum Temperature[c]
HYDROCARBONS			
Apiezon L	B, S, P; II-V(0)	B, T	300
Apiezon M	B, S, P; II-V(0)	B, T	275
Apiezon N	III, IV	B, T	300

Refs. p. 405. (Continued) Suppliers p. 403.

Stationary Phase	Use[a]	Solvent[b]	Maximum Temperature[c]
Asphalt	IV, V	B	300
Benzylbiphenyl	II, IV	A	100
n-Hexadecane	IV, V	B, T	50
Paraffin wax	N; IV(0)	C	200
Squalane	B, N, P; III-V; gases	T	140
Squalene	V(0)	T	150
ESTERS			
Beeswax	Essential oils	C	200
Butanediol succinate (BDS)[d]	N; II, IV; sugars	C	225
Castorwax	I; essential oils	C	200
Dibutyl phthalate	P; III, IV	A, M	100
Diethylene glycol adipate (DEGA)	III	A	200
Diethylene glycol succinate(DEGS)	P, S; II-IV	A, C	200
Dinonyl phthalate	N; II-IV	A, C	150
Ethylene glycol succinate (EGS)	P; II-IV	C	200
Tricresyl phosphate (TCP)	S; II-V; gases	A, M	125
POLYGLYCOLS			
Carbowax 400	N; I-IV	C, M	125
Carbowax 1500	N, S; II, III	A, M	200
Carbowax 4000	II-IV; sugars	A, M	200
Carbowax 20M	P, S; I-V	C	250
Ucon 50 HB 280X	I-IV	C	200
Ucon 50 LB 550X	III-V	A	200
Ucon 50 HB 2000	II-V	A, M	200
AMIDES			
Hallcomid M18[f]	I-III	A, C	150
Hallcomid M180L[g]	I-III	M	150
Versamid 900	N; II-IV	h	350
Versamid 940	N; II-IV	h	275
SILICONES[i]			
DC 200 (oil)	P, S; III-IV	T	250
DC 550 (oil)	B, N, (0); II-IV; gases	A, T	275
DC 710 (oil)	N, P, (0); IV, V	A, C	300
SF 96	Si; IV, V	T	250
SE 30[j]	N, P, S, (0); II-V; gases	C, T	350
XE 60 (a cyano derivative)	N, P, S; II-V	A	250

(Continued)

Refs. p. 405.

Suppliers p. 403.

Stationary Phase	Use[a]	Solvent[b]	Maximum Temperature[c]
MISCELLANEOUS			
Bentone 34[k]	IV	B	200
FFAP	S, Si, (0); I-IV	CH_2Cl_2	275
β,β'-Iminodipropionitrile	IV	M	100
Liquid crystals[l]	Depends on specific liquid; see reference 20		
β,β'-Oxydipropionitrile	IV, V	A, M	100
Porous polymers (e.g. Porapak; see p. 398)			
$AgNO_3$-benzyl cyanide[m]	Olefins	M	50
Dexsil 300 GC[n]	High boilers	B, C	500

[a]Recommended applications, indicated in two ways by code. (1) Compounds of a specific element (B, S, Si, P, etc.); a phase used for inorganic and organometallic compounds is designated (0). (2) According to the Ewell Classification (see reference 14a): Class I--compounds forming extensive H-bond networks (water, glycols, aminoalcohols, dibasic acids, hydroxyacids, polyhydroxybenzenes); Class II--compounds with H-bond donor (O, F, N) and active-H (alcohols, phenols, primary and secondary amines, fatty acids, oximes, RCN and RNO_2 with α-H, NH_3, HF, N_2H_4, HCN); Class III--compounds with H-bond donor, but no active H (aldehydes, ketones, esters, ethers, tert amines, RNO_2 and RCN with no α-H); Class IV--compounds with "active" H but no donor (chlorocarbons, some heterocycles, aromatic and olefinic hydrocarbons); Class V--neutral compounds (no H-bond tendency: saturated hydrocarbons, mercaptans, sulfides, CS_2, perchloro-compounds). For another classification scheme, see the Supelco Catalog [use of Kováts indices and Rohrschneider constants; L. Rohrschneider, Advan. Chromatog., 4, 333 (1967)].

[b]Useful solvents for preparing column (heating may be necessary in some cases). A = acetone, B = benzene, C = chloroform, M = methanol, T = toluene.

[c]Upper column T (°C) permissible.

[d]Craig polyester.

[e]Polar, water-soluble type designated H or HB; nonpolar, water-insoluble designated LB.

[f]Dimethylstearamide.

[g]Dimethyloleylamide.

[h]Chloroform/1-butanol (1/1).

[i]DC refers to Dow Corning; SF to silcone fluid, SE and XE to gum rubber (General Electric Co.).

[j]Probably the most versatile phase in use.

[k]An organic derivative of an aluminosilicate clay, usually admixed with an ester; separates o-, m-, p-isomers.

[l]Available from Aldrich; Eastman; and Princeton Organics, Route 206, Princeton, N. J. 08540.

[m]Saturated solutions of $AgNO_3$ in benzyl cyanide, ethylene glycol, glycerol, and others, are highly efficient for olefin separation, including cis-trans isomers. Benzyl cyanide is the preferred liquid (nonhygroscopic).

[n]A new polycarboranesiloxane (Olin Corp., New Haven, Conn.) available from Analabs; very good for high-temperature work.

X. CHROMATOGRAPHY SUPPLY DIRECTORY

The multitudinous types and sources of chromatography supplies and equipment may be found in several annual guides: A.C.S. Laboratory Guide (published by Analytical Chemistry); Guide to Scientific Instruments (published by the AAAS journal, Science); International Chromatography Guide (published by J. Chromatog. Sci., Preston Technical Abstract Co., 2101 Dempster Street, Evanston, Ill. 60201; formerly J. Gas Chromatog.). The following listing gives manufacturers and some key distributors; the usual laboratory supply houses (Fisher Scientific, e.g.) may carry some items, as well as their own brands of common supplies (such as alumina and silica gel). In using this listing, refer to the specific chromatography section for detailed information.

A. Guide

Product	Addresses	Product	Addresses
	LIQUID COLUMN CHROMATOGRAPHY (Adsorption, partition, gel)		
Alumina	1,10,48,52,70,72	Molecular sieves	17,41
Bio-Gels	10	Polyamide	68,70
Carbon	23,52	Porous glass	10,16,70
Cellulose	11,44,59,63,71	Porous polymers	10,16,40,70
Diatom. earth	40	Prepacked columns	14,57,70
Gels	10,56,70	Sephadex	56
Magnesium silicate	2,10,29,68,70	Silica gel	6,10,11,17,45,48,70,72
Merck-o-gels	48		
	PAPER CHROMATOGRAPHY		
	(See p. 380)		
	THIN LAYER CHROMATOGRAPHY		
Alumina	1,10,11,12,48,70,72	Ion exchange	10,11,31,44,63,65
Cellulose	10,11,32,44,59,70	Polyamides	11,32,46,70,72
Gels	10,56,70	Precoated plates	5,7,11,12,27,33,45,46,48,57
General supplies	6,11,12,19,31,32,48, 68,72	Silica gel	11,12,48,70,72
	ION EXCHANGE CHROMATOGRAPHY		
	(See p. 384)		
	GEL FILTRATION AND PERMEATION		
	(See p. 387)		

(Continued)

Refs. p. 405.

Product	Addresses	Product	Addresses
	ELECTROPHORESIS		
(For suppliers of the usual media, such as papers and gels, refer to the manufacturers listed under Column, Paper, Thin Layer, and Gel Chromatography.)			
Disk apparatus	(see Gel apparatus)	Immunoelectrophoresis	13,49
Free flow (Tiselius, etc.)	3,8,11,53,58	Low-voltage apparatus (Zone)	8,12,26,32,38, 51,66
Gel apparatus	10,13,26,62,66	Preparative apparatus	66
General supplies	12,13,32,49,51,66	Thin layer apparatus	60,66
High-voltage equipment	12,62		
	VAPOR PHASE CHROMATOGRAPHY		
Apiezons	9	Silicones	24,33
General supplies	4,6,37,52,69	Stationary phases	4,6,37,68,69
Liquid crystals	See p. 47	Supports	See p. 397
Porous glass, polymers	10,16,40,70	Syringes	36

B. Suppliers

1. Aluminum Company of America, 1501 Alcoa Building, Pittsburgh, Pa. 15219.
2. Alupharm Chemicals, 608-612 Commercial Place, P.O. Box 30628, New Orleans, La. 70130.
3. American Instrument Co., 8030 Georgia Avenue, Silver Spring, Md. 20910.
4. Analabs, Inc., 80 Republic Drive, North Haven, Conn. 06473.
5. Analtech, Inc., Blue Hen Industrial Park, South Chapel Street Extension, Newark, Del. 19711.
6. Applied Science Labs., Inc., P.O. Box 440, State College, Pa. 16801.
7. J. T. Baker Chemical Co., 222 Red School Lane, Phillipsburgh, N.J. 08865 (J. T. Baker Chemikalien, Flemingfeld, West Germany).
8. Beckman Instruments, 2500 Harbor Boulevard, Fullerton, Calif. 92634.
9. James G. Biddle Co., Township Line & Jolly Roads, Plymouth Meeting, Pa. 19462.
10. Bio-Rad Labs, 32nd and Griffin Avenues, Richmond, Calif. 94804 (Bio-Rad Laboratories, München-Gräfelfing, am Kirchenholze 6, West Germany).
11. Brinkmann Instruments, Inc., Cantiague Road, Westbury, N.Y. 11743.
12. Camag, 4132 Muttenz, Homburgerstrasse 24, Switzerland (in the United States: Camag, Inc., 11830 West Ripley Avenue, Milwaukee, Wis. 53226).
13. Canalco Inc., 5635 Fisher Lane, Rockville, Md. 20852.
14. Chromatronix, Inc., 2743 Eighth Street, Berkeley, Calif. 94710.
15. Coast Engineering Lab., Chromatography Supply, Inc. 3755 Inglewood Avenue, Redondo Beach, Calif. 90277.
16. Corning Glass Works, 3910 Crystal Street, Corning, N.Y. 14830.
17. Davidson Chemical Division, W. R. Grace and Co., 101 North Charles Street, Baltimore, Md. 21203.

Refs. p. 405.

18. Debton Company P.W.B.A., Langeleem Straat 259, Antwerp, Belgium.
19. C. Desaga GmbH, Maass Strasse 26/28, P. O. Box 407, Heidelberg, West Germany.
20. Diamond Shamrock Chemical Co., 1901 Spring Street, Redwood City, Calif. 94063.
21. Disc Instruments, Inc., 2701 South Halladay Street, Santa Ana, Calif. 92705 (Disc Instruments, Limited, Ebberns Road, Hamel Hempstead, Hertfordshire, England).
22. Distillation Products, Eastman Kodak, Rochester, N.Y. 14603.
23. Dow Chemical Co., 2030 Abbott Road Court, Midland, Mich. 48640.
24. Dow Corning Corp. Engineering Products Division, Midland, Mich. 48641.
25. Eagle-Picher Co., American Building, Cincinnati, Ohio 45201.
26. E-C Apparatus Corp., 220 South 40th Street, Philadelphia, Pa. 19104.
27. Eastman Kodak Co., Eastman Organic Chemicals, 343 State Street, Rochester, N.Y. 14650.
28. Eaton-Dikeman Co., Mt. Holly Springs, Pa. 17065.
29. Floridin Co., Three Penn Center, Pittsburgh, Pa. 15234.
30. The Fluorocarbon Co., 1754 South Clementine Street, Anaheim, Calif. 92803.
31. Gallard-Schlesinger Co., 580 Mineola Avenue, Carle Place, N.Y. 11514.
32. Gelman Instrument Co., 600 South Wagner Road, Ann Arbor, Mich. 48106.
33. General Electric Co., Silicone Products Department, Waterford, N.Y. 12188.
34. Griffin and George, Ltd., London, England.
35. R. Grutzmacher, Berlin, Germany.
36. Hamilton Co., 12440 East Lambert Road, Whittier, Calif. 90608; Precision Sampling Corp., Box 15119, Baton Rouge, La. 70815; Shandon Scientific Co., Inc., 515 Broad Street, Sewickley, Pa. 15143; Unimetrics Universal Corp., 1853 Raymond Avenue, Anaheim, Calif. 92801.
37. Hewlett-Packard Avondale Division, Route 41, Avondale, Pa. 19311 (Hewlett-Packard S.A., Meyrin-Geneva, Switzerland).
38. Instrumentation Specialities Co., 5624 Seward Ave., Lincoln, Neb. 68507.
39. Ionac Chemical Co., Birmingham, N.J. 08011.
40. Johns-Manville, J-M Celite Division, 22 East 40th Street, New York, N.Y. 10016.
41. Linde Air Products, East Park Drive and Woodward Avenue, Tonawanda, N.Y. 14150.
42. LKB Instruments, 12221 Parklawn Drive, Rockville, Md. 20852.
43. 3M Co., 3M Center, St. Paul, Minn. 55101.
44. Macherey, Nagel & Co., Werkstrasse 6-8, P.O. Box 307, D-516 Dueren, West Germany (in the United States, see Brinkmann).
45. Mallinckrodt Chemical Works, Science Products, 3600 North Second Street, St. Louis, Mo. 63160.
46. Mann Research Laboratories (now called Schwarz-Mann Bioresearch), 136 Liberty Street, New York, N.Y. 10006.
47. May and Baker, Ltd., Dagenham, Essex, England (in the United States, Rhodia, Inc., 600 Madison Avenue, New York, N.Y., and Gallard-Schlesinger Co.).
48. E. Merck A.G., Chemische Fabrik, Darmstadt, Germany (in the United States, see Brinkmann).
49. Metaloglass Inc., 466 Blue Hill Avenue, Boston, Mass. 02121.
50. National Aluminate Corp., 6216 West 66th Place, Chicago, Ill.
51. Ortec Inc., 100 Midland Road, Oak Ridge, Tenn. 37830.
52. Perco Supplies, P.O. Box 201, San Gabriel, Calif. 91778.
53. Perkin-Elmer Corp., Main Avenue, Norwalk, Conn. 06851.
54. The Permutit Co., East 49 Midland Avenue, Paramus, N.J.

Refs. p. 405.

55. The Permutit Co., Ltd., Pemberton House, 632/652 London Road, Isleworth, Middlesex, England.
56. Pharmacia Fine Chemicals, Inc., 800 Centennial Avenue, Piscataway, N.J. 08854 (Pharmacia Fine Chemicals AB, Uppsala, Sweden).
57. Quantum Industries, 341 Kaplan Drive, Fairfield, N.J. 07006.
58. Quickfit, Inc., 7 Just Road, Fairfield, N.J. 07006.
59. H. Reeve Angel, 9 Bridewell Place, Clifton, N.J. 07014 (14 New Bridge Street, London E.C. 4, England).
60. Research Specialties Co., 200 South Garrard Boulevard, Richmond, Calif.
61. Rohm and Haas, Independence Mall West, Philadelphia, Pa. 19105.
62. Savant Instruments Inc., 221 Park Avenue, Hicksville, N.Y. 11801. For additional high-voltage equipment: Gilson Medical Electronics, Middelton, Wis.; Servonuclear Corp., Long Island City, N.Y.
63. Schleicher & Schuell, Inc., 543 Washington Street, Keene, N.H. 03431 (also Dassel, Krs. Einbeck, Germany).
64. Seravac Labs., (PTY) Ltd., Holyport Maidenhead, Berkshire, England.
65. Serva Entwicklungslabor, Heidelberg, Romerstrasse 118, Germany (in the United States, Gallard-Schlesinger).
66. Shandon Scientific Co., Sewickley, Pa. 15143.
67. Sterchamol-Werke, Wuelfratch, West Germany (in the United States, see Alupharm).
68. Supelco, Inc., Supelco Park, Bellefonte, Pa. 16823.
69. Varian Aerograph, 2700 Mitchell Drive, Walnut Creek, Calif. 94598.
70. Waters Associates, Inc., 61 Fountain Street, Framingham, Mass. 01701 (Waters Messtechnik GmbH, 6 Frankfurt/Main, Savignystrasse 51, West Germany).
71. Whatman, Wand R. Balston, Ltd., Maidstone Kent, England (distributors, Reeve Angel).
72. M. Woelm, 344 Eschwege, Postfach 850, West Germany (in the United States, Waters Associates).
73. Zerolit Ltd., Gunnersburg Avenue, London W. 4, England.

XI. REFERENCES

1. For complete details on theory and practice, the following are recommended. (a) Chromatography, E. Heftmann, Ed., 2nd ed., Rheinhold, New York, 1967. (b) O. Mikeš, Laboratory Handbook of Chromatographic Methods, Van Nostrand, London, 1970. (c) R. Stock and C. Rice, Chromatographic Methods, 2nd ed., Chapman and Hall, London, 1967. (d) An excellent, concise treatment of column, tlc, and glc methods is J. M. Bobbitt, A. E. Schwarting, and R. J. Gritter, Introduction to Chromatography, Reinhold, New York, 1968.
2. Current developments are described in the following periodicals and series. From Elsevier (Amsterdam)--Journal of Chromatography, Chromatographic Reviews, Gas Chromatography Abstracts. From Marcel Dekker (New York)--Separation Science, Advances in Chromatography. From Preston Technical Abstracts (Evanston, Ill.)--Journal of Chromatographic Science (formerly Journal of Gas Chromatography), G. C. Abstract Service. From A.C.S. (Washington, D.C.)--Analytical Chemistry. Also, the series Progress in Separation and Purification, Interscience, New York., Vol. 1, 1968; Vol. 2, 1969.
3. L. R. Snyder, J. Chromatog., 12, 488 (1963). Also, Floridin Technical Data and Product Specifications, Floridin Co., 375 Park Avenue, New York, N.Y. 10022.

4. L. R. Snyder, J. Chromatog., 28, 300 (1967).
5. A. Kiselev, Advan. Chromatog., 4, 113 (1967); J. Hassler, Activated Carbon, Chemical Publishing Co., New York, 1963.
6. L. R. Snyder, Principles of Adsorption Chromatography, Marcel Dekker, New York, 1968, Ch. 7; also reference 1a, pp. 54-57.
7. Considerable details for specific compound types, especially natural products, are given in the following publications. (a) G. M.-Bèttolo, Ed., Thin-Layer Chromatography, Elsevier, Amsterdam, 1964. (b) G. Pataki, Techniques of TLC in Amino Acid and Peptide Chemistry, Ann Arbor Science Publishers, Inc., Ann Arbor, Mich., 1968 (German edition: W. de Gruyter and Co., Berlin, 1966). (c) Reference 1b. (d) For extensive detail, E. Stahl, TLC. A Laboratory Handbook, Springer-Verlag, New York, 1969. (e) For extensive bibliographies, consult Thin Layer Chromatography. Cumulative Bibliography I and II, D. Jänchen, Ed., CAMAG, Muttenz, Switzerland and New Berlin, Wis.
8. For a fairly comprehensive list with applications, write for TLC Visualization Reagents and Chromatographic Solvents, Eastman Organic Chemicals, Rochester, N.Y. 14650 (Kodak Publication No. JJ-5).
9. For extensive details, see the treatise Paper Chromatography by I. Hais and K. Macek, Academic Press, New York, 1963, and references 1a, 1b, and 1c and references cited therein. A recent treatise by G. Zweig is also excellent: Paper Chromatography, Vol. I of Paper Chromatography and Electrophoresis, Academic Press, New York, 1967.
10. Adapted from reference 9.
11. For recent reviews, see (a) D. D. Bly, Science, 168, 527 (1970), and (b) J. Cazes, J. Chem. Educ., 47, A461 and A505 (1970).
12. For extensive coverage, see H. Determann, Gel Chromatography, Springer-Verlag, New York, 1968.
13. Bibliographies and general applications are found in P. Flodin, Dextran Gels and their Applications in Gel Filtration, AB Pharmacia, Uppsala, Sweden, 1962; H. Determann, Angew. Chem., 76, 635 (1964); and in periodic abstracts and indexes obtainable from the Pharmacia (Sephadex) and Bio-Rad (Bio-Gel) companies.
14. There are several good books on theory and practice, including (a) H. M. McNair and E. J. Bonelli, Basic Gas Chromatography, Varian Aerograph (2700 Mitchell Drive, Walnut Creek, Calif. 94598, and Pelikanweg 2, 4002 Basel, Switzerland), 1967; (b) S. Dal Nogare and R. S. Juvet, Jr., Gas Liquid Chromatography, Interscience, New York, 1963.
15. For a review, see E. Kováts, Advan. Chromatog., 1, 229 (1965). Collections of Kováts indices are found in W. O. McReynolds, Gas Chromatographic Retention Data, Preston Technical Abstracts, Evanston, Ill., 1966, and O. E. Schupp III and J. S. Lewis, Eds., Compilation of Gas Chromatographic Data, ASTM Publication DS 25a, American Society for Testing and Materials, Philadelphia (a computer compilation series). A very comprehensive Gas Chromatographic Data Compilation has been published recently by the ASTM as Publication AMD 25A S-1, which updates and considerably improves the data in the first edition. Order by Publication Code No. 10-025011-39.
16. M. Beroza, Accts. Chem. Res., 3, 33 (1970); M. Beroza and R. Coad, J. Gas. Chromatog., 199 (1966).
17. C. Phillips and P. Timms, J. Chromatog., 5, 131 (1961).
18. G. Feldman, M. Maude, and A. Windeler, American Laboratory, February 1970, p. 61; L. Condal-Bosch, J. Chem. Educ., 41, A235 (1964); H. L. Pardue, M. F. Burke, and J. R. Barnes, ibid., 44, 694 (1967). For an excellent review, see H. W. Johnson, Jr., Advan. Chromatog., 5, 175 (1968). G. W. Ewing has also reviewed various types laboratory integrators in J. Chem. Educ., 49, A333 (1972).

19. The various GC abstract services (2); some books such as references 1b and 14a and J. S. Lewis, Compilation of Gas Chromatographic Data, ASTM STP No. 343, 1963.
20. H. Kelker and E. von Schivizhoffen, Advan. Chromatog., 6, 247 (1968)--use of liquid crystals in vpc.
21. J. R. Whitaker, Electrophoresis in Stabilizing Media, Vol. 1 of Paper Chromatography and Electrophoresis, Academic Press, New York, 1967.
22. I. Smith, Zone Electrophoresis, Vol. II of Chromatographic and Electrophoretic Techniques, W. Heinemann-Medical Books, London and Interscience, New York, 1960.
23. B. J. Haywood, Electrophoresis-Technical Applications. A Bibliography of Abstracts, Ann Arbor-Humphrey Science Publishers, Ann Arbor, Mich., and London, England, 1969.
24. Polyacrylamide gel electrophoresis (PAGE) has been reviewed recently by Chrambach and Rodbard, Science, 172, 440 (1971).
25. Recent developments in gel electrophoresis are covered in Ann. N.Y. Acad. Sci., Vol. 121, Art. 2, pp. 305-650, 1964.
26. For a recent review on "Retention Volume Theories for Gas Chromatography," see J. F. Parcher, J. Chem. Educ., 49, 472 (1972).

EXPERIMENTAL TECHNIQUES

I. PROPERTIES OF COMMON LABORATORY MATERIALS

It is often necessary in laboratory work to assemble experimental apparatus from hundreds of different types of materials. This may require construction from raw materials as in machining of metals or blowing of glass. More commonly, it requires the assembling of commercially prepared components, which may be as simple as fitting together two glass joints. At both extremes it is necessary to be familiar with at least some of the properties of the materials used. It is with this in mind that this section was compiled. Much additional information can be obtained from the many available texts and manuals dealing with laboratory techniques and procedures. Some of the most important of these are listed in references 1 to 8; a particularly useful book by Brunner and Batzer (17) provides many data and much discussion on uses of various materials, especially in connection with vacuum techniques.

Related to the actual materials is the question of the types of tools available in the laboratory; in addition to the usual collection, readers may be interested in less familiar Hard to Find Tools (Brookstone Co., Dept. C, 5 Brookstone Building, Peterborough, N.H. 03458) and unique tools (National Camera, Dept. JSB, Englewood, Colo. 80110).

A. Glass

Unquestionably the most common laboratory material is glass, not only because of its versatility but also because of its excellent chemical resistance. A very good book describing the techniques of both scientific and hobby glassblowing is Creative Glass Blowing by Hammesfahr and Strong (9). For extensive details on properties, see reference 10. Optical properties of some glasses are described on pages 179 to 182 in this book.

Because of its high thermal coefficient of expansion, very slow heating and cooling are required when working with soft glass. Almost all labware, however, is made of borosilicate glass and can be heated or cooled very rapidly, provided that there are no unusual strains or flaws. Borosilicate glass can easily be distinguished from other types by immersion of the clean, dry piece in a 16/84 (weight) mixture of methanol/benzene. Because of its refractive index, it will be practically invisible.

1. Properties of Some Commercial Glasses

The data below are taken from various catalogs and from references 5 and 10. It should be noted, however, that different methods of manufacture may result in slightly different compositions and thus slightly different properties.

	Soda-Lime (Soft; Flint; TEKK; EXAX)	Borosilicate (Kimax; Pyrex)	96% Silica (Vycor)	Fused Silica
Strain point[a] (°C)	480	510	820	1020
Annealing point[b] (°C)	520	550	910	1120
Softening point[c] (°C)	700	820	1500	1710
Working point[d] (°C)	1010	1245		

(Continued)

Refs. p. 428.

	Soda-Lime (Soft; Flint; TEKK; EXAX)	Borosilicate (Kimax; Pyrex)	96% Silica (Vycor)	Fused Silica
Linear coefficient of expansion[e]	9.3	3.3	0.8	0.6
Refractive index (n_D)	1.52	1.474	1.458	1.54
Density (g/cm^3)	2.5	2.23	2.18	2.65

[a]Point at which strain is relieved after 4 hr.
[b]Point at which strain is relieved in 15 min.
[c]Point at which sagging begins.
[d]Point at which glass can be easily manipulated and blown by mouth.
[e]Parts per million per °C between 0 and 300°C.

Several specialty glasses are available, such as "Pyroceram" (used for hot-plate tops and other very high temperature work) and "COREX" (fortified glass used in unbreakable pipets, etc.), both from Corning; others include alkali-resistant glass (low in B_2O_3; very stable to alkali and acid) and "low actinic glass" (filters out visible light). For information on glass filters, see page 360.

2. Joints, Stopcocks, and Stoppers

The National Bureau of Standards has set standards for taper-ground joints and stoppers (₸, standard taper), spherical-ground joints (Ş, spherical joint), stoppers, and stopcocks in order to guarantee their interchangeability. The data in these tables are taken from reference 11. A fairly successful method for loosening "frozen" ground glass joints involves preparation of the following solution: 10 parts chloral hydrate, 5 parts glycerine, 5 parts water, and 3 parts concentrated hydrochloric acid; paint the solution onto the joints or immerse them in the solution and let stand.

a. Taper-Ground Joints. Designated by two numbers: (A) gives the approximate diameter in mm at the large end of the ground zone; (B) gives the approximate length in mm of the ground zone.

Full-Length Taper-Ground Joints		Short-Length Taper-Ground Joints		Medium-Length Taper-Ground Joints	
Size (A/B)	Approximate Diameter of Small End (mm)	Size (A/B)	Approximate Diameter of Small End (mm)	Size (A/B)	Approximate Diameter of Small End (mm)
7/25	5.0	5/8	4.2	5/12	3.8
10/30	7.0	7/10	6.5	7/15	6.0
12/30	9.5	10/7	9.3	10/18	8.2
14/35	11.0	10/10	9.0	12/18	10.7
19/38	15.0	12/10	11.5	14/20	12.5

(Continued)

Full-Length Taper-Ground Joints		Short-Length Taper-Ground Joints		Medium-Length Taper-Ground Joints	
Size (A/B)	Approximate Diameter of Small End (mm)	Size (A/B)	Approximate Diameter of Small End (mm)	Size (A/B)	Approximate Diameter of Small End (mm)
24/40	20.0	14/10	13.5	19/22	16.6
29/42	25.0	19/10	17.8	24/25	21.5
34/45	30.0	24/12	22.8	29/26	26.6
40/50	35.0	29/12	28.0	34/28	31.7
45/50	40.0	34/12	33.3	40/35	36.5
50/50	45.0	40/12	38.8		
55/50	50.0	45/12	43.8		
60/50	55.0	50/12	48.8		
71/60	65.0	55/12	53.8		
103/60	97.0	60/12	58.8		
		71/15	69.5		

b. Spherical-Ground Joints. Designated by two numbers: (A) gives the outside diameter of the ball member in mm; (B) gives the inside diameter of the joint in mm. Standard sizes are as follows (A/B):

7/1 12/1 12/1.5 12/2 12/3 12/5 18/7 18/9 28/12
28/15 35/20 35/25 40/25 50/30 65/40 75/50 102/75

c. Taper-Ground Stopcocks. Designated by one number: the diameter (mm) of the bore hole in the plug.

Single straight-bore stopcock No.	1-M	1	1 1/2	2	3	4	5	6	7	10
Diameter of plug at center-line of bore (mm)	7	12	12	12	17	17	20	20	25	35

(The designation numbers used for single oblique-bore (2-way) and double oblique-bore (3-way) stopcocks are 1, 1 1/2, 2, 3, 4, with the same meaning as above.)

d. Taper-Ground Stoppers. Designated by one number: the diameter (mm) at the large end of the ground zone. "Reagent Bottle Stoppers" are interchangeable with taper-ground joints, whereas "Flask Stoppers" are not.

Reagent Bottle Stoppers		Flask Stoppers	
No.	Length of Ground Zone (mm)	No.	Length of Ground Zone (mm)
14	20 ± 1.5	8	10.0 ± 1.0
19	22 ± 1.5	9	14.0 ± 1.0
24	30 ± 2.0	13	14.0 ± 1.0
29	35 ± 2.0	16	15.0 ± 1.0
34	40 ± 2.0	19	17.0 ± 1.0
45	47 ± 2.0	22	20.5 ± 1.5
		27	21.5 ± 1.5
		32	21.5 ± 1.5
		38	30.0 ± 2.0

Refs. p. 428.

3. Joining Glass to Metal

For many applications, glass may be joined to other materials simply with glue or epoxy cement (see below). It is also possible to solder glass to metal using solder glasses as described in the first issue of Chem. Tech. (12). This technique involves using a low-melting, high-lead glass as solder. There are two types of solder glasses. The first is a thermosetting glass or devitrifying glass (called Pyroceramcement by Corning Glass Works). These thermosetting glasses melt between 350 and 650°C and set to rigid by becoming crystalline between 400 and 750°C. These devitrifying glasses form opaque, hermetic, mechanically strong seals. They are available in powdered form (from Corning) and can be applied as a slurry in volatile liquid, which is then burned off during glazing. The second type of solder glass is simply a thermoplastic glass (stable glass) with low enough flow temperature to permit forming a seal without melting or distorting the glass piece being sealed. These joints are often transparent. Both types of solder glasses are available with expansion coefficients to allow seals to either borosilicate or soft glass. Of course, the expansion coefficient of the metal must be taken into account and should be less than 500 ppm different from that of the glass over the temperature range involved (generally sealing temperature to room temperature). That is, $\Delta E \cdot \Delta T$ should be less than 500, where ΔE is the difference in expansion coefficients in ppm/°C of the two joined pieces and ΔT is the temperature range to which the seal will be subjected. For the expansion coefficients of some common metals, see below. Of course, it is also possible to use a graded seal of several different glasses with varying expansion coefficients.

Conventional solder, too, can be used by forming a thin film of metal onto the glass being sealed. One method (13) for doing this consists of dipping the glass into a solution of chloroplatinic acid ($H_2PtCl_6 \cdot 6H_2O$) made by mixing 0.2 g of the acid, 5 ml ethanol, and 5 ml ethyl ether and adding 4 or 5 drops of turpentine. Burn away the liquid to leave a film of platinum to which a solder joint can be made. If done carefully, this makes a gas tight joint. Of course, the same expansion considerations as above must be made.

Note that copper (expansion coefficient 16.8 ppm/°C) can be sealed to any glass, because of the metal's superior ductility. Some alloys have expansion coefficients similar to those of glass, such as Kovar (Westinghouse Electric Co.) and Fernico (General Electric Co.) (alloys of Ni, Co, and Fe); Sealmet (Higrade Sylvania Corp.); and Dumet (Cu coated Ni-Fe alloy). See reference 9, page 175, for more details.

B. Plastics

Plastics are available in an almost limitless variety of properties, each with its advantages and disadvantages. They are, however, generally not so strong as metals, not so chemically resistant as glass, and restricted to a relatively lower temperature limit. On the other hand, these disadvantages may be absent, and plastics are increasingly becoming one of the principal structural materials. Only a very few of the many available types of plastics have been included below; however, most of the common plastics used in the laboratory are covered.

Refs. p. 428.

Refs. p. 428.

1. Properties of Common Plastics

The data in this table are taken from numerous sources including various catalogs. Plastics are not precisely defined substances, hence the properties reported may vary somewhat depending on the method of production.

Plastics	Composition (Monomer)	Temperature Limit (°C)	Transparency	Flexibility	Machinability	Tensile Strength (1000 psi)	Other Properties
Kel-F (3M Company)	Chlorotrifluoroethylene	-200 to +200	Clear	Good	Good	4.6-5.7	Almost as chemically resistant as teflon but even more expensive
Nylon	Various polyamides	+150	Opaque	Rigid	Good	10-12	Tough; abrasion resistant; weak acid and strong alkali resistant but attacked by oxidizing agents
Plexiglas [Acrilan; Lucite; Perspex (England)]	Methyl Methacrylate	+70	Clear	Rigid	Good		Pieces can be glued together using trichloroethylene as a solvent. High electrical resistivity and optical quality
Polycarbonate (e.g., BPA: "Lexan")	Carbonate of 2,2-bis(4-hydroxyphenyl) propane and similar bis-(phenols)	-70 to +135	Clear	Rigid		9-10.5	Has high impact strength and extreme toughness. Sensitivity to basic solutions should be considered when choosing cleaning solution
Polyethylene (conventional)	C_2H_4	-100 to +80	Translucent	Extreme	Poor	1.4-2.5	Lower density type usually found in squeeze-type wash bottles

Polyethylene (linear)	C_2H_4	-100 to +120	Opaque	Rigid	Fair	2.9-4.0	Much less permability to O_2 than conventional polyethylene
Polypropylene	C_3H_6	-30 to +135	Translucent	Rigid	Fair	5.0	Harder and more rigid than polyethylene
Polystyrene (styrofoam)	$C_6H_5C_2H_3$	+75	Clear	Rigid	Fair	5.0-8.0	Can be made biologically inert
Teflon FEP (DuPont)	Fluorinated ethylene-propylene copolymer	-270 to +205	Translucent	Extreme	Good	2.7-3.1	FEP differs from TFE only in that it is more easily molded and is not as thermally stable
Teflon TFE (DuPont)	Tetrafluoroethylene	-265 to +315	Opaque	Extreme	Good	2.5-3.5	Has excellent chemical resistance and is very easy to clean. Has extremely low coefficient of friction and excellent wear resistance but scratches easily. Chemical intertness makes it almost impossible to bond to other materials
Tefzel (DuPont)	A close analog of Teflon	+180	Translucent	Moderate	Good	6.5	
TPX (transparent polymer X; Imperial Chemicals, England)	4-methyl-1-pentene	0 to +180	Clear	Rigid		4.0	Has high O_2 permeability
Tygon (Saran) (Norton Plastics, et al.)	Polyvinyl Chloride (PVC)	+75	Clear	Extreme		1.0-3.5 (3.0-5.0)	Tygon tubing contains plasticizers that are dissolved by most solvents
Tyril (Dow Chemical)	Styrene-acrylonitrile copolymer	+80	Clear	Rigid		11.0	

Refs. p. 428.

2. Chemical Resistance of Plastics[a]

	Kel-F	Plexi-glass	Polycar-bonate	Poly-ethylene	Poly-propylene	Poly-styrene	Teflon	TPX	Tygon	Tyril
Acids, inorganic	E	P/G[b]	E	E	E	NR	E	E	G	E
Acids, organic	E	G[b]	G	E	E	G	E	E	G	E
Alcohols	E	E	G	E	E	G	E	E	G	G
Aldehydes	E	NR	P/G	G	G	NR	E	G	P/G	P/G
Amines	E	NR	NR	G	G	G	E	G	NR	G
Bases	E	P/G[b]	NR	E	E	G	E	E	E	E
Dimethyl sulfoxide	E	NR	NR	E	E	NR	E	E	NR	NR
Esters	E	NR	NR	E	G	NR	E	E	P	NR
Ethers	E	NR	P	G	G	P	E	G	P	NR
Hydrocarbons, aliphatic	E	P	P	G	G	NR	E	G	P	E
Hydrocarbons, aromatic or halogenated	E	NR	NR	G	G	NR	E	NR	NR	NR
Ketones	E	NR	NR	G	G	NR	E	G	NR	NR
Oils, essential	E	P	G	G	G	NR	E	G	NR	P
Oils, lubricating	E	E	G	E	E	G	E	E	E	G
Oils, vegetable	E	E	E	E	E	G	E	E	E	E
Salts	E	E	E	E	E	E	E	E	E	E
Silicones	E	G	E	G	E	G	E	E	G	G
Water	E	E	E	E	E	E	E	E	E	E

[a] E = excellent; G = good; P = poor; NR = not recommended.
[b] Attacked by oxidizing acids, strong alkali, and formic acid.

Refs. p. 428.

C. Rubber Compounds

There are so many different types of rubber compounds that any collection of properties must necessarily be selective. Many rubber compounds are formulated for a particular application. Nevertheless, the following data of some of the more common rubber compounds, both synthetic and natural, may be helpful in choosing the one most suited for your needs. These data are taken from reference 14.

Rubber	Specific Gravity	Tensile Strength (1000 psi)	Hardness (Shore)[a]	Maximum Temperature (°C) Continuous Use	Effect of Sunlight
IIR (butyl, GR-1)	0.91	2.3-3.0	40-70	120-150	None
NR (natural rubber)	0.93	3.0-4.5	20-100	80	Deteriorates
Hard rubber	1.2-1.95	4.0-10.0	65-95	100	
CR (neoprene, GR-M)	1.25	2.0-3.5	30-90	120	None
SBR (styrene, Buna S)	0.94	1.6-3.7	35-90	80	Deteriorates
NBR (nitrile, Buna N)	0.99	0.5-4.0	40-90	120	Slight
Polysulfide (Thiokol ST)	1.35	0.7-1.25	50-80	100	None
Silicone (high temperature)		0.7-0.8	45-65	300	None
Fluoroelastomers (Viton A)	1.85	2.0-3.0	60-95	260	None
Polyisoprene	0.93	2.0-3.0	40-80	80	Deteriorates
Polybutadiene	0.94	2.5	45-80	80	Deteriorates

[a]Shore hardness is based on the degree of rebound of a diamond-tipped hammer dropped from a fixed height, as compared to a standard of high carbon steel (shore hardness = 100).

D. Silicone Oils

Dow Corning silicone oils are some of the most useful liquids for oil baths, diffusion pumps, lubrication, and similar purposes. The properties listed here are the more common ones, although many more are available. For instance, viscosities ranging from 0.65 to 100,000 centistokes (cSt) can be purchased. Most of these oils are soluble in hydrocarbon and chlorinated solvents as well as aromatics. These data are taken from literature made available by Dow Corning. More information

Refs. p. 428.

can be obtained by writing to Dow Corning Corp., Midland, Mich. 48640, or to any local sales office. The temperature range is based on volatility, flash point, and freezing point or pour point. The viscosity is in centistokes at 25°C.

No.	Viscosity	Flash Point (°C)	Temperature Range (°C)	Specific Gravity (25°)	Viscosity-Temperature Coefficient[a]	Comments
200	50	280	-45 to 200	0.96	0.59	Available in viscosities from 0.65 to 60,000
	350	315	-50 to 200	0.97	0.62	
330	50	290	-80 to 200	0.97	0.57	Has excellent dielectric properties
510	100	275	-50 to 220	0.99	0.62	Available also in viscosities 50 and 500
	1000	275	-50 to 250	0.99	0.63	
550	100-150	300	-40 to 220	1.07	0.76	
555	10-30	120	-35 to 100	1.06		Miscible with many organic solvents
710	475-525	300	-15 to 220	1.10	0.84	

[a] 1-(viscosity at 99°/viscosity at 37.8°).

E. Metals and Alloys

Metals are among the strongest, most durable construction materials. They are usually unaffected by temperatures 1000°C or higher. Their excellent conductive properties for both heat and electricity are valuable, but at the same time might be a disadvantage. Metals and alloys vary in properties depending on method of manufacture. The properties below are for the most commonly available form. The data are taken from various sources including references 5, 13, and 14.

Substance	MP (°C)	Density	Brinell Hardness[a]	Tensile Strength (1000 psi)	Coefficient of Expansion[b]	Welda-bility	Comments
Aluminum	660	2.7	16	8.5	26.7	Poor	Sensitive to strong bases
Brass (67% Cu, 33% Zn)	940	8.4	145	66.8	20.2	Fair	Easily soldered with either soft or silver solder

Refs. p. 428.

Copper	1082	8.5		33	17.8	Fair	Easily soldered
Duralumin[c]	≃540	2.9	125	88.2	26		Resistant to corrosion from acids and sea water
Gold	1063	19.3		36			Very good chemical resistance
Inconel[d]							Very good chemical resistance. Good for elevated temperatures
Invar[e]	1495	8.0			0.9(20°)		Very low thermal expansion
Iron, cast	≃1200	≃7.5		15-18	10.6(40°)	Fair	High chemical sensitivity; rusts easily
Iron, wrought	1510	7.7		42-52	11.4(0-100°)		High chemical sensitivity; rusts easily
Lead	327	11.3	≃6	3.0	29(18-100°)	Good	
Monel (67% Ni, 33% Cu)	≃1350	≃8.8	130-300	75-170	15	Good	Very good chemical resistance, even to acids and fluorine
Nichrome[f]							Common electrical resistance wire
Nickel	1455	8.9	110-300	75-170	14	Good	Very resistant to strong alkali, HCl, H_2SO_4
Permalloy[g]							Unusual magnetic properties
Platinum	1773	21.5	64	48	8.99(40°)		Very good chemical resistance

(Continued)

Refs. p. 428.

	MP (°C)	Density	Brinell Hardness[a]	Tensile Strength (1000 psi)	Coefficient of Expansion[b]	Welda-bility	Comments
Silver	961	10.5		40	18.8(20°)		Very good chemical (especially acid) resistance
Steel, carbon	1430	7.8	400	200	10.5(0-100°)	Good	High chemical sensitivity; rusts easily
Steel, mild	1430	7.8		50-60		Good	High chemical sensitivity; rusts easily
Steel, stainless	≃1500	≃7.8	≃500	250-300	17-20	Varies	Excellent structural strength

[a]Brinell harness number (15): "A hard spherical indenter of diameter D mm is pressed into the metal surface under a load W kg and the mean chordal diameter of the resultant indentation measured (d mm). The Brinell hardness number (Bhn) is defined as:

$$\text{Bhn} = \frac{W}{\text{curved area of indentation}} = \frac{2W}{\pi D(D - D^2 - d^2)}$$

and is expressed in kg/mm^3."

[b]Parts per million per °C in range of 0 to 300°C unless another temperature range or specific temperature is stated in parentheses.

[c]A durable aluminum alloy containing 4% Cu, 0.5% Mg, 0.25-1.0% Mn, and small amounts of Fe and Si.

[d]78% Ni, 15% Cr, 6% Fe.

[e]Alloy of 36% Ni, 64% steel (0.2% C in steel); used in delicate measuring instruments.

[f]Alloy of 60% Ni, 24% Fe, 16% Cr, 0.1% C; resistivity = 113 μΩ-cm at 0° C.

[g]Alloy of iron and >30% Ni. Very high magnetic permeability and electrical resistivity.

Refs. p. 428.

F. Solders and Fluxes (14,17,18)

Soft solders are mainly alloys of Sn and Pb and are used for joining relatively low-melting-point, nonferrous metals using a flame or soldering gun. Hard solders (silver solder, brazing compounds) are applied by flame. The fluxes used for soft soldering are aqueous solutions or petroleum jelly pastes of NH_4Cl-$ZnCl_2$ mixtures; for electrical work, alcohol solutions of various resins are employed (rosin-core solder, e.g.), since ordinary fluxes are quite acidic and leave damaging traces of acid. For hard solders, borax-boric acid fluxes are used; other types are also available. Most fluxes are corrosive and serve to break up oxide layers on the metals to be joined; they also prevent oxidation of solder and metal surfaces during heating.

The melting range in the tables below is that over which the alloy changes from a complete solid (solidus) to a complete liquid (liquidus); for eutectics, the liquidus and solidus occur at the same temperature. For vacuum-tight joints, narrow melting range alloys are preferred.

There are various Federal, ASTM, and SAE specifications for all types of solders (17). In many of the solders listed below, small amounts ($\leq$1%) of such elements as Sb, Cu, and Bi are usually present, but only the principal ones are mentioned here. For soldering Al, an alloy of 60% Sn-40% Zn is useful.

1. Soft Solders

Type	Composition (%)		Melting Range (C°)
	Lead	Tin	
--	100	-	327
Plumber's	70	30	177-260
Fine	60	40	177-238
Ordinary	50	50	177-216
Eutectic	38	62	183
--	-	100	232

2. Hard Solders

Solders composed mainly of Cu and Zn are referred to as brazing solders; silver solders are mixtures of Ag, Cu, and Zn. For additional information, write for catalog and bulletins from Handy and Harman Brazing Products, 82 Fulton Street, New York, N.Y.

The comments below as well as some of the data are taken from reference 17.

Composition (%)				Melting Range (°C)	Comments
Ag	Cu	Zn	Other		
-	100	-	-	1082	
-	45-55	55-45	-	854-882	Ordinary brazing solder

(Continued)

Refs. p. 428.

Composition (%)				Melting	
Ag	Cu	Zn	Other	Range (°C)	Comments
-	50	47-45	3-5 Sn	857-882	Ordinary brazing solder
72	28	-	-	779	Good for Cu to Cu and for stainless and Inconel
60	30	-	10 Sn	600-715	Hard and strong; good for ferrous and non-ferrous metals
50	34	16	-	693-774	
50	15.5	16.5	18 Cd	620-636	Good flow properties; for ferrous and non-ferrous metals
15	80	-	5 P	641-704	Mainly for Cu or Cu alloys, but useful for Mo, W, and Ag; Forms brittle joints on ferrous metals
10	52	38	-	821-871	

G. Adhesives

One of the most common methods for connecting dissimilar materials is simply to glue them together. Only general comments on some of the more useful of the many different types of available adhesives are made here. For more specific information, reference 16 can be consulted.

White Glue. This is perhaps the most common type of adhesive, partly because of its low cost and ease of use (e.g., "Elmer's" glue.) It is excellent for gluing almost all porous materials such as wood, paper, and leather. It is nontoxic and nonflammable and often forms joints much stronger than the material being glued. It is fairly sensitive to water over a long period of time.

Epoxy. This includes a large class of adhesives capable of joining almost any two materials. Exceptions are soft plastics (e.g., polyethylene, PVC) and rubbers (e.g., natural, butyl). Epoxys come in two parts: an epoxide-based resin and a curing agent. When mixed, a glue is formed that sets to a very strong, rigid bond. The various available properties include high chemical resistance, very low residual vapor pressure, high or low thermal conductivity, and high temperature resistance (>250°C). Typical products are "Bipax" (prepackaged, easy to mix components from TRA-CON, 55 North Street, Medford, Mass. 02155), "Epoxi-Patch Kit" (excellent for vacuum and long-term stability and inertness, from Hysol Division, The Dexter Corp., Olean, N.Y. 14760), and "Duro Epoxy" (Woodhill Chemical Corp., Cleveland, Ohio 44128). Extensive technical details can be found in "Epoxy Resins," Advan. Chem., 92, ACS, Washington, D.C., 1970 (cf. 19).

RTV (Room-Temperature Vulcanizing) Silicone Rubber. Liquids or pastes that cure to strong, durable, resilient silicone rubber, used for sealing,

Refs. p. 428.

bonding, encapsulating, and mold fabrication. They are available in either one- or two-component form. The one-package silicones form bonds that are flexible at -120°C and withstand temperatures up to 300°. They bond well to most materials, with the exception of some rubbers and flexible plastics. Setting time might be as long as 2 to 3 days depending on thickness, and full strength is not reached for several days. More information can be obtained from General Electric, Silicone Products Department, Waterford, N.Y. 12188. Similar resins are available from Dow Corning, Midland, Mich. 48640.

Sauereisen. The Sauereisen Cement Co. (RIDC, Industrial Park, Pittsburgh, Pa. 15238) sells a variety of very versatile ceramic-based cements. Sauereisen No. 1 Paste, for example, is an acid-proof (except HF) adhesive that resists most solvents, and temperatures as high as 1100°C. It is useful for glass, porcelain, metal, asbestos, wood, and so on, and hardens into an unglazed porcelain body without the use of kilns.

Contact Cement. These instant-sticking adhesives are good for most materials (not Cu metal) and are useful for rush jobs or those that cannot be clamped. Seals are too weak for heavily loaded or stressed parts; the cement may damage some plastics. Examples of products are "Duro Plastic Contact Cement," "Mr. Fix Contact Cement," and "Weldwood All Purpose Contact Cement."

Cyanoacrylate. Cyanoacrylate monomer (such as Eastman 910) polymerizes when squeezed between two surfaces and sets rapidly. It is extremely versatile; it adheres to skin rather well, but is not toxic. An example is "Permabond" sold by Laboratory Supplies Co., 29 Jefry Lane, Hicksville, N.Y. 11801. The hardened seal is soluble in acetone, DMF, and nitromethane. Another new cyanoacrylate is "Zipbond," which cures in about 1 min. without applied heat or pressure; it joins most materials (Instrument Division of Tescom Corp., 2633 South East 4th Street, Minneapolis, Minn. 55414).

Litharge. Very useful in cementing glass to metal. Grind in a mortar, PbO (260 g) to which diluted glycerine (67 cc glycerin/33 cc water) is added. The mixture may be worked like putty and sets in a day.

Miscellaneous. Various special application adhesives, such as acid-proof cement, De Khotinsky cement (shellac-pine tar mixture good for sealing), and cupric oxide cement (metal to metal or glass), are described in the "Arts and Recipes" section of old editions of the Handbook of Chemistry and Physics (e.g., 37th ed.). Among general-purpose adhesives, one of the most common and useful, especially for joining plastics, is "Duco" (DuPont).

H. Stopcock Greases and Related Materials

Various greases, oils, and waxes for different laboratory applications are available from many of the usual supply houses (Fisher, Curtin, A. H. Thomas, etc.). The data below summarize the properties of the most common and useful types. See reference 17 for additional information.

1. Greases

Most of the greases are water insoluble and inert to routine chemicals; the fluorocarbon-based greases are the most inert. Silicone and other greases are leached from joints after prolonged exposure to hydrocarbons

Refs. p. 428.

and ethers, and decompose somewhat under uv irradiation. The greases are suitable for O-ring (elastomer) as well as glass seals.

Type[a]	Mp (°C)	Vp at rt (mm Hg)	Recommended Temperature Range (°C)		Comments
			Stopcocks	General	
Apiezon H	b	10^{-9}	5 to 150	-15 to 250	c
Apiezon L	47	10^{-11}	10 to 30	10 to 30	d
Apiezon M	44	10^{-8}	10 to 30	10 to 30	Good general use[d]
Apiezon N	43	10^{-9}	10 to 30	10 to 30	Designed for stopcocks[c]
Apiezon T	125	10^{-8}	10 to 80	0 to 120	c
Cello-Grease	120	(Fisher Scientific; heavier and darker than Cello-Seal; high-vacuum use)			
Cello-Seal	100	(Fisher Scientific; gas-tight seals between glass and rubber tubing; coating outside of rubber tubing to prevent diffusion)			
DC[e] Silicone		$<10^{-5}$	Constant viscosity from -40° to 200°		
DC[e] High-Vacuum		$<5 \times 10^{-6}$	Constant viscosity from -40° to 200°		
Kel-F Grease		$<10^{-3}$	-18 to 180	(made by 3M and M. W. Kellogg Co.)	
Lubriseal	40				General use
Lubriseal, high-vacuum	50	3×10^{-6}			
Nonaq		(Fisher; particularly good for contact with ketones, aldehydes, halocarbons, and hydrocarbons, but not hydroxylic solvents)			

[a]Apiezon greases are high-purity lubricants; they are made in England, but are distributed in the United States by James G. Biddle Co., Plymouth Meeting, Pa. 19462, from whom detailed data are available (Bulletin 43).
[b]Softens at 250° but does not melt to a free-flowing liquid.
[c]Has rubbery nature, forming cushion between mating surfaces.
[d]Also used as stationary liquid phase in glc.
[e]Dow Corning.

Other products include the "Versilubes" (effective from -73° to 230°C), "Insulgreases" (nonlubricating, dielectric greases, useful from about -70° to 200°C), and "General Purpose Compounds" (corrosion protection, heat sink), all available from General Electric (Silicone Products, Waterford, N.Y.). "Fluoro-Glide" is a colloidal Teflon in aerosol form, for spray coating a lubricating Teflon film. There are also two new Apiezon greases--AP 100 (softening point 30°C; high-vacuum use) and AP 101 (softening point 185°C; for moderate vacuum)--designed especially to prevent joints, etc., from seizing ("freezing"). They are remarkably stable and are not leached by agressive solvents, yet are easily removed by washing with aqueous detergent.

Refs. p. 428.

2. Sealing Compounds and Waxes

The following Apiezon products are available (see footnote a in the preceding table):

Sealing Compound Q--a low cost, putty-like sealing medium used to seal joints, fill holes, seal around edges of unground joints, and so on. Good for moderate vacuum. Soluble in kerosene and other hydrocarbons.

Waxes--used to seal high-vacuum joints that are not required to be moved. Wax W is used for most "permanent" joint applications; it comes in black stick form and is applied like other hard waxes by warming the joints or surfaces involved and rubbing the wax over the hot surface. The waxes are soluble in benzene and carbon tetrachloride. [See also "Cello-Seal" (Fisher) in the table on greases, and DeKhotinsky cement under Adhesives above.]

Type	Vp at 20°C (mm Hg)	Temperature Data (°C)		
		Softening Point	Apply at	Useful up to
Compound Q	10^{-5}	45	Rt	30
Wax W	10^{-8}	85	100	80
Wax W 100	10^{-7}	55	80	50
Wax W 40	10^{-7}	45	40-50	30
(DeKhotinsky[a,b]	10^{-4}	140	150	100)
(Glyptal[a,c]	10^{-5}	-	Rt	125)

[a]Not an Apiezon product.
[b]Slightly soluble in acetone.
[c]Alkyd-type resins (red color) (General Electric Co., 1 River Road, Schenectady 5, N.Y.); useful for pipe threads. Partly soluble in acetone, MEK, or xylene.

3. Diffusion Pump Oils

Organic pump oils are superior to mercury (vp 10^{-4} mm at 0°C; 10^{-2} mm at 50°C; 5 mm at 150°C). They are generally esters (E), hydrocarbons (H), and silicones (S); newer types made of polyphenyl ethers (P) have extremely low vapor pressure and excellent thermal stability. See reference 17 for additional information.

Type	BP (°C)[a]	Chemical Structure	Vapor Pressure (mm Hg, 20°C)
Apiezon A	190	H	10^{-6}
Apiezon B	220	H	5×10^{-8}
Apiezon C	255	H	5×10^{-9}
Convalex		P	Very low
DC[b] 702	175	S	5×10^{-8}
DC 703	180	S	2×10^{-8}
DC 704	190	S	1×10^{-8}
DC 705	245	S	Very low
Octoil	183	E[c]	2×10^{-7}
Octoil S	199	E[d]	5×10^{-8}
Santovac 5		P[e]	Very low

Refs. p. 428.

[a]Apiezons at 1 mm Hg, all others at 0.5 mm Hg.
[b]Dow Corning.
[c]Bis(2-ethylhexyl)phthalate.
[d]Bis(2-ethylhexyl)sebacate.
[e]Monsanto Co., 800 North Lindbergh Boulevard, St. Louis, Mo. 63166.

I. Filter Characteristics

The most common filter paper is made of cellulose, comes in many sizes and shapes, and has various retention (speed) qualities. Newer types, made of borosilicate glass fibers, offer fine particle retention with good filtering speed, resistance to corrosive solvents, and thermal stability (to 475°C). A more recent type is "phase separating paper," a water-repellent paper used in place of separatory funnels; the paper passes organic solvents but retains water (and particulate matter). For extensive details, write for the Whatman catalog (Reeve Angel, 9 Bridewell Place, Clifton, N.J. 07014). Another type of filter is a Teflon (TFE) membrane which permits filtration of corrosive liquids from -260 to +280°C (available from Ace Scientific, 1420 East Linden Avenue, Linden, N.J. 07036, and others). More information on papers is found under "Paper Chromatography," page 380.

1. Standard Filter Paper Sizes

Standard Conical Funnels (58 or 60°)		Buchner Funnels (Coor No. 490)	
Funnel Diameter (mm)	Paper Size (cm)	Funnel Number	Paper Size (cm)
35	5.5	0000	1.4
45	7.0	0	4.25
55	9.0	1	5.5
65	11.0	1A	7.0
75	12.5	2	7.5
90	15.0	2A	9.0
100	18.5	3	11.0
160 (pint funnel)	24.0	4	12.5
180 (quart funnel)	32.0	4A	15.0
220	40.0	5	18.5
260 (gallon funnel)	50.0	6	24.0

2. Filtering Aids

Filtering aids expedite the filtration and washing of gelatinous precipitates, and help to retain very fine precipitates ($BaSO_4$, etc.) and particulate matter (activated charcoal, catalyst preparations, etc.). One type is made of ashless paper pulp in the form of a compressed tablet or a bulky, fibrous material; both forms are used to form a retentive filter bed over customary paper (or sintered glass funnels). Another very common aid is diatomaceous earth ("Celite," Johns-Manville, 22 East 40th Street, New York, N.Y. 10016).

Refs. p. 428.

3. Millipore Filters

Millipore filters are thin, porous structures made of pure and biologically inert cellulose esters or related polymeric substances; they are available in more than 20 pore sizes ranging from 14 μm to 25 mm (0.025 μm), in disks from 13 to 293 mm in diameter. Their characteristics include very uniform pore size, high porosity and flow rate, high chemical stability (even toward concentrated acids and bases), negligible residue after incineration, transparency when pores are filled with an immersion oil of matching refractive index. Write for Catalog MC/1, Millipore Corporation, Ashby Road, Bedford, Mass. 01730. The accompanying size reference scale is reproduced with permission of Millipore Corporation.

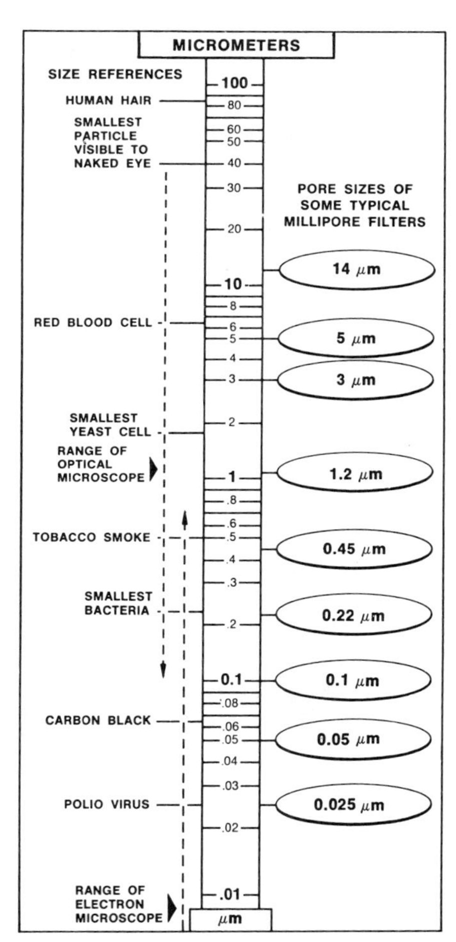

Scale comparing Millipore filter pore sizes with microbes and micro-particles.

4. Amicon Filters

The Amicon Corporation (21 Hartwell Avenue, Lexington, Mass. 02173) sells a variety of materials for "ultrafiltration" for the concentration, separation, and purification of colloids and macromolecules. These include the DIAFLO ultrafiltration membranes, the DIAPOR microporous filters, and

Refs. p. 428.

special equipment for their use. Write for "Ultrafiltration for Laboratory and Clinical Uses," Publication No. 403, for details including literature references.

J. REFERENCES

1. K. B. Wiberg, Laboratory Technique in Organic Chemistry, McGraw-Hill, 1960.
2. "Laboratory and Workshop Notes 1959-1961," R. Lang, Ed., Edward Arnold, Ltd., 1962.
3. A. I. Vogel, Practical Organic Chemistry, Wiley, New York, 1956.
4. A. I. Vogel, Elementary Practical Organic Chemistry, Parts 1, 2, and 3, Wiley, New York, 1966.
5. MIT Dept. of Chem. Research Manual (D. Traficante, personal commun.).
6. A. Weissberger, Technique of Organic Chemistry, Vols. 1-14, Wiley, 1951-1969.
7. L. F. Fieser and M. Fieser, Reagents for Organic Synthesis, Vols. 1 and 2, Wiley-Interscience, 1967, 1969.
8. L. F. Fieser, Experiments in Organic Chemistry, D. C. Heath and Co., 1957.
9. J. E. Hammesfahr and C. L. Strong, Creative Glass Blowing, W. H. Freeman, San Francisco, 1968.
10. Corning Glass Works, Properties of Selected Commercial Glasses, Bulletin B-83, 1963.
11. Commercial Standard 21-58, Interchangeable Taper-Ground Joints, Stopcocks, Stoppers and Spherical-Ground Joints, published by U.S. Department of Commerce and available from The Clearing House, 5285 Port Royal Road, Springfield, Va.
12. R. E. Hogan, Chem. Tech., 1, 41 (1971).
13. Handbook of Chemistry and Physics, 41st ed., Chemical Rubber Company, Cleveland, 1960.
14. Perry's Chemical Engineering Handbook, 4th ed., McGraw-Hill, New York, 1963.
15. Natl. Bur. Standards (U.S.) Circ. C447, 1943.
16. Handbook of Adhesives, Irving Skeist, Ed., Reinhold, 1962.
17. W. H. Brunner and T. H. Batzer, Practical Vacuum Techniques, Van Nostrand Reinhold Co., New York, 1965.
18. Lange's Handbook of Chemistry, 10th ed., McGraw-Hill, New York, 1966.
19. Epoxy Resin Handbook, Noyes Data Corp. (Park Ridge, N.J. 07656), 1972.

II. Standard Glassware Cleaning Solutions

Routine, simple cleaning is conveniently accomplished by brushing with detergent or soap solution; special products are available (such as "Alconox") that act as wetting agents. Laboratory glassware washing machines are also sold, as well as ultrasonic cleaning baths. However, hard-to-clean glassware is most often treated with the solutions described below. It is economically advisable to buy technical grade chemicals for their preparation.

A. Chromic Acid

For routine cleaning (burets, pipets, etc.), add slowly with stirring 800 cc concentrated H_2SO_4 to a solution of 92 g $Na_2Cr_2O_7 \cdot 2H_2O$ in 458 cc H_2O.

Alternatively, add 1 liter of acid to ≈35 cc of a saturated aqueous sodium dichromate solution; the latter solution is obviously more concentrated and is used for very difficult cases. When the red-brown solution turns green or becomes too diluted after use, discard. Potassium dichromate may also be used, but is considerably less soluble than the sodium compound. Very many (>10) washings of the glassware with demineralized water are necessary to remove traces of chromium ions if desired. For use in chelometric titrations and the like, treat the acid-washed glassware as follows for a more effective metal "decontamination": soak for a couple of hours with a solution (aqueous) containing 1% disodium ethylenediaminetetraacetate and 2% NaOH; rinse several times with demineralized water.

A ready-made dichromate concentrate is sold, which is added to a standard 9-lb bottle of H_2SO_4 to make cleaning mixture ("Chromerge"; Manostat Corp., 20 North Moore Street, New York, N.Y.). Other products are available, such as ready-to-use "Dichrol" (Scientific Products, 1210 Leon Place, Evanston, Ill. 60201).

B. Alcoholic Sodium (Potassium) Hydroxide

Add about 1 liter of 95% ethanol to 120 cc water containing 120 g NaOH (or KOH). A very powerful cleaning medium, but prolonged exposure will damage (by etching) ground glass joints.

C. "Nochromix"

This inorganic oxidizer contains no metallic ions. The white powder is dissolved in water and mixed with concentrated H_2SO_4, giving a solution that is reportedly more strongly oxidizing than chromic acid. The solution is clear, but turns orange as the oxidizer is used up. Available from Godax Laboratories, 6 Varick Street, New York, N.Y. 10013.

D. Sulfuric-Fuming Nitric Acids

A mixture of the two is especially good for very greasy, dirty glassware. Alternatively, grease may be removed by soaking in warm soap solution for ≈1 hr; after rinsing with H_2O soak in concentrated HCl.

E. Trisodium Phosphate

To remove carbon residues, soak for several minutes in the following solution and brush off the incrustation: trisodium phosphate (Na_3PO_4)--57 g; sodium oleate--28.5 g, water--470 cc. Carbonaceous material may also be removed with 10 to 15% aqueous NaOH or KOH.

III. PURIFICATION OF COMMON SOLVENTS

A. Introduction

1. Purity

The requirements for degree of solvent purity will vary, of course, with the solvent's use. There are no adequate experimental criteria for ideal purity, and the achievement of a "100%" pure solvent can only be approximated. A practical definition has been stated as follows: "A material is sufficiently pure if it does not contain impurities of such nature and in such quantity as to interfere with the use for which it is intended" (1). For example, a so-called "electronic grade" of some solvents is sold for use in applications where trace metal contamination is undesirable; removal of trace inorganic impurities from organic substances requires special techniques (e.g., see Y. Okamoto, A. J. Gordon et al., "Effects of Impurities on the Electrical Conductivity of Simple Polycyclic Aromatic Hydrocarbons," in Organic Semiconductors, Eds., J. J. Brophy and J. W. Buttrey, Macmillan, New York, 1962, pp. 100-108).

2. General Precautions

Several don'ts and do's associated with purification and handling of solvents must be emphasized:

(a) Never--under any circumstances--use Na, other active metals, or metal hydrides to dry acidic or halogenated liquids or compounds that may act as oxidizing agents.

(b) Never attempt to use a powerful drying agent (such as Na, CaH, $LiAlH_4$, H_2SO_4, P_2O_5) before gross predrying with ordinary agents (Na_2SO_4, e.g.), unless assured of low water content.

(c) Always test for and destroy peroxides in suspicious ethers or other solvents before attempting to distill and/or dry them. Also, because of the peroxide danger, most ethers should not be exposed to light and air for prolonged periods.

(d) Remember that many solvents are toxic and may be cumulative poisons (benzene, etc.); always avoid inhalation of vapors. Also, avoid flames except for known noninflammable solvents (CCl_4, $CHCl_3$); diethyl ether and CS_2 are particularly dangerous.

(e) Highly purified solvents are best stored in sealed glass containers under an inert atmosphere (dry, O_2-free N_2 usually). If sealing is impractical, a positive pressure of an inert gas over the liquid should be maintained. With some solvents, sealing a stoppered container with paraffin wax enables long-term storage.

B. Purification Directions

By following the techniques described below, sufficient purification can be achieved for general use of a solvent in most chemical and physical experiments (synthesis, kinetics, spectroscopy, dipole moment, etc.). It is assumed that the experimenter begins with a reasonably pure commercial grade, rather than very impure "technical" grades. Directions for purification procedures for very specific applications (conductance, ionization, thermodynamic, and other measurements) are given in reference 1. Unless indicated, all distillations should be conducted at atmospheric pressure. Unless directions are given for crystallization of a solvent from another liquid as host solvent, crystallization implies freezing of the solvent-to-be-purified, with up to 20% of the unsolidified liquid discarded. Additional data on the drying and deoxygenation of solvents can be found on page 438; information on detection and removal of peroxides is located on page 437. Unless other references are cited, most of the material below is from references 1 to 5. The number following each solvent name refers to the entry in the table beginning on page 2. The introductory remarks to that table should also be consulted for comments on commercially available "pure" solvents and for references to further information. In addition to the techniques given below, a method known as "adsorptive filtration" using activated alumina is recommended for many applications; write for Application Note 105 from Waters Associates, Inc., 61 Fountain Street, Framingham, Mass. 01701.

1. Acetic Acid--47 (Bp 118)

Commercial glacial acetic acid (∿99.5%) contains carbonyl impurities, removed by refluxing with 2 to 5% (wt.) $KMnO_4$ or excess Cr_2O_3, followed by distillation. Traces of water are removed by treatment with triacetyl borate, made by heating boric acid and Ac_2O (1 part to 5 by wt.) at 60°; cool and filter off the crystals. Heat acetic acid with a two to threefold excess of borate estimated for reaction with the water present; distillation gives anhydrous acid. Alternatively, distillation from P_2O_5 may suffice.

2. Acetone--68 (Bp 56.2)

Acetone is very difficult to dry; many of the usual drying agents cause reaction, even $MgSO_4$. Molecular sieve 4A works well, as does K_2CO_3. Distillation from $KMnO_4$ (small amount) destroys aldehydes, e.g. For very high purity acetone, saturate with dry NaI at 25-30°, decant the solution and cool it to -10°; crystals of the NaI complex form, which are filtered then warmed to 30°, the resulting liquid finally being distilled.

3. Acetonitrile--40 (Bp 81.6)

The recommended method for general use is as follows (1). Predry if wet, then stir with CaH until gas evolution ceases. Distill from P_2O_5 ($\leq$5 g/liter) in an all-glass apparatus, using high reflux ratio. Reflux the distillate over CaH (5 g/liter) for at least an hour, then distill slowly, discarding the first 5 and last 10% of distillate so as to reduce acrylonitrile content. If benzene is an impurity (uv absorption at λ260, strong tailing λ220 nm), remove by azeotropic distillation with water before P_2O_5 treatment.

4. Ammonia--220 (Bp -33.4)

For handling of liquid NH_3 cylinders, see reference 5, page 116. Very dry NH_3 is prepared by dissolving Na in it, followed by distillation. The process is best handled in a vacuum line of a metal high-pressure system. At room temperature, ammonia has a vapor pressure of about 10 atm.

5. Aromatic Hydrocarbons

Very high purity benzene--137 (bp 80.1°)--is prepared by fractional crystallization (mp 5.53° if ultrapure) from ethanol or methanol, followed by distillation. More routine purification is achieved by shaking or stirring benzene with concentrated sulfuric acid (∿100 ml/liter benzene), removing the acid layer and repeating until darkening becomes slight. Decant the benzene into a clean flask for distillation. Any thiophene present, as well as olefinic impurities and water will be removed by the H_2SO_4 procedure.

Toluene--159 (bp 110.6°)--and the xylenes--172, 173, 174--may be similarly purified; however, keep cool (below 30°) during H_2SO_4 treatment due to the greater reactivity of these hydrocarbons toward sulfonation.

Refs. p. 436.

In lieu of H_2SO_4, $CaCl_2$ drying is recommended, although just distillation should suffice, since all the hydrocarbons mentioned form an azeotrope with water or have a significantly higher boiling point.

6. tert-Butyl Alcohol--104 (Bp 82)

Distillation from CaO followed by multiple crystallization of the distillate yields very high purity alcohol (mp 25.4°).

7. Carbon Disulfide--25 (Bp 46.2)

This extremely flammable and toxic liquid must be handled with caution. Distill very carefully, using a water bath heated not too much above the bp of CS_2. To remove sulfur impurities, shake with Hg; then with a cold, saturated $HgCl_2$ solution; finally with cold, saturated $KMnO_4$; dry over P_2O_5 and distill.

8. Carbon Tetrachloride--7 (Bp 76.5)

Any CS_2 that may be present is removed by stirring hot CCl_4 containing about 10% by volume of concentrated alcoholic KOH. Wash with H_2O after several treatments, dry over $CaCl_2$, then distill from P_2O_5.

9. Chloroform--12 (Bp 61.2)

Most $CHCl_3$ contains about 1% ethanol as stabilizer to suppress air oxidation to phosgene. Purification by any one of the following is recommended, followed by storage in the dark under a N_2 atmosphere.

(a) Shake with concentrated H_2SO_4, wash with water, dry over $CaCl_2$ or K_2CO_3, and distill.
(b) Pass through a column of Grade I activated alumina (about 25 g/500 ml $CHCl_3$).
(c) Shake several times with one-half its volume of water, dry with $CaCl_2$, and distill from P_2O_5.

10. Diethyl Ether--101 (Bp 34.5)

Unless beginning with "absolute" grade, the ether should be checked for peroxides and treated accordingly. Heed the additional warnings given for CS_2. Ether can be dried adequately by storage over and distillation from Na wire. However, distillation from $LiAlH_4$ (or CaH) is by far the most efficient method. Ether should be distilled from a hydride just prior to use for best results.

11. Diglyme (Diethyleneglycol Dimethyl Ether)--151 (Bp 161)

As with other glycol ethers, purification is accomplished by heating with Na, followed by distillation over Na.

12. N,N-Dimethylformamide--75 (Bp 152)

N,N-DMF may contain formic acid as well as water. Stir or shake with KOH and distill from CaO or BaO.

Refs. p. 436.

13. Dimethyl Sulfoxide--56 [Bp 189(d)]

DMSO may contain dimethyl sulfide and sulfone in addition to water. Store overnight or longer over freshly activated alumina, Drierite, BaO, or NaOH. Distill at reduced pressure (∿2-3 mm Hg, bp 50°) from NaOH pellets or BaO and store over molecular sieve 4A.

14. 1,4-Dioxane--92 (Bp 102)

Dioxane may contain many impurities and is difficult to purify thoroughly. Many published methods are now known to be ineffective due to decomposition of the liquid; a long and detailed, but proved method is described in reference 1 (p. 708). However, for routine purification, the following procedure (3,4) is adequate. Reflux a mixture of 300 ml H_2O, 40 ml concentrated HCl, and 3 liters dioxane for 12 hr, maintaining a slow stream of N_2 (entrains the acetaldehyde formed from hydrolysis of a glycol acetal impurity). Cool the solution, add KOH pellets until no more dissolve and a separate layer forms. Decant the dioxane (upper) layer and dry with fresh KOH. Reflux the dried dioxane over Na overnight or until the metal remains bright. Distill from Na and store in the dark under N_2.

It is not a good practice to use $LiAlH_4$ with dioxane, since the hydride may decompose at the bp. To keep dioxane oxygen and peroxide free, the use of benzophenone ketyl is recommended (see p. 438).

15. Ethanol--55 (Bp 78.3)

Pure ethanol is commercially available as "absolute" grade, which contains about 0.1 to 0.5% water plus (by law) anywhere from 0.5 to 10% of a denaturant (acetone, benzene, diethyl ether, or methanol, e.g.). A more common and less expensive form is the azeotrope containing 4.5% water, referred to as 95% ethanol or rectified spirit (bp 78.2°), which is preferable as a uv solvent (no benzene or other denaturant, if reagent grade or USP). Pure ethanol is extremely hygroscopic and readily picks up moisture during transfer; therefore, a carefully controlled system must be used to prepare very dry ethanol.

A recommended procedure for removing nearly all the water from absolute ethanol is as follows. Reflux a mixture of 60 ml absolute ethanol, 5 g Mg turnings, and a few drops of CCl_4 or $CHCl_3$ (catalyst) until all the Mg is converted to the ethoxide. Add 900 ml more of absolute ethanol, reflux for 1 hr, and distill. The highly volatile ethyl bromide can be used as catalyst instead of $CHCl_3$ or CCl_4 if halogen compounds must be excluded. Down to 0.05% water in ethanol can be detected by formation of a voluminous precipate upon admixture of a benzene solution of aluminum ethoxide (6). Storage over molecular sieve 3A helps maintain alcohol dry to about 0.005% water.

Most of the water can be removed from 95% ethanol by refluxing over fresh lime (CaO) for several hours, followed by distillation. Alternatively, water can be distilled off as one component of a ternary azeotrope, such as benzene-ethanol-water, bp 64.48°. After the water is removed, benzene can be distilled off as the benzene-ethanol azeotrope (bp 68.24°).

16. Ethyl Acetate--93 (Bp 77.1)

The water, ethanol, and acid impurities in most commercial ethyl acetate are removed by washing with 5% aqueous sodium carbonate, then with saturated calcium chloride, followed by drying over anh potassium carbonate, and finally distilling from P_2O_5.

Refs. p. 436.

17. Ethylene Dichloride--43 (1,2-Dichloroethane) (Bp 83.8)

First shake with basic alumina (1 g/liter) or store over NaOH to remove acid impurities (especially HCl, formed on exposure to light). Distillation from P_2O_5 gives dry solvent.

18. Hexamethylphosphoramide--156 [Bp 66 (0.5 mm); 127 (20 mm)]

This solvent may contain basic impurities in addition to water. However, most reported purification methods call for vacuum distillation from BaO or CaO; storage over molecular sieve 4A is also recommended.

19. Methanol--23 (Bp 64.5)

In addition to water, the impurities are principally C-1 to C-4 carbonyl and hydroxy compounds, but they are present only in trace amounts in reagent grade solvent. Any acetone present is removed as iodoform after treatment with NaOI. Most water is removed by distillation since no azeotrope forms. Very dry methanol is obtained by storage over or passage through a column of molecular sieve 3A or 4A, followed by storage over lumps of calcium hydride. Drierite is not recommended as a drying agent for methanol! Residual water can also be removed with magnesium methoxide as follows: 50 ml methanol, 5 g Mg turnings and 0.5 g sublimed iodine are refluxed until the iodine color disappears, and hydrogen evolution occurs. After methoxide formation add 1 liter methanol and reflux about 1/2 hr, followed by careful distillation.

20. Methylene Chloride--16 (Bp 40.8)

Wash with concentrated sulfuric acid, then with aqueous carbonate and water, followed by drying over calcium chloride. Distillation from P_2O_5 provides pure, dry solvent.

21. Morpholine--99 (Bp 128)

Drying with Drierite followed by careful fractional distillation is satisfactory. Distillation from, and storage over sodium is also recommended.

22. Nitroalkanes

The commercially available C-1 through C-3 compounds are satisfactorily purified by drying over calcium chloride or P_2O_5 followed by careful distillation. Nitromethane can also be highly purified by fractional crystallization (mp -28.6°).

23. Nitrobenzene--136 (Bp 211)

When purified by repeated fractional crystallization (mp 5.76°) and distillation from P_2O_5, nitrobenzene is colorless. If impure, it becomes colored rapidly over P_2O_5; otherwise hardly any color is evident even after prolonged contact with P_2O_5.

24. Piperidine--117 (Bp 106)

Careful distillation from NaOH or KOH pellets or from sodium metal have been suggested. Partial crystallization (mp -10.5°) of the distillate,

Refs. p. 436.

followed by redistillation of the crystallizate, provides even purer liquid.

25. 1-Propanol--77 (Bp 97.4)

Allyl alcohol is a common impurity and is removed by adding about 1.5 ml Br_2/liter of alcohol, followed by distillation from anh K_2CO_3. For very dry alcohol, use the Mg method described for methanol. Note that 1-propanol forms an azeotrope with water (72% alcohol, bp 88°).

26. 2-Propanol--78 (Bp 82.4)

This alcohol forms an azeotrope with water (9% water, bp 80.3°) which can be dehydrated by refluxing over and distilling from lime. 2-Propanol may form peroxides, which are conveniently destroyed by refluxing with solid $SnCl_2$. In general, distillation from anhydrous calcium sulfate produces satisfactory material; use the Mg method described for methanol for very dry solvent.

27. Pyridine--111 (Bp 115.3)

Dry over KOH pellets for prolonged periods, followed by distillation from BaO. Care is required to exclude moisture from this very hygroscopic solvent (forms a hydrate, bp 94.5°).

28. Saturated Hydrocarbons

The ordinary compounds, such as pentane, hexane, heptane, cyclohexane, methylcyclohexane, their isomers and mixtures (petroleum ethers, etc.), all may contain olefinic and/or aromatic impurities; some may contain sulfur compounds. Shake the solvent with a mixture of concentrated sulfuric and nitric acids, repeat two or three times; if a positive permanganate test for olefins is obtained, use a concentrated solution of $KMnO_4$ in 10% sulfuric acid until the test is negative. Wash thoroughly with water, dry over calcium chloride, and distill. Passage through a column of activated alumina removes any residual nonsaturated compounds. Note that this procedure is not adequate to separate closely boiling isomers (say, C-6 compounds) if so desired; see reference 1 for details. A further note of caution--various petroleum ethers (low boiling, high boiling, etc.) are commonly used as purchased for crystallization solvents or chromatography eluants. It is recommended that for such uses the solvents be stored over a drying agent (calcium sulfate, etc.) and flash-distilled before use; the low-boiling petroleum ethers particularly, often contain high-boiling "impurities" which contaminate chromatographic fractions.

29. Sulfolane--94 (Bp 283)

Acid impurities are removed by passage through activated alumina. Repeated vacuum distillation from NaOH has been recommended (7) until a 1 ml sample does not become colored within 5 min after addition of 1 ml of 100% sulfuric acid. A final distillation from CaH will remove any residual water.

Refs. p. 436.

30. Sulfur Dioxide--223 (Bp -10.1)

Bubbling through concentrated sulfuric acid removes water and sulfur trioxide impurities. A vacuum line or other closed system is recommended for handling.

31. Sulfuric Acid--218 (Bp ∿305)

As described by Jolly (5), 100% acid is conveniently prepared by adding sufficient fuming sulfuric acid to standard 96% acid to convert all the water to sulfuric acid. To determine the endpoint, blow a gust of moist air across the acid with a small rubber syringe. If the acid contains excess SO_3, a fog appears; if less than 100%, no fog appears. This procedure allows the composition to be adjusted to within 0.02% of pure acid! Protect the extremely hygroscopic acid from the atmosphere.

32. Tetrahydrofuran--90 (Bp 66)

Test for peroxides and treat accordingly; traces of peroxide are removed by refluxing a 0.5% suspension of cuprous chloride for 1/2 hr, followed by distillation before proceeding. Predry over KOH pellets and finally reflux over, and distill from lithium aluminum hydride or calcium hydride for very dry solvent.

33. Miscellaneous

Cellosolves and carbitols are purified by drying over calcium sulfate and distilling. Acid anhydrides are purified by fractional distillation from the fused sodium salt of the corresponding acid; the high molecular weight anhydrides (C-6, etc.) decompose if distilled at atmospheric pressure.

C. References

1. J.A. Riddick and W. B. Bunger, Organic Solvents, Volume 2 of Techniques of Chemistry, Wiley-Interscience, New York, 1971.
2. A. I. Vogel, A Textbook of Practical Organic Chemistry, 3rd ed., Wiley, New York, 1966, pp. 163-179.
3. K. B. Wiberg, Laboratory Technique in Organic Chemistry, McGraw-Hill, New York, 1960, pp. 240-251.
4. L. F. Fieser, Experiments in Organic Chemistry, 3rd ed., D. C. Heath, Boston, 1955, pp. 281-292.
5. W. L. Jolly, The Synthesis and Characterization of Inorganic Compounds, Prentice-Hall, Englewood Cliffs, N.J., 1970, pp. 114-121.
6. F. Henle, Chem. Ber., 53, 719 (1920).
7. E. M. Arnett and C. F. Douty, J. Amer. Chem. Soc., 86, 409 (1964).

IV. DETECTION OF PEROXIDES AND THEIR REMOVAL

For recent general reviews, see H. L. Jackson et al., J. Chem. Educ., 47, A175 (1970), and R. M. Johnson and I. W. Siddiqi, The Determination of Organic Peroxides, Pergamon Press, New York, N.Y., 1970.

A. Rapid Tests for Detecting Peroxides in Liquids

1. The most sensitive test (detects down to 0.001%) involves conversion of colorless ferrothiocyanate to red ferrithiocyanate upon admixing a drop of the liquid with the reagent. The reagent is prepared as follows. Dissolve 9 g $FeSO_4 \cdot 7H_2O$ in 50 ml 18% HCl. Add a little granular Zn, then 5 g sodium thiocyanate; when the transient red color fades, add 12 more g sodium thiocyanate and decant the solution from the unused Zn into a clean storage bottle.
2. Place a few milliliters into a glass-stoppered flask. Add 1 ml of freshly prepared 10% aqueous KI, shake, and let stand for 1 min. Appearance of a yellow color indicates peroxide. A more rapid method is as follows: add about 1 ml of the liquid to an equal volume of glacial acetic acid which contains about 100 mg NaI or KI. A yellow color indicates a low concentration, and brown a high concentration of peroxides.
3. For water-insoluble liquids, add several milliliters to a solution containing about 1 mg sodium dichromate, 1 ml water, and 1 drop diluted H_2SO_4. A blue color in the organic layer is a positive test (perchromate ion).
4. Shake a portion of the liquid with a drop of clean mercury; in the presence of peroxides a black film of oxidized mercury will form.

B. Removal of Peroxides (Especially from Ethers)

1. Large amounts may be removed by storing over alumina or by passage through a short alumina column. If activated alumina is used, the solvent is also dried in the process. Caution: do not allow the alumina to go dry; elute or wash out the adsorbed peroxides with 5% aqueous $FeSO_4$, for example (see below).
2. For water-insoluble solvents, shake with a concentrated solution of a ferrous salt (100 g ferrous sulfate, 42 ml concentrated HCl, 85 ml water). With some ethers, this treatment may produce small amounts of aldehydes, which can be removed by washing with 1% $KMnO_4$, then with 5% aqueous NaOH, and then with water.
3. One of the most effective reagents is aqueous sodium pyrosulfite (also called metabisulfite, $Na_2S_2O_5$), reacting quickly with peroxides in essentially stoichiometric amounts [A. C. Hemstead, Ind. Eng. Chem., 56, 37 (1964)].
4. Cold washing with 25% by weight (of ether) of triethylenetetramine completely removes high concentrations of peroxide (Hemstead, loc. cit.).
5. Stannous chloride is the only inorganic reagent effective in the solid form (Hemstead, loc. cit.).
6. For water-soluble ethers, a convenient methods is to reflux the ether with 0.5% (wt.) cuprous chloride, followed by distillation.

V. CHEMICAL METHODS FOR DEOXYGENATING GASES AND LIQUIDS

Liquids are often deoxygenated (degassed in general) by high-vacuum and/or inert atmosphere distillation; the use of freeze-thaw degas cycles is also convenient for some substances. However, there are several convenient and efficient preparations which will remove most or all oxygen from a fluid by employing the appropriate treatment. For example, passage of a gas through any of the following is relatively easy. Some of the methods listed below also remove water and are excellent for purifying solvents (method F, e.g.).

For special situations and more detail, consult The Manipulation of Air-Sensitive Compounds, by D. F. Shriver (McGraw-Hill, New York, 1969). A recent article on inert atmosphere enclosures (dry-box techniques) is also useful [L. F. Druding, J. Chem. Educ., 47, A815 (1970)]; a simple, inexpensive variation using a disposable polyethylene glove is described by Srinivasan and Wiles [J. Chem. Educ., 43, 348 (1971)].

A. Pyrogallol

Strongly alkaline aqueous solution of pyrogallol (1,2,3-trihydroxybenzene). Comment: Absorbs O_2 rapidly; difficult to detect visibly when solution is no longer active.

B. Fieser's Solution

To 20 g KOH in 100 ml H_2O add 2 g sodium anthraquinone-β-sulfonate and 15 g sodium hydrosulfite ($Na_2S_2O_4$--also called sodium dithionite); keep warm until blood-red solution is obtained. Comment: The cool solution is effective until hydrosulfite is exhausted (color turns to brown or brown-red, or a ppt appears). Capacity of fresh solution is about 800 ml O_2. (H_2S may be produced, but is removed by bubbling through saturated aqueous lead acetate.)

C. Sodium Dithionite

Several seconds vigorous shaking of a 10% aqueous $Na_2S_2O_4$ solution with an organic liquid will effectively remove most of the dissolved O_2 in the liquid. Comment: Solution unstable and must be used fresh.

D. Chromous (Cr^{2+})

Make a 0.4 M aqueous $Cr(ClO_4)_2$(anhydrous) solution (13.5 g/100 ml H_2O); add 25 ml concentrated HCl. To this solution, add ∿33 g Zn that has been amalgamated as follows: wash Zn for 30 sec with 3M HCl; amalgamate with $HgCl_2$ for 3 min (10 g $HgCl_2$/100 ml H_2O) and wash the Zn thoroughly with distilled water. Comment: This dark blue solution is most effective for removing traces of O_2; when solution turns light blue to green, it is inactive. ($CrCl_2$ and $CrSO_4$ may also be used, but the latter is not as satisfactory.)

E. $LiAlH_4$- Benzopinacolone

0.5 g Lithium aluminum hydride and 10 g benzopinacolone (Ph_3CCOPh) in 50 ml pyridine. Comment: Solution is effective as long as it remains red. It is particularly good for removing both H_2O and O_2 and thus avoids a possible necessary drying step in methods A to D above. Entrained pyridine can be removed (if necessary) by bubbling through concentrated H_2SO_4.

F. Benzophenone Ketyl

A very effective method for removing H_2O and O_2 from solvents is as follows. Predry and remove any acids present by stirring the warm solvent over NaOH (KOH); filter. Reflux the solvent over metallic Na for at least an hour. Add a quantity of solid benzophenone to the mixture to give a deep blue-purple solution. Comment: Distillation under N_2 gives O_2 and H_2O free solvent; the ketyl mixture can also be used to purify any nonreacting gas. The ketyl is exhausted when the color fades or turns pale.

G. "BTS Catalyst"

O_2 as well as H_2, H_2S, CO, COS, and vinyl chloride are effectively removed from gases and liquids by this finely dispersed Cu-on-activated-reagents. It is supplied (by BASF Co., 866 Third Avenue, New York, N.Y. 10022) in the oxidized state and may be used directly to remove reducing agents; for O_2 removal it must be reduced at 120 to 200° with H_2 or CO. For details, write the company for BTS Catalyst brochure.

H. Copper Filings

The gas is passed through a bed of Cu filings heated at 500 to 600°.

I. "Deoxo," "Oxisorb," and "Ridox"

"Deoxo" is a commercial, catalytic purifier (unit attaches directly to gas cylinder) designed to remove O_2 from H_2 at room temperature. (Available from most gas supply companies.) "Oxisorb" contains a replaceable cartridge which removes O_2 from many gases by chemisorption on active metal, and also removes H_2O by absorption on molecular sieves (Analabs, Inc., 80 Republic Drive, New Haven, Conn. 06473). "Ridox" is a recently introduced highly active granular reagent which removes up to 99% of oxygen in certain gases (Fisher Scientific; write for Bulletin No. 199). Comment: These devices, as well as the other solid purifiers (methods G and H), are not as efficient as methods D to F in removing the last traces of O_2; also, it is difficult to detect when the system becomes inactive.

VI. SIMPLE CHEMICAL METHODS FOR DETECTING SPECIFIC GASES

It is often of interest to know if a specific gas is liberated from a reaction (e.g., CO in photochemical decarbonylation). A convenient method for monitoring the effluent from such a reaction is to pass the vapors (by means of an inert gas stream, e.g.) through detection reagents, some of which are outlined below. The following general references were useful:

N. H. Furman, Ed., Scott's Standard Methods of Chemical Analysis, Vol. 2, 5th ed., D. Van Nostrand, New York, 1947, pp. 2336-2433.
F. Feigl, Spot Tests, Vol. 1, Elsevier, Amsterdam, 1954.

Gas	Test and Observations
H_2	Pass the gas over cold, solid $PdCl_2$; in the absence of CO and olefins, HCl is evolved which is easily detected (see below).
HF	Turns zirconium-alizarin paper yellow. To make paper, soak dry filter paper in a 5% solution of $Zr(NO_3)_4$ in 5% aqueous HCl; then place the paper in a 2% aqueous solution of sodium alizarin sulfonate to produce the red-violet test paper which is washed and dried. Moisten the paper before use with 50% acetic acid.
HCl, HBr, HI	Hold an open bottle of concentrated ammonia near the effluent; dense, white fumes form. Absorption in saturated $AgNO_3$ also good (silver halide precipitates).
HCN	Same as $AgNO_3$ test for HCl. Also, when filter paper is impregnated with solutions of Cu(II) acetate and benzidine acetate contacts HCN, an immediate blue color appears (works also for HBr and HI).
H_2CO	Formaldehyde can be precipitated quantitatively as methylenebisdimedon (mp 188°) by reaction with dimedon (5,5-dimethyl-1,3-cyclohexanedione). Optimum reagent conditions are [Yoe and Reid, Anal. Chem., 13, 238 (1941)]: 6.4 ml saturated aqueous dimedon (~4 mg/ml) + 250 ml of Walpole's acetate buffer for pH 4.6; prepare buffer by combining 25.5 ml 0.2M HOAc (12 g/liter), 24.5 ml 0.2M NaOAc (16.4 g/liter), and 50 ml H_2O. Alternatively: H_2CO gives a purple color with chromotropic acid. Dissolve 0.1 g chromotropic acid (4,5-dihydroxy-2,7-naphthalenedisulfonic acid disodium salt; Eastman) in 1 ml H_2O; add 10 ml concentrated H_2SO_4 (Cheronis and Ma, Organic Functional Group Analysis, Interscience, New York, 1964, p. 505).
H_2S	Filter paper moistened with aqueous Pb(II) acetate turns black.
O_2	Turns colorless potassium pyrogallate solution red-brown (5 g pyrogallic acid in 100 ml H_2O containing 120 g KOH). See also page 438.
O_3	Liberates I_2 from 4% (wt.) aqueous KI. Also, turns Ag foil black and $MnCl_2$ paper brown. Gas has odor of clover or new-mown hay.
CO	Passage through aqueous Pd(II) chloride precipitates black Pd (other reducing gases will do the same). Also, a dilute aqueous $KMnO_4/AgNO_3$ solution acidified with HNO_3 is decolorized (MnO_2) by CO. The standard absorption medium for CO is ammoniacal or acidic (HCl) Cu(I) chloride or sulfate; however, there is hardly any detectable visible change.

(Continued)

Gas	Test and Observations
CO	Precipitates as carbonate from $Ca(OH)_2$ or $Ba(OH)_2$ solution.
Cl_2, Br_2, I_2	Starch-iodide paper turns blue. Also, filter paper impregnated with aqueous fluorescein will turn red on contact with Br_2 or I_2 (not Cl_2).
NH_3	Dense, white fumes near open bottle of concentrated HCl and characteristic odor.
SO_2	Instantly decolorizes aqueous fuchsin or malachite green solution, as well as starch-I_2 paper.
Acetylene	Turns ammoniacal Cu(I) chloride red-brown (resulting mixture explosive when dry; keep moist).
Ethylene	Decolorizes diluted Br_2/CCl_4.

VII. SIMPLE PREPARATION OF SOME DRY GASES

Some dry gases are easily prepared by heating certain dry solids (see, e.g., R. E. Dodd and P. L. Robinson, Experimental Inorganic Chemistry, Elsevier, New York, 1954, Ch. 3). The following are taken from W. L. Jolly, The Synthesis and Characterization of Inorganic Compounds, Prentice-Hall, N.J., 1970, page 545.

Gas	Solids
H_2	UH_3 (heat to 650°)
H_2S (H_2Se)	S(Se) + paraffin wax
CO_2	$MgCO_3$; dry ice
$(CN)_2$	$Hg(CN)_2$
C_2F_4	Teflon
BF_3	$CaF(BF_4)$ or B_2O_3 + KBF_4
N_2	NaN_3; $(NH_4)_2Cr_2O_7$
NO	63.8g KNO_2, 25.2 g KNO_3, 76 g Cr_2O_3, 120 g Fe_2O_3, all thoroughly admixed.
N_2O	NH_4NO_3
Cl_2	$CuCl_2$
SiF_4	$BaSiF_6$
GeF_4	$BaGeF_6$
SO_2	$Na_2S_2O_5$
SO_3	$Fe_2(SO_4)_3$

VIII. COMMON SOLVENTS FOR CRYSTALLIZATION

Solvents are listed in approximate order of decreasing polarity. Number refers to entry in table, "Properties of Solvents and Common Liquids," page 2, which can be consulted for further data. In choosing a second solvent for a mixture, trial and error are usually required. However, there are some generally successful mixtures, such as diethyl ether-methanol (or ethanol) for highly associated solids (especially amides, alcohols) and many natural products, and diethyl ether-petroleum ether (or benzene) for polar compounds (especially esters, alcohols) and hydrocarbons.

Solvent	No.	Bp (°C)	Flammability[a]	Toxicity[a]	Good for	Second Solvent for Mixture	Comments[b]
Water	218	100	0	0	Salts, amides, some carboxylic acids	Acetone, alcohols, dioxane, acetonitrile	Precipitates dry slowly
Acetic acid	47	118	+	++	Salts, amides, some carboxylic acids	Water	Difficult to remove; pungent odor
Acetonitrile	40	81.6	+++	+++	Polar compounds	Water, ethyl ether, benzene	
Methanol	23	64.5	+++	+	General, esters, nitro and bromo compounds	Water, ethyl ether, benzene	
Ethanol	55	78.3	+++	0	General, esters, nitro and bromo compounds	Water, hydrocarbons, ethyl acetate	
Acetone	68	56	+++	+	General, nitro and bromo compounds, osazones	Water, hydrocarbons, ethyl ether	Should be dried if not used with water
Methyl cellosolve	79	124	++	++	Sugars	Water, benzene, ethyl ether	
Pyridine	111	115.6	+++	++	High-melting, insoluble compounds	Water, methanol, hydrocarbons	Solvent power depends on dryness; difficult to remove
Methyl acetate	71	57	++++	++	General, esters	Water, ethyl ether	

Ethyl acetate	93	77.1	+++	+	General, esters	Ethyl ether, hydrocarbons, benzene	
Methylene chloride	16	40	0	++	General, low-melting compounds	Ethanol, hydrocarbons	Easily removed and dried
Ethyl ether	101	34.5	++++	++	General, low-melting compounds	Acetone, hydrocarbons	May creep up side of flask and deposit precipitate
Chloroform	12	61.7	0	++++	General, acid chlorides	Ethanol, hydrocarbons	Easily removed and dried
Dioxane	92	102	+++	++	Amides	Water, benzene, hydrocarbons	May form solvates with some ethers
Carbon Tetrachloride	7	76.5	0	++++	Nonpolar compounds, acid chlorides, anhydrides	Ethyl ether, benzene, hydrocarbons	Reacts with some nitrogen bases; cumulative poison
Toluene	159	110.6	+++	++	Aromatics, hydrocarbons	Ethyl ether, ethyl acetate, hydrocarbons	
Benzene	137	80.1	+++	+++	Aromatics, hydrocarbons, molecular complexes	Ethyl ether, ethyl acetate, hydrocarbons	Cumulative poison
Ligroin	-	90-110	+++	+	Hydrocarbons	Ethyl acetate, benzene, methylene chloride	Trace of ethanol sometimes greatly increases solvent power
Petroleum ether (ACS)	-	35-60	++++	+	Hydrocarbons	Any solvent on this list from ethanol down	
n-Pentane	121	36.1	++++	+	Hydrocarbons	Any solvent on this list from ethanol down	
n-Hexane	146	69	++++	+	Hydrocarbons	Any solvent on this list from ethanol down	
Cyclohexane	143	80.7	+++	+	Hydrocarbons	Any solvent on this list from ethanol down	
n-Heptane	170	98.4	+++	+	Hydrocarbons	Any solvent on this list from ethanol down	

[a] More + means more flammable or toxic.
[b] Comments apply principally to the main solvent, not a possible mixture.

IX. SOLVENTS FOR EXTRACTION OF AQUEOUS SOLUTIONS[a]

Emulsions may form during the extraction of aqueous solutions by organic solvents, making good separation very difficult, if not impossible. Their formation is especially liable to occur if the solution is alkaline; addition of dilute sulfuric acid (if permissible) may break up such an emulsion. The following are general methods for breaking up emulsions: saturation of the aqueous phase with a salt (NaCl, Na_2SO_4, etc.); addition of several drops of alcohol or ether (especially when $CHCl_3$ is the organic layer); centrifugation of the mixture, one of the most successful techniques.

Solvent	No.[b]	Bp (°C)	Flammability[c]	Toxicity[c]	Comments
Benzene	137	80.1	+++	+++	Prone to emulsion; good for alkaloids and phenols from buffered solutions
2-Butanol	102	99.5	+	+++	High boiling; good for highly polar water-soluble materials from buffered solution
Carbon Tetrachloride	7	76.5	0	++++	Easily dried; good for nonpolar materials
Chloroform	12	61.7	0	++++	May form emulsion; easily dried
Diethyl ether	101	34.5	++++	++	Absorbs large amounts of water; good general solvent
Diisopropyl ether	-	69	+++	++	May form explosive peroxides on long storage; good for acids from phosphate buffered solutions
Ethyl acetate	93	77.1	+++	+	Absorbs large amount of water; good for polar materials
Freon 11	-	24	0	+	Freons good for volatile nonpolar compounds; are fairly expensive
Freon 113	-	47.7	0	+	
Methylene chloride	16	40	0	++	May form emulsions; easily dried
n-Pentane	121	36.1	++++	+	Hydrocarbons are easily dried; are poor solvents for polar compounds
n-Hexane	146	69	++++	+	
n-Heptane	170	98.4	+++	+	

[a]Data in this table are taken mostly from Introduction to Organic Research, an unpublished manual of the Department of Chemistry, Massachusetts Institute of Technology (D. Traficante, personal communication) and from Craig and Craig, "Extraction and Distribution," in Vol. 3 of Weissberger, Ed., Technique in Organic Chemistry, New York, Interscience, 1950, p. 301.

[b]No. refers to entry in table, Properties of Solvents and Common Liquids, page 2.

[c]More + means more toxic or flammable.

X. DRYING AGENTS

A. Dehydrating Agents for Liquids

1. The best dehydrating agents are those that react rapidly and irreversibly with water (and not with the solvent or solutes); they are also the most dangerous and should be used only after gross predrying with a less vigorous drying agent (see below). These agents are almost always used only to dry a solvent prior to and during distillation (see also Methods for Deoxygenating Fluids, p. 438). Although $MgClO_4$ is one of the most efficient drying agents, it is not recommended because of its solubility in many solvents and because it can cause explosions if mishandled.

Agent	Products Formed with H_2O	Comments
Na[a]	NaOH, H_2	Excellent for saturated hydrocarbons and ethers; do not use with any halogenated compounds.
CaH	$Ca(OH)_2$, H_2	One of the best agents; slower than $LiAlH_4$ but just as efficient and safer. Use for hydrocarbons, ethers, amines, esters, C_4 and higher alcohols (not C_1, C_2, C_3 alcohols). Do not use for aldehydes and active carbonyl compounds.
$LiAlH_4$[b]	LiOH, $Al(OH)_3$, H_2	Use only with inert solvents (hydrocarbons, aryl (not alkyl) halides, ethers); reacts with any acidic hydrogen and most functional groups (halo, carbonyl, nitro, etc.). Use caution; excess may be destroyed by slow addition of ethyl acetate.
BaO or CaO	$Ba(OH)_2$ or $Ca(OH)_2$	Slow but efficient; good mainly for alcohols and amines, but should not be used with compounds sensitive to strong base.
P_2O_5	HPO_3, H_3PO_4, $H_4P_2O_7$	Very fast and efficient; very acidic. Predrying recommended. Use only with inert compounds (especially hydrocarbons, ethers, halides, acids, and anhydrides).

[a] J. T. Baker Co. sells an alloy of 10% Na, 90% Pb called "Dri-Na"; this dry, granular agent reacts only slowly with air, but is as efficient as Na wire for drying ethers, etc. See Fieser and Fieser, Reagents, Vol. 2, Wiley, New York, 1969, p. 385.

[b] A less dangerous, but equally efficient alternative is $Na(CH_3OCH_2CH_2O)_2$-AlH_2, known as "Vitride" (Realco Chemical Company, 783 Jersey Avenue, New Brunswick, N.J. 08902; available from Eastman Kodak).

2. For drying solutions and/or solvents, the following common adsorbing/absorbing agents are used.

Agent	Capacity[a]	Speed[b]	Comments
$CaSO_4$	1/2 H_2O	Very fast (1)	Sold commercially as "Drierite" with or without a color indicator; very efficient. When dry, the indicator ($CoCl_2$) is blue, but turns pink as it takes on H_2O (capacity $CoCl_2 \cdot 6H_2O$); useful in temperature range -50° to +86°. Some organic solvents leach out, or change the color of $CoCl_2$ (acetone, alcohols, pyridine, etc.).
$CaCl_2$	6 H_2O	Very fast (2)	Not very efficient; use only for hydrocarbons and alkyl halides (forms solvates, complexes, or reacts with many N and O compounds).
$MgSO_4$	7 H_2O	Fast (4)	Excellent general agent; very inert but may be slightly acidic (avoid with very acid-sensitive compounds). May be soluble in some organic solvents.
Molecular Sieve 4A	High	Fast (30)	Very efficient; predrying with a more common agent recommended (see below for details on molecular sieves). Sieve 3A also excellent.
Na_2SO_4	10 H_2O	Slow (290)	Very mild, inefficient, slow, inexpensive, high capacity; good for gross predrying, but do not warm the solution.
K_2CO_3	2 H_2O	Fast	Good for esters, nitriles, ketones and especially alcohols; do not use with acidic compounds.
NaOH, KOH	Very high	Fast	Powerful, but used only with inert solutions in which agent is insoluble; especially good for amines.
H_2SO_4	Very high	Very fast	Very efficient, but use limited to saturated or aromatic hydrocarbons or halides (will remove olefins and other "basic" compounds).

(Continued)

Agent	Capacity[a]	Speed[b]	Comments
Alumina or Silica Gel	Very high	Very fast	Especially good for hydrocarbons. Should be finely divided; can be reactivated after use by heating (300° for SiO_2, 500° for Al_2O_3).

[a]Moles H_2O per mole of agent (maximum).
[b]Relative rating. For first five entries, number in parentheses is relative drying speed for benzene--low number, rapid drying; order may change for slow agents with change in solvent [B. Pearson and J. Ollerenshaw, Chem. Ind., 370 (1966)].

B. Agents for Desiccators and Drying Tubes

1. For drying solids (drying pistol or desiccator), liquids or oils, (desiccator) or gases (bubbler or drying tube); or for keeping atmospheric moisture out of vessel (drying tube): $CaCl_2$, $CaSO_4$, KOH, P_2O_5, H_2SO_4. Commonly used is P_2O_5, but fresh (white) surface must be exposed; H_2SO_4 should not be used at low pressure. For details concerning dry-box atmosphere control, see L. F. Druding, J. Chem. Educ., 47, A815 (1970).
2. To keep out atmospheric CO_2, a drying tube of "Ascarite" (finely divided NaOH) is useful; dry CaO may also be used.
3. Paraffin wax shavings are useful to dry a sample wet with hydrocarbon solvents.

C. Molecular Sieves

Synthetic zeolites (aluminosilicates) with pore sizes capable of selectively adsorbing molecules of narrow size range. [see Fieser and Fieser, Reagents for Organic Synthesis, Vol. 1, Wiley, New York, 1967, p. 703, and D. W. Brack, J. Chem. Educ., 41, 768 (1964)]. Available from Fisher Scientific Co., Matheson, Union Carbide, and others.

Type	Pore Size (Å)	Comments
3A	3	Adsorbs H_2O, NH_3 (not C_2H_6); good for drying polar liquids.
4A	4	Adsorbs H_2O, EtOH, H_2S, CO_2, SO_2, C_2H_4, C_2H_6, C_3H_6 (not C_3H_8 and higher hydrocarbons); good for drying nonpolar liquids and gases.
5A	5	Adsorbs n-C_4H_9OH and n-C_4H_{10}, not isocompounds and rings C_4 or greater.
10X	8	Adsorbs highly brached hydrocarbons and aromatics; useful for purification and drying of gases.
13X	10	Adsorbs di-n-butylamine, not tri-n-butylamine; especially good for drying hexamethylphosphoramide, $[(CH_3)_2N]_3PO$.

D. Karl Fischer Titration and Water Analysis

Karl Fischer titration is one of the most sensitive techniques available for the actual measurement of very small quantities of water in organic liquids. The Karl Fischer Reagent is a solution of iodine, sulfur dioxide, and pyridine usually in methanol as a solvent; methanol may be replaced by methyl cellosolve, dioxane, or glacial acetic acid, but pyridine is essential.

$$H_2O + I_2 + SO_2 + Py\ (excess) \longrightarrow 2Py \cdot HI + Py \cdot SO_3$$

$$Py \cdot SO_3 + CH_3OH \longrightarrow Py^{+}H\ CH_3OSO_2^{-}$$

For vinyl ethers, methanol interferes with the determination and methyl cellosolve should be used instead. Reagent solutions are available commercially (Fisher Scientific; Matheson; A. H. Thomas) as well as a special titration unit (A. H. Thomas). For more detail, see N. D. Cheronis and T. S. Ma, Organic Functional Group Analysis, Interscience, New York, 1964, pp. 472-475. An extensive study of the method can be found in J. Mitchell and D. Smith, Aquametry, Interscience, New York, 1948.

A more rapid and versatile determination of water by glc has been developed which uses a "Poropak" column [J. Hogan, R. Engel, and H. Stevenson, Anal. Chem., 42, 249 (1970)]. For details of the technique, write for "Trace Water Analysis in Organics" from Fisher Scientific Company; analyses of 1 to 10 ppm water are accomplished with high precision and accuracy.

Other methods of water analysis are reviewed in Organic Solvents, Vol. 2 of Techniques of Chemistry, Wiley-Interscience, New York, 1971.

XI. SOLVENTS AND BATHS FOR HEATING AND COOLING

It is often necessary to maintain a temperature different from that of the laboratory. In cases where precise control is needed, a liquid bath is usually superior to other methods. For temperatures above that of the room, any liquid that is nonviscous at the temperature desired and boils at least 20 to 30° above that temperature can usually be used. Of course, other factors such as expense, toxicity, and flammability are important. The table below lists some commonly used liquids for heating baths, as well as several not so common media for high temperatures or wide ranges of use. It is also possible to use many of these liquids as solvents for high-temperature reactions. The second table lists some mixtures of common laboratory liquids that can be used to very low temperatures, yet remain liquid at room temperature. These mixtures can also be used as solvents at these low temperatures. For temperatures down to about -50°, there are many common organic liquids that can be used.

While most laboratories have several means for achieving and maintaining temperatures above room temperature (e.g., bunsen burners and hot plates), cooling apparatus is much less common. The last two tables in this section offer convenient ways to cool a system to a specified temperature without the need for expensive apparatus.

A. Liquid Media for Heating Baths

The data in this table are taken mostly from R. Egly, "Heating and Cooling," in Vol. 3, Part 2, of Weissberger, ed., *Technique in Organic Chemistry*, Interscience, New York, 1957, p. 152. The percentages refer to weight percent. For other useful high-temperature baths see fused salts, page 40.

Medium	Mp (°C)	Bp (°C)	Useful Range (°C)[a]	Flash Point (°C)	Comments
H_2O	0	100	0-80	None	Ideal in limited range
Ethylene glycol	-12	197	-10-180	115	Cheap; flammable; difficult to remove from apparatus
DC[b] 330 silicone oil	<-148	-	30-280	290	Becomes viscous at low temperature
20% H_3PO_3, 80% H_3PO_4	<20	-	20-250	None	Water soluble; nonflammable; corrosive; steam evolved at high temperature
Triethylene glycol	-5	287	0-250	156	Water soluble; stable
DC[b] 550 silicon oil	-60	-	0-250	310	Noncorrosive; can be used to 400° in N_2 atmosphere
Glycerol	18	290	-20-260	160	Supercools; water soluble; nontoxic; viscous
Paraffin	∿50	-	60-300	199	Flammable
H_2SO_4	10	330	20-300	None	Corrosive and hazardous; nonflammable
33% H_3PO_3, 67% H_3PO_4	-	-	125-340	None	Corrosive and hazardous; nonflammable
Dibutyl phthalate	-	340	150-320	-	Viscous at low temperature
Wood's metal (50% Bi, 25% Pb, 12.5% Sn, 12.5% Cd)	70	-	70-350	-	Oxidizes if used at >250° for long period of time
Mercury	-40	357	-35-350	None	Nonflammable; stable; vapor highly toxic
Tetracresyl silicate	<-48	∿440	20-400	-	Noncorrosive; nontoxic; fire resistant; expensive

(Continued)

Medium	Mp (°C)	Bp (°C)	Useful Range (°C)[a]	Flash Point (°C)	Comments
Sulfur	113	445	120-400	207	Flammable; highly viscous
51.3% KNO_3, 48.7% $NaNO_3$	219	-	230-500	None	Stable in air; strong oxidant; nonflammable
40% $NaNO_2$, 7% $NaNO_3$, 53% KNO_3	142	-	150-500	None	Noncorrosive; nonflammable, oxidizes slowly at high temperatures
Lead	327	1613	350-800	None	High specific gravity; vapor toxic
40% NaOH, 60% KOH	167	-	200-1000	None	Corrosive and hazardous

[a]Useful and safe range for bath open to atmosphere. [b]Dow Corning.

B. Special Liquids for Low Temperature

Special solvent mixtures with very low freezing points but with boiling points above or near room temperature are contained in this table. These mixtures are good solvents at low temperatures and when used as such often freeze substantially lower than the value reported here. Since most of these substances are highly chlorinated and/or can be obtained deuterated, they make excellent low-temperature nmr solvents.

Substance	Vol. %	Mp (°C)	Substance	Vol. %	Mp (°C)
Carbon tetrachloride Chloroform	49 51	-81	Chloroform Methylene chloride Ethyl bromide trans-1,2-Dichloroethylene Trichloroethylene	14.5 25.3 33.4 10.4 16.4	-150
Chloroform Trichloroethylene	31 69	-100			
Chloroform Methylene chloride Carbon tetrachloride	27 60 13	-111	Methyl chloride Dimethyl ether	25 75	-154
Chloroform trans-1,2-Dichloroethylene Trichloroethylene Ethyl bromide	20 14 21 45	-139	n-Pentane Methyl cyclohexane n-Propanol	64.5 24.4 11.1	<-180

C. Low-Temperature Baths

Two types of systems are shown in this table. One involves pouring liquid nitrogen (bp 196°) into a solvent by stirring until a slush is formed. The temperature may be maintained by periodically adding nitrogen to maintain the slush. These data are taken from R. E. Randeau, J. Chem. Eng. Data, 11, 124 (1966). The second system involves addition of small lumps of dry ice to the solvent until a slight excess of dry ice coated with frozen solvent remains. Again, temperature may be maintained by periodically adding more dry ice. These data are taken for the most part from A. M. Phipps and D. N. Hume, J. Chem. Educ., 45, 664 (1968). The latter reference describes the use of the dry ice method in mixtures of ortho and meta xylene of varying composition. Their plot of temperature versus volume fraction of o-xylene is reproduced at the end of this table.

System	°C	System	°C
p-Xylene/N_2	13	Carbitol acetate/CO_2	-67
p-Dioxane/N_2	12	t-Butyl amine/N_2	-68
Cyclohexane/N_2	6	Ethanol/CO_2	-72
Benzene/N_2	5	Trichloroethylene/N_2	-73
Formamide/N_2	2	Butyl acetate/N_2	-77
Aniline/N_2	-6	Acetone/CO_2	-77
Cycloheptane/N_2	-12	Isoamyl acetate/N_2	-79
Benzonitrile/N_2	-13	Acrylonitrile/N_2	-82
Ethylene glycol/CO_2	-15	Sulfur dioxide/CO_2	-82
o-Dichlorobenzene/N_2	-18	Ethyl acetate/N_2	-84
Tetrachloroethane/N_2	-22	Ethyl methyl ketone/N_2	-86
Carbon tetrachloride/N_2	-23	Acrolein/N_2	-88
Carbon tetrachloride/CO_2	-23	Nitroethane/N_2	-90
m-Dichlorobenzene/N_2	-25	Heptane/N_2	-91
Nitromethane/N_2	-29	Cyclopentane/N_2	-93
o-Xylene/N_2	-29	Hexane/N_2	-94
Bromobenzene/N_2	-30	Toluene/N_2	-95
Iodobenzene/N_2	-31	Methanol/N_2	-98
Thiophene/N_2	-38	Diethyl ether/CO_2	-100
3-Heptanone/CO_2	-38	n-Propyl iodide/N_2	-101
Acetonitrile/N_2	-41	n-Butyl iodide/N_2	-103
Pyridine/N_2	-42	Cyclohexene/N_2	-104
Acetonitrile/CO_2	-42	Isooctane/N_2	-107
Chlorobenzene/N_2	-45	Ethyl iodide/N_2	-109
Cyclohexanone/CO_2	-46	Carbon disulfide/N_2	-110
m-Xylene/N_2	-47	Butyl bromide/N_2	-112
n-Butyl amine/N_2	-50	Ethyl bromide/N_2	-119
Diethyl carbitol/CO_2	-52	Acetaldehyde/N_2	-124
n-Octane/N_2	-56	Methyl cyclohexane/N_2	-126
Chloroform/CO_2	-61(-77)	n-Pentane/N_2	-131
Chloroform/N_2	-63	1,5-Hexadiene/N_2	-141
Methyl iodide/N_2	-66	i-Pentane/N_2	-160

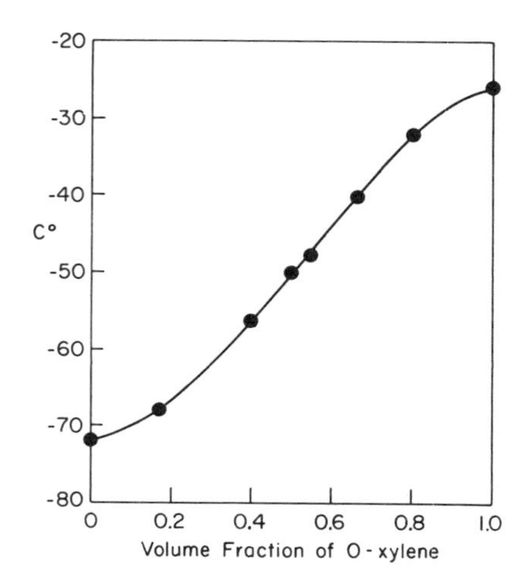

Steady state temperature of dry-ice—o-xylene, *m*-xylene mixtures.

D. Salt-Ice Cooling Mixtures

This table lists salt/ice cooling mixtures that can be obtained by mixing the salt (at about room temperature) with water or ice at the specified temperature in the amount noted. In actual practice, these temperatures are often difficult to reach and may depend on rate of stirring and on how finely crushed the ice is.

Substance	Initial Temperature (°C)	g/100g H_2O	Final Temperature (°C)
Na_2CO_3	-1 (ice)	20	-2.0
NH_4NO_3	20	106	-4.0
$NaC_2H_3O_2$	10.7	85	-4.7
NH_4Cl	13.3	30	-5.1
$NaNO_3$	13.2	75	-5.3
$Na_2S_2O_3 \cdot 5H_2O$	10.7	110	-8.0
$CaCl_2 \cdot 6H_2O$	-1 (ice)	41	-9.0
KCl	0 (ice)	30	-10.9
KI	10.8	140	-11.7
NH_4NO_3	13.6	60	-13.6
NH_4Cl	-1 (ice)	25	-15.4
NH_4NO_3	-1 (ice)	45	-16.8
NH_4SCN	13.2	133	-18.0
NaCl	-1 (ice)	33	-21.3
$CaCl_2 \cdot 6H_2O$	0 (ice)	81	-21.5
H_2SO_4 (66.2%)	0 (ice)	23	-25
NaBr	0 (ice)	66	-28
H_2SO_4 (66.2%)	0 (ice)	40	-30
C_2H_5OH (4°)	0 (ice)	105	-30
$MgCl_2$	0 (ice)	85	-34
H_2SO_4 (66.2%)	0 (ice)	91	-37
$CaCl_2 \cdot 6H_2O$	0 (ice)	123	-40.3
$CaCl_2 \cdot 6H_2O$	0 (ice)	143	-55

XII. MOLECULAR WEIGHT DETERMINATION

The most accurate method, but requiring the most expensive instrumentation, is mass spectrometry. Most determinations are still made by methods taking advantage of freezing and melting point depression, and boiling point elevation of a solvent (ebulliometry), to which the unknown has been admixed. Molecular weight is calculated in each case from:

$$mw = \frac{100Kw}{W\ \Delta t}$$

where K = molal depression or elevation constant of the solvent, w = weight of solute, W = weight of solvent, and Δt = depression or elevation (degrees). As little as 20 mg of solute is sufficient. The Rast method uses a solid "solvent," such as camphor, to determine the mw of nonvolatile solids by melting point depression; it is easier to perform and more accurate than the cryoscopic method which depends on freezing point depression of a liquid solvent. The following data are taken from Organic Structure Determination, by Pasto and Johnson, Prentice-Hall, Englewood Cliffs, N.J., 1969, pages 73ff, and The Synthesis and Characterization of Inorganic Compounds, by Jolly, Prentice-Hall, Englewood Cliffs, N.J., 1970, page 284; these references discuss the general topic of molecular weight measurements.

A recent review of molecular weight determination techniques, including the newly developed mass chromatography method, is found in C. E. Bennett et al., American Laboratory, May 1971, page 67.

Rast Solvent	Mp (°C)	K (°C)
Benzophenone	48	9.8
Biphenyl	70	8.0
Borneol	202	35.8
Bornylamine	164	40.6
Camphene	49	31.08
Camphoquinone	190	45.7
Camphor[a]	178	39.7
Cyclopentadecanone	65.6	21.3
Naphthalene	80.2	6.9
Perylene	276	25.7
Tribromophenol	96.0	20.4

Cryoscopic Solvent	Fp (°C)	K (°C)
Acetic acid	16.7	3.9
Ammonia	-77.7	0.957
Benzene	5.5	5.12
Cyclohexane	6.5	20.0
Sulfuric acid	10.5	6.81
Water	0.0	1.86

[a]Value of K decreases if concentration is less than ≈0.2M in solute.

Ebullioscopic Solvent	Bp (°C)	K (°C)
Acetic acid	118	3.07
Acetone	56	1.71
Benzene	80	2.53
Carbon tetrachloride	76	5.03
Chlorobenzene	132	4.15
Chloroform	60	3.63
Cyclohexane	81	2.79
Ethanol	78	1.22
Nitrobenzene	210	5.24
n-Octane	126	4.02
Toluene	110	3.33
Water	100	0.512

In the isopiestic method, a solution of a standard substance, s, is equilibrated in a closed system at constant temperature with a solution of the unknown, x, until the vapor pressures above the respective solutions are equalized. At equilibrium, the mole fractions of s and x must be equal in the solutions. The mw is calculated from the volume change. If M = molecular weight, W = weight, V_s, V_x = final volume of the two solutions, and d = solvent density, then

$$\frac{W_x/M_x}{W_x M_x + V_x d/M_{solv}} = \frac{W_s/M_s}{W_s/M_s + V_s d/M_{solv}}$$

A related technique, the vapor phase osmometric method, depends on the vaporization-condensation transfer of pure solvent to a separate solution of the unknown in the same solvent. This transfer is adiabatic, producing a temperature difference between pure solvent and solution due to the heat of vaporization generated during the process. The difference is proportional to the vapor pressure difference between solvent and solution, hence is proportional to the solute concentration and dependent on solvent type. For details, write for information on the Mechrolab Vapor Pressure Osmometer (Hewlett-Packard, F and M Scientific Division, Route 41, Avondale, Pa. 19311).

MATHEMATICAL AND NUMERICAL INFORMATION

I. APPROVED INTERNATIONAL UNITS SYSTEM AND GENERAL CONSTANTS

In addition to the references cited below, the reader is referred to the thorough work by M. L. McGlashan: Physico-Chemical Quantities and Units, Royal Institute of Chemistry, London, 1968, and to the definitive IUPAC work, Manual of Symbols and Terminology for Physicochemical Quantities and Units, Butterworth, London, 1970.

A. SI Units

The Systēme Internationale d'Unities (International System of Units or SI) has been adopted (at least in principle) by most of the world's organizations of authority in units. For a detailed review, see E. A. Mechtly, The International System of Units, NASA Publication, NASA SP-7012 (1970), which may be ordered for $0.55 from the U. S. Government Printing Office, Washington, D. C. 20402, as document NAS 1.21:7012/2. The National Bureau of Standards has a similar publication which can be ordered as S.D. Catalog No. C 13.10:330 (cost $0.50); brief discussions are found in G. Socrates, J. Chem. Educ., 46, 710 (1969), and in J. Chem. Educ., 48, 569 (1971). As the standard metric system is referred to as CGS, SI is called MKSA (meter, kilogram, second, ampere).

1. Basic and Derived Units

Quantity	Unit	Symbol	Quantity	Unit	Symbol
Length	meter	m	Luminous intensity	candela	cd
Mass	kilogram	kg	Electric current	ampere	A
Time	second	s	Plane angle	radian	rad
Thermodynamic temperature	kelvin	K	Solid angle	steradian	sr
Acceleration		m/s^2	Luminance		cd/m^2
Angular acceleration		rad/s^2	Luminous flux	lumen	lm
Angular velocity		rad/s	Magnetomotive force	ampere	A
Area		m^2	Magnetic field strength		A/m
Density		kg/m^3	Magnetic flux	weber	Wb
Dynamic viscosity		Ns/m^2	Magnetic flux density	tesla	T
Electric capacitance	farad	F	Power	watt	W
Electric charge	coulomb	C	Pressure		N/m^2
Electric field strength		V/m	Radiant intensity		W/sr
Electric resistance	ohm	Ω	Radioactivity (counts)		s^{-1}
Energy (work, heat)	joule	J	Specific heat		J/kg K
Entropy		J/K	Thermal conductivity		W/mK
Force	newton	N	Velocity		m/s
Frequency	hertz	Hz	Volume		m^3
Illumination	lux	lx	Voltage (EMF)	volt	V
Inductance	henry	H	Wave number		m^{-1}
Kinematic viscosity		m^2/s			

2. Recommended Rules of Usage

a. Periods are not used after abbreviations (g, cm, m; not g., cm., m.).
b. Plurals are not used (10 kg; not 10 kgs).
c. Group digits in threes about the decimal point (or decimal comma) using a space, not a comma (2 457 623.049 2; not 2,457,623.0492).
d. Avoid use of a solidus (/); if needed, use only one (J/K mol; not J/K/mol).
e. Omit the degree sign for kelvin scale (50 K; not 50° K).

3. Defined Values of Basic Units

Meter (m): 1 650 763.73 wavelengths of 2 p_{10}-5 d_5 transition in ^{86}Kr.
Second (s): the duration of 9 192 631 770 periods of the radiation corresponding to the transition between the two hyperfine levels of the fundamental state of ^{133}Cs. The "ephemeris" second is 1/31 556 925.974 7 of the tropical year 1900.
Kilogram (kg): mass of the international kilogram (Sèvres, France), a cylinder of platinum-iridium alloy.
Kelvin (K): unit of thermodynamic temperature; 273.16 K ≡ triple point of water.
Liter (l): 0.001 m^3 = 1000 cm^3.
Mole (mol): amount of matter having the same number of formula units as atoms in 0.012 kg of ^{12}C.
Unified atomic mass unit (u): 1/12 the mass of an atom of ^{12}C = 1.660 43 x 10^{-24} g.
Standard free fall acceleration (g_n): 9.806 65 ms^{-2} = 980.665 cm s^{-2}.
Normal atmospheric pressure (atm): 101 325 N m^{-2} = 1 013 250 dyn cm^{-2}.
Thermochemical calorie (cal): 4.1840 J = 4.1840 x 10^7 erg.
International Steam Table calorie (cal_{IT}): 4.1868 J = 4.1868 x 10^7 erg.
Inch (in.): 0.0254 m = 2.54 cm.
Pound (avdp): 0.453 592 37 kg = 453.592 37 g.

4. Prefixes for Fractions and Multiples of Units[a,b]

Prefix	Symbol	Factor (x in 10^x)	Prefix	Symbol	Factor (x in 10^x)
tera	T	12	centi[c]	c	-2
giga	G	9	milli	m	-3
mega	M	6	micro[d]	μ	-6
kilo	k	3	nano	n	-9
hecto[c]	h	2	pico	p	-12
deka[c]	da	1	femto	f	-15
deci[c]	d	-1	atto	a	-18

[a]Use only one prefix for a unit (1000 kg = 1 Mg not 1 k kg).
[b]Prefixes used should differ from a unit in steps of 10^3.
[c]Use only if recommended prefixes[b] are inconvenient.
[d]The symbol μ is no longer used for the length unit micron (10^{-6} m), but is used only as a prefix. The old 1 micron is now 1 μm (as used, e.g., for infrared wavelengths).

B. Physical and Chemical Constants

1. Recommended Values

The following values are considered to be the most accurate and up to date, based on a paper by Taylor, Parker and Langenberg, Rev. Mod. Phys., July 1969. See Mechtly, op. cit., page 7.

Constant (Symbol)	Value	Units[a] SI	CGS
Avogadro constant (N_A)	6.022 169	$x10^{23}$ mol^{-1}	$x10^{23}$ mol^{-1}
Bohr radius (a_o)	5.291 771 5	$x10^{-11}$ m	$x10^{-9}$ cm
Bohr magneton (μ_B)	9.274 096	$x10^{-24}$ JT^{-1}	$x10^{-21}$ erg G^{-1}[c]
Boltzmann constant (k)[b]	1.380 622	$x10^{-23}$ JK^{-1}	$x10^{-16}$ erg K^{-1}
Electron rest mass (m_e)	9.109 558	$x10^{-31}$ kg	$x10^{-28}$ g
Elementary charge (e)	5.485 97	$x10^{-4}$ u	$x10^{-4}$ u
	1.602 191 7	$x10^{-19}$ C	$x10^{-20}$ $cm^{1/2}$ $g^{1/2}$[c]
	4.802 98	--	$x10^{-10}$ $cm^{3/2}$ $g^{1/2}$ s^{-1}[d]
Electron charge/ mass (e/m)	1.758 802 8	$x10^{11}$ C kg^{-1}	$x10^{7}$ $cm^{1/2}$ $g^{-1/2}$[c]
	5.272 74	--	$x10^{17}$ $cm^{3/2}$ $g^{-1/2}$ s^{-1}[d]
Electron radius (r_e)	2.817 939	$x10^{-15}$ m	$x10^{-13}$ cm
($r^2{}_e$)	7.940 780	$x10^{-30}$ m^2	$x10^{-26}$ cm^2
Electron volt (1 eV)	1.602 10	$x10^{-19}$ J	$x10^{-12}$ erg
Faraday constant (F)	9.648 670	$x10^{4}$ C mol^{-1}	$x10^{3}$ $cm^{1/2}$ $g^{1/2}$ mol^{-1}[c]
Fine structure constant (α)	7.297 351	$x10^{-3}$	$x10^{-3}$
Gas constant (R)[e]	8.314 34	$x10^{0}$ JK^{-1} mol^{-1}	$x10^{7}$ erg K^{-1} mol^{-1}
Gravitational constant (G)	6.673 2	$x10^{-11}$ Nm^2 kg^{-2}	$x10^{-8}$ dyn cm^2 g^{-2}
Gyromagnetic ratio of protons in water (γ) (corrected for diamagnetism of H_2O)	2.675 196 5	$x10^{8}$ rad s^{-1} T^{-1}	$x10^{4}$ rad s^{-1} G^{-1}[c]
Molar volume (ideal gas) (V_o)	2.241 36	$x10^{-2}$ m^3 mol^{-1}	$x10^{4}$ cm^3 mol^{-1}

(Continued)

Constant (Symbol)	Value	Units[a]	
		SI	CGS
Neutron rest mass (m_n)	1.674 920	$\times 10^{-27}$ kg	$\times 10^{24}$ g
	1.008 665 20	$\times 10^{0}$ u	$\times 10^{0}$ u
Nuclear magneton (μ_N)	5.050 951	$\times 10^{-27}$ JT^{-1}	$\times 10^{-24}$ erg G^{-1} [c]
Planck constant (h)[f]	6.626 196	$\times 10^{-34}$ Js	$\times 10^{-27}$ erg s
($h/2\pi$)	1.054 591 9	$\times 10^{-34}$ Js	$\times 10^{-27}$ erg s
Proton moment (μ_p)	1.410 620 3	$\times 10^{-26}$ JT^{-1}	$\times 10^{-23}$ erg G^{-1} [c]
Proton rest mass (m_p)	1.672 614	$\times 10^{-27}$ kg	$\times 10^{-24}$ g
	1.007 276 61	$\times 10^{0}$ u	$\times 10^{0}$ u
Rydberg constant (R_∞)	1.097 373 12	$\times 10^{7}$ m^{-1}	$\times 10^{5}$ cm^{-1}
Speed of light in vacuum (c)	2.997 925 0	$\times 10^{8}$ m s^{-1}	$\times 10^{10}$ cm s^{-1}
Stefan-Boltzmann constant (σ)	5.669 61	$\times 10^{-8}$ Wm^{-2} K^{-4}	$\times 10^{-5}$ erg cm^{-2} s^{-1} K^{-4}
Wien displacement constant (b)	2.897	$\times 10^{-3}$ mK	$\times 10^{-1}$ cm K
Zeeman splitting constant (μ_B/hc)	4.668 59	$\times 10^{1}$ m^{-1} T^{-1}	$\times 10^{-5}$ cm^{-1} G^{-1} [c]

[a] C = coulomb, G = gauss, J = joule, K = °Kelvin, N = newton, T = tesla, u = atomic mass unit, W = watt.

[b] Other useful value: 3.298×10^{-24} cal K^{-1} $molecule^{-1}$.

[c] Electromagnetic system.

[d] Electrostatic system.

[e] Other useful values: 1.9872 cal K^{-1} mol^{-1}; 8.2053×10^{-2} ℓ atm K^{-1} mol^{-1}.

[f] Other useful values: 9.534×10^{-14} k cal-s/mol = 1.583×10^{-34} cal-s/molecule.

2. Miscellaneous Constants and Units

Natural log base (e)	2.71828
pi (π)	3.14159
curie	3.70×10^{10} disintegrations per s (dps) = 2.22×10^{12} dpm

(Continued)

roentgen	1 esu charge per cm^3 (0.001293 g) dry air at STP; 1.61 x 10^{12} ion pairs/g air; charge equivalent to 2 x 10^9 electrons/cm^3 air
rad	100 erg/g
rep	93 erg/g absorbed in tissue
1 barn	10^{-24} cm^2
96,500 coulombs	6.02 x 10^{23} electrons
1 gamma	10^{-6} g (μg)
1 lambda (λ)	10^{-6} ℓ (10^{-3} mℓ)

3. Size Scale of the Universe: Orders of Magnitude

The following scale shows in cm the approximate sizes of objects and distances in the universe. For an expanded scale of the microscopic range, see page 427.

Atomic nucleus	10^{-11}
Atom	10^{-8}
Bacterial virus	10^{-5}
Grain of sand	10^{-2}
Mouse	10
Man	10^2
Skyscraper	10^4
Mt. Everest	10^6
Earth diameter	10^9
Sun diameter	10^{11}
Distance from earth to sun	10^{13}
Distance from earth to nearest star	10^{19}
Diameter of a galaxy	10^{23}
Distance to farthest observable galaxy	10^{28}

II. USEFUL CONVERSION FACTORS

The tables in Sections A, B, and C below are to be used as illustrated in the following example from Section A: 1 cm (from column on far left) is equivalent to 0.3937 inch or 3.281 x 10^{-2} foot; expressed another way, there is 0.3937 in./cm or 0.003281 ft/cm.

A. Length, Area, Volume

	m	cm	mm	nm[a]	Å	in.	ft
1 m	1	10^2	10^3	10^9	10^{10}	39.37	3.281
1 cm	10^{-2}	1	10	10^7	10^8	0.3937	$3.281x10^{-2}$
1 mm	10^{-3}	10^{-1}	1	10^6	10^7	$3.937x10^{-2}$	$3.281x10^{-3}$
1 nm	10^{-9}	10^{-7}	10^{-6}	1	10	$3.937x10^{-8}$	$3.281x10^{-9}$
1 Å	10^{-10}	10^{-8}	10^{-7}	10^{-1}	1	$3.937x10^{-9}$	$3.281x10^{-10}$
1 in.	0.0254	2.54	25.40	$2.54x10^7$	$2.54x10^8$	1	$8.333x10^{-2}$
1 ft	0.3048	30.48	304.8	$3.048x10^8$	$3.048x10^9$	12	1

[a]Nanometer; formerly called a millimicron (mμ).

	m^2	cm^2	mm^2	$in.^2$	ft^2	mi^2
1 m^2	1	10^4	10^6	$1.55x10^3$	10.764	$3.861x10^{-7}$
1 cm^2	10^{-4}	1	10^2	0.155	$1.076x10^{-3}$	$3.861x10^{-11}$
1 mm^2	10^{-6}	10^{-2}	1	$1.55x10^{-3}$	$1.076x10^{-5}$	$3.861x10^{-13}$
1 $in.^2$	$6.452x10^{-4}$	6.452	645.2	1	$6.944x10^{-3}$	$2.49x10^{-10}$
1 ft^2	$9.29x10^{-2}$	929.0	$9.29x10^4$	144	1	$3.587x10^{-8}$
1 mi^2	$2.59x10^6$	$2.59x10^{10}$	$2.59x10^{12}$	$4.01x10^9$	$2.788x10^7$	1
1 Acre	$4.05x10^3$	$4.05x10^7$	$4.05x10^9$	$6.28x10^6$	$4.36x10^4$	$1.56x10^{-3}$

	m^3	cm^3	mm^3	$in.^3$	ft^3	liter
1 m^3	1	10^6	10^9	$6.10x10^4$	35.31	10^3
1 cm^3	10^{-6}	1	10^3	$6.10x10^{-2}$	$3.53x10^{-5}$	10^{-3}
1 mm^3	10^{-9}	10^{-3}	1	$6.10x10^{-5}$	$3.53x10^{-8}$	10^{-6}
1 $in.^3$	$1.639x10^{-5}$	16.39	$1.639x10^4$	1	$5.79x10^{-4}$	$1.639x10^{-2}$

(Continued)

	m^3	cm^3	mm^3	in^3	ft^3	liter
1 ft^3	2.832×10^{-2}	2.832×10^4	2.832×10^7	1728	1	28.316
1 liter	10^{-3}	10^3	10^6	61.02	3.53×10^{-2}	1
1 pint[a]	4.73×10^{-4}	473.18	4.73×10^5	28.88	1.67×10^{-2}	0.4731
1 quart[a]	9.46×10^{-4}	946.35	9.46×10^5	57.75	3.34×10^{-2}	0.9463
1 gallon[a]	3.78×10^{-3}	3.78×10^3	3.78×10^6	231	0.1334	3.785
1 ounce[a]	2.96×10^{-5}	29.57	2.96×10^4	1.805	1.04×10^{-3}	2.957×10^{-2}

[a]U.S., fluid. 1 British pint, quart, or gallon = 1.20 U.S. pints, quarts, or gallons; 1 British ounce = 0.9608 U.S. ounce.

B. Mass

	g	kg	mg	grain	ounce[a]	pound[a]
1 g	1	10^{-3}	10^3	15.432	3.527×10^{-2}	2.205×10^{-3}
1 kg	10^3	1	10^6	1.543×10^4	35.27	2.205
1 mg	10^{-3}	10^{-6}	1	1.543×10^2	3.527×10^{-5}	2.205×10^{-6}
1 grain	6.48×10^{-2}	6.48×10^{-5}	64.8	1	2.29×10^{-3}	1.43×10^{-4}
1 ounce[a]	28.3495	2.83×10^{-2}	2.83×10^4	437.5	1	6.25×10^{-2}
1 pound[a]	453.592	0.45359	4.53×10^5	7000	16	1
1 ton[b]	10^6	10^3	10^9	1.543×10^7	3.527×10^4	2.205×10^3

[a]Avoirdupoir (avdp.).
[b]Metric. "Short" ton (U.S.) = 2000 pounds = 907.18 kg.

C. Energy, Frequency, Wavelength

For useful equations relating energy to frequency and wavelength, see page 166.

	cm^{-1}	J/mol	eV[a]	cal/mol[a]	MHz
1 m	10^2	8.35911	8.0658×10^5	3.4977×10^2	4.3294×10^{-3}
1 cm^{-1}	1	11.963	1.2398×10^{-4}	2.859	2.9979×10^4
1 erg/ molecule	5.0348×10^{15}	6.025×10^{16}	6.2421×10^{11}	1.43956×10^{16}	1.5094×10^{20}

(Continued)

	cm^{-1}	J/mol	eV[a]	cal/mol[a]	MHz
1 J/mol	8.3591×10^{-2}	1	1.036×10^{-5}	0.239006	2.506×10^{3}
1 eV[a]	8.0658×10^{3}	9.6525×10^{4}	1	2.306×10^{4}	2.4181×10^{8}
1 cal/mol[a]	0.34975	4.1840	4.3361×10^{-5}	1	1.0485×10^{4}
1 MHz	3.3356×10^{-5}	3.9904×10^{-4}	4.1355×10^{-9}	9.537×10^{-5}	1
1 kT (1 K)	0.69501	8.3144	8.6169×10^{-5}	1.9872	2.0836×10^{4}

[a]Atomic units of energy (au), known as hartrees, are converted as follows: 0.627709 au = 1 cal/mol; 27.2097 au = 1 eV.

D. Temperature

The units of temperature are Celsius (or centigrade), °C, and Kelvin, °K, in the metric system, and Fahrenheit, °F, and Rankine, °R, in the English system. Kelvin and Rankine units represent "absolute temperature" scales, derived from thermodynamics and the concept of an absolute zero. Celsius and Fahrenheit units are the basis of empirical temperature scales, based on the ice point (0°C, 32°F) and steam point (100°C, 212°F) of water.

$$°C = \frac{5}{9}(°F - 32) \qquad °K = \frac{5}{9}(°R)$$

$$°F = \frac{9}{5}(°C) + 32 \qquad °R = \frac{9}{5}(°K)$$

$$°K = °C + 273.15 = \frac{5}{9}(°F) + 255.37$$

$$°R = °F + 459.67 = \frac{9}{5}(°C) + 491.67$$

E. Pressure, Force, Power

1 atm = 1.01325 bars = 760 mm Hg (0°) = 1.01325×10^6 dynes/cm^2 = 14.696 psi

1 torr = 1 mm Hg (0°) = 133.322 N/m^2

1 bar = 105 N/m^2

1 N = 10^5 dynes = 0.22481 pounds (lb) = 7.2330 poundals = 101.971 g

1 g-cm = 9.30113×10^{-8} Btu = 2.34385×10^{-5} cal = 980.665 dyne-cm = 7.23301×10^{-5} ft-lb = 3.65303×10^{-11} hp-hr = 9.80665 Nm

1 watt (W) = 3.41443 Btu/hr = 860.421 cal/hr = 10^7 erg/s = 1 J/s

1 hp = 745.700 W

III. WAVELENGTH-WAVENUMBER CONVERSION TABLE

The table is reprinted by permission from K. Nakanishi, Infrared Absorption Spectroscopy-Practical, Holden-Day, San Francisco, 1967, and Nankodo Company Ltd., Tokyo, 1967. An example of the use of this table is as follows: a wavenumber of 3185 cm^{-1} is equivalent to 3.14 μm.

Wavelength (μm)	Wave-number (cm^{-1})									
	0	1	2	3	4	5	6	7	8	9
2.0	5000	4975	4950	4926	4902	4878	4854	4831	4808	4785
2.1	4762	4739	4717	4695	4673	4651	4630	4608	4587	4566
2.2	4545	4525	4505	4484	4464	4444	4425	4405	4386	4367
2.3	4348	4329	4310	4292	4274	4255	4237	4219	4202	4184
2.4	4167	4149	4132	4115	4098	4082	4065	4049	4032	4016
2.5	4000	3984	3968	3953	3937	3922	3906	3891	3876	3861
2.6	3846	3831	3817	3802	3788	3774	3759	3745	3731	3717
2.7	3704	3690	3676	3663	3650	3636	3623	3610	3597	3584
2.8	3571	3559	3546	3534	3521	3509	3497	3484	3472	3460
2.9	3448	3436	3425	3413	3401	3390	3378	3367	3356	3344
3.0	3333	3322	3311	3300	3289	3279	3268	3257	3247	3236
3.1	3226	3215	3205	3195	3185	3175	3165	3155	3145	3135
3.2	3125	3115	3106	3096	3086	3077	3067	3058	3049	3040
3.3	3030	3021	3012	3003	2994	2985	2976	2967	2959	2950
3.4	2941	2933	2924	2915	2907	2899	2890	2882	2874	2865
3.5	2857	2849	2841	2833	2825	2817	2809	2801	2793	2786
3.6	2778	2770	2762	2755	2747	2740	2732	2725	2717	2710
3.7	2703	2695	2688	2681	2674	2667	2660	2653	2646	2639
3.8	2632	2625	2618	2611	2604	2597	2591	2584	2577	2571
3.9	2564	2558	2551	2545	2538	2532	2525	2519	2513	2506
4.0	2500	2494	2488	2481	2475	2469	2463	2457	2451	2445
4.1	2439	2433	2427	2421	2415	2410	2404	2398	2392	2387
4.2	2381	2375	2370	2364	2358	2353	2347	2342	2336	2331
4.3	2326	2320	2315	2309	2304	2299	2294	2288	2283	2278
4.4	2273	2268	2262	2257	2252	2247	2242	2237	2232	2227
4.5	2222	2217	2212	2208	2203	2198	2193	2188	2183	2179
4.6	2174	2169	2165	2160	2155	2151	2146	2141	2137	2132
4.7	2128	2123	2119	2114	2110	2105	2101	2096	2092	2088
4.8	2083	2079	2075	2070	2066	2062	2058	2053	2049	2045
4.9	2041	2037	2033	2028	2024	2020	2016	2012	2008	2004
5.0	2000	1996	1992	1988	1984	1980	1976	1972	1969	1965
5.1	1961	1957	1953	1949	1946	1942	1938	1934	1931	1927
5.2	1923	1919	1916	1912	1908	1905	1901	1898	1894	1890
5.3	1887	1883	1880	1876	1873	1869	1866	1862	1859	1855
5.4	1852	1848	1845	1842	1838	1835	1832	1828	1825	1821
5.5	1818	1815	1812	1808	1805	1802	1799	1795	1792	1789
5.6	1786	1783	1779	1776	1773	1770	1767	1764	1761	1757
5.7	1754	1751	1748	1745	1742	1739	1736	1733	1730	1727
5.8	1724	1721	1718	1715	1712	1709	1706	1704	1701	1698
5.9	1695	1692	1689	1686	1684	1681	1678	1675	1672	1669
	0	1	2	3	4	5	6	7	8	9

Wavelength (μm)	Wave-number (cm^{-1})									
	0	1	2	3	4	5	6	7	8	9
6.0	1667	1664	1661	1658	1656	1653	1650	1647	1645	1642
6.1	1639	1637	1634	1631	1629	1626	1623	1621	1618	1616
6.2	1613	1610	1608	1605	1603	1600	1597	1595	1592	1590
6.3	1587	1585	1582	1580	1577	1575	1572	1570	1567	1565
6.4	1563	1560	1558	1555	1553	1550	1548	1546	1543	1541
6.5	1538	1536	1534	1531	1529	1527	1524	1522	1520	1517
6.6	1515	1513	1511	1508	1506	1504	1502	1499	1497	1495
6.7	1493	1490	1488	1486	1484	1481	1479	1477	1475	1473
6.8	1471	1468	1466	1464	1462	1460	1458	1456	1453	1451
6.9	1449	1447	1445	1443	1441	1439	1437	1435	1433	1431
7.0	1429	1427	1425	1422	1420	1418	1416	1414	1412	1410
7.1	1408	1406	1404	1403	1401	1399	1397	1395	1393	1391
7.2	1389	1387	1385	1383	1381	1379	1377	1376	1374	1372
7.3	1370	1368	1366	1364	1362	1361	1359	1357	1355	1353
7.4	1351	1350	1348	1346	1344	1342	1340	1339	1337	1335
7.5	1333	1332	1330	1328	1326	1325	1323	1321	1319	1318
7.6	1316	1314	1312	1311	1309	1307	1305	1304	1302	1300
7.7	1299	1297	1295	1294	1292	1290	1289	1287	1285	1284
7.8	1282	1280	1279	1277	1276	1274	1272	1271	1269	1267
7.9	1266	1264	1263	1261	1259	1258	1256	1255	1253	1252
8.0	1250	1248	1247	1245	1244	1242	1241	1239	1238	1236
8.1	1235	1233	1232	1230	1229	1227	1225	1224	1222	1221
8.2	1220	1218	1217	1215	1214	1212	1211	1209	1208	1206
8.3	1205	1203	1202	1200	1199	1198	1196	1195	1193	1192
8.4	1190	1189	1188	1186	1185	1183	1182	1181	1179	1178
8.5	1176	1175	1174	1172	1171	1170	1168	1167	1166	1164
8.6	1163	1161	1160	1159	1157	1156	1155	1153	1152	1151
8.7	1149	1148	1147	1145	1144	1143	1142	1140	1139	1138
8.8	1136	1135	1134	1133	1131	1130	1129	1127	1126	1125
8.9	1124	1122	1121	1120	1119	1117	1116	1115	1114	1112
9.0	1111	1110	1109	1107	1106	1105	1104	1103	1101	1100
9.1	1099	1098	1096	1095	1094	1093	1092	1091	1089	1088
9.2	1087	1086	1085	1083	1082	1081	1080	1079	1078	1076
9.3	1075	1074	1073	1072	1071	1070	1068	1067	1066	1065
9.4	1064	1063	1062	1060	1059	1058	1057	1056	1055	1054
9.5	1053	1052	1050	1049	1048	1047	1046	1045	1044	1043
9.6	1042	1041	1040	1038	1037	1036	1035	1034	1033	1032
9.7	1031	1030	1029	1028	1027	1026	1025	1024	1022	1021
9.8	1020	1019	1018	1017	1016	1015	1014	1013	1012	1011
9.9	1010	1009	1008	1007	1006	1005	1004	1003	1002	1001
	0	1	2	3	4	5	6	7	8	9

Wavelength (μm)	Wave-number (cm^{-1})									
	0	1	2	3	4	5	6	7	8	9
10.0	1000.0	999.0	998.0	997.0	996.0	995.0	994.0	993.0	992.1	991.1
10.1	990.1	989.1	988.1	987.2	986.2	985.2	984.3	983.3	982.3	981.4
10.2	980.4	979.4	978.5	977.5	976.6	975.6	974.7	973.7	972.8	971.8
10.3	970.9	969.9	969.0	968.1	967.1	966.2	965.3	964.3	963.4	962.5
10.4	961.5	960.6	959.7	958.8	957.9	956.9	956.0	955.1	954.2	953.3
10.5	952.4	951.5	950.6	949.7	948.8	947.9	947.0	946.1	945.2	944.3
10.6	943.4	942.5	941.6	940.7	939.8	939.0	938.1	937.2	936.3	935.5
10.7	934.6	933.7	932.8	932.0	931.1	930.2	929.4	928.5	927.6	926.8
10.8	925.9	925.1	924.2	923.4	922.5	921.7	920.8	920.0	919.1	918.3
10.9	917.4	916.6	915.8	914.9	914.1	913.2	912.4	911.6	910.7	909.9
11.0	909.1	908.3	907.4	906.6	905.8	905.0	904.2	903.3	902.5	901.7
11.1	900.9	900.1	899.3	898.5	897.7	896.9	896.1	895.3	894.5	893.7
11.2	892.9	892.1	891.3	890.5	889.7	888.9	888.1	887.3	886.5	885.7
11.3	885.0	884.2	883.4	882.6	881.8	881.1	880.3	879.5	878.7	878.0
11.4	877.2	876.4	875.7	874.9	874.1	873.4	872.6	871.8	871.1	870.3
11.5	869.6	868.8	868.1	867.3	866.6	865.8	865.1	864.3	863.6	862.8
11.6	862.1	861.3	860.6	859.8	859.1	858.4	857.6	856.9	856.2	855.4
11.7	854.7	854.0	853.2	852.5	851.8	851.1	850.3	849.6	848.9	848.2
11.8	847.5	846.7	846.0	845.3	844.6	843.9	843.2	842.5	841.8	841.0
11.9	840.3	839.6	838.9	838.2	837.5	836.8	836.1	835.4	834.7	834.0
12.0	833.3	832.6	831.9	831.3	830.6	829.9	829.2	828.5	827.8	827.1
12.1	826.4	825.8	825.1	824.4	823.7	823.0	822.4	821.7	821.0	820.3
12.2	819.7	819.0	818.3	817.7	817.0	816.3	815.7	815.0	814.3	813.7
12.3	813.0	812.3	811.7	811.0	810.4	809.7	809.1	808.4	807.8	807.1
12.4	806.5	805.8	805.2	804.5	803.9	803.2	802.6	801.9	801.3	800.6
12.5	800.0	799.4	798.7	798.1	797.4	796.8	796.2	795.5	794.9	794.3
12.6	793.7	793.0	792.4	791.8	791.1	790.5	789.9	789.3	788.6	788.0
12.7	787.4	786.8	786.2	785.5	784.9	784.3	783.7	783.1	782.5	781.9
12.8	781.3	780.6	780.0	779.4	778.8	778.2	777.6	777.0	776.4	775.8
12.9	775.2	774.6	774.0	773.4	772.8	772.2	771.6	771.0	770.4	769.8
13.0	769.2	768.6	768.0	767.5	766.9	766.3	765.7	765.1	764.5	763.9
13.1	763.4	762.8	762.2	761.6	761.0	760.5	759.9	759.3	758.7	758.2
13.2	757.6	757.0	756.4	755.9	755.3	754.7	754.1	753.6	753.0	752.4
13.3	751.9	751.3	750.8	750.2	749.6	749.1	748.5	747.9	747.4	746.8
13.4	746.3	745.7	745.2	744.6	744.0	743.5	742.9	742.4	741.8	741.3
13.5	740.7	740.2	739.6	739.1	738.6	738.0	737.5	736.9	736.4	735.8
13.6	735.3	734.8	734.2	733.7	733.1	732.6	732.1	731.5	731.0	730.5
13.7	729.9	729.4	728.9	728.3	727.8	727.3	726.7	726.2	725.7	725.2
13.8	724.6	724.1	723.6	723.1	722.5	722.0	721.5	721.0	720.5	719.9
13.9	719.4	718.9	718.4	717.9	717.4	716.8	716.3	715.8	715.3	714.8
14.0	714.3	713.8	713.3	712.8	712.3	711.7	711.2	710.7	710.2	709.7
14.1	709.2	708.7	708.2	707.7	707.2	706.7	706.2	705.7	705.2	704.7
14.2	704.2	703.7	703.2	702.7	702.2	701.8	701.3	700.8	700.3	699.8
14.3	699.3	698.8	698.3	697.8	697.4	696.9	696.4	695.9	695.4	694.9
14.4	694.4	694.0	693.5	693.0	692.5	692.0	691.6	691.1	690.6	690.1
14.5	689.7	689.2	688.7	688.2	687.8	687.3	686.8	686.3	685.9	685.4
14.6	684.9	684.5	684.0	683.5	683.1	682.6	682.1	681.7	681.2	680.7
14.7	680.3	679.8	679.3	678.9	678.4	678.0	677.5	677.0	676.6	676.1
14.8	675.7	675.2	674.8	674.3	673.9	673.4	672.9	672.5	672.0	671.6
14.9	671.1	670.7	670.2	669.8	669.3	668.9	668.4	668.0	667.6	667.1
	0	1	2	3	4	5	6	7	8	9

Wavelength (μm)	Wave-number (cm^{-1})									
	0	1	2	3	4	5	6	7	8	9
15.0	666.7	666.2	665.8	665.3	664.9	664.5	664.0	663.6	663.1	662.7
15.1	662.3	661.8	661.4	660.9	660.5	660.1	659.6	659.2	658.8	658.3
15.2	657.9	657.5	657.0	656.6	656.2	655.7	655.3	654.9	654.5	654.0
15.3	653.6	653.2	652.7	652.3	651.9	651.5	651.0	650.6	650.2	649.8
15.4	649.4	648.9	648.5	648.1	647.7	647.2	646.8	646.4	646.0	645.6
15.5	645.2	644.7	644.3	643.9	643.5	643.1	642.7	642.3	641.8	641.4
15.6	641.0	640.6	640.2	639.8	639.4	639.0	638.6	638.2	637.8	637.3
15.7	636.9	636.5	636.1	635.7	635.3	634.9	634.5	634.1	633.7	633.3
15.8	632.9	632.5	632.1	631.7	631.3	630.9	630.5	630.1	629.7	629.3
15.9	628.9	628.5	628.1	627.7	627.4	627.0	626.6	626.2	625.8	625.4
16.0	625.0	624.6	624.2	623.8	623.4	623.1	622.7	622.3	621.9	621.5
16.1	621.1	620.7	620.3	620.0	619.6	619.2	618.8	618.4	618.0	617.7
16.2	617.3	616.9	616.5	616.1	615.8	615.4	615.0	614.6	614.3	613.9
16.3	613.5	613.1	612.7	612.4	612.0	611.6	611.2	610.9	610.5	610.1
16.4	609.8	609.4	609.0	608.6	608.3	607.9	607.5	607.2	606.8	606.4
16.5	606.1	605.7	605.3	605.0	604.6	604.2	603.9	603.5	603.1	602.8
16.6	602.4	602.0	601.7	601.3	601.0	600.6	600.2	599.9	599.5	599.2
16.7	598.8	598.4	598.1	597.7	597.4	597.0	596.7	596.3	595.9	595.6
16.8	595.2	594.9	594.5	594.2	593.8	593.5	593.1	592.8	592.4	592.1
16.9	591.7	591.4	591.0	590.7	590.3	590.0	589.6	589.3	588.9	588.6
17.0	588.2	587.9	587.5	587.2	586.9	586.5	586.2	585.8	585.5	585.1
17.1	584.8	584.5	584.1	583.8	583.4	583.1	582.8	582.4	582.1	581.7
17.2	581.4	581.1	580.7	580.4	580.0	579.7	579.4	579.0	578.7	578.4
17.3	578.0	577.7	577.4	577.0	576.7	576.4	576.0	575.7	575.4	575.0
17.4	574.7	574.4	574.1	573.7	573.4	573.1	572.7	572.4	572.1	571.8
17.5	571.4	571.1	570.8	570.5	570.1	569.8	569.5	569.2	568.8	568.5
17.6	568.2	567.9	567.5	567.2	566.9	566.6	566.3	565.9	565.6	565.3
17.7	565.0	564.7	564.3	564.0	563.7	563.4	563.1	562.7	562.4	562.1
17.8	561.8	561.5	561.2	560.9	560.5	560.2	559.9	559.6	559.3	559.0
17.9	558.7	558.3	558.0	557.7	557.4	557.1	556.8	556.5	556.2	555.9
18.0	555.6	555.2	554.9	554.6	554.3	554.0	553.7	553.4	553.1	552.8
18.1	552.5	552.2	551.9	551.6	551.3	551.0	550.7	550.4	550.1	549.8
18.2	549.5	549.1	548.8	548.5	548.2	547.9	547.6	547.3	547.0	546.7
18.3	546.4	546.1	545.9	545.6	545.3	545.0	544.7	544.4	544.1	543.8
18.4	543.5	543.2	542.9	542.6	542.3	542.0	541.7	541.4	541.1	540.8
18.5	540.5	540.2	540.0	539.7	539.4	539.1	538.8	538.5	538.2	537.9
18.6	537.6	537.3	537.1	536.8	536.5	536.2	535.9	535.6	535.3	535.0
18.7	534.8	534.5	534.2	533.9	533.6	533.3	533.0	532.8	532.5	532.2
18.8	531.9	531.6	531.3	531.1	530.8	530.5	530.2	529.9	529.7	529.4
18.9	529.1	528.8	528.5	528.3	528.0	527.7	527.4	527.1	526.9	526.6
19.0	526.3	526.0	525.8	525.5	525.2	524.9	524.7	524.4	524.1	523.8
19.1	523.6	523.3	523.0	522.7	522.5	522.2	521.9	521.6	521.4	521.1
19.2	520.8	520.6	520.3	520.0	519.8	519.5	519.2	518.9	518.7	518.4
19.3	518.1	517.9	517.6	517.3	517.1	516.8	516.5	516.3	516.0	515.7
19.4	515.5	515.2	514.9	514.7	514.4	514.1	513.9	513.6	513.3	513.1
19.5	512.8	512.6	512.3	512.0	511.8	511.5	511.2	511.0	510.7	510.5
19.6	510.2	509.9	509.7	509.4	509.2	508.9	508.6	508.4	508.1	507.9
19.7	507.6	507.4	507.1	506.8	506.6	506.3	506.1	505.8	505.6	505.3
19.8	505.1	504.8	504.5	504.3	504.0	503.8	503.5	503.3	503.0	502.8
19.9	502.5	502.3	502.0	501.8	501.5	501.3	501.0	500.8	500.5	500.3
	0	1	2	3	4	5	6	7	8	9

Wavelength (μm)	Wave-number (cm⁻¹) 0	1	2	3	4	5	6	7	8	9
20.0	500.0	499.8	499.5	499.3	499.0	498.8	498.5	498.3	498.0	497.8
20.1	497.5	497.3	497.0	496.8	496.5	496.3	496.0	495.8	495.5	495.3
20.2	*495.0*	*494.8*	*494.6*	*494.3*	*494.1*	*493.8*	*493.6*	*493.3*	*493.1*	*492.9*
20.3	492.6	492.4	492.1	491.9	491.6	491.4	491.2	490.9	490.7	490.4
20.4	490.2	490.0	489.7	489.5	489.2	489.0	488.8	488.5	488.3	488.0
20.5	487.8	487.6	487.3	487.1	486.9	486.6	486.4	486.1	485.9	485.7
20.6	485.4	485.2	485.0	484.7	484.5	484.3	484.0	483.8	483.6	483.3
20.7	*483.1*	*482.9*	*482.6*	*482.4*	*482.2*	*481.9*	*481.7*	*481.5*	*481.2*	*481.0*
20.8	480.8	480.5	480.3	480.1	479.8	479.6	479.4	479.2	478.9	478.7
20.9	478.5	478.2	478.0	477.8	477.6	477.3	477.1	476.9	476.6	476.4
21.0	476.2	476.0	475.7	475.5	475.3	475.1	474.8	474.6	474.4	474.2
21.1	473.9	473.7	473.5	473.3	473.0	472.8	472.6	472.4	472.1	471.9
21.2	*471.7*	*471.5*	*471.3*	*471.0*	*470.8*	*470.6*	*470.4*	*470.1*	*469.9*	*469.7*
21.3	469.5	469.3	469.0	468.8	468.6	468.4	468.2	467.9	467.7	467.5
21.4	467.3	467.1	466.9	466.6	466.4	466.2	466.0	465.8	465.5	465.3
21.5	465.1	464.9	464.7	464.5	464.3	464.0	463.8	463.6	463.4	463.2
21.6	463.0	462.7	462.5	462.3	462.1	461.9	461.7	461.5	461.3	461.0
21.7	*460.8*	*460.6*	*460.4*	*460.2*	*460.0*	*459.8*	*459.6*	*459.3*	*459.1*	*458.9*
21.8	458.7	458.5	458.3	458.1	457.9	457.7	457.5	457.2	457.0	456.8
21.9	456.6	456.4	456.2	456.0	455.8	455.6	455.4	455.2	455.0	454.8
22.0	454.5	454.3	454.1	453.9	453.7	453.5	453.3	453.1	452.9	452.7
22.1	452.5	452.3	452.1	451.9	451.7	451.5	451.3	451.1	450.9	450.7
22.2	*450.5*	*450.2*	*450.0*	*449.8*	*449.6*	*449.4*	*449.2*	*449.0*	*448.8*	*448.6*
22.3	448.4	448.2	448.0	447.8	447.6	447.4	447.2	447.0	446.8	446.6
22.4	446.4	446.2	446.0	445.8	445.6	445.4	445.2	445.0	444.8	444.6
22.5	444.4	444.2	444.0	443.9	443.7	443.5	443.3	443.1	442.9	442.7
22.6	442.5	442.3	442.1	441.9	441.7	441.5	441.3	441.1	440.9	440.7
22.7	*440.5*	*440.3*	*440.1*	*439.9*	*439.8*	*439.6*	*439.4*	*439.2*	*439.0*	*438.8*
22.8	438.6	438.4	438.2	438.0	437.8	437.6	437.4	437.3	437.1	436.9
22.9	436.7	436.5	436.3	436.1	435.9	435.7	435.5	435.4	435.2	435.0
23.0	434.8	434.6	434.4	434.2	434.0	433.8	433.7	433.5	433.3	433.1
23.1	432.9	432.7	432.5	432.3	432.2	432.0	431.8	431.6	431.4	431.2
23.2	*431.0*	*430.8*	*430.7*	*430.5*	*430.3*	*430.1*	*429.9*	*429.7*	*429.6*	*429.4*
23.3	429.2	429.0	428.8	428.6	428.4	428.3	428.1	427.9	427.7	427.5
23.4	427.4	427.2	427.0	426.8	426.6	426.4	426.3	426.1	425.9	425.7
23.5	425.5	425.4	425.2	425.0	424.8	424.6	424.4	424.3	424.1	423.9
23.6	423.7	423.5	423.4	423.2	423.0	422.8	422.7	422.5	422.3	422.1
23.7	*421.9*	*421.8*	*421.6*	*421.4*	*421.2*	*421.1*	*420.9*	*420.7*	*420.5*	*420.3*
23.8	420.2	420.0	419.8	419.6	419.5	419.3	419.1	418.9	418.8	418.6
23.9	418.4	418.2	418.1	417.9	417.7	417.5	417.4	417.2	417.0	416.8
24.0	416.7	416.5	416.3	416.1	416.0	415.8	415.6	415.5	415.3	415.1
24.1	414.9	414.8	414.6	414.4	414.3	414.1	413.9	413.7	413.6	413.4
24.2	*413.2*	*413.1*	*412.9*	*412.7*	*412.5*	*412.4*	*412.2*	*412.0*	*411.9*	*411.7*
24.3	411.5	411.4	411.2	411.0	410.8	410.7	410.5	410.3	410.2	410.0
24.4	409.8	409.7	409.5	409.3	409.2	409.0	408.8	408.7	408.5	408.3
24.5	408.2	408.0	407.8	407.7	407.5	407.3	407.2	407.0	406.8	406.7
24.6	406.5	406.3	406.2	406.0	405.8	405.7	405.5	405.4	405.2	405.0
24.7	*404.9*	*404.7*	*404.5*	*404.4*	*404.2*	*404.0*	*403.9*	*403.7*	*403.6*	*403.4*
24.8	403.2	403.1	402.9	402.7	402.6	402.4	402.3	402.1	401.9	401.8
24.9	401.6	401.4	401.3	401.1	401.0	400.8	400.6	400.5	400.3	400.2
	0	1	2	3	4	5	6	7	8	9

IV. MULTIPLES OF ELEMENT AND GROUP WEIGHTS

The following table may be used for the calculation of molecular weight and percent elemental composition. As an example, take $C_{15}H_{30}O_2$.

	Weight	Log
C_{15}	180.17	2.25567
H_{30}	30.24	1.48058
O_2	31.998	1.50512
	242.41	

From a four-place log table, log 242.41 = 2.3845.

$$\% \ C = \frac{180.17}{242.41} \times 100 = 100 \times \text{antilog}\ (2.2557 - 2.3845)$$
$$= 100 \times \text{antilog}\ (-0.1288)$$
$$= 100 \times \text{antilog}\ (9.8712 - 10)$$
$$= 100 \times 0.7432 = 74.32$$

(and so forth for H and O).

	Weight	Log		Weight	Log		Weight	Log
H	1.008	0.00346	H_{23}	23.18	1.36519	H_{45}	45.36	1.65667
H_2	2.016	0.30449	H_{24}	24.19	1.38367	H_{46}	46.37	1.66622
H_3	3.024	0.48058	H_{25}	25.20	1.40140	H_{47}	47.38	1.67558
H_4	4.032	0.60552	H_{26}	26.21	1.41843	H_{48}	48.38	1.68470
H_5	5.040	0.70243	H_{27}	27.22	1.43482	H_{49}	49.39	1.69366
H_6	6.048	0.78161	H_{28}	28.22	1.45062	H_{50}	50.40	1.70243
H_7	7.056	0.84856	H_{29}	29.23	1.46585	H_{51}	51.41	1.71103
H_8	8.064	0.90655	H_{30}	30.24	1.48058	H_{52}	52.42	1.71946
H_9	9.072	0.95770	H_{31}	31.25	1.49482	H_{53}	53.42	1.72774
H_{10}	10.08	1.00346	H_{32}	32.26	1.50861	H_{54}	54.43	1.73585
H_{11}	11.09	1.04485	H_{33}	33.26	1.52197	H_{55}	55.44	1.74382
H_{12}	12.10	1.08264	H_{34}	34.27	1.53494	H_{56}	56.45	1.75165
H_{13}	13.10	1.11774	H_{35}	35.28	1.54753	H_{57}	57.46	1.75934
H_{14}	14.11	1.14959	H_{36}	36.29	1.55976	H_{58}	58.46	1.76689
H_{15}	15.12	1.17955	H_{37}	37.30	1.57166	H_{59}	59.47	1.77431
H_{16}	16.13	1.20758	H_{38}	38.30	1.58324	H_{60}	60.48	1.78161
H_{17}	17.14	1.23339	H_{39}	39.31	1.59453	H_{61}	61.49	1.78879
H_{18}	18.14	1.25873	H_{40}	40.32	1.60552	H_{62}	62.50	1.79585
H_{19}	19.15	1.28221	H_{41}	41.33	1.61624	H_{63}	63.50	1.80280
H_{20}	20.16	1.30449	H_{42}	42.34	1.62671	H_{64}	64.51	1.80964
H_{21}	21.17	1.32567	H_{43}	43.34	1.63693	H_{65}	65.52	1.81637
H_{22}	22.18	1.34588	H_{44}	44.35	1.64691			

(Continued)

	Weight	Log
Li	6.94	0.84136
Li_2	13.9	1.14239
Li_3	20.8	1.31848
Li_4	27.8	1.44342
B	10.81	1.03383
B_2	21.62	1.33486
B_3	32.43	1.51095
B_4	43.24	1.63589
B_5	54.05	1.73280
B_6	64.86	1.81198
C	12.011	1.079579
C_2	24.022	1.380609
C_3	36.033	1.556700
C_4	48.044	1.681639
C_5	60.055	1.778549
C_6	72.066	1.857730
C_7	84.077	1.924677
C_8	96.088	1.982669
C_9	108.10	1.033822
C_{10}	120.11	2.079579
C_{11}	132.12	2.120972
C_{12}	144.13	2.158760
C_{13}	156.14	2.193523
C_{14}	168.15	2.225707
C_{15}	180.17	2.255670
C_{16}	192.18	2.283699
C_{17}	204.19	2.310028
C_{18}	216.20	2.334852
C_{19}	228.21	2.358333
C_{20}	240.22	2.380609
C_{21}	252.23	2.401798
C_{22}	264.24	2.422002
C_{23}	276.25	2.441307
C_{24}	288.26	2.459790
C_{25}	300.28	2.477519
C_{26}	312.29	2.494553
C_{27}	324.30	2.510943
C_{28}	336.31	2.526737
C_{29}	348.32	2.541977
C_{30}	360.33	2.556700
C_{31}	372.34	2.570941
C_{32}	384.35	2.584729
C_{33}	396.36	2.598093
C_{34}	408.37	2.611058
C_{35}	420.39	2.623647
C_{36}	432.40	2.635882
C_{37}	444.41	2.647781
C_{38}	456.42	2.659363
C_{39}	468.43	2.670644
C_{40}	480.44	2.681639
C_{41}	492.45	2.692363
C_{42}	504.46	2.702828
C_{43}	516.47	2.713048
C_{44}	528.48	2.723032
C_{45}	540.50	2.732792
C_{46}	552.51	2.742337
C_{47}	564.52	2.751677
C_{48}	576.53	2.760820
C_{49}	588.54	2.769775
C_{50}	600.55	2.778549
N	14.007	1.146345
N_2	28.014	1.447375
N_3	42.021	1.623466
N_4	56.028	1.748405
N_5	70.035	1.845315
N_6	84.042	1.924496
O	15.999	1.204093
O_2	31.998	1.505123
O_3	47.997	1.681214
O_4	63.996	1.806153
O_5	79.995	1.903063
O_6	95.994	1.982244
O_7	111.99	2.049191
O_8	127.99	2.107183
O_9	143.99	2.158335
O_{10}	159.99	2.204093
F	18.998	1.278708
F_2	37.996	1.579738
F_3	56.994	1.755829
Na	22.990	1.361539
Na_2	45.980	1.662569
Na_3	68.970	1.838660
Si	28.09	1.44855
Si_2	56.18	1.74958
Si_3	84.27	1.92567
Si_4	112.4	2.05061
Si_5	140.5	2.14752
Si_6	168.5	2.22670
P	30.974	1.490997
P_2	61.948	1.792027
P_3	92.922	1.968119
P_4	123.90	2.09306
S	32.06	1.50596
S_2	64.12	1.80699
S_3	96.18	1.98308
S_4	128.2	2.10802
Cl	35.453	1.549653
Cl_2	70.906	1.850683
Cl_3	106.36	2.026774
Cl_4	141.81	2.151713
Cl_5	177.27	2.248623
K	39.10	1.59218
K_2	78.20	1.89321
K_3	117.30	2.06930
Cu	63.55	1.80312
Cu_2	127.1	1.10415
Zn	65.4	1.81558
Zn_2	131	2.11661
Se	79.0	1.89763
Se_2	158	2.19866
Br	79.904	1.902570
Br_2	159.81	2.203607
Br_3	239.71	2.379688
Br_4	319.62	2.504636
Ag	107.868	2.032893
Ag_2	215.736	2.333923
Sn	118.7	2.07445
Sn_2	237.4	2.37548
I	126.905	2.103479
I_2	253.810	2.404509
I_3	380.715	2.580600
Pt	195.1	2.29026
Pt_2	390.2	2.59129
Hg	200.6	2.30233
Hg_2	401.2	2.60336
Pb	207.2	2.31639
Pb_2	414.4	2.61750

(Continued)

	Weight	Log		Weight	Log
$(H_2O)_{1/2}$	9.0075	0.954604	$OCOCH_3$	59.044	1.771176
H_2O	18.015	1.255634	$(OCOCH_3)_2$	118.09	2.072206
$(H_2O)_{1\,1/2}$	27.023	1.431725	$(OCOCH_3)_3$	177.13	2.248297
$(H_2O)_2$	36.030	1.55666	$(OCOCH_3)_4$	236.18	2.373236
$(H_2O)_3$	54.045	1.732756	$(OCOCH_3)_5$	295.22	2.470146
$(H_2O)_4$	72.060	1.857694	$(OCOCH_3)_6$	354.26	2.549327
$(H_2O)_5$	90.075	1.954604	$(OCOCH_3)_7$	413.31	2.616274
$(H_2O)_6$	108.09	1.033786	$(OCOCH_3)_8$	472.35	2.674266
			$(OCOCH_3)_9$	531.40	2.725418
OCH_3	31.034	1.491838	$(OCOCH_3)_{10}$	590.44	2.771176
$(OCH_3)_2$	62.068	1.792867			
$(OCH_3)_3$	93.102	1.968959	$(CH_3)_3Si$	73.195	1.864481
$(OCH_3)_4$	124.14	2.093897	$[(CH_3)_3Si]_2$	146.39	2.165511
$(OCH_3)_5$	155.17	2.190807	$[(CH_3)_3Si]_3$	219.59	2.341603
$(OCH_3)_6$	186.20	2.26999	$[(CH_3)_3Si]_4$	292.78	2.466541
$(OCH_3)_7$	217.24	2.336936	$[(CH_3)_3Si]_5$	365.98	2.563451
$(OCH_3)_8$	248.27	2.394928	$[(CH_3)_3Si]_6$	439.17	2.642633
OC_2H_5	45.061	1.653801	C_6H_5	77.106	1.887088
$(OC_2H_5)_2$	90.122	1.954831	$(C_6H_5)_2$	154.21	2.188118
$(OC_2H_5)_3$	135.18	2.130922	$(C_6H_5)_3$	231.32	2.364209
$(OC_2H_5)_4$	180.24	2.255861	$(C_6H_5)_4$	308.42	2.489148
$(OC_2H_5)_5$	225.31	2.352771	$(C_6H_5)_5$	385.53	2.586058
			$(C_6H_5)_6$	462.64	2.665239
			$(C_6H_5)_7$	539.74	2.732186
			$(C_6H_5)_8$	616.85	2.790178

V. MOLECULAR SYMMETRY: DEFINITIONS AND COMMON SYSTEMS

This section is meant to serve as a guide to point group classification of molecules (objects). The symbols are those of the Schoenflies system. For more thorough treatment and special topics, the following bibliography is offered.

F. A. Cotton, Chemical Applications of Group Theory, Interscience, New York, 1963 (2nd ed., 1971).
H. Jaffé and M Orchin, Symmetry in Chemistry, Wiley, New York, 1965.
M. Orchin and H. Jaffé, Symmetry, Point Groups and Character Tables, J. Chem. Educ., 47, 246, 372, 510 (1970).
R. L. Carter, A Flow-Chart Approach to Point Group Classification, J. Chem. Educ., 45, 44 (1968). See also J. Donohue, ibid., 46, 27 (1969).
J. L. Carlos, Jr., Molecular Symmetry and Optical Inactivity, J. Chem. Educ., 45, 248 (1968).
E. L. Muetterties, Topological Representation of Stereoisomerism, J. Amer. Chem. Soc., 91, 1936 (1969) [cf. ibid., 91, 3098 (1969)].
K Mislow and M. Raban, "Stereoisomeric Relationships of Groups in Molecules," in Topics in Stereochemistry, Vol. 1, Interscience, New York, 1967.

A. Definitions and Symbols

Symmetry operation--movement of a body in such a way, that after the movement has been performed every point of the body is coincident with an equivalent point (maybe the same point) in the original orientation. In other words, the body has been moved into a configuration indistinguishable from (but not necessarily identical to) the original. For example, rotation by 180° about the axis shown in an equilateral triangle produces a triangle indistinguishable from the original with respect to the viewer, but not strictly identical if the corners have an imaginary label:

Symmetry element--a line, plane, or point about which (with reference to which) one or more symmetry operations may be carried out.

Symmetry Element	Symbol	Symmetry Operation(s)
Identity	E	None.
n-Fold axis (proper axis)	C_n	Rotate object about such an axis through $2\pi/n$.
Mirror plane	σ	Reflect object in the plane.
Center of symmetry (center of inversion)	i	Inversion of all atoms or groups through the center.
n-Fold alternating axis (improper axis; rotation-reflection axis)	S_n	Rotate object by $2\pi/n$ about axis, followed by reflection in plane perpendicular to the axis.

B. Classification of Shapes

Objects are conveniently classified within the following scheme (K. Mislow, Introduction to Stereochemistry, Benjamin, New York, 1966).

1. No Reflection Symmetry:Dissymmetry (object and its mirror image are nonsuperimposable)

a. No C_n, therefore asymmetric--lacking any symmetry elements (except the trivial onefold axis; point group $\underline{C}_1$).
b. One or more C_n (point groups $\underline{C}_n$ and $\underline{D}_n$, therefore dissymmetric).

2. Reflection Symmetry

a. A σ but no C_n (in point group $\underline{C}_s$).
b. No σ (group $\underline{S}_n$).

c. Both σ and C_n (groups $\underline{C}_{nv}$, $\underline{C}_{nh}$, $\underline{D}_{nd}$, $\underline{D}_{nh}$, $\underline{T}_d$, $\underline{O}_h$, $\underline{I}_h$, $\underline{K}_h$). (The symbols stand for the following: $\underline{C}$ = cylindrical; $\underline{D}$ = dihedral; $\underline{T}$ = tetrahedral; $\underline{O}$ = octahedral; $\underline{I}$ = icosahedral; $\underline{K}_h$ = a sphere, which possesses all possible symmetry elements).

3. Flow Chart for Classifying Molecular Symmetry into Point Groups

The following is reproduced from R. L. Carter, J. Chem. Educ., 45, 44 (1968) by permission of the Division of Chemical Education, ACS.

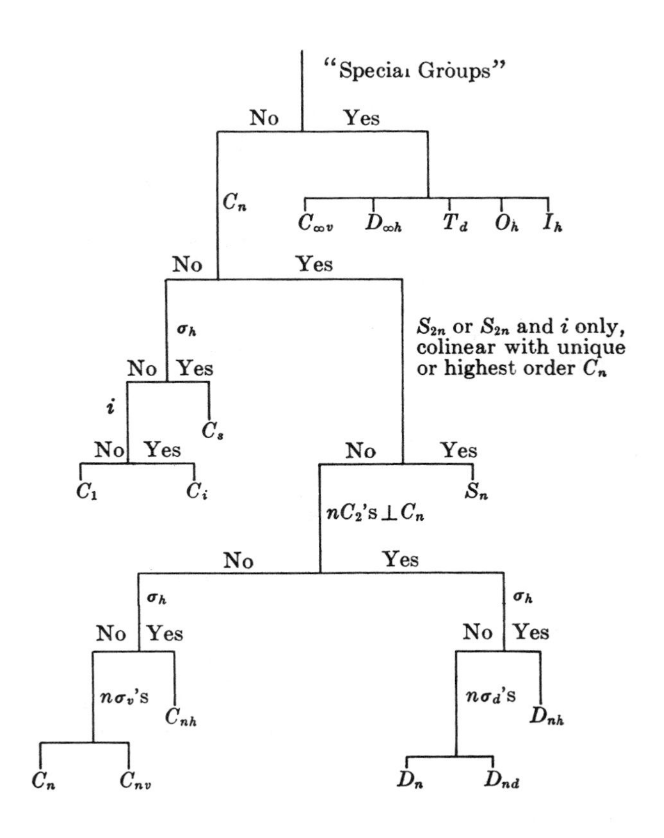

C. Examples of Common Point Groups

Not all symmetry elements may be identified in the illustrations.

Group	Definition	Examples
$\underline{C}_n$	Has only a C_n (∴ dissymmetric)	A hand (C_1, asymmetric); XCH=C=CHX ($\underline{C}_2$); hexahelicene ($\underline{C}_2$);

($\underline{C}_2$)

(Continued)

Group	Definition	Examples
$\underline{D}_n$	Has one C_n ⊥ which there are n-C_2's : $\underline{D}_n \equiv C_n + nC_2$	($\underline{D}_2$) (doubly bridged biphenyls) 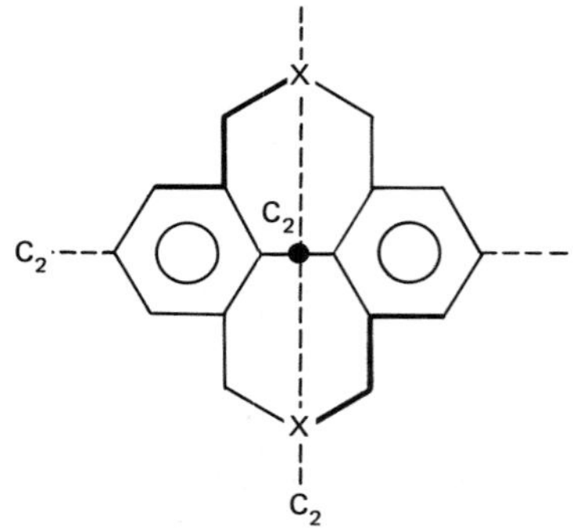 ($\underline{D}_3$) 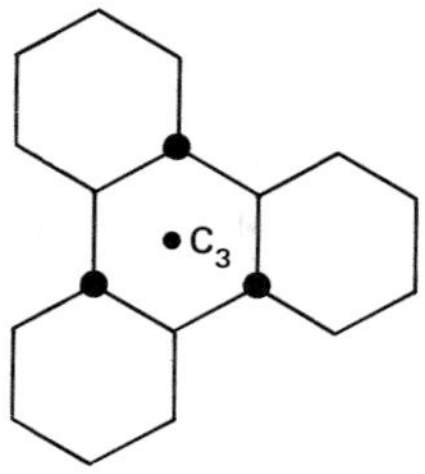 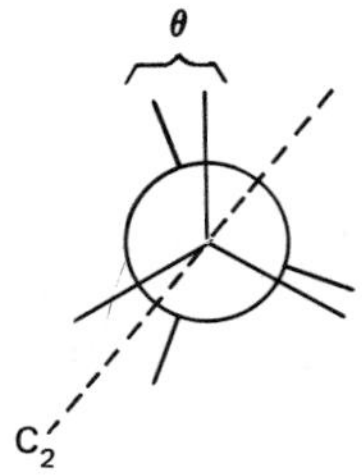 ($\underline{D}_3$) Skew conformation of ethane (0°<Θ<60°)
$\underline{C}_s$	Has one σ but no C_n.	CH_2=CHCl; CH_2=CBrCl; bromocyclopropane; CH_3OH in skew or eclipsed conformation; 2-fluoronaphthalene.
$\underline{S}_n$	A rare group. Has <u>no</u> σ, but an S_n which passes <u>through</u> the molecule. $\underline{S}_2$ is same as $\underline{C}_i$ (center of <u>inversion</u>).	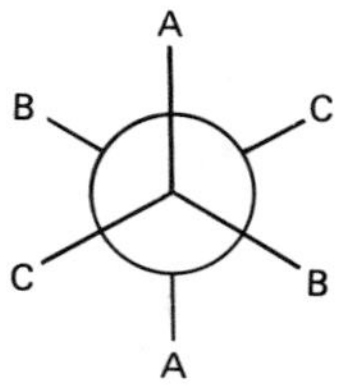 ($\underline{S}_2 \equiv \underline{C}_i$) 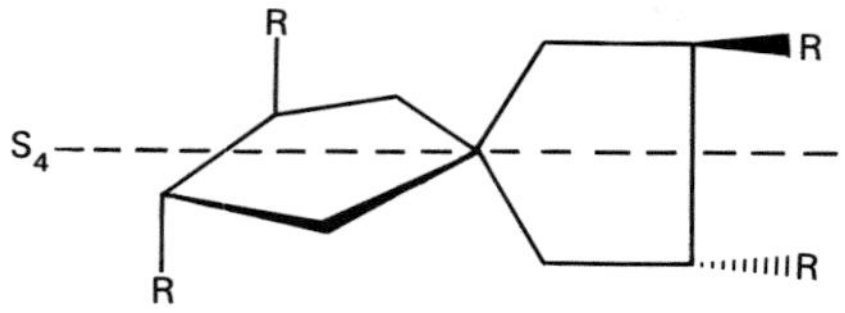 ($\underline{S}_4$)

(Continued)

Group	Definitions	Examples
$\underline{C}_{nv}$	One C_n plus n-σ containing the C_n: $\underline{C}_{nv} \equiv C_n + n\sigma$. $\underline{C}_{1v}$ is same as $\underline{C}_s$ and C_{1h} (see below). A special case is $C_{\infty v}$ which applies to linear molecules only and is referred to as conical symmetry; e.g., HBr, HCN, FC≡CH, CO.	$H_2O(\underline{C}_{2v})$; $CH_2{=}CF_2(\underline{C}_{2v})$; $CH_2Cl_2(\underline{C}_{2v})$; $H_2CO(\underline{C}_{2v})$; $NH_3(\underline{C}_{3v})$; $CHCl_3(\underline{C}_{3v})$; eclipsed $CH_3CF_3(\underline{C}_{3v})$; planar-cisoid 1,3-butadiene $(\underline{C}_{2v})$.
$\underline{C}_{nh}$	One C_n plus one "horizontal" plane $\perp$ C_n: $C_{nh} \equiv C_n + \sigma_n$. Note: all $\underline{C}_{2h}$ objects also have i.	Planar-transoid 1,3-butadiene $(\underline{C}_{2h})$. 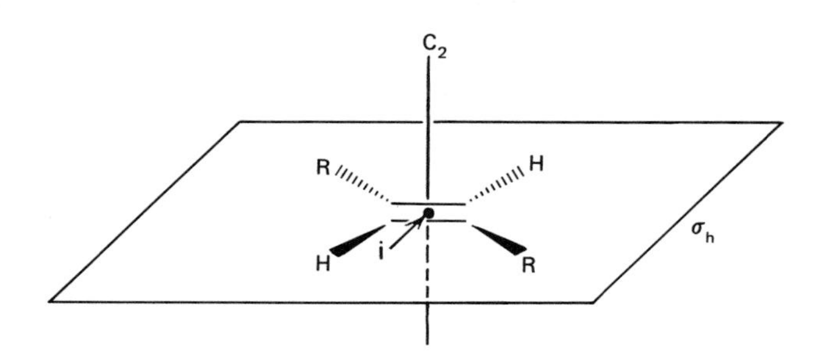
$\underline{D}_{nd}$	Has all symmetry elements of D_n plus n-diagonal planes bisecting the angles between the n-C_2's. $\underline{D}_{nd} \equiv (C_n + nC_2) + n\sigma_d$.	Cyclo-octatetraene (tub-shaped) (D_{2d}); allene $(\underline{D}_{2d})$; chair cyclohexane $(\underline{D}_{3d})$; staggered ferrocene $(\underline{D}_{5d})$. Staggered ethane $(\underline{D}_{3d})$:
$\underline{D}_{nh}$	Has all symmetry elements of $\underline{D}_n$ and $\underline{C}_{nv}$ plus one $\sigma \perp C_n$: $\underline{D}_{nh} \equiv (C_n + nC_2) + n\sigma_v + \sigma_h$. A special case is $\underline{D}_{\infty h}$ (referred to as cylindrical symmetry), containing a C_∞, ∞-σ_v's, and one σ_h; e.g., acetylene, CO_2, OCCCO (carbon suboxide). Note: $\underline{D}_{\infty h}$ molecules also contain ∞- C_2's perpendicular to the molecular axis.	Naphthalene $(\underline{D}_{2h})$; benzene $(\underline{D}_{6h})$; CH_2CH_2 $(\underline{D}_{2h})$; eclipsed ethane $(\underline{D}_{3h})$; planar CH_3^+ $(\underline{D}_{3h})$; cyclopropane $(\underline{D}_{3h})$; planar cyclopentane $(\underline{D}_{5h})$; eclipsed ferrocene $(\underline{D}_{5h})$.

(Continued)

Group	Definitions	Examples
T_d	Has 3 mutually perpendicular C_2's + 4 C_3's + 6 σ's. (The C_2's coincide with 3 S_4's).	Methane; adamantane; $SiCl_4$.
O_h	Has 6 C_4's + 8 C_3's + 9 C_2's (3 of which are coincident with C_4) + i + 9 σ.	Regular octahedron, as in $Fe(CN)_6^{\equiv}$; $Cr(CO)_6$.

VI. CHARACTER TABLES FOR COMMON SYMMETRY GROUPS

C_s	E	σ_h		
A'	1	1	x, y, R_z	x^2, y^2, z^2, xy
A"	1	-1	z, R_x, R_y	yz, xz

C_{2v}	E	C_2	$\sigma_v(xz)$	$\sigma_v'(yz)$		
A_1	1	1	1	1	z	x^2, y^2, z^2
A_2	1	1	-1	-1	R_z	xy
B_1	1	-1	1	-1	x, R_y	xz
B_2	1	-1	-1	1	y, R_x	yz

C_{3v}	E	$2C_3$	$3\sigma_v$		
A_1	1	1	1	z	$x^2 + y^2$, z^2
A_2	1	1	-1	R	
E	2	-1	0	(x, y)(R_x, R_y)	($x^2 - y^2$, xy)(xz, yz)

C_{4v}	E	$2C_4$	C_2	$2\sigma_v$	$2\sigma_d$		
A_1	1	1	1	1	1	z	$x^2 + y^2$, z^2
A_2	1	1	1	-1	-1	R_z	
B_1	1	-1	1	1	-1		$x^2 - y^2$
B_2	1	-1	1	-1	1		xy
E	2	0	-2	0	0	(x, y)(R_x, R_y)	(xz, yz)

$C_{\infty v}$	E	$2C_\infty^\Phi$	...	$\infty\sigma_v$		
$A_1 \equiv \Sigma^+$	1	1	...	1	z	$x^2 + y^2$, z^2
$A_2 \equiv \Sigma^-$	1	1	...	-1	R_z	
$E_1 \equiv \Pi$	2	$2 \cos \Phi$	...	0	(x, y); (R_x, R_y)	(xz, yz)
$E_2 \equiv \Delta$	2	$2 \cos 2\Phi$	...	0		($x^2 - y^2$, xy)
$E_3 \equiv \Phi$	2	$2 \cos 3\Phi$	...	0		
...	...	...	...	...		

C_{2h}	E	C_2	i	σ_h		
A_g	1	1	1	1	R_z	x^2, y^2, z^2, xy
B_g	1	-1	1	-1	R_x, R_y	xz, yz
A_u	1	1	-1	-1	z	
B_u	1	-1	-1	1	x, y	

D_{2h}	E	$C_2(z)$	$C_2(y)$	$C_2(x)$	i	$\sigma(xy)$	$\sigma(xz)$	$\sigma(yz)$		
A_g	1	1	1	1	1	1	1	1		x^2, y^2, z^2
B_{1g}	1	1	-1	-1	1	1	-1	-1	R_z	xy
B_{2g}	1	-1	1	-1	1	-1	1	-1	R_y	xz
B_{3g}	1	-1	-1	1	1	-1	-1	1	R_x	yz
A_u	1	1	1	1	-1	-1	-1	-1		
B_{1u}	1	1	-1	-1	-1	-1	1	1	z	
B_{2u}	1	-1	1	-1	-1	1	-1	1	y	
B_{3u}	1	-1	-1	1	-1	1	1	-1	x	

D_{3h}	E	$2C_3$	$3C_2$	σ_h	$2S_3$	$3\sigma_v$		
A_1'	1	1	1	1	1	1		$x^2 + y^2$, z^2
A_2'	1	1	-1	1	1	-1	R_z	
E'	2	-1	0	2	-1	0	(x, y)	($x^2 - y^2$, xy)
A_1''	1	1	1	-1	-1	-1		
A_2''	1	1	-1	-1	-1	1	z	
E"	2	-1	0	-2	1	0	(R_x, R_y)	(xz, yz)

D_{4h}	E	$2C_4$	C_2	$2C_2'$	$2C_2''$	i	$2S_4$	σ_h	$2\sigma_v$	$2\sigma_d$		
A_{1g}	1	1	1	1	1	1	1	1	1	1		$x^2 + y^2$, z^2
A_{2g}	1	1	1	-1	-1	1	1	1	-1	-1	R_z	
B_{1g}	1	-1	1	1	-1	1	-1	1	1	-1		$x^2 - y^2$
B_{2g}	1	-1	1	-1	1	1	-1	1	-1	1		xy
E_g	2	0	-2	0	0	2	0	-2	0	0	(R_x, R_y)	(xz, yz)
A_{1u}	1	1	1	1	1	-1	-1	-1	-1	-1		
A_{2u}	1	1	1	-1	-1	-1	-1	-1	1	1	z	
B_{1u}	1	-1	1	1	-1	-1	1	-1	-1	1		
B_{2u}	1	-1	1	-1	1	-1	1	-1	1	-1		
E_u	2	0	-2	0	0	-2	0	2	0	0	(x, y)	

D_{6h}	E	$2C_6$	$2C_3$	C_2	$3C_2'$	$3C_2''$	i	$2S_3$	$2S_6$	σ_h	$3\sigma_d$	$3\sigma_v$		
A_{1g}	1	1	1	1	1	1	1	1	1	1	1	1		x^2+y^2, z^2
A_{2g}	1	1	1	1	-1	-1	1	1	1	1	-1	-1	R_z	
B_{1g}	1	-1	1	-1	1	-1	1	-1	1	-1	1	-1		
B_{2g}	1	-1	1	-1	-1	1	1	-1	1	-1	-1	1		
E_{1g}	2	1	-1	-2	0	0	2	1	-1	-2	0	0	(R_x,R_y)	(xz,yz)
E_{2g}	2	-1	-1	2	0	0	2	-1	-1	2	0	0		$(x^2-y^2$, xy)
A_{1u}	1	1	1	1	1	1	-1	-1	-1	-1	-1	-1		
A_{2u}	1	1	1	1	-1	-1	-1	-1	-1	-1	1	1	z	
B_{1u}	1	-1	1	-1	1	-1	-1	1	-1	1	-1	1		
B_{2u}	1	-1	1	-1	-1	1	-1	1	-1	1	1	-1		
E_{1u}	2	1	-1	-2	0	0	-2	-1	1	2	0	0	(x,y)	
E_{2u}	2	-1	-1	2	0	0	-2	1	1	-2	0	0		

$D_{\infty h}$	E	$2C_\infty^\Phi$	···	$\infty\sigma_v$	i	$2S_\infty^\Phi$	···	∞C_2		
Σ_g^+	1	1	···	1	1	1	···	1		$x^2 + y^2$, z^2
Σ_g^-	1	1	···	-1	1	1	···	-1	R_z	
Π_g	2	2 cos Φ	···	0	2	-2 cos Φ	···	0	(R_x, R_y)	(xz, yz)
Δ_g	2	2 cos 2Φ	···	0	2	2 cos Φ	···	0		$(x^2 - y^2$, xy)
···	···	···	···	···	···	···	···	···		
Σ_u^+	1	1	···	1	-1	-1	···	-1	z	
Σ_u^-	1	1	···	-1	-1	-1	···	1		
Π_u	2	2 cos Φ	···	0	-2	2 cos Φ	···	0	(x, y)	
Δ_u	2	2 cos 2Φ	···	0	-2	-2 cos 2Φ	···	0		
···	···	···	···	···	···	···	···	···		

D_{2d}	E	$2S_4$	C_2	$2C_2'$	$2\sigma_d$		
A_1	1	1	1	1	1		$x^2 + y^2$, z^2
A_2	1	1	1	-1	-1	R_z	
B_1	1	-1	1	1	-1		$x^2 - y^2$
B_2	1	-1	1	-1	1	z	xy
E	2	0	-2	0	0	(x, y); (R_x, R_y)	(xz, yz)

D_{3d}	E	$2C_3$	$3C_2$	i	$2S_6$	$3\sigma_d$		
A_{1g}	1	1	1	1	1	1		$x^2 + y^2$, z^2
A_{2g}	1	1	-1	1	1	-1	R_z	
E_g	2	-1	0	2	-1	0	(R_x, R_y)	$(x^2 - y^2, xy)$, (xz, yz)
A_{1u}	1	1	1	-1	-1	-1		
A_{2u}	1	1	-1	-1	-1	1	z	
E_u	2	-1	0	-2	1	0	(x, y)	

D_{4d}	E	$2S_8$	$2C_4$	$2S_8^3$	C_2	$4C_2'$	$4\sigma_d$		
A_1	1	1	1	1	1	1	1		$x^2 + y^2$, z^2
A_2	1	1	1	1	1	-1	-1	R_z	
B_1	1	-1	1	-1	1	1	-1		
B_2	1	-1	1	-1	1	-1	1	z	
E_1	2	$\sqrt{2}$	0	$-\sqrt{2}$	-2	0	0	(x, y)	
E_2	2	0	-2	0	2	0	0		$(x^2 - y^2, xy)$
E_3	2	$-\sqrt{2}$	0	$\sqrt{2}$	-2	0	0	(R_x, R_y)	(xz, yz)

T_d	E	$8C_3$	$3C_2''$	$6S_4$*	$6\sigma_d$		
A_1	1	1	1	1	1		$x^2 + y^2 + z^2$
A_2	1	1	1	-1	-1		
E	2	-1	2	0	0		$(2z^2 - x^2 - y^2, x^2 - y^2)$
T_1	3	0	-1	1	-1	(R_x, R_y, R_z)	
T_2	3	0	-1	-1	1	(x, y, z)	(xy, xz, yz)

*The x-, y-, and z-axes.

O_h	E	$8C_3$	$6C_2$	$6C_4$*	$3C_2''$ ($\equiv 3C_4^2$)	i	$6S_4$	$8S_6$	$3\sigma_h$	$6\sigma_d$		
A_{1g}	1	1	1	1	1	1	1	1	1	1		$x^2 + y^2 + z^2$
A_{2g}	1	1	-1	-1	1	1	-1	1	1	-1		
E_g	2	-1	0	0	2	2	0	-1	2	0		$(2z^2 - x^2 - y^2, x^2 - y^2)$
T_{1g}	3	0	-1	1	-1	3	1	0	-1	-1	(R_x, R_y, R_z)	
T_{2g}	3	0	1	-1	-1	3	-1	0	-1	1		(xz, yz, xy)
A_{1u}	1	1	1	1	1	-1	-1	-1	-1	-1		
A_{2u}	1	1	-1	-1	1	-1	1	-1	-1	1		
E_u	2	-1	0	0	2	-2	0	1	-2	0		
T_{1u}	3	0	-1	1	-1	-3	-1	0	1	1	(x, y, z)	
T_{2u}	3	0	1	-1	-1	-3	1	0	1	-1		

*For x-, y-, and z-axes.

VII. COMPUTER PROGRAMS

The increasing use of computers in all areas of chemistry has made necessary the establishment of libraries of computer programs. The largest of these and the one containing the widest selection of programs is the Quantum Chemistry Program Exchange, Chemistry Department, Room 204, Indiana University, Bloomington, Ind. 47401. A membership fee and a charge for each program requested is required; however, in cases where no funds are available the charges may be waived. Programs in the following categories are available from QCPE:

- Matrix, algebraic, and arithmetic utility
- Expansions and special functions
- Quantum mechanical integrals
- Eigenvalues and eigenvectors
- Symmetry analysis and allied numerical quantities
- Self-consistent field programs
- Other programs based on electronic energy
- Other treatments of chemical systems (spectral and rate data, etc.)
- Miscellaneous programs (sort, etc.)

Listings of several computer programs of general interest can be found in the series, Computer Programs for Chemistry, D. F. DeTar, Ed., Vols. 1-3, Benjamin, New York, 1968-1969. These programs are also available on tape from Benjamin.

A large number of computer programs can be obtained, often without charge, directly from their authors. Many of these can be found by checking the notes in the Journal of Chemical Education. The ones mentioned in this source are oriented toward educational usage but many have direct research application.

A library of programs for X-ray crystallography can be found in I U Cr World List of Crystallographic Computer Programs, 2nd ed., D. P. Shoemaker, Ed., 1966, available from Polycrystal Book Service, P. O. Box 11567, Pittsburgh, Pa. 15238, or A. Oosterhoek's Vitgevers Mij. N. U., Domstraat 11-13, Utrecht, Netherlands. Another source of programs for X-ray crystallography is the University of Maryland Computer Science Center. For further information, contact Dr. James M. Stewart in care of the University, College Park, Md. 20742.

Listings of several programs as well as general discussions of the use of computers in chemistry appear in the references below. It should be noted, however, that copying a program from these books may be in violation of copyright and permission of the publisher should be obtained.

T. L. Isenhour, Introduction to Computer Programming for Chemists, Allyn and Bacon, Boston, 1972.
T. R. Dickson, The Computer and Chemistry, W. H. Freeman, San Francisco, 1968.
R. S. Ledley, Use of Computers in Biology and Medicine, McGraw-Hill, New York, 1965.
K. Wiberg, Computer Programming for Chemists, Benjamin, New York, 1965.

VIII. STATISTICAL TREATMENT OF DATA

Measurement of physical or chemical data necessarily involves a certain amount of error. This section offers some guidelines for the expression and treatment of this error. Almost any analytical chemistry text will contain a section on the statistical treatment of data. The following references are recommended for detailed discussions:

H. A. Neidig et al., Chemistry, 40, No. 10, p. 28, and No. 11, p. 28, 1967 (a brief introduction covering concepts and procedures).
P. D. Clark, B. R. Craven, and R. C. L. Bosworth, The Handling of Chemical Data, Pergamon Press, New York, 1968.
J. D. Hinchen, Practical Statistics for Chemical Research, Methuen, London, 1969.
E. L. Bauer, A Statistical Manual for Chemists, Academic Press, New York, 1971.

A. Significant Figures

The digits in a number that are "significant" are comprised of those that are known with certainty, plus the first digit whose value is in doubt. For example, if three successive weighings of a sample yield the values 0.656, 0.658, and 0.662, the calculated average weight would be 0.658666... . Obviously the weighings are not reliable in the third decimal place, so that the measurement contains three significant figures (two certain digits and one about which there is some doubt). The average, therefore, should contain the same number of significant figures and should be rounded off to 0.659. This rounding off is done according to the following rule: if the digit following the last significant figure is greater than 5, the significant figure is raised by 1; if less than 5, no change is made; if equal to 5, the last significant figure should be left even. For example, 0.66050 would be 0.660 (three significant figures). Zeros following a number after the decimal are counted as significant figures (4.250 has four significant figures). Zeros preceeding a number, or following a number before the decimal, are not counted. Thus, both 0.066 and 66,000 have only two significant figures, but 1660. and 660.0 have four.

A method for indicating significant figures is to record the last digit as a subscript. In the first example above, 0.659 may be written as 0.65_9.

When making calculations involving measured values, results must be expressed so that they contain only the number of significant figures justified by the certainty of the original measurement. For example, addition or subtraction results are rounded off to the position of the number containing the least accurately known value: 13.4 + 1478.224 = 1491.624, rounded off to 1491.6. Multiplication or division results are expressed with the same number of significant figures as the least certain original value used in the calculation: 31 x 350.1 = 10,853.1, rounded off to 11,000.

B. Reliability of Measurements

The result of a single experimental measurement is the measured value often represented as X. The difference between X and the true value X_t is the error of that particular measurement. Where only random error is encountered (see below) and an infinite number of measured values are obtained, a plot of these measured values versus the frequency of observation of each value results in a symmetrical curve (referred to as a normal or Gaussian distribution curve.

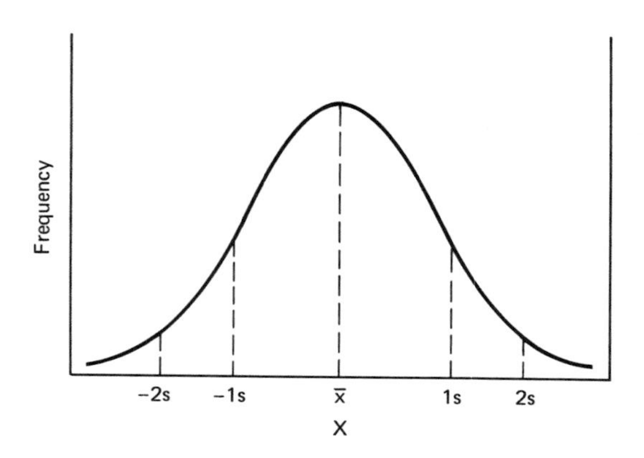

In such a plot, the true value, X_t, would appear most often and would be equal to the arithmetic mean or average, $\bar{X}$, obtained by dividing the sum of all the measurements by the number of such measurements (N): $\bar{X} = \frac{1}{N}\Sigma X_i$. In practice, of course, an infinite number of measurements cannot be made, nor is a perfectly random error seen. Thus $\bar{X}$ only approaches X_t as the number of measurements increases.

The width of the curve above is indicative of the reproducibility or precision of the measurements. Precision can be expressed in several ways, one being the magnitude of the deviations of the measurements, where deviation is the difference between a given measured value and the arithmetic mean: $D = |X_i - \bar{X}|$ (other methods for expressing precision are discussed below).

It should be noted that a constant "reproducible" error may be present when a high degree of precision is obtained. Every effort must be made to eliminate such errors, since they are undetectable by statistical analysis and, of course, result in poor accuracy. Accuracy is the agreement between a measured value and the true value.

C. Errors

Errors can be classified as either determinate or indeterminate. Determinate errors are those whose magnitude can be detected and thus corrected for. Examples of determinate errors are the use of an inaccurate balance or poorly calibrated equipment; poor technique, such as loss of part of the sample or presence of impurities in a standard; errors in the method of determination, such as an analysis based on an incomplete reaction.

Indeterminate errors are random deviations from the true value that arise, for example, from inability to determine the exact position of a pointer between calibration lines, the exact peak position in a spectrum, and the exact point of color change for an indicator. These indeterminate errors are caused by random variations in an instrument, line width, operator, and so on, and thus should be completely random. It is this type of error that statistical analysis is designed to treat.

D. Treatment of Errors

1. Average Deviation

If the true value is known, the error of a single measurement is simply the difference between it and the true value. However, as discussed above, the true value can be estimated only by the arithmetic mean; hence only the deviation of each measurement ($D = |X_i - \overline{X}|$) can be determined. If the deviations of all measurements are averaged, another expression for precision is obtained: the average deviation, $\overline{D}$, often called the average error. Thus for N values of X:

$$\overline{D} = \frac{\Sigma |X_i - \overline{X}|}{N}$$

2. Standard Deviation

Precision is most often expressed in terms of the standard deviation, s, which is the square root of the second moment around the mean. The determination of s is based on an analysis of the error curve shown above, and is thus a more realistic measure of precision. It can be shown that for a random distribution of errors (or, more correctly, deviations) one standard deviation, s, approximates the limits above and below the arithmetic mean between which there is a 68.26% probability of finding any given measurement. The standard deviation can be calculated using the following formulae:

$$s = \sqrt{\frac{\Sigma (X_i - \overline{X})^2}{N - 1}} = \sqrt{\frac{(\Sigma X_i^2) - [(\Sigma X_i)^2/N]}{N - 1}}$$

The meaning of s is expressed as follows, referring to the curve shown above. In a set of observations (data), 68.26% of them will show deviation from the mean by no more than ±1 standard deviation (the area under the curve between ±1s is 68.26% of the total). Similarly, 95.46% of the observations will show deviation of no more than ±2s; 99.73% for ±3s. The square of the standard deviation, s^2, is called the variance (V) of the data which is the basic measure of deviation and is yet another expression of the precision of the method; V is not as useful as s, since its units are those of the square of the measured units.

The following example illustrates the calculation of the standard deviation:

X	X^2
26.2	686.44
26.0	676.00
27.5	756.25
25.6	655.36
25.2	635.04
24.9	620.01
$\Sigma X = 155.4$	$\Sigma X^2 = 4029.10$

$$\text{Mean} = \frac{155.4}{6} = 25.9$$

$$(\Sigma X)^2 = 155.4^2 = 24{,}149.16$$

$$s = \sqrt{\frac{4029.10 - (24{,}149.16/6)}{5}}$$

$$= \sqrt{0.848}$$

$$= 0.92$$

This standard deviation is an expression of the precision of the method (spread of the data). As stated above, it is the limits ($\pm s$) within which there is a 68% probability of finding the next value. It is not the standard deviation of the mean (s_m). The mean is often reported as $\pm$ the standard deviation (in our example, 25.9 ± 0.9); however, a more proper expression of the true value is the mean ± the standard deviation of the mean, s_m, calculated by the equation

$$s_m = \frac{s}{\sqrt{N}}$$

where N is the number of measurements (s_m is also called the standard error). This now is an expression of the limits ($\pm s_m$) within which there is a 68% probability of finding the true value. Thus the results in the example can be reported as $25.9 \pm 0.92/\sqrt{6} = 25.9 \pm 0.376$, rounded off for significant figures to 25.9 ± 0.4.

It can be said that 25.9 ± 0.4 represents the true value with a confidence of 68%; the values (25.9 + 0.4) and (25.9 - 0.4) are referred to as the confidence limits of the mean, while 68% is the level of confidence, or confidence coefficient (in this example, for one standard deviation; for $2s_m$, it is 95.46%; etc.). Expressed another way, 68% of the time, a measured value of X would lie between 25.5 and 26.3. A general practice is to use 2 standard deviations (25.9 ± 0.8) as a cutoff point for acceptable data (1 part in 22, or about 5%); that is, if one reading from a set is 3s from the mean, it is discarded as an improbable occurrence (1 chance in 370). Other tests for data rejection are discussed below.

3. The t-Distribution

When the sample size is small ($N < 30$), values of s fluctuate considerably and the distribution of the values is no longer a standard, normal distribution. In order to calculate confidence limits for any probability (not just 1s, 2s, etc.) in such cases, use is made of t-values, which are correction factors applied to the value of s and are derived from what is called the Student (or Gosset) distribution. The Student distribution is similar to the normal (Gaussian) distribution but is spread out more (higher probability in the tails, less in the center).

Significance Limits of the Student Distribution: t-Values

(N = number of measurements in sample)

Degrees of Freedom (N-1)	Level of Confidence		
	90%	95%	99%
1	6.314	12.706	63.657
2	2.920	4.303	9.925
3	2.353	3.182	5.841
4	2.132	2.776	4.604
5	2.015	2.571	4.032
6	1.943	2.447	3.707
7	1.895	2.365	3.499
8	1.860	2.306	3.355
9	1.833	2.262	3.250
10	1.812	2.228	3.169
11	1.796	2.201	3.106
12	1.782	2.179	3.055
13	1.771	2.160	3.012
14	1.761	2.145	2.977
15	1.753	2.131	2.947
20	1.725	2.086	2.845
25	1.708	2.060	2.787
30	1.697	2.042	2.750
50	1.678	2.009	2.678
100	1.660	1.984	2.626
∞[a]	1.645	1.960	2.576

[a]Values same as for normal distribution.

Thus, for the example above, there would be a 95% probability of finding the true value in the range

$$25.9 \pm (0.376 \times 2.571) = 25.9 \pm 1.0$$

E. Rejection of Data

When one measurement from a set appears to deviate significantly from the others or from the mean, a decision must be made whether that measurement should be rejected. There is no one accepted method for making this decision and the following methods should be used with discretion.

1. The $4\bar{D}$ Method

This method uses the concept of average deviation, $\bar{D}$ (see above). Disregarding the questionable measurement, the mean and a new average deviation are calculated for the remaining measurements. If the disregarded measurement differs from the mean by more than four times the average deviation, the measurement is rejected. In the example used above, the value of 27.5 might be questioned. The mean for the remaining five numbers is 25.6 with an average deviation $\bar{D} = 0.4$. The questioned measurement (27.5) differs from this new average by 1.9, which is more than $4\bar{D}$, and thus by this criterion the value can be rejected. If the number of measurements is very large ($\gtrsim 30$), this method results in approximately a 99% probability that one is justified in rejecting that particular measurement, which is perhaps the result of other than random error.

A probability of this magnitude (99%) is often not required and a lesser confidence factor might by used. If $2.5\bar{D}$ is used, the probability that the rejection is justified is about 20 to 1, which is adequate for most purposes in which experimental error is ~5%. In our example, the questioned measurement would, of course, be rejected, since 1.9 is much greater than 2.5 x 0.4.

2. Use of t-Values

This method is based on the use of the standard deviation. In the sample above, if again the 27.5 is rejected, a mean of 25.6 with a standard deviation s = 0.5 is obtained. With five values (therefore 4 degrees of freedom), the t-factor is 2.776 for a 95% level of confidence, and any measurements outside the limits of 25.6 ± (0.5 x 2.776) = 25.6 ± 1.4 could be discarded. The 27.5 is beyond these limits and rejection is again justified (for a 95% confidence level).

3. The Q-Test

This method is the preferred one when less than ten measurement are involved. The total spread of the data (maximum value minus minimum value), w, is divided into the difference between the questionable value and its numerically nearest neighbor:

$$Q = \frac{|X_m - X_n|}{w}$$

The result is compared with the following standard Q-values:

N	Q	N	Q
3	0.94	7	0.51
4	0.76	8	0.47
5	0.64	9	0.44
6	0.56	10	0.41

If it is larger, the measurement is rejected. In the example, for the value of 27.5

$$Q = \frac{27.5 - 26.2}{27.5 - 24.9} = \frac{1.3}{2.6} = 0.5$$

This value is less than the value from the table (0.56) for N = 6; therefore, by this criterion, rejection of 27.5 is not justified.

F. Tests of Significance

Several statistical tests may be used to evaluate different sets of data from the same or related experiments. For example, two different methods may be used to obtain a physical constant such as solubility; an appropriate test will provide statistical evidence that any observed difference in the results from the two sets of data is, in fact, significant or not (at some desired confidence level). The specific test used depends greatly on the desired comparison and the nature of the data.

The most common tests are the Student t-test and the chi-square (χ^2) test. For details, refer to any of the references already given.

1. Student t-test. Used specifically to analyze differences between means from different sets of data.
2. Chi-square (χ^2) test. To determine the goodness of fit between observed and expected (theoretical) distribution of data. Tables of χ^2-values provide an answer to the question: What is the probability (p) of obtaining a fit, due only to chance, that is as poor as or worse than that observed?

G. Regression Analysis: Method of Least Squares

Many experiments in chemistry involve the measurement of some dependent, random variable (Y) as a function of an independent, controlled variable (X). Common examples are found in kinetics, where Y may be a product (or reactant) concentration (or some measure of concentration such as a titration value, pressure, or a meter reading) and X is usually time, a parameter known with great accuracy. Analysis of such data is called regression analysis, and the line or other plot obtained is referred to as the line (or plot) of regression.

1. Linear Least-Squares Plot

More often than not, a straight-line (linear) relationship is sought between Y and X, in which case the function (regression line) describing the data would take the form

$$Y = mX + b$$

Knowledge of the best values of slope, m, and Y-intercept, b, would allow prediction of Y for any X with a given probability of error. It can be shown that the line having the highest probability of being the true line is the one for which the sum of the squares of deviations of Y-values (defined by this "best" line) from Y-values measured experimentally, is at a minimum. This linear least-squares treatment gives values of m and b which describe the line of most accurate fit to the data:

$$m = \frac{N\Sigma X_i Y_i - (\Sigma X_i)(\Sigma Y_i)}{N\Sigma X_i^2 - (\Sigma X_i)^2}$$

$$b = \overline{Y} - m\overline{X} = \frac{(\Sigma Y_i)(\Sigma X_i^2) - (\Sigma X_i)(\Sigma X_i Y_i)}{N\Sigma X_i^2 - (\Sigma X_i)^2}$$

where Y_i is an actual measured value at X_i, and N is the number of "points." In a first-order kinetic situation, it is customary to construct a semilogarithm plot, whereby Y becomes ln Y (or 2.303 log Y) for the calculation.

2. Standard Error of Estimate

Just as the standard deviation measures the variation of scatter about the arithmetic mean, the standard error of estimate is a measure of variation or scatter about a regression line. It is merely the root-mean-square of the Y-deviations about the fitted curve, and as such is a measure of the precision of a least-squares fit. The Y-deviations, d (d = Y-actual minus Y-theoretical), are often called the residuals around the regression line, and S_Y is referred to as the standard deviation of residuals:

$$S_Y = \sqrt{\frac{\Sigma d^2}{N}}$$

The denominator becomes (N - 1) if $N < 30$ points. S_Y is used in the same manner as s; one S_Y includes 68% of the cases when measured above and below the regression line (i.e., individual Y_i-values are assumed to follow a normal or approximately normal distribution about the regression line). Application of t-values is also appropriate, as already discussed.

3. A Least-Squares Computer Program

The calculation of a least-squares line is commonly done with the aid of a desk calculator. However, with access to a computer the chore is considerably simplified. The Fortran IV program presented below was written for operation via a time-sharing terminal connected to a PDP-10 computer (Digital Equipment Corporation). Modifications for use on another system will be few and straightforward.

The test values (X_i, Y_i) for the input (N = 9) were chosen at random. The output lists the following information:

(a) Variance ($=S_Y^2$) of the regression line.

(b) Slope (m) and intercept (b) along with their respective standard deviations, defined as follows:

$$S_m^2 = S_Y^2/\Sigma(X_i - \overline{X})^2 = S_Y^2/[\Sigma X_i^2 - (\Sigma X_i)^2/N]$$

$$S_b^2 = S_Y^2 \Sigma X_i^2/N\Sigma(X_i - \overline{X})^2 = S_Y^2 \Sigma X_i^2/[N\Sigma X_i^2 - (\Sigma X_i)^2]$$

(c) Observed X_i and Y_i (input data), calculated Y_i-values, and the residual (d_i) for each point.

The d_i-values are useful in determining whether rejection of any point is justified by any of the tests already described. A listing of the program follows the input and output; statements on the listings preceded by a C are comments inserted here for instructional purposes only and are not an integral part of the program.

a. Input (N = 9)

```
    TEST DATA FØR LEAST SQUARES PRØGRAM
 9
1.0,7.63
5.0,7.11
10.0,6.34
15.0,5.74
20.0,5.11
25.0,4.52
30.0,3.86
35.0,3.19
40.0,2.55
```

b. Output

```
    TEST DATA FØR LEAST SQUARES PRØGRAM

THE DATA HAVE A VARIANCE ØF  0.0018
THE RESULTING SLØPE IS    -0.1292 WITH A STD DEVIATIØN ØF  0.0011
THE RESULTING INTERCEPT IS   7.7156 WITH A STD DEVIATIØN ØF  0.0261

ØBSERVED X    ØBSERVED Y    CALCULATED Y    DIFFERENCE

  1.00000       7.63000       7.58634        0.04366
  5.00000       7.11000       7.06943        0.04057
 10.00000       6.34000       6.42330       -0.08330
 15.00000       5.74000       5.77716       -0.03716
 20.00000       5.11000       5.13103       -0.02103
 25.00000       4.52000       4.48489        0.03511
 30.00000       3.86000       3.83875        0.02125
 35.00000       3.19000       3.19262       -0.00262
 40.00000       2.55000       2.54648        0.00352
END ØF FILE ØN DSK
```

c. Program Listing

```
      DIMENSIØN LABEL(14),X(50),Y(50)
C
C     THE NEXT 8 STATEMENTS ALLØW VARIATIØN ØF INPUT AND ØUTPUT
C     DEVICES.  IF ALL INPUT AND ØUTPUT WILL BE FRØM THE SAME
C     DEVICE, THEN NIN AND NØUT CAN BE SET EQUAL TØ THE DEVICE
C     NUMBERS ØR CAN BE REPLACED BY THE SYMBØLIC NØTATIØN FØR
C     THE DEVICE (I.E. LPT FØR LINEPRINTER) AND THE NEXT 8
C     STATEMENTS ELIMINATED.
C
```

(Continued)

```
      TYPE 100
  100 FØRMAT(35H1TYPE NUMBER ØF UNIT USED FØR INPUT ,/)
      ACCEPT 200,NIN
  200 FØRMAT(13)
      TYPE 300
  300 FØRMAT(36H TYPE NUMBER ØF UNIT USED FØR ØUTPUT ,/)
      ACCEPT 400,NØUT
  400 FØRMAT(13)
C
C     READ IN LABEL
C
    1 READ(NIN,500)(LABEL(I),I=1,14)
  500 FØRMAT(14A5)
C
C     READ IN NUMBER ØF X,Y VALUES.  MAXIMUM = 50.
C
      READ(NIN,600)N
  600 FØRMAT(12)
C
C     READ IN X,Y VALUES, ØNE VALUE ØF EACH (SEPARATED BY A CØMMA)
C     ØN EACH LINE.
C
      READ(NIN,700)(X(I),Y(I),I=1,N)
  700 FØRMAT(2F)
      SUMX = 0.0
      SUMY = 0.0
      SUMX2 = 0.0
      SUMXY = 0.0
      SUMD2 = 0.0
C
C     CALCULATIØN ØF SLØPE AND INTERCEPT
C
      DØ 2 I=1,N
      SUMX = SUMX + X(I)
      SUMY = SUMY + Y(I)
      SUMX2 = SUMX2 + X(I)*X(I)
    2 SUMXY = SUMXY + X(I)*Y(I)
      DIV = N*SUMX2 - SUMX*SUMX
      SLØPE = (N*SUMXY - SUMX*SUMY)/DIV
      YINTR = (SUMY*SUMX2 - SUMX*SUMXY)/DIV
C
C     CALCULATIØN ØF VARIANCE ØF Y AND STD DEV ØF SLØPE AND INTERCEPT
C
      DØ 3 I=1,N
      DEV = SLØPE*X(I) + YINTR - Y(I)
    3 SUMD2 = SUMD2 + DEV*DEV
      VARY = SUMD2/(N-1)
      DSLØPE = SQRT(N*VARY/DIV)
      DYINTR = SQRT(SUMX2*VARY/DIV)
C
C     ØUTPUT RESULTS
C
      WRITE(NØUT,800)(LABEL(I),I=1,14),VARY,SLØPE,DSLØPE,YINTR,DYINTR
  800 FORMAT(1H1,14A5,//,' THE DATA HAVE A VARIANCE ØF',F8.4,/,
     1' THE RESULTING SLØPE IS',F10.4,' WITH A STD DEVIATION ØF',
     2F8.4,/,' THE RESULTING INTERCEPT IS',F10.4,' WITH STD DEVIATIØN
     3 ØF',F8.4,///)
```

(Continued)

```
C
C     THE ØBSERVED AND CALCULATED VALUES ØF Y FØR EACH X ARE PRINTED
C     ØUT ALØNG WITH THEIR DEVIATIØNS.  IF THIS ØUTPUT IS NØT
C     DESIRED, THE NEXT 7 STATEMENTS MAY BE DELETED.
C
      WRITE(NØUT,900)
  900 FØRMAT(' ØBSERVED X   ØBSERVED Y   CALCULATED Y   DIFFERENCE',
     1//)
      DØ 4 I=1,N
      YCALC = SLØPE*X(I)+YINTR
      DIF = Y(I)-YCALC
    4 WRITE (NØUT,1000)X(I),Y(I),YCALC,DIF
 1000 FØRMAT(4(F10.5,3X))
      GØ TØ 1
      END
```

4. Weighted Linear Least-Squares Calculations

Most least-squares treatments incorporate a constant, absolute error in each point. Thus suppose that the standard deviation for Y is 0.05; for a constant, absolute error, Y-values would all be characterized as Y ± 0.05 over the whole range of Y-values. However, in practice experimental conditions usually dictate a constant percentage error, in which case Y-values must be weighted in a least-squares calculation. Then the sum of the weighted squares of residuals is minimized. The techniques for modifying the ordinary least-squares treatment to accommodate this factor are discussed in the following papers in J. Chem. Educ., 44 (1967):

K. P. Anderson and R. L. Snow, pp. 756-757.
E. D. Smith and O. M. Mathews, pp. 757-759.

More discussion is found in K. A. Brownlee, Statistical Theory and Methodology in Science and Engineering, Wiley, New York, 1960, pp 308-312.

H. Correlation Coefficient

A correlation analysis is similar to a regression analysis. Regression analysis is usually concerned with predicting one variable from a knowledge of the fixed (accurately known) independent variable. A correlation, however, deals with the general problem of finding a relationship and its statistical meaning between two or more random variables. For example, in the study of a reaction mechanism (or whatever) it may be of interest to determine what influence on rate (concentration change, Y) any or all of the following (X) will have: pressure, viscosity, concentration of any component, light intensity, and so on. A plot of Y versus X is called a scatter diagram; if the points follow closely a straight line with positive slope, a high positive correlation exists between the two variables (for negative slope, a high negative correlation). For a strictly random pattern, there is a zero correlation (no clear linear relationship exists between X and Y). We are restricting discussion here to linear relationships (keeping in mind that X and/or Y may be expressed as their logarithms).

In quantitative terms, a measure of the linear relationship between X and Y (any two random variables) is defined as the linear correlation coefficient, r, a comparative index of association between X and Y:

$$r^2 = 1 - \frac{S_Y^2}{s_Y^2}$$

Values of r^2 lie between 0 and 1; thus r ranges from -1 to +1. If $S_Y = 0$ (perfect straight line), r will be -1 or +1. Note that a value of $r^2 = r = 0$ implies a lack of linearity but not necessarily a lack of any kind of association.

Too much interpretation should not be placed on r; two values of, say, 0.4 and 0.8 mean that there are two positive correlations, one stronger than the other (not twice as good). However, for r = 0.8 it can be said that (100 r^2)% or 64% of the variation in random variable Y is accounted for by differences in the variable X.

MISCELLANEOUS

I. IMPORTANT CHEMISTRY REFERENCE SOURCES: A BIBLIOGRAPHY

In addition to the many key sources cited throughout this book, there are many others dealing extensively with general and specialized chemical needs. These are listed here. No attempt is made to discuss the usual and well known references such as the CRC and Lange Handbooks, the Merck Index, the Landolt-Börnstein Series, International Critical Tables, and various other chemical and biochemical dictionaries and handbooks. Rather, this is a collection of useful and recommended, but relatively unknown sources. For a more complete collection dealing mainly with spectral properties of matter, see page 342. An extensive list of the people and places of chemistry is found in the International Chemistry Directory, Benjamin, New York, 1969. A useful guide to sources of information appears in C. R. Burman, How to Find Out in Chemistry, 2nd ed., Pergamon Press, London, 1966. Another general reference of interest is D. W. G. Ballentyne and D. R. Lovett, A Dictionary of Named Effects and Laws in Chemistry, Physics and Mathematics, 3rd ed., Chapman and Hall, New York (available in the United States from Barnes and Noble), 1970.

A. Sources of Numerical Data

1. Continuous Numerical Data Projects

Compete descriptions of references to specific kinds of data are found in Continuous Numerical Data Projects, A Survey and Analysis, NAS-NRC, Printing and Publishing Office, NAS, 2101 Constitution Avenue, Washington, D. C. 20418 ($5.00), and in the recent International Compendium of Numerical Data Projects. A Survey and Analysis, by CODATA (the Committee on Data for Science and Technology of the International Council of Scientific Unions), Springer-Verlag, New York, Heidelberg, and Berlin, 1969. These are the best references to determine where specific thermodynamic, metallurgical, physicochemical, nuclear, and other data may be located. An abstract of the NAS-NRC book is given in the CRC Handbook of Chemistry and Physics, 50th ed., p. F-281.

2. National Standard Reference Data System (NSRDS)

The NSRDS was originally established in 1963 as a program of the National Bureau of Standards to provide a definitive source of critically evaluated data in the physical sciences. Until 1972, the series was published by NBS, and publications can be purchased from Superintendent of Documents, U.S. Government Printing Office, Washington, D.C. 20402. For a description, ask for NSRDS-NBS 1 (cost: 15 cents). However, beginning in 1972, the data compilations and reviews appear in a new periodical, Journal of Physical and Chemical Reference Data, published jointly by the American Chemical Society and American Institute of Physics.

3. Biochemistry

A book containing useful laboratory information is Data for Biochemical Research, R. M. C. Dawson et al., Eds., Oxford University Press, London, 1969. Other up-to-date sources are The Handbook of Biochemistry and Biophysics and its companion volume Methods and References in Biochemistry and Biophysics, H. C. Damm et al., Eds., World Publishing Company, New York, 1966, and later editions.

4. Inorganic Chemistry

A new book by R. B. Heslop, Numerical Aspects of Inorganic Chemistry, American Elsevier, New York, 1970, is a compact, inexpensive collection of essential data and techniques for the inorganic chemist.

5. Organometallic Chemistry

The series Organometallic Compounds, M. Dub, Ed., Springer-Verlag, New York, provides a comprehensive source of information on the methods of synthesis, physical constants, and reactions of transition element compounds (Vol. 1, 1966); Ge, Sn, and Pb compounds (Vol. 2, 1967); and As, Sb, and Bi compounds (Vol. 3, 1968). A formula index to the series is also available (1970). Another fairly comprehensive collection of known organometallic compounds and their mp or bp is Handbook of Organometallic Compounds, M. Hagihara, M. Kumada, and R. Okawara, Eds., Benjamin, New York, 1968.

6. Organic Chemistry

A good reference is Physico-Chemical Constants of Pure Organic Compounds, by J. Timmermens, American Elsevier, New York, Vol. 1, 1950, and Vol. 2, 1965. Another systematic tabulation of accurate and extensive data mainly for hydrocarbons and derivatives is Physical Properties of Chemical Compounds, by R. R. Dreisbach, Advances in Chemistry Series, No. 15 (1955; Vol. I), No. 22 (1959; Vol. II), No. 29 (Vol. III; 1961), ACS, Washington, D.C. A collection of mp and bp for many thousands of compounds is found in the Handbook of Tables for Organic Compound Identification, 3rd ed., Z. Rappoport, Ed., CRC Press, Cleveland, 1967. Critical constants and vapor pressure data for organic compounds are found in Chem. Rev., 68, 659 (1968), and J. E. Jordan, Vapor Pressure of Organic Compounds, Wiley-Interscience, New York, 1954. These data and more are collected in the 51st edition of the CRC Handbook of Chemistry and Physics, pp. D-146 to D-178. A definitive treatment on thermodynamic properties is The Chemical Thermodynamics of Organic Compounds, by D. R. Stull, E. F. Westrum, Jr., and G. C. Sinke, Wiley-Interscience, New York, 1969. A similar book is J. D. Cox and G. Pilcher, Thermochemistry of Organic and Organometallic Compounds, Academic Press, London and New York, 1970.

7. Thermophysical Properties of Matter (TPRC Data Series)

This is a 13-volume series compiling and evaluating the most comprehensive data ever collected on thermal conductivity, specific heat, thermal radiative properties, diffusivity, viscosity, and thermal conductivity. IFI/Plenum Data Corporation, 227 West 17th Street, New York, N.Y. 10011.

B. Directories of Chemicals and Equipment

1. Chem Sources

This is an annual publication listing organic and inorganic chemicals commercially available in the United States and places of purchase (Directories Publishing Co., Box 422, Flemington, N.J. 08822).

2. Radioisotopes

There are three very useful publications giving technical data as well as procurement information.

(a) Radioisotope Directory, an annual tear-out insert, usually of the December issue of Nuclear News. Available for $3.00 from American Nuclear Society, 244 East Ogden Avenue, Hinsdale, Ill. 60521 (Keith Voigt, 312-325-1991).

(b) Isotope User's Guide, ORNL-IIC-19. Available free from Oak Ridge National Laboratory, Isotope Information Center, Box X, Oak Ridge, Tenn. 37830.

(c) Special Sources of Information on Isotopes, ORNL-IIC-12. Available for $3.00 from National Technical Information Service (see paragraph C2, below). This document is an extensive bibliography.

For special information, contact Dr. John N. Maddox, Division of Isotopes Development, AEC, Washington, D.C. 20545 (202-973-4071).

3. "Directory of Chemical Producers"

A continuously updated and thorough subscription service, this compendium lists information on all chemical companies and their products; a separate alphabetical chemical index is also provided, giving the producers and plant locations. Available from Stanford Research Institute, Menlo Park, Calif. (John R. Strickland) at $200 for the first year and $120 for each additional year.

4. Reagents for Organic Synthesis

By L. F. Fieser and M. F. Fieser, Wiley-Interscience, New York, Vol. 1, 1967; Vol. 2, 1969; Vol. 3, 1971. An anthology of organic chemicals and organic synthesis.

5. Guide to Scientific Instruments

An annual issue of Science (AAAS, 1515 Massachusetts Avenue, N.W., Washington, D.C. 20005).

6. Laboratory Guide to Instruments, Equipment, and Chemicals

An annual publication of the ACS as an issue of Analytical Chemistry.

7. Chemical Week Buyers' Guide

An annual issue of Chemical Week (McGraw-Hill), providing a "complete guide to sources for chemicals and packaging," with a very extensive company directory.

8. International Chromatography Guide

An annual publication of J. Chromatog. Sci. (Preston Technical Abstract Co., 2101 Dempster Street, Evanston, Ill. 60201).

9. NBS Standard Reference Materials

The National Bureau of Standards designs, prepares, and distributes reference materials in all areas of science: pH standards; calibrated and certified thermometers; instrument calibration standards; standards for accurate measurement of air pollutants for trace elements in tissue, and for other measurements; and many others. For a recent description, see American Laboratory, June 1971, page 10, for an article by J. P. Cali and R. W. Seward (Office of Standard Reference Materials, NBS, Washington, D.C. 20234). Test and calibration work are also performed for a fee; for details, write for Calibration and Test Services of the NBS (includes fees), S.N. 0303-0801, from U.S. Government Printing Office ($2.00).

C. Science Information Centers

There are several information centers that provide specialized information, usually at no charge. For a discussion, see V. F. Maturi et al., J. Chem. Educ., 43, 605 (1966).

1. National Referral Center for Science and Technology

Located at the Library of Congress, Washington, D.C., 20540, it functions as an intermediary, directing the inquirer to key people or sources. The Center publishes highly useful, but apparently little known books, giving names, addresses, and telephone numbers for such sources of information on specific subjects as libraries, companies, and associations. They may be obtained by ordering from the U.S. Government Printing Office the following: A Directory of Information Resources in the United States: Physical Sciences and Engineering (2900 sources) and Biological Sciences (2200 sources), National Referral Center for Science and Technology, 1971. For special inquiries, contact John Price, Head, Reference and Referral Section, Library of Congress (101-426-5670).

2. National Technical Information Service

This service, formerly known as the Clearinghouse for Scientific and Technical Information is located at U.S. Department of Commerce, Springfield, Va. 22151 (703-321-8543). Supplies unclassified technical reports on United States government research and development, as well as other literature and awareness services.

3. Science Information Exchange

Located at 1730 M Street, N.W., Washington, D.C. 20036, this activity of the Smithsonian Institution concerns itself with current, unpublished research and provides information on who is doing what research and where.

4. Chemical Abstracts [Box 1378, Columbus, Ohio 43216]

Magnetic tapes used in compiling Chemical Titles are available for searches of author, bibliography, keyword, and combinations of these indexes. Custom searching is performed on the basis of a submitted interest profile. A fee is charged.

5. Science Data Bank, Inc. [18901 Cranwood Parkway, Cleveland, Ohio 44128]

This company sells a variety of scientific data bases on magnetic tape; the data come principally from the CRC Handbook of Chemistry and Physics and include physical properties, ir spectra, and Wiswesser line notations. The appropriate computer programs for storage and retrieval are included.

D. Nomenclature

Current developments in nomenclature are covered in the new series, Notes on Nomenclature, a feature of the Journal of Chemical Education. The first article [48, 433 (1971)] provides a discussion and bibliography of definitive rules, and should be consulted for guidance to the literature. The most recent definitive volumes from the IUPAC are Nomenclature of

Inorganic Chemistry, 2nd ed., 1972, and Nomenclature of Organic Chemistry, Sections A, B, and C, 3rd ed., 1971, published by Butterworth and Co., London.*

Most chemical nomenclature questions (inorganic, organic, biochemical) can be answered by referring to the following key sources, which contain detailed bibliographies or actual nomenclature rules:

1. K. L. Loening, "International Cooperation on Scientific Nomenclature," J. Chem. Docum., 10, 231 (1970).
2. R. F. Trimble, "Bibliography of Rules of Chemical Nomenclature in Various Languages. I. Inorganic Chemistry," ibid., 10, 227 (1970). "II. Organic Chemistry," ibid., to be published.
3. R. S. Cahn, Ed., Handbook for Chemical Society Authors, 2nd ed., Special Publication No. 14, the Chemical Society, London, 1961. An extensive collection of authoritative rules.
4. Chemical Abstracts Service Publications. Many, mostly free booklets on IUPAC and other, new recommended nomenclature rules are available. Write for a catalog (see address below).
5. The Ring Index, 2nd ed., 1960, and supplements, ACS, Special Issue Sales Department, Washington D.C. Naming and numbering of cyclic systems.
6. SOCMA Handbook. Commercial Organic Chemical Names, ACS, Special Issue Sales Department, Washington, D.C., 1966.
7. W. Gardner, Chemical Synonyms and Trade Names, CRC Press, London, 6th ed., 1968; 7th ed., 1971 (E. I. Cooke, Ed.).
8. E. Eliel, Recent Advances in Stereochemical Nomenclature, J. Chem. Educ., 48, 163 (1971), and references cited.

For special nomenclature problems, call or write to K. L. Loening, Director of Nomenclature, Chemical Abstracts Service, Ohio State University, Columbus, Ohio 43210, telephone 614-422-7109. Mr. Loening is also Chairman of the ACS Committee on Nomenclature.

E. Experimental Technique

The standard treatises in English are the Wiley-Interscience multivolume series, Technique of Inorganic Chemistry (H. B. Jonassen and A. Weissberger, Eds.) and Technique of Organic Chemistry (A. Weissberger, Ed.). The first four volumes and their subvolumes in Houben-Weyl, Methoden der Organischen Chemie, Georg Thieme Verlag, Stuttgart, cover in thorough detail laboratory techniques in nearly all aspects of chemistry, including analytical, physical, radiochemical, and spectroscopic methods. Below are listed some recommended concise sources for more specific treatments of experimental detail.

1. The several books by A. I. Vogel (published by Wiley, New York): Textbook of Quantitative Inorganic Analysis, 3rd ed., 1962; Textbook of Macro and Semimacro Qualitative Inorganic Analysis, 4th ed., 1954; Textbook of Practical Organic Chemistry, 3rd ed., 1966; Elementary Practical Organic Chemistry, 2nd ed., 1966, in three separate parts (volumes): Part 1, Small Scale Preparations; Part 2, Qualitative Organic Analysis; Part 3, Quantitative Organic Analysis.

*Information on other volumes available, including free appendixes to existing titles, is available from IUPAC Secretariat, Bank Court Chambers, 2-3 Pound Way, Cowley Center, Oxford OX43YF, U.K.

2. The Synthesis and Characterization of Inorganic Compounds, by W. L. Jolly, Prentice-Hall, N.J., 1970. Contains very useful sections on practical experimental techniques. For review of this book, and references to similar books, see G. B. Kauffman, J. Chem. Educ., 48, A461 (1971).
3. Laboratory Technique in Organic Chemistry, by K. Wiberg, McGraw-Hill, New York, 1960.
4. Experiments in Organic Chemistry, by L. F. Fieser, D. C. Heath, Boston, 3rd ed., 1957, or any of the older or new Fieser laboratory manuals.
5. The Manipulation of Air-Sensitive Compounds, by D. F. Shriver, McGraw-Hill, New York, 1969.
6. Practical Vacuum Techniques, by W. H. Brunner and T. H. Batzer, Van Nostrand-Rheinhold, New York, 1965.
7. Creative Glass Blowing, by J. E. Hammesfahr and C. L. Strong, W. H. Freeman, San Francisco, 1968.

F. Safety

One of the standard works is the Handbook of Laboratory Safety, 2nd ed., CRC Press, Cleveland, 1970. Another useful collection is Safety in the Chemical Laboratory, Vol. 2, N. V. Steere, Ed., Chemical Education Publishing Co., Easton, Pa. 1971. The latter is a reprint of all of the articles from February 1967-January 1970 in the regular column of the same name in J. Chem. Educ.; this continuing column can be consulted for current developments.

A very thorough Manual of Hazardous Chemical Reactions (2350 entries) is available for $3.25 from the National Fire Protection Association, 60 Battery-March Street, Boston, Mass. 02110 (telephone: 617-482-8755). A "chemical transportation safety index" in the form of a slide rule is available from RSMA, 181 East Lake Shore Drive, Chicago, Ill. 60611 ($3.50 each). Recently, the Manufacturing Chemists Association has established the Chemical Transportation Emergency Center (Washington, D.C.), known as CHEMTREC. Anyone nationwide needing help in a transportation emergency involving chemicals can call 800-424-9300 toll-free any time (24 hour service).

II. ATOMIC AND MOLECULAR MODELS*

This section describes commercially available model sets; a representative list of suppliers is included. The uses and limitations of some model systems are discussed in references 1a and 1b. In addition to commercial sets, many homemade systems are described in the literature (1c,2-11).

A. Atom and Orbital Models

These are used to demonstrate details of isolated atoms and atomic or hybrid orbitals. Discussions of atomic and molecular orbitals that are helpful in connection with the use of such models are found in reference 12a.

*Most of this section is taken from A. J. Gordon, J. Chem. Educ., 47, 30 (1970), by permission of the copyright owner (Division of Chemical Education, ACS).

Refs. p. 505.

Suppliers p. 504.

Type	Construction	Description	Suppliers
Atom	Ni-plated metal	Bohr-Sommerfeld model; uses magnets; 15 in. tall.	4
Atom	Plastic-coated cardboard	Foldout type showing elementary particles (H-Ne).	4
Orbital	Plastic template	For drawing AO's, MO's.	4
Orbital	"Sturdy"	For making models of all s, p, and d atomic, hybrid, and molecular orbitals with snap fasteners; 8 in. tall.	4
Orbital	Drilled balls; metal frame	Demonstrate angular distribution of all s, p, d, and f orbitals; about 10 in. tall. For description, see reference 12b.	10
Atom, orbital	Plastic	Prefolded strips; illustrate hybridized orbitals on metal stands. About 7 in. on an edge.	5, 6

B. Polyhedra Models

A variety of regular and irregular polyhedra and other solid shapes are available in opaque or transparent plastic, wood, and other materials; kits for construction of such shapes are also available (from suppliers 4 and 5, among others).

C. Crystal Lattice Models

Type	Description	Supplier
Crystal plane model	Heavy coated cardboard on metal frame; illustrates geometry of crystal faces and Miller indices.	4
Crystal systems	Many types including preassembled or kit form and ball-and-stick; made of plastic, foam, wood, and/or metal. Illustrate the six basic crystal forms.	4,5,7,10 11,20, 21
Crystal lattices	Preassembled or in kit form (ball-and-spoke; foam; etc.) these demonstrate specific crystal structures (e.g., NaCl, graphite, wurtzite); some are space-filling (supplier 4).	4,5,7,10, 11,20, 21

Refs. p. 505.

Suppliers p. 504.

D. Molecular Model Sets

Many of the types described in the table are available as complete sets (such as general chemistry, and organic chemistry) and as separate elements; price and descriptive details can be obtained from the supplier or in current catalogs. The characteristics of each type are indicated in the table by a key, defined as follows:

s = space filling [i.e., component atoms occupy a volume to represent magnitude of relative size (electronic density)]
r = bond lengths of different bonds to scale; followed in parentheses by scale used (in cm/Å)
a = bond angles to scale
v = covalent and/or van der Waals radii to scale
m = metal
pf = firm plastic or rubber
pc = compressible or flexible plastic or rubber
t = special tool or aid used for model assembly and/or disassembly

Unless stated otherwise: all atoms are attached by snap-on connectors of some type; most, if not all the usual atoms are available in most of their oxidation (coordination) states (H, C, N, O, halogens, P, S, metals); atoms are color coded. References to the literature describing special uses of some models are given for each type.

Commercially Available Molecular Model Sets

Type	Characteristics	Comments	Suppliers
Cenco-Petersen (1b)	r(5.0),a,m,pc	Flexible neoprene atoms with threaded aluminum inserts; steel bonds of variable lengths and scales. Accurate. Strained angles accommodated.	22
Corey-Pauling-Koltun (CPK) (13)	s,r(1.25),a,v, pf,t	Designed for bio molecules (RNA, DNA, etc.); many sets available including one for steroids.	16
Courtauld (14)	s,r(2.0),a,v, pf,t	Has distortable links (for strained bonds, etc.); good for stereochemical problems.	16
Dreiding (15)	r(2.5),a,m	Connection by tubes and rods, very easy to use, provides uncluttered, accurate view of dimensions of molecular skeleton.	8,14,15, 18

(Continued)

Refs. p. 505.

Suppliers p. 504.

Type	Characteristics	Comments	Suppliers
Fieser stereomodels (16)	r(5.2),a,pf,m	Design based on Dreiding models, but made of Lexan, polyethylene, and aluminum. Inexpensive, good for student use, limited variety of functional groups and structural types. Only C, N, and O atoms color coded.	14
Fisher-Hirschfelder-Taylor (FHT) (17)	s,r(1.0),a,v, pf,t	Stuart-type; accurate stereomodels of large variety. Also comes as a metal-coordination model kit. Not useful for highly strained angles. Sturdy; very versatile (very similar to Leybold models--see below).	2,8,15
Framework molecular models (FMM) (18)	r(variable),a, m,pc	Each element a metal-pronged cluster to which color-coded plastic tubing (cut to desired size) is attached; π and d orbital frames can be constructed. Very adaptable; inexpensive. Good for student use.	13
Frey Fused Molecule Sets	r(∿4.9),a,pf	Large, lightweight; balls coded in six colors, not to atomic scale; collars represent multiple bonds; bond lengths not strictly to scale.	9
Godfrey-Bronwill			
Molecular Models	s,r(2.5),a,v, pc+pf	Made of pliable PVC magnets to illustrate H-bonding. Versatile.	3,13,15 19
Stereomodels (19)	r(2.5),a,pf	For visualizing skeletal structure and dimensions; limited number of functional groups. Tetrahedra made by snapping slotted "half-atoms" together.	3,12,15, 19
HGS (Benjamin-Maruzen)	r(2.5),a,pf	Made of various polyhedra (ABS resin) with holes for straight and curved tubing; sturdy, versatile, inexpensive. Only C-C and C-H bonds to scale. Good for student use.	17

(Continued)

Refs. p. 505.

Suppliers p. 504.

Type	Characteristics	Comments	Suppliers
Kendrew	r(2),a,m,t	Makes skeletal models of brass rods joined by small barrel with screws; many angles (gauge available) and lengths possible. For accurate dimension measurement.	16
Lande-Edmund (20)	s,r(0.5),a,v, pf	White polystyrene-rubber plastic connected by steel pins or sockets. Designed for biomacromolecules. Series of pregrouped clusters. Not color coded.	5
Leybold (Stuart-Briegleb)	s,r(1.5),a,v, m+pf,t	Accurate stereomodels of large variety. Not suitable for highly strained angles. Sturdy; very versatile.	11
PMDM	r(variable),a, v,pf	These polyhedra molecular demonstration models are good for classroom use; bond lengths adjustable (5 to 15 in.).	13
Scale Atoms	s,r($\sim$1.5),a,v, pf	Similar to Leybold type, but less expensive; atoms come as "bowls" and removable "caps" and are joined by a peg arrangement. Contains C, H, N, O, S, P, Cl, Br.	24
SRM	s,r(3.3),v, w+pf,t	Very large stereomodels of accurate dimensions; magnetic connectors for H-bonding. Set contains C, H, N, O, S, Cl, F, lone pair.	4,16
SRP	r($\sim$3),v,m+pf	Similar to ball-and-stick type, but atomic units snap together directly; bond sticks and fasteners also included. Limited variety of units.	4

E. Miscellaneous

1. Ball-and-Stick. Many types; most have drilled wooden balls for atoms, wooden sticks and springs for bonds. Among the least expensive, but least accurate and versatile models. Supplier 5 offers polystyrene and styrofoam with threaded metal dowels as connectors. Suppliers: 3,5,6,9,20.

2. Kennard-Doré Stereo-Model Construction Set. A sophisticated and elaborate set of tools and materials (plastic, metal, brazing alloy, etc.) for construction of accurate, custom-made models. Supplier: 16.

3. "Visual Motion Device." This battery-operated item is not strictly classified with models; the E.M.E. Molecular Motion Demonstrator is used to simulate atomic and molecular motion to classroom-size groups (Educational Materials and Equipment Co., Box 63, Bronxville, N.Y. 10708). A similar device, "The Molecular Dynamics Simulator," is available from Holt, Rinehart and Winston, Inc., 383 Madison Avenue, New York, N.Y. 10017.

4. Models to Demonstrate Atomic Inversion. One style is designed to illustrate the Walden inversion mechanism; described in reference 21. Fairly expensive. Supplier: 23. For others (homemade) see reference 15b and H. Nyquist, J. Chem. Educ., 42, 103 (1965), as well as D. Hamon, ibid., 47, 398 (1970), and J. S. Meek, ibid., 48, 112 (1971).

5. Models in Structural Inorganic Chemistry. A model-building kit of several hundred plastic polyhedra, balls, connectors, etc., in various colors and sizes for inorganic structures. Described in A. F. Wells, Models in Structural Inorganic Chemistry; book and models available from Oxford Press, New York, N.Y., 1970.

F. Suppliers

1. Ace Scientific Supply Company, 1420 East Linden Avenue, Linden, N.J. 07036.
2. Arthur H. Thomas Co., Vine Street at Third, P.O. Box 779, Philadelphia, Pa. 19105.
3. Curtin Scientific Apparatus and Chemicals, 12340 Parklawn Drive, Rockville, Md. 20852.
4. Science Related Materials, Inc., P.O. Box 1009, Evanston, Ill. 60204.
5. Edmund Scientific Co., 400 Edscorp Building, Barrington, N.J. 08007.
6. E. H. Sargent and Co., 4647 West Foster Avenue, Chicago, Ill. 60630.
7. E. Leybold's Nachfolger, Köln-Bayental, Germany.
8. Fisher Scientific Company, 52 Fadem Road, Springfield, N.J. 07081.
9. Frey Scientific Co., 465 South Diamon Street, Mansfield, Ohio 44903.
10. Klinger Scientific Apparatus Corp., 83-45 Parsons Boulevard, Jamaica, N.Y. 11432.
11. Lapine Scientific Co., 6001 South Knox Avenue, Chicago, Ill. 60629.
12. Matheson Scientific, 5922 Triumph Street, Los Angeles, Calif. 90022.
13. Prentice-Hall, Inc., Englewood Cliffs, N.J. 07632.
14. Rinco Instrument Co., 503 South Prairie Street, P.O. Box 167, Greenville, Ill. 62246.
15. Scientific Glass Apparatus Co., 735 Broad Street, Bloomfield, N.J. 07003.
16. The Ealing Corporation, 2225 Massachusetts Avenue, Cambridge, Mass. 02140.
17. W. A. Benjamin, Inc., 2 Park Avenue, New York, N.Y. 10016.
18. W. Büchi Scientific Apparatus, Flawil, Switzerland.
19. Will Scientific Inc., Box 1050, Rochester, N.Y. 14603.
20. Science Labs, Box 737, El Cerrito, Calif. 94530.
21. LeMont Scientific Inc., Pike Street, Lemont, Pa. 16851.

Refs. p. 505.

22. Central Scientific Co., 1700 Irving Park Road, Chicago, Ill.
23. S. Wylie Cc., Ltd., 430 Perth Avenue, Trading Estate, Slough, Buck, England.
24. Incentive Learning Systems, P.O. Box 43027, S-100 72, Stockholm 43, Sweden. Available in the U.S.A. from, among others, Scott Engineering Sciences, 1400 South West 8th Street, Pompano Beach, Fla. 33060.

G. References

1. (a) For example, see K. Mislow, Introduction to Stereochemistry, Benjamin, New York, 1965, pp. 42-46.
   (b) Q. Petersen, J. Chem. Educ., 47, 24 (1970).
   (c) H. Bassow, Construction and Uses of Atomic and Molecular Models, Pergamon Press, New York, 1968.
2. C. Conrad and H. Bent, "A Jig for Making Styrofoam Molecular Models," J. Chem. Educ., 46, 492 (1969).
3. R. G. Gymer, "Fluid-Flow Simulation of Molecular Orbitals," J. Chem. Educ., 46, 493 (1969).
4. M. J. Nye, "Wooden Models of Asymmetric Structures," J. Chem. Educ., 46, 175 (1969).
5. N. C. Craig, "Molecular Symmetry Models," (lucite and brass), J. Chem. Educ., 46, 23 (1969).
6. F. Rodriguez, "Simple Models for Polymer Stereochemistry" (use of plastic sheeting), J. Chem. Educ., 45, 507 (1968), and H. Kaye, "Disposable Models for the Demonstration of Configuration of Vinyl Polymers" (use of plastic coated clothesline wire and pipe clearners), ibid., 48, 201 (1971).
7. S. Yamana, "An Easily Constructed Tetrahedral Model" (paper), J. Chem. Educ., 45, 245 (1968), and B. H. Freeland and R. J. O'Brien, "Easily Constructed Paper Stereomodels," ibid., 48, 771 (1971).
8. R. C. Olsen, "Crystal Models" (cork balls and plexiglas), J. Chem. Educ., 44, 729 (1967).
9. W. J. Sheppard, "The Construction of Solid Tetrahedral and Octahedral Models" (cardboard sheets), J. Chem. Educ., 44, 683 (1967).
10. G. O. Larson, "Atomic and Molecular Models Made from Vinyl Covered Wire," J. Chem. Educ., 41, 219 (1964).
11. G. Brumlik, "Molecular Models Featuring Molecular Orbitals" (Adiprene L-100, a flexible plastic; scale 1.3 cm/Å), J. Chem. Educ., 38, 502 (1961).
12. (a) I. Cohen and T. Bustard, "Atomic Orbitals. Limitations and Variations," J. Chem. Educ., 43, 187 (1966); S. Berry, "Atomic Orbitals," J. Chem. Educ., 43, 283 (1966); A. C. Wahl, "Molecular Orbital Densities: Pictorial Studies," Science, 151, 961 (1966).
    (b) A. Norbury, Educ. Chem., 5, 1 (1968).
13. W. Koltun, Biopolymers, 3, 665 (1965).
14. Robinson, Disc. Far. Soc., 16, 125 (1954).
15. (a) W. Chipman, J. Chem. Educ., 46, 119 (1969) (describes use in conjunction with FMM Set).
    (b) J. Gootjes and G. Bakuwel, "Dreiding Model of a Reversible Nitrogen Atom," J. Chem. Educ., 42, 407 (1965).
16. L. F. Fieser, J. Chem. Educ., 42, 409 (1965); L. F. Fieser, Chemistry in Three Dimensions, Rinco Instrument Co., 1963.
17. H. Hendrickson and P. Srere, "Molecular Models of Metal Chelates to Illustrate Enzymatic Reactions," J. Chem. Educ., 45, 539 (1968).

18. G. Brumlik, E. Barrett and R. Baumgarten, J. Chem. Educ., 41, 221 (1964); E. Barrett, "Models Illustrating d Orbitals involved in Multiple Bonding," J. Chem. Educ., 44, 147 (1967); C. Bumgardner and G. Wahl, Jr., "Framework Molecular Models to Illustrate Linnett's Double Quartet Theory," J. Chem. Educ., 45, 347 (1968).
19. J. C. Godfrey, J. Chem. Educ., 42, 404 (1965).
20. S. Lande, "Conformations of Peptides. Speculation Based on Molecular Models," J. Chem. Educ., 45, 587 (1968).
21. R. Berry and C. Botterill, Educ. Chem., 4, 139 (1967).

III. ADDRESSES OF PUBLISHERS THAT DEAL WITH CHEMISTRY

Abelard-Schuman, 6 West 57th Street, New York, N.Y. 10019.
The Aberdeen University Press Ltd., Farmers Hall, Aberdeen, Scotland.
Academic Press, 111 Fifth Avenue, New York, N.Y. 10003.
Addison-Wesley Publishing Co., Inc., Reading Mass. 01867.
Allyn & Bacon, Inc., 470 Atlantic Avenue, Boston, Mass. 02210 (also, Rockleigh, N.J. 07647).
American Association for the Advancement of Science, Publications Department, 1515 Massachusetts Avenue, N.W., Washington D.C. 20005.
American Chemical Society, Special Issue Sales, 1155 Sixteenth Street, N.W., Wash., D.C. 20036.
American Elsevier Publishing Co., Inc., 52 Vanderbilt Avenue, New York, N.Y. 10017.
American Institute of Physics, 335 East 45th Street, New York, N.Y. 10017.
American Library Association, 50 East Huron Street, Chicago, Ill. 60611.
American Society of Biological Chemists, Inc., 428 East Preston Street, Baltimore, Md. 21202.
American Society for Testing and Materials (ASTM), 1916 Race Street, Philadelphia, Pa. 19103.
Annual Reviews, Inc., 4139 El Camino Way, Palo Alto, Calif. 94306.
Atheneum Publishers, 122 East 42nd Street, New York, N.Y. 10017.
Barnes & Noble, Inc., 105 Fifth Avenue, New York, N.Y. 10003 (United States distributors for Methuen, Ltd., and Chapman and Hall, Ltd.).
Basic Books, Inc., Publishers, 404 Park Avenue South, New York, N.Y. 10016.
W. A. Benjamin, Inc., 2 Park Avenue, New York, N.Y. 10016.
Blaisdell Publishing Company, (A Division of Ginn & Co.) 275 Wyman Street, Waltham, Mass. 02154.
Clark Boardman Co., Ltd., 435 Hudson Street, New York, N.Y. 10014.
Burgess Publishing Co., 426 S. Sixth Street, Minneapolis, Minn. 55415.
Butterworth & Co. (Publishers) Ltd., 88 Kingsway, London, WC2B 6AB, England.
Butterworth & Co., Canada, Ltd., 1367 Danforth Avenue, Toronto 6, Ont., Canada.
Cambridge University Press, 32 East 57th Street, New York, N.Y. 10022.
Central Book Co., 850 Dekalb Avenue, Brooklyn, N.Y. 11221.
Chapman & Hall, Ltd., see Barnes & Noble, Inc.
Chemical Abstracts Service, The Ohio State University, Columbus, Ohio 43210.
Chemical Education Publishing Co., 20th and Northampton Streets, Easton, Pa. 18042.
Chemical Publishing Co., Inc., 200 Park Avenue South, Department 512, New York, N.Y. 10003.
Chemical Rubber Co., 18901 Cranwood Parkway, Cleveland, Ohio 44128.
The Chemical Society, Burlington House, Piccadilly, London, W1V GBN.
The Clarendon Press, see Oxford University Press.

Clearinghouse for Federal Scientific and Technical Information, Springfield, Va. 22151.
Collier Books, see The Macmillan Company.
Commission on Undergraduate Education in the Biological Sciences (CUEBS), 3900 Wisconsin Avenue, N.W., Washington, D.C. 20016.
Consultants Bureau, see Plenum Publishing Corp.
Cornell University Press, 124 Roberts Place, Ithaca, N.Y. 14850.
The Council of Chief State School Officers; order from Ginn and Co.
Marcel Dekker, Inc., 95 Madison Avenue, New York, N.Y. 10016.
Doubleday and Company, Inc., 501 Franklin Avenue, Garden City, N.Y. 11530.
Dover Publications, Inc., 180 Varick Street, New York, N.Y. 10014.
E. P. Dutton and Co., Inc., 201 Park Avenue, New York, N.Y. 10003.
Elsevier Publishing Co., P.O. Box 211, 110-112 Spuistraat, Amsterdam, Holland.
Free Press, see The Macmillan Company.
W. H. Freeman and Company, 660 Market Street, San Francisco, Calif. 94104.
Ginn and Co., Statler Building, Park Square, Boston, Mass. 02154.
Willard Grant Press, 53 State Street, Boston, Mass. 02109.
Hafner Publishing Company, see Stechert-Hafner, Inc.
Harcourt, Brace & World, Inc., 757 Third Avenue, New York, N.Y. 10017.
Harper & Row Publishers, 49 East 33rd Street, New York, N.Y. 10016.
Harvard University Press, 79 Garden Street, Cambridge, Mass. 02138.
D. C. Heath & Co. (Division of Raytheon Education Co.) 285 Columbus Avenue, Boston, Mass. 02116 (also, 2700 North Richardt Avenue, Indianapolis, Ind. 46219).
Heyden and Son, Ltd., Spectrum House, Alderton Crescent, London N.W.4, England.
Holden-Day, Inc., 500 Sansome Street, San Francisco, Calif. 94111.
Holt, Rinehart & Winston, Inc., 383 Madison Avenue, New York, N.Y. 10017.
Houghton Mifflin Co., 110 Tremont Street, Boston, Mass. 02107.
Imperial Chemical Industries Ltd., Millbank, London, S.W.1, England.
Intercontinental Medical Book Corporation, 381 Park Avenue South, New York, N.Y. 10016.
Interscience Publishers, see John Wiley & Sons, Inc.
Walter J. Johnson, Inc., 111 Fifth Avenue, New York, N.Y. 10003.
S. Karger AG, Albert J. Phiebig: American Representative, P.O. Box 352, White Plains, N.Y. 10602.
E. & S. Livingston, Ltd., 15-16-17 Treviot Place, Edinburgh 1, Scotland (United States distributor, William & Wilkins Co.).
Longmans, Green and Co., Ltd., 6 and 7 Clifford St., London, W1, England (also 119 West 40th Street, New York, N.Y. 10018).
McGraw-Hill Book Company, Inc., 330 West 42nd Street, New York, N.Y. 10036.
Mack Publishing Company, 20th and Northampton Streets, Easton, Pa. 18043.
Macmillan (Journals) Ltd., Brunel Road, Basingstoke, Hampshire, England.
The Macmillan Company (Subsidiary of Crowell-Collier & Macmillan, Inc.), 866 Third Avenue, New York, N.Y. 10022 (also handle Collier Books and Free Press).
Merck & Co., Inc., Rahway, N.J. 07065.
Methuen and Co., Ltd., 11 New Fetter Lane, London EC4, England.
Nankado Co., Ltd., 42-6 Hongo, 3-Chome, Bunkyo-ku, Tokyo, Japan.
National Research Council, National Academy of Sciences, 2101 Constitution Avenue, Washington, D.C. 20418.
National Research Council of Canada, Ottawa 7, Canada.
National Translations Center, John Crerar Library, 35 West 33rd Street, Chicago, Ill. 60616.
New American Library (Subsidiary of the Times Mirror Co.), 1301 Avenue of the Americas, New York, N.Y. 10019.

North Holland Publishing Co., P.O. Box 103, Amsterdam C, Holland.
Nutrition Foundation, Inc., 99 Park Avenue, New York, N.Y. 10016.
Oxford University Press, 16-00 Pollitt Drive, Fair Lawn, N.J. 07410 (also handle the Clarendon Press).
Penguin Books, Inc., 7110 Ambassador Road, Baltimore, Md. 21207.
Pergamon Press, Inc., Maxwell House, Fairview Park, Elmsford, N.Y. 10523.
Plenum Publishing Corporation (Consultants Bureau), 227 West 17th Street, New York, N.Y. 10011.
Prentice-Hall, Inc., Englewood Cliffs, N.J. 07632.
Preston Technical Abstracts Co., 909 Pitner Avenue, Evanston, Ill. 60202.
Reinhold Book Corporation, see Van Nostrand Reinhold Co.
The Ronald Press Company, 79 Madison Avenue, New York, N.Y. 10016.
The Royal Australian Chemical Institute (from Plenum Publishing Corp.).
St. Martin's Press, 175 Fifth Avenue, New York, N.Y. 10010.
Sadtler Res. Labs., Inc., 3316 Spring Garden Street, Philadelphia, Pa. 19104.
W. B. Saunders Company, West Washington Square, Philadelphia, Pa. 19105.
Scientific American, Inc., 415 Madison Avenue, New York, N.Y. 10017.
Springer-Verlag New York, Inc., 175 Fifth Avenue, New York, N.Y. 10010.
Stechert-Hafner, Inc., 31 East 10th Street, New York, N.Y. 10003.
Stipes Publishing Co., 10-12 Chester Street, Champaign, Ill. 61820.
Georg Thieme Verlag, Stuttgart N. Herdweg 63, Germany.
Charles C. Thomas, Publisher, 301-327 East Lawrence Avenue, Springfield, Ill. 62703.
The University of Chicago Press, 5750 Ellis Avenue, Chicago, Ill. 60637.
U.S. Government Printing Office, Washington, D.C. 20402.
University of Michigan Press, 615 East University, Ann Arbor, Mich. 48106.
Van Nostrand Reinhold Company, 450 West 33rd Street, New York, N.Y. 10001 (also 300 Pike Street, Cincinnati, Ohio 45202).
Varian Associates, 611 Hansen Way, Palo Alto, Calif. 94306.
Verlag-Chemie—GmbH, 694 Weinheim/Bergstrasse, Postfach 129/149, Germany.
Verlag Helvetica Chimica Acta, 4000 Basel 7, Switzerland.
John Wiley & Sons, Inc. (Wiley-Interscience) 605 Third Avenue, New York, N.Y. 10016.
William & Wilkins Co., 428 East Preston Street, Baltimore, Md. 21202.
Yale University Press, 92A Yale Station, New Haven, Conn. 06520.

IV. COMBUSTION MICROANALYSIS AND OTHER CUSTOM-ANALYTICAL SERVICES

This section contains a representative list of companies that offer analytical services on a custom basis. For routine combustion analysis (C,H, etc.) about 3 to 5 mg of sample per element is usually required; most companies performing elemental organic analyses do C, H, N, O, S, P, halogens, and ash (as oxide or sulfate), while several do nonroutine elements (B, Si, etc.).

A highly recommended monograph on microanalysis and related general microchemical techniques is that by G. Tölg, Ultramicro Elemental Analysis, Interscience, New York, 1970.

Analysis	Abbreviation	Companies (see below)
Elemental (organic)	O	1,4-6,8,10,12,14,16,18-20,23,24,27-30
Elemental (inorganic, metals)	I	6-8,10,12,14,25,28,29,32,35
Molecular Weight (other than by mass spectroscopy)	M	1,6,8,9,16,18,27-30
Functional groups (alkoxyl, active H, etc.; neutralization and saponification equivalents, I-number, etc.)	F	1,6,16,18,27-30
Thermal (DTA, TGA, DSC)	T	7,9,28,33,37
X-ray	X	3,8,28
Infrared	ir	8,9,28,29,32
Ultraviolet-visible	uv	9,28,29,32
Mass spectra	ms	11,12,22,37
Magnetic resonance	nmr, esr	21,28,31,34,37
Gas chromatography (prep.)	gc	8,9,12,14,28,35,37
Liquid chromatography (adsorption, gel, etc.)	lc	8,9,14,35,36,37
Separations (distillation, extraction, ion exchange, precipitation)	S	8,14,35,36
Miscellaneous (Mössbauer, fluorescence, deuterium, emission, flame photometry, neutron activation, electrophoresis, radiochemistry, electron microscopy, surface area, particle size, atomic absorption, vicosity, density)		2,8,13-15,17,23,26,28,29,32,35

1. Alfred Bernhardt, 433 Mülheim (Ruhr), West Germany. O, M, F
2. Andonian Associates, Inc., 26 Thayer Road, Waltham, Mass. 02154. Esr, Mössbauer, fluorescence
3. Cerac, Inc., Box 597, Butler, Wis. 53007. X
4. Chemalytics, Inc., 2330 South Industrial Park Drive, Tempe, Ariz. 85281. O
5. Childers Microanalytical Labs., Box 517, Milford, N.J. 08848. O
6. Clark Microanalytical Lab., Box 69, Urbana, Ill. 61801. O, M, F, I
7. Coors Spectro-Chemical Lab., Box 5865, Denver, Colo. 80217. I (ceramics), T
8. Crobaugh Labs., 3800 Perkins Avenue, Cleveland, Ohio, 44114. O, M, F, I, ir, gc, X, particle size, pesticide residue, lc, S
9. DeBell and Richardson, Inc., Hazardville, Conn. 06036. T; ir, uv, gc (of plastics); M (distribution by gel chrom.); lc
10. Galbraith Labs., Inc., Box 4187, Knoxville, Tenn. 37921. O, I

11. GCA Technology Div. (GCCA Corp.), Bedford, Mass. ms
12. Gollob Analytical Service, Inc., 46 Industrial Road, Berkeley Heights, N.J. 07922. O, I, ms-gc
13. Gulf General Atomic, Inc., Box 608 San Diego, Calif. 92112. Neutron activation, radiochemistry
14. Harris Labs., Inc., 624 Peach Street, Box 427, Lincoln, Neb. 68501. O, I, gc, lc, S, electrophoresis, radiochemistry
15. Holbert Electron Microscopy Lab., 80 Highland Avenue, Glen Ridge, N.J. 07028 and Structure Probe, Inc., 535 East Gay Street, West Chester, Pa. Electron microscopy
16. Huffman Labs., Inc., Wheatridge, Colo. 80033. O, M, F
17. ICN Chemical and Radioisotope Div., 2727 Campus Drive, Irvine, Calif. 92664. Mössbauer spectra
18. M-H-W Labs., Box 326, Garden City, Mich. 48135. O, M, F
19. Micro-Analysis, Inc., Box 5088, Marshallton, Wilmington, Del. 19808. O
20. Micro-Tech Labs., Inc., 4117 Oakton Street, Skokie, Ill. 60076.
21. Midwest Research Institute, 425 Volker Boulevard, Kansas City, Mo. 64110. Nmr, esr
22. Morgan-Shaffer Corp., 5110 Courtrai Avenue, Montreal, Que. (Canada). Ms
23. Josef Nemeth, 303 West Washington Street, Urbana, Ill. Deuterium
24. Par Alexander Labs., Box 4137, South Daytona, Fla. O
25. Lucius Pitkin, Inc., 50 Hudson Street, New York, N.Y. I, trace analysis
26. Quantachrome Corp., 337 Glen Cove Road, Greenvale, N.Y. Surface area, particle size
27. George I. Robertson, 52 West End Avenue, Florham Park, N.J. O, M, F
28. Sadtler Research Labs., Inc., 3328 Spring Garden Street, Philadelphia, Pa. 19104. O, M, ir, uv, nmr, gc, X, T, emission spectroscopy, flamephotometry, neutron activation
29. Schwarzkopf Microanalytical Lab., Inc., 56-19 37th Avenue, Woodside, N.Y. 11377. O, M, F, I, ir, uv, emission spectrophotometry
30. Spang Microanalytical Lab., Box 1111, Ann Arbor, Mic. 48106. O, M, F
31. Spectratec, Inc., Box 17081, Dulles International Airport, Washington, D.C. 20041. Nmr
32. Stewart Labs., Inc., 820 Tulip Avenue, Knoxville, Tenn. 37921. Uv, ir, emission, atomic absorption
33. Thermal Analysis Services, Box 1652, Bridgeport, Conn. 06601. T
34. Varian, Analytical Inst. Div., 611 Hansen Way, Palo Alto, Calif. 94303. Nmr (220 Mhz)
35. WHARF Inst., Inc., 506 North Walnut Street, Madison, Wis. 53701. I, lc, gc, physical properties (viscosity,density, etc.), S
36. Waters Associates, Inc., 61 Fountain Street, Framingham, Mass. 01701. lc, S
37. West Coast Technical Service Inc., 1049 South San Gabriel Boulevard, San Gabriel, Calif. 91776. Ir, nmr, esr, ms-gc, T

V. HAZARDS OF COMMON CHEMICALS

The following table summarizes various hazards of some common chemicals. It is reproduced by permission from A Condensed Laboratory Handbook, E. I. duPont deNemours and Co., Wilmington, Del. 19898, 1971, pp. 20-27. Many of the data were condensed from Fire Protection Guide on Hazardous

Materials, National Fire Protection Association (see p. 499). Threshold limit values for the various substances are accepted to mean that an average person can be exposed to that concentration of vapor for 8 hr/day, 5 days/week without harm. The values were taken from Threshold Limit Values of Airborne Contaminants, adopted in 1969 by American Conference of Governmental Industrial Hygienists, 1014 Broadway, Cincinnati, Ohio 45202. Department of Transportation (DOT) label classifications are part of "shipping regulations," details of which are given in Agent T.C. George's Tariff #23, Publishing Hazardous Materials Regulations of the Department of Transportation, Including Specifications for Shipping Containers, T. C. George, Agent, 2 Penn Plaza, New York, N.Y. 10001 (effective since September 1969); useful information is also found in recent editions of the Condensed Chemical Dictionary, Reinhold, New York, and in a series of articles by N. V. Steeve in J. Chem. Educ. [e.g., 48, A709 (1971)].

Most governmental and industrial authorities use the same categories to classify labels:

Category	Color of Label	Abbreviation
Compressed flammable gas	Red	F.G.
Compressed nonflammable gas	Green	Nonf. G.
Flammable solid	Yellow	F.S.
Flammable liquid	Red	F.L.
Corrosive liquid	White	Cor. L.
Oxidizing material	Yellow	Oxy. M.
Poison labels		
Class A gas or liquid		Pois. A
Class B liquid or solid		Pois. B
Class C (tear gas)		Pois. C
Class D (radioactive)	Red or blue	Pois. D
Explosives labels		
Classes A, B, C (A is most, C least dangerous)		Expl. A, Expl. B, Expl. C

Note the following key for the tables below: * vapors form explosive mixtures with air; ** milligrams per cubic meter.

NAME	PHYSICAL STATE Liquid (L)—Solid (S) or Gas (G)	FLASH POINT °F. (Closed Cup)	IGNITION TEMP. °F.	BOILING POINT °F.	FLAMMABLE LIMITS % by Volume		LIFE HAZARDS				THRESHOLD LIMIT—p.p.m.	EXPLOSION HAZARD	FIRE-FIGHTING MATERIALS								STORAGE	Protect from			D.O.T. LABEL CLASSIFICATION
					Lower	Upper	Skin Irritant	Eye Irritant	Respir. Irritant	Toxic			Water	Water Spray	No Water	"Alcohol" Foam	Regular Foam	Carbon Dioxide	Dry Chemical	Other Powder	Isolate (I) or Separate (S)	Physical Damage	Heat and Sparks	Moisture	
Acetaldehyde	L	−36	365	70	4.1	55	X	X	X	X	200	*	X	X		X		X	X		I	X	X		F.L.
Acetic Acid (Glacial)	L	109	800	245	5.4	16 @ 212°	X	X	X	X	10	*		X		X		X	X		S	X			Cor. L.
Acetic Anhydride	L	129	734	284	2.9	10.3	X	X	X	X	5	*		X		X		X	X		S	X	X	X	Cor. L.
Acetone (Dimethyl Ketone)	L	0	1000	134	2.6	12.8	X	X	X	X	1000	*		X		'X		X	X		S	X			F.L.
Acetyl Chloride	L	40	734	124			X	X	X	X	5?	*			X			X	X			X	X	X	Cor. L.
Acetyl Peroxide (25% Solution)	L						X	X	X	X		X			X				X		I	X	X		Oxy. M.
Ammonia (Anhydrous)	G		1204	−28	16	25	X	X	X	X	50			X				X	X		S	X	X		Nonf. G.
Ammonium Nitrate	S							X	X	X		X	X								S	X			Oxy. M.
Ammonium Perchlorate	S									X		X	X								I	X	X		Oxy. M.
Aniline	L	158	1418	364	1.3		X	X	X	X	5	*		X		X		X	X		S	X			Pois. B.
Barium Peroxide	S									X	5**	X			X					X	S	X		X	Oxy. M.
Benzene	L	12	1044	176	1.3	7.1	X	X	X	X	25	*		X			X	X	X		I	X			F.L.
Benzoyl Peroxide	S									X		X	X								I	X			Oxy. M.
Beryllium (Dust or Powder)	S						X	X	X	X	.002**	*			X					X	I	X		X	Pois. B.
Bromine	L			139			X	X	X	X	0.1		X								S	X		X	Cor. L.
Butyl Alcohol	L	84	689	243	1.4	11.2	X	X		X	100	*				X		X	X			X			F.L.
Calcium Carbide	S						X	X	X	X		*			X					X	I	X		X	
Calcium Cyanide	S						X	X	X	X	5**				X						S	X		X	Pois. B.
Calcium Hypochlorite	S						X	X	X	X				X							S	X		X	Oxy. M.
Carbon Disulfide	L	−22	212	115	1.3	4.4	X	X	X	X	10	*						X	X		I	X	X		F.L.
Cellulose Nitrate	S	55						X		X		X	X	X		X					I	X	X		F.L. or Expl. A
Chlorine	G						X	X	X	X	1	X			X						S	X			Nonf. G.
Chlorosulfonic Acid	L						X	X	X	X		X			X					X	I	X			Cor. L.
Cresol (Ortho)	L	178	1110	376	1.4 @ 300°		X			X	5	*		X			X	X	X		S	X			
Cresol (Meta) Cresol (Para)	L	202	1038	395	1.1 @ 302°		X			X	5	*		X X			X X	X X	X X		S S	X X			
Cyanamide	S	285 melts @ 113					X	X	X	X		X						X	X			X		X	
Cyclohexane	L	−4	500	179	1.3	8			X	X	400	*					X	X	X			X			F.L.
Diborane	G		100-125		0.9	98			X	X	0.1				X			X			I	X			F.G.
Dioxane-1,4	L	54	356	214	2	22	X	X	X	X	100	*	X			X		X	X		I	X			
Ether (Ethyl Ether; Diethyl Ether)	L	−49	356	95	1.9	48			X	X	400	X			X	X		X	X		I	X			F.L.
Ethyl Acetate	L	24	800	171	2.5	9.0	X	X	X	X	200	*		X		X	X	X				X			F.L.
Ethyl Alcohol	L	55	739	173	4.3	19	X	X	X	X	1000		X	X		X		X	X			X			F.L.
Ethyl Chloride	L	−58	966	54	3.8	15.4	X	X	X	X	1000	*		X				X	X			X			F.L.
Fluorine	G						X	X	X	X	0.1	X									I	X	X		F.G.
Hexane (Normal)	L	−7	502	156	1.1	7.5		X		X	500	*					X	X	X			X	X		F.L.
Hydrazine (Anhydrous)	L	100	75-518	236	4.7	100		X	X	X	1	*	X			X		X	X		S	X	X		Cor. L.
Hydrochloric Acid	L						X	X	X	X	5	X	X							X	S	X			Cor. L.
Hydrogen Cyanide	L	0	1000	79	6	41	X		X	X	10	*						X	X		I	X	X		Pois. A.
Hydrogen Fluoride	L			67			X	X	X	X	3	X	X								S	X			Cor. L.
Hydrogen Peroxide (27.5 to 52% by wt.)	L						X	X	X	X	1	X	X								I	X			Oxy. M.
Hydrogen Sulfide	G		500	−76	4.3	45		X	X	X	10	*						X	X		I	X	X		F.G.
Lead Nitrate	S									X		X	X								S	X		X	Oxy. M.

NAME	PHYSICAL STATE Liquid (L)—Solid (S) or Gas (G)	FLASH POINT °F. (Closed Cup)	IGNITION TEMP. °F.	BOILING POINT °F.	FLAMMABLE LIMITS % by Volume		LIFE HAZARDS				THRESHOLD LIMIT—p.p.m.	EXPLOSION HAZARD	FIRE-FIGHTING MATERIALS								STORAGE				D.O.T. LABEL CLASSIFICATION
					Lower	Upper	Skin Irritant	Eye Irritant	Respir. Irritant	Toxic			Water	Water Spray	No Water	"Alcohol" Foam	Regular Foam	Carbon Dioxide	Dry Chemical	Other Powder	Isolate (I) or Separate (S)	Protect from: Physical Damage	Protect from: Heat and Sparks	Protect from: Moisture	
Magnesium (including alloys)	S											X			X					X	I	X		X	F.S.
Methyl Alcohol	L	52	867	147	7.3	36			X	X	200	*	X	X		X		X	X			X	X		F.L.
Methylamine-Mono	G		806	21	4.9	20.7	X	X	X	X		*						X	X						F.G.
Di-methylamine	G		806	45	2.8	14.4	X	X	X	X		*						X	X						F.G.
Tri-methylamine	G		374	38	2.0	11.6	X	X	X	X		*						X	X						F.G.
Methyl Methacrylate	L	50 (oc)		212			X	X	X	X		*					X	X	X		S	X	X		F.L.
Monochlorobenzene	L	84	1180	270	1.3	7.1	X	X	X	X	75	*					X	X	X			X			
Naphthalene	S	174	979	424	0.9	5.9	X	X	X	X		*	X				X	X	X		S	X	X		
Nitric Acid	L						X	X	X	X	2	X		X							S	X			Cor. L.
Nitroaniline	S	390								X	1			X			X	X	X			X	X	X	
Nitrobenzene	L	190	900	412	1.8 @ 200°			X		X	1	*		X			X	X	X		S	X	X		Pois. B.
Perchloric Acid	L						X	X	X	X		X		X							S				Cor. L.
Phenol (Carbolic Acid)	S	175	1319	358			X	X	X	X	5	X		X		X	X	X	X		S	X	X		Pois. B.
Phosphorus, Red	S		500	538				X		X		X	X								S	X			F.S.
Phosphorus (White or Yellow)	S		86	549			X	X	X	X	0.1**	X	X								S	X			F.S.
Picric Acid	S		Less than 572					X	X	X	0.1**	X	Do not fight								I	X	X		Expl. A.
Potassium	S						X	X	X	X		X			X				X	X	I	X	X		F.S.
Potassium Chlorate	S									X		X	X								S	X			Oxy. M.
Potassium Cyanide	S						X			X	5**										S	X		X	Pois. B.
Potassium Hydroxide	S-L						X	X	X	X			X								S	X		X	Cor. L.
Potassium Nitrate	S							X		X		X	X								S	X		X	Oxy. M.
Potassium Permanganate	S							X				X	X								S	X			Oxy. M.
Silver Nitrate	S						X	X	X	X			X								S	X			Oxy. M.
Sodium	S						X	X	X	X		X			X				X	X	I	X	X	X	F.S.
Sodium Hydroxide	S						X	X	X	X	2**		X								S	X		X	
Sodium Peroxide	S						X	X	X	X		X			X				X		S	X		X	Oxy. M.
Sodium Sulfide	S						X	X		X		X	X								S	X			F.S.
Strontium Nitrate	S									X			X								S	X			Oxy. M.
Styrene	L	90	914	295	1.1	6.1	X	X	X	X	100	*		X			X	X	X			X	X		
Sulfuric Acid	L						X	X	X	X	1**								X		S	X		X	Cor. L.
Tetraethyl Lead	L	200					X	X	X	X	.075**	*	X				X	X	X		I	X	X		Pois. B.
Toluene	L	40	997	231	1.4	6.7		X	X	X	200	*		X			X	X	X			X	X		F.L.
Trinitrobenzene	S								X	X		X	Do not fight								I	X	X		Expl. A.
Vinyl Chloride	G		882	7	4	22	X			X	500	*						X	X			X	X		F.G.
Xylene (Ortho)	L	90	867	292	1.0	6.0	X		X	X	200	*		X			X	X	X			X	X		F.L.
Xylene (Meta)	L	84	982	282	1.1	7.0	X		X	X	200	*		X			X	X	X			X	X		F.L.
Xylene (Para)	L	81	984	281	1.1	7.0	X		X	X	200	*		X			X	X	X			X	X		F.L.

SUPPLIERS INDEX

SUBJECT INDEX

FOUR-PLACE MANTISSAS FOR COMMON LOGARITHMS

N	0	1	2	3	4	5	6	7	8	9	Proportional Parts								
											1	2	3	4	5	6	7	8	9
10	0000	0043	0086	0128	0170	0212	0253	0294	0334	0374	*4	8	12	17	21	25	29	33	37
11	0414	0453	0492	0531	0569	0607	0645	0682	0719	0755	4	8	11	15	19	23	26	30	34
12	0792	0828	0864	0899	0934	0969	1004	1038	1072	1106	3	7	10	14	17	21	24	28	31
13	1139	1173	1206	1239	1271	1303	1335	1367	1399	1430	3	6	10	13	16	19	23	26	29
14	1461	1492	1523	1553	1584	1614	1644	1673	1703	1732	3	6	9	12	15	18	21	24	27
15	1761	1790	1818	1847	1875	1903	1931	1959	1987	2014	*3	6	8	11	14	17	20	22	25
16	2041	2068	2095	2122	2148	2175	2201	2227	2253	2279	3	5	8	11	13	16	18	21	24
17	2304	2330	2355	2380	2405	2430	2455	2480	2504	2529	2	5	7	10	12	15	17	20	22
18	2553	2577	2601	2625	2648	2672	2695	2718	2742	2765	2	5	7	9	12	14	16	19	21
19	2788	2810	2833	2856	2878	2900	2923	2945	2967	2989	2	4	7	9	11	13	16	18	20
20	3010	3032	3054	3075	3096	3118	3139	3160	3181	3201	2	4	6	8	11	13	15	17	19
21	3222	3243	3263	3284	3304	3324	3345	3365	3385	3404	2	4	6	8	10	12	14	16	18
22	3424	3444	3464	3483	3502	3522	3541	3560	3579	3598	2	4	6	8	10	12	14	15	17
23	3617	3636	3655	3674	3692	3711	3729	3747	3766	3784	2	4	6	7	9	11	13	15	17
24	3802	3820	3838	3856	3874	3892	3909	3927	3945	3962	2	4	5	7	9	11	12	14	16
25	3979	3997	4014	4031	4048	4065	4082	4099	4116	4133	2	3	5	7	9	10	12	14	15
26	4150	4166	4183	4200	4216	4232	4249	4265	4281	4298	2	3	5	7	8	10	11	13	15
27	4314	4330	4346	4362	4378	4393	4409	4425	4440	4456	2	3	5	6	8	9	11	13	14
28	4472	4487	4502	4518	4533	4548	4564	4579	4594	4609	2	3	5	6	8	9	11	12	14
29	4624	4639	4654	4669	4683	4698	4713	4728	4742	4757	1	3	4	6	7	9	10	12	13
30	4771	4786	4800	4814	4829	4843	4857	4871	4886	4900	1	3	4	6	7	9	10	11	13
31	4914	4928	4942	4955	4969	4983	4997	5011	5024	5038	1	3	4	6	7	8	10	11	12
32	5051	5065	5079	5092	5105	5119	5132	5145	5159	5172	1	3	4	5	7	8	9	11	12
33	5185	5198	5211	5224	5237	5250	5263	5276	5289	5302	1	3	4	5	6	8	9	10	12
34	5315	5328	5340	5353	5366	5378	5391	5403	5416	5428	1	3	4	5	6	8	9	10	11
35	5441	5453	5465	5478	5490	5502	5514	5527	5539	5551	1	2	4	5	6	7	9	10	11
36	5563	5575	5587	5599	5611	5623	5635	5647	5658	5670	1	2	4	5	6	7	8	10	11
37	5682	5694	5705	5717	5729	5740	5752	5763	5775	5786	1	2	3	5	6	7	8	9	10
38	5798	5809	5821	5832	5843	5855	5866	5877	5888	5899	1	2	3	5	6	7	8	9	10
39	5911	5922	5933	5944	5955	5966	5977	5988	5999	6010	1	2	3	4	5	7	8	9	10
40	6021	6031	6042	6053	6064	6075	6085	6096	6107	6117	1	2	3	4	5	6	8	9	10
41	6128	6138	6149	6160	6170	6180	6191	6201	6212	6222	1	2	3	4	5	6	7	8	9
42	6232	6243	6253	6263	6274	6284	6294	6304	6314	6325	1	2	3	4	5	6	7	8	9
43	6335	6345	6355	6365	6375	6385	6395	6405	6415	6425	1	2	3	4	5	6	7	8	9
44	6435	6444	6454	6464	6474	6484	6493	6503	6513	6522	1	2	3	4	5	6	7	8	9
45	6532	6542	6551	6561	6571	6580	6590	6599	6609	6618	1	2	3	4	5	6	7	8	9
46	6628	6637	6646	6656	6665	6675	6684	6693	6702	6712	1	2	3	4	5	6	7	7	8
47	6721	6730	6739	6749	6758	6767	6776	6785	6794	6803	1	2	3	4	5	5	6	7	8
48	6812	6821	6830	6839	6848	6857	6866	6875	6884	6893	1	2	3	4	4	5	6	7	8
49	6902	6911	6920	6928	6937	6946	6955	6964	6972	6981	1	2	3	4	4	5	6	7	8
50	6990	6998	7007	7016	7024	7033	7042	7050	7059	7067	1	2	3	3	4	5	6	7	8
51	7076	7084	7093	7101	7110	7118	7126	7135	7143	7152	1	2	3	3	4	5	6	7	8
52	7160	7168	7177	7185	7193	7202	7210	7218	7226	7235	1	2	2	3	4	5	6	7	7
53	7243	7251	7259	7267	7275	7284	7292	7300	7308	7316	1	2	2	3	4	5	6	6	7
54	7324	7332	7340	7348	7356	7364	7372	7380	7388	7396	1	2	2	3	4	5	6	6	7
N	0	1	2	3	4	5	6	7	8	9	1	2	3	4	5	6	7	8	9

* Interpolation in this section of the table is inaccurate.